Annals of Mathematics Studies

Number 177

Hypoelliptic Laplacian and Orbital Integrals

Jean-Michel Bismut

PRINCETON UNIVERSITY PRESS

PRINCETON AND OXFORD

2011

Copyright © 2011 by Princeton University Press

Published by Princeton University Press,
41 William Street, Princeton, New Jersey 08540

In the United Kingdom: Princeton University Press,
6 Oxford Street, Woodstock, Oxfordshire OX20 1TW

Library of Congress Cataloging-in-Publication Data

Bismut, Jean-Michel.
  Hypoelliptic Laplacian and Orbital Integrals / Jean-Michel Bismut
    p. cm.
  Includes bibliographical references and index.
  ISBN-13: 978-0-691-15129-8 (alk. paper)
  ISBN-13: 978-0-691-15130-4 (pbk. : alk. paper)
  1. Hypoelliptic equations. 2. Index theory and related fixed point theorems.

British Library Cataloging-in-Publication Data is available

This book has been composed in LᴬTEX

The publisher would like to acknowledge the author of this volume for providing
the camera-ready copy from which this book was printed.

Printed on acid-free paper. ∞

press.princeton.edu

Printed in the United States of America

10 9 8 7 6 5 4 3 2 1

Est-ce à votre cocher, Monsieur,
ou bien à votre cuisinier, que vous
voulez parler? car je suis l'un et l'autre.

MOLIÈRE, *L'avare*

Contents

Acknowledgments

The author is indebted to Yves Benoist, Laurent Clozel, Patrick Gérard, François Labourie, Gilles Lebeau, Yves Le Jan, Werner Müller, and Yuri Tschinkel for very helpful discussions and remarks. I am especially grateful to Gilles Lebeau for many heated, fruitful, if nonacademic, discussions. Without his support, this work would not have been completed. I am also indebted to Xiaonan Ma and Weiping Zhang for their critical comments. Finally, I thank Courant Institute and New York University for supporting my stay during the spring terms of 2007, 2008, and 2009.

Introduction

The purpose of this book is to use the hypoelliptic Laplacian to evaluate semisimple orbital integrals, in a formalism that unifies index theory and the trace formula.

0.1 The trace formula as a Lefschetz formula

Let us explain how to think formally of such a unified treatment, while allowing ourselves a temporarily unbridled use of mathematical analogies. Let X be a compact Riemannian manifold, and let Δ^X be the corresponding Laplace-Beltrami operator. For $t > 0$, consider the trace of the heat kernel $\mathrm{Tr}\left[\exp\left(t\Delta^X/2\right)\right]$. If L_2^X is the Hilbert space of square-integrable functions on X, $\mathrm{Tr}\left[\exp\left(t\Delta^X/2\right)\right]$ is the trace of the 'group element' $\exp\left(t\Delta^X/2\right)$ acting on L_2^X.

Suppose that L_2^X is the cohomology of an acyclic complex R on which Δ^X acts. Then $\mathrm{Tr}\left[\exp\left(t\Delta^X/2\right)\right]$ can be viewed as the evaluation of a Lefschetz trace, so that cohomological methods can be applied to the evaluation of this trace. In our case, R will be the fibrewise de Rham complex of the total space $\widehat{\mathcal{X}}$ of a flat vector bundle over X, which contains TX as a subbundle. The Lefschetz fixed point formulas of Atiyah-Bott [ABo67, ABo68] provide a model for the evaluation of such cohomological traces.

The McKean-Singer formula [McKS67] indicates that if $\square^R$ is a Hodge like Laplacian operator acting on R and commuting with Δ^X, for any $b > 0$,

$$\mathrm{Tr}^{L_2^X}\left[\exp\left(t\Delta^X/2\right)\right] = \mathrm{Tr_s}^R\left[\exp\left(t\Delta^X/2 - t\square^R/2b^2\right)\right]. \qquad (0.1)$$

In (0.1), $\mathrm{Tr_s}$ is our notation for the supertrace. Note that the formula involves two parameters: t is a parameter in a Lie algebra, and $1/b^2$ is a genuine time parameter. For $b \to 0$, the right-hand side of (0.1) obviously converges to the left-hand side.

To establish the Atiyah-Bott formulas, the heat equation method of Gilkey [Gi73, Gi84] and Atiyah-Bott-Patodi [ABoP73] consists in making $b \to +\infty$ in (0.1), and to show that the local supertrace in the right-hand side of (0.1) localizes on the fixed point set of the isometry $\exp\left(t\Delta^X/2\right)$, while exhibiting the nontrivial local cancellations anticipated by McKean-Singer [McKS67]. One should obtain formulas this way that are analogous to the fixed point formulas of [ABo67, ABo68].

The present book is an attempt to make sense of the above, in the case where X is a compact locally symmetric space of noncompact type. In this case, the Selberg trace formula should be thought of as the evaluation of a Lefschetz trace. Contrary to what happens in Atiyah-Bott [ABo67, ABo68], the operator $\mathcal{L}_b^X = \Delta^X/2 - \Box^R/2b^2$ is not elliptic, but just hypoelliptic.

0.2 A short history of the hypoelliptic Laplacian

Let us now give the proper rigorous background to the present work. Let X be a compact Riemannian manifold, let $\mathcal{X}, \mathcal{X}^*$ be the total spaces of its tangent and cotangent bundle. In [B05], we introduced a deformation of the classical Hodge theory of X. The corresponding Laplacian $L_b^X, b > 0$ is a hypoelliptic operator acting over $\mathcal{X}^*$. It is essentially the weighted sum of the harmonic oscillator along the fibre and of the generator of the geodesic flow. Arguments given in [B05] showed that as $b \to 0$, the operator L_b^X should converge in the proper sense to the Hodge Laplacian $\Box^X/2$ of X via a collapsing mechanism, and that as $b \to +\infty$, L_b^X converges to the generator of the geodesic flow.

The program outlined in [B05] was carried out in Bismut-Lebeau [BL08], at least for bounded values of b. In [BL08], it was shown that in a very precise way, L_b^X converges to $\Box^X/2$. A consequence of the results of [BL08] is that given $t > 0$, $\exp\left(-tL_b^X\right)$ is trace class, and that as $b \to 0$, its trace converges to the trace of $\exp\left(-t\Box^X/2\right)$. The spectral theory of L_b^X was also studied in [BL08], as well as its local index theory. An important result of [BL08] is that if F is a flat vector bundle on X, the Ray-Singer metric on $\det H^\cdot(X, F)$ one can attach to L_b^X coincides with the classical elliptic Ray-Singer metric [RS71, BZ92]. This paves the way to a possible proof using the hypoelliptic Laplacian of the Fried conjecture [Fri86, Fri88] concerning the relation of the Ray-Singer torsion to special values of the dynamical zeta function of the geodesic flow.

In [B08a], we gave a deformation of the classical Dirac operator, the deformed Dirac operator acting over $\mathcal{X}$, its square still being hypoelliptic, and having the same analytic structure as the operator L_b^X. Results similar to the ones in [BL08] were established in [B08a] for Quillen metrics.

As a warm-up to the present book, if G is a compact connected semisimple Lie group with Lie algebra $\mathfrak{g}$, we produced in [B08b] a deformation of the Casimir operator of G to a hypoelliptic Laplacian over $G \times \mathfrak{g}$ acting on smooth sections of $\Lambda^\cdot(\mathfrak{g}^*)$, and we showed that the supertrace of its heat kernel coincides with the trace of the scalar heat kernel of G. In particular, the spectrum of the Casimir operator is embedded as a fixed part of the spectrum of the hypoelliptic Laplacian. By making $b \to +\infty$, we recovered known formulas [Fr84] expressing the heat kernel of G as Poisson sums over its coroot lattice. The deformation of the Casimir operator in [B08b] was obtained via a deformation of the Dirac operator of Kostant [Ko76, Ko97], whose square coincides, up to a constant, with the Casimir operator.

The question arises of knowing whether, in the case of locally symmetric spaces of noncompact type, a deformation of the Casimir operator to a hypoelliptic operator is possible, which would have the same interpolation properties as before, and would produce a version of the Selberg trace formula. Such a construction would provide a justification for the formal considerations we made in section 0.1. The purpose of this book is to show that this question has a positive answer.

In this book, the spectral side of the trace formula will be essentially ignored. Our main result, which is given in chapter 6, is an explicit local formula for certain semisimple orbital integrals in a reductive group. The Selberg trace formula expresses the trace of certain trace class operators as a sum of orbital integrals. What is done in the book is to evaluate these orbital integrals individually, by a method that is inspired by index theory.

0.3 The hypoelliptic Laplacian on a symmetric space

First, we will explain the construction of the hypoelliptic Laplacian that is carried out in the present book.

Let G be a reductive Lie group with Lie algebra $\mathfrak{g}$, let K be a maximal compact subgroup of G with Lie algebra $\mathfrak{k}$, let B be an invariant nondegenerate bilinear form on $\mathfrak{g}$, and let $\mathfrak{g} = \mathfrak{p} \oplus \mathfrak{k}$ be the Cartan decomposition of $\mathfrak{g}$. Let $U(\mathfrak{g})$ be the enveloping algebra. Let $C^{\mathfrak{g}} \in U(\mathfrak{g})$ be the Casimir element.

Set $X = G/K$. Then X is contractible. Moreover, if $\rho^E : K \to \mathrm{Aut}(E)$ is a unitary representation of K, E descends to a Hermitian vector bundle F on X.

Then $\mathfrak{p}, \mathfrak{k}$ descend to vector bundles TX, N, TX being the tangent bundle of X. Moreover, $TX \oplus N$ can be canonically identified with the trivial vector bundle $\mathfrak{g}$ over X, and so it is equipped with a canonical flat connection. Let $\pi : \mathcal{X} \to X, \widehat{\pi} : \widehat{\mathcal{X}} \to X$ be the total spaces of $TX, TX \oplus N$.

Let $\widehat{c}(\mathfrak{g})$ be the Clifford algebra of $(\mathfrak{g}, -B)$. Following Kostant [Ko76, Ko97], in Definition 2.7.1, we introduce the Dirac operator $\widehat{D}^{\mathfrak{g}} \in \widehat{c}(\mathfrak{g}) \otimes U(\mathfrak{g})$, whose square coincides, up to a constant, with the negative of the Casimir.

In Definition 2.13.1, from $-\frac{1}{2}\widehat{D}^{X,2}$, we obtain the elliptic operator $\mathcal{L}^X$ acting on $C^\infty(X, F)$. Up to a constant, $\mathcal{L}^X$ coincides with the action of $\frac{1}{2}C^{\mathfrak{g}}$ on $C^\infty(X, F)$, so that $\mathcal{L}^X$ is an elliptic operator.

Also, the Dirac operator $\widehat{D}^{\mathfrak{g}}$ descends to an operator $\widehat{D}^{\mathfrak{g},X}$ acting on

$$C^\infty\left(\widehat{\mathcal{X}}, \widehat{\pi}^*\left(\Lambda^{\cdot}(T^*X \oplus N^*) \otimes F\right)\right).$$

In Definition 2.9.1 and in section 2.12, for $b > 0$, we introduce the operator $\mathfrak{D}_b^X$ that acts on the above vector space. It is given by the formula

$$\mathfrak{D}_b^X = \widehat{D}^{\mathfrak{g},X} + ic\left([Y^N, Y^{TX}]\right) + \frac{1}{b}\left(\mathcal{D}^{TX} + \mathcal{E}^{TX} - i\mathcal{D}^N + i\mathcal{E}^N\right). \quad (0.1)$$

In (0.1), c denotes the natural action of the Clifford algebra of $(TX \oplus N, B)$ on $\Lambda^{\cdot}(T^*X \oplus N^*)$. Also the operator $\mathcal{D}^{TX} + \mathcal{E}^{TX} - i\mathcal{D}^N + i\mathcal{E}^N$ is some version of the standard Dirac operator along the Euclidean fibre $TX \oplus N$ introduced by Witten [Wi82].

In Definition 2.13.1, we define the operator $\mathcal{L}_b^X$, which is given by

$$\mathcal{L}_b^X = -\frac{1}{2}\widehat{D}^{\mathfrak{g},X,2} + \frac{1}{2}\mathfrak{D}_b^{X,2}. \tag{0.2}$$

As an aside, we suggest the reader compare the operator that appears in the right-hand side of (0.1) with the right-hand side of (0.2).

The following formula is established in section 2.13,

$$\mathcal{L}_b^X = \frac{1}{2}\left|\left[Y^N, Y^{TX}\right]\right|^2 + \frac{1}{2b^2}\left(-\Delta^{TX\oplus N} + |Y|^2 - m - n\right) + \frac{N^{\Lambda^{\cdot}(T^*X\oplus N^*)}}{b^2}$$

$$+ \frac{1}{b}\left(\nabla_{Y^{TX}}^{C^\infty\left(TX\oplus N, \pi^*\left(\Lambda^{\cdot}(T^*X\oplus N^*)\widehat{\otimes}F\right)\right)} + \widehat{c}\left(\mathrm{ad}\left(Y^{TX}\right)\right)\right.$$

$$\left. - c\left(\mathrm{ad}\left(Y^{TX}\right) + i\theta\mathrm{ad}\left(Y^N\right)\right) - i\rho^E\left(Y^N\right)\right). \tag{0.3}$$

Let us just say that in (0.3), $\Delta^{TX\oplus N}$ is the Euclidean Laplacian along the fibre $TX \oplus N$. Also observe that in $\mathcal{L}_b^X$, the harmonic oscillator along the fibres of the Euclidean vector bundle $TX \oplus N$ appears with the factor $1/b^2$. By results of Hörmander [Hö67], the operator $\frac{\partial}{\partial t} + \mathcal{L}_b^X$ is hypoelliptic. Its structure is very close to the structure of the hypoelliptic Laplacian in [B05, BL08, B08a].

By adding a trivial matrix operator A to $\mathcal{L}^X, \mathcal{L}_b^X$, we obtain operators $\mathcal{L}_A^X, \mathcal{L}_{A,b}^X$. In the context of the present book, the operator $\mathcal{L}_{A,b}^X$ will be called a hypoelliptic Laplacian.

0.4 The hypoelliptic Laplacian and its heat kernel

In chapter 11, the proper functional analytic machinery is developed, in order to obtain a chain of Sobolev spaces on which the hypoelliptic Laplacian acts as an unbounded operator. This is done by inspiring ourselves from our previous work with Lebeau [BL08, chapter 15], which is valid for the case where the base manifold is compact. Also regularizing properties of its resolvent and of its heat operator are obtained. The heat operator is shown to be given by a smooth kernel.

A probabilistic method is also given to construct the heat operator on $\mathcal{X}$ or $\widehat{\mathcal{X}}$. The fact that the functional analytic and probabilistic constructions coincide is proved using the Itô calculus. The probabilistic construction of the heat kernel is relatively easy, but does not give the refined properties on the resolvent that one obtains by the functional analytic machinery.

In the remainder of the book, most of the hard analysis is done via the probabilistic construction of the heat kernel, while the functional analytic estimates do not play a significant role. This is because contrary to the situation in [BL08], where it was essential to obtain proper understanding of the spectral properties of the hypoelliptic Laplacian, here, this aspect can be essentially disregarded.

0.5 Elliptic and hypoelliptic orbital integrals

Let $\gamma \in G$ be semisimple. In chapter 4, we introduce the heat operators $\exp\left(-t\mathcal{L}_A^X\right)$, $\exp\left(-t\mathcal{L}_{A,b}^X\right)$, and we define the corresponding orbital integrals $\mathrm{Tr}^{[\gamma]}\left[\exp\left(-t\mathcal{L}_A^X\right)\right]$, $\mathrm{Tr}_{\mathrm{s}}^{[\gamma]}\left[\exp\left(-t\mathcal{L}_{A,b}^X\right)\right]$. These orbital integrals are said to be respectively elliptic and hypoelliptic. As the notation suggests, the elliptic orbital integrals are generalized traces, while the hypoelliptic orbital integrals are generalized supertraces. While the existence of the elliptic orbital integrals follows from standard Gaussian estimates for the heat kernel for $\exp\left(-t\mathcal{L}_A^X\right)$ on X, the existence of the orbital integrals for $\exp\left(-t\mathcal{L}_{A,b}^X\right)$ relies on a nontrivial estimate on the hypoelliptic heat kernel $q_{b,t}^X\left((x,Y),(x',Y')\right)$. This estimate is stated in Theorem 4.5.2. It says that given $\epsilon > 0, M > 0, \epsilon \leq M$, there exist $C > 0, C' > 0$ such that for $0 < b \leq M, \epsilon \leq t \leq M$, and $(x,Y),(x',Y') \in \widehat{\mathcal{X}}$, then

$$\left|q_{b,t}^X\left((x,Y),(x',Y')\right)\right| \leq C \exp\left(-C'\left(d^2\left(x,x'\right) + |Y|^2 + |Y'|^2\right)\right). \quad (0.1)$$

0.6 The limit as $b \to 0$

In Theorem 4.6.1, we prove that for any $b > 0, t > 0$,

$$\mathrm{Tr}_{\mathrm{s}}^{[\gamma]}\left[\exp\left(-t\mathcal{L}_{A,b}^X\right)\right] = \mathrm{Tr}^{[\gamma]}\left[\exp\left(-t\mathcal{L}_A^X\right)\right]. \quad (0.1)$$

Equation (0.1) is closely related to a corresponding identity established for ordinary traces over a compact Lie group G in [B08b].

The proof of (0.1) consists of two steps. The fact that the left-hand side of (0.1) does not depend on $b > 0$ is proved by a method very closely related to the proof of the McKean-Singer formula [McKS67] in index theory. It is in this sense that the book unifies index theory and the evaluation of orbital integrals.

The proof of (0.1) is then reduced to showing that as $b \to 0$, the left-hand side converges to the right-hand side. Proving this fact is obtained by a nontrivial analysis of the heat kernel for $\exp\left(-t\mathcal{L}_{A,b}^X\right)$. The uniform estimate (0.1) plays a crucial role in the proof. In section 0.8, we will give more details on the analytic arguments used in the book.

0.7 The limit as $b \to +\infty$: an explicit formula for the orbital integrals

Our final formula is obtained by making $b \to +\infty$ in (0.1). Let $d_\gamma(x) = d(x, \gamma x)$ be the displacement function associated with γ [BaGSc85], which is known to be a convex function. Let $X(\gamma) \subset X$ be its critical set, which is a totally geodesic submanifold of X. Then $X(\gamma)$ is the symmetric space associated with the centralizer $Z(\gamma) \subset G$ of γ. As $b \to +\infty$, the analysis of the hypoelliptic orbital integral $\mathrm{Tr_s}^{[\gamma]}\left[\exp\left(-t\mathcal{L}_{A,b}^X\right)\right]$ localizes near $X(\gamma)$. More precisely, in chapter 9, to obtain the asymptotics of $\mathrm{Tr_s}^{[\gamma]}\left[\exp\left(-t\mathcal{L}_{A,b}^X\right)\right]$, we choose $x \in X(\gamma)$, and we take the suitable expansion of $\mathcal{L}_{A,b}^X$ near the geodesic connecting x and γx. This expansion involves a rescaling of coordinates, and also a corresponding Getzler rescaling [Ge86] of the Clifford variables $\widehat{c}$.

Ultimately, the limit operator is a hypoelliptic operator acting on $\mathfrak{p} \times \mathfrak{g}$. The existence of a canonical flat connection over $TX \oplus N$ plays a critical role in the computations. Combining the existence of this flat connection with the existence of the central Casimir operator introduces two major differences with respect to what was done in [B05, BL08, B08a].

After conjugation, we may assume that $\gamma = e^a k^{-1}, a \in \mathfrak{p}, k \in K$, and $\mathrm{Ad}(k)a = a$. Let $\mathfrak{k}(\gamma) \subset \mathfrak{k}$ be the Lie algebra of the centralizer of γ in K. Our explicit formula for $\mathrm{Tr}^{[\gamma]}\left[\exp\left(-t\mathcal{L}_{A,b}^X\right)\right]$ is stated in Theorem 6.1.1. It is given by an explicit integral over $\mathfrak{k}(\gamma)$. In chapter 6, we show how to derive corresponding formulas for arbitrary kernels, which include the wave kernel. This is all the more remarkable, since, contrary to $\mathcal{L}_A^X$, $\mathcal{L}_{A,b}^X$ does not have a wave kernel.

0.8 The analysis of the hypoelliptic orbital integrals

In the analysis of the hypoelliptic orbital integrals, there is some overlap with the analysis of the hypoelliptic Laplacian in [BL08]. In [BL08], the Riemannian manifold X was assumed to be compact, and genuine traces or supertraces were considered. Here X is noncompact, and the orbital integrals that appear in (0.1) are defined using explicit properties of the corresponding heat kernels like the estimate in (0.1). Such estimates do not follow from [BL08].

In [BL08], the limit as $b \to 0$ of hypoelliptic supertraces was studied by functional analytic methods involving semiclassical pseudodifferential operators. Chapter 17 in [BL08] is entirely devoted to this question. Since here X is noncompact, and since we deal explicitly with the kernels of the considered operators, the results of [BL08] cannot be used as such.

Finally, the limit as $b \to +\infty$ involves questions that were not addressed in [BL08]. Again uniform estimates are needed.

In the present book, these analytic questions are dealt with by a combination of probabilistic and analytic methods. In probability, we use the Itô calculus, and also the stochastic calculus of variations, or Malliavin calculus [M78, St81b, B81a, M97].

0.9 The heat kernel for bounded b and the Malliavin calculus

Estimates like (0.1) are essentially obtained in three steps:

1. In chapter 12, we obtain rough estimates on scalar heat kernels associated with a scalar version of $\mathcal{L}_b^X$. By rough estimates, we mean uniform bounds on the kernels and their derivatives of arbitrary order. Such bounds are obtained using the Malliavin calculus. Also we study the limit as $b \to 0$ of the scalar hypoelliptic heat kernel.

2. In chapter 13, using the semigroup property of the scalar heat kernel combined with the rough bounds, we establish decay estimates similar to (0.1) for the scalar heat kernel.

3. In chapter 14, the estimates for the scalar heat kernel are transferred to the kernel $q_{b,t}^X$.

We will briefly explain why probabilistic methods are relevant for step (1). Note that the geodesic flow on X is a differential equation. Also, by (0.3), $\mathcal{L}_{A,b}^X$ is a differential operator of order 1 in the variables in X, while being of order 2 in the variables in $TX \oplus N$. To the scalar part of $\mathcal{L}_{A,b}^X$, one can associate a stochastic differential equation on $\widehat{\mathcal{X}}$, which projects to a differential equation on X. This differential equation is a perturbation of the geodesic flow. The heat equation semigroup for the scalar part of $\mathcal{L}_{A,b}^X$ describes the probability law in $\widehat{\mathcal{X}}$ of the corresponding diffusion process at a given time t.

The Malliavin calculus consists in exploiting the structure of the stochastic differential equation. More precisely, the properties of the heat kernel are obtained by using the fact that the scalar heat kernel is the image by the stochastic differential equation map Φ of a classical Brownian measure. Integration by parts on Wiener space can then be used to control the derivatives of the heat kernel. Estimates on heat kernels are ultimately obtained via the estimation of the Malliavin covariance matrix $\Phi'\Phi'^*$.

For bounded b, estimating the covariance matrix is essentially equivalent to the proper uniform control of an action functional depending on $b > 0$. For $b > 0$, if $x_s, 0 \le s \le t$ is a smooth curve with values in X with fixed $(x_0, \dot{x}_0), (x_t, \dot{x}_t)$, set

$$H_{b,t}(x.) = \frac{1}{2} \int_0^t \left(|\dot{x}|^2 + b^4 |\ddot{x}|^2 \right) ds. \tag{0.1}$$

This action functional was introduced in [B05] for smooth curves in X, and the corresponding variational problem was studied by Lebeau [L05]. Still problems remained because of the possible nonsmoothness of the solution of the associated Hamilton-Jacobi equation. However, it turns out that the estimate of the Malliavin covariance matrix represents a tangent variational problem, which can be controlled by the solution of a related variational problem, where X is replaced by the Euclidean vector space $\mathfrak{p}$. This problem had precisely been studied by Lebeau in [L05] as a warm-up to a full understanding of the variational problem on X.

This is why, prior to chapter 12, we devote chapter 10 to a detailed study of the above variational problem on an Euclidean vector space. The results of chapter 10 are used in chapter 12 to obtain a control of the integration by parts formula. Besides, when properly interpreted, chapter 10 can be viewed as an explicit verification of the soundness of our method of proof, when G is an Euclidean vector space. The fact that in this case, the intermediate steps can be made completely explicit is of special interest.

In chapter 12, we also obtain the limit as $b \to 0$ of the scalar hypoelliptic heat kernel. As explained before, when the base manifold is compact, a functional analytic version of this problem was solved in Bismut-Lebeau [BL08, chapter 17]. Here, this result is reobtained by probabilistic methods for the noncompact manifold X.

In chapter 13, we obtain a Gaussian decay of the scalar heat kernel similar to (0.1) using the rough bounds in chapter 12, and also by exploiting the semigroup property. Probabilistic methods are still used, but they are more elementary than in chapter 12. One difficulty is to show that as $b \to 0$, in spite of the fact that the energy of the underlying diffusion in X tends to $+\infty$ as $b \to 0$, this diffusion does not escape to infinity in X, the energy being absorbed by random fluctuations. Ultimately, the estimates follow easily from the rough bounds, and from Mehler's formula for the heat kernel of the harmonic oscillator.

In chapter 14, the transfer of the estimates for the scalar heat kernel to estimates on $q_{b,t}^X$ is obtained using a matrix version of the Feynman-Kac formula. In the case where F is nontrivial, the symmetric space associated with the complexification $K_{\mathbf{C}}$ of K plays an important role.

Let us point out that many of our estimates are still valid in the case of variable curvature. In particular, the techniques developed in the present book can be used to give a different proof of some of results established in [BL08], with explicit estimates on the heat kernels.

Finally, let us explain in more detail how we use our estimates. The uniform bounds on the heat kernels are needed to prove that the orbital integrals are well defined, and also to show that dominated convergence can be applied to the integrand defining the orbital integral when $b \to 0$ and $b \to +\infty$. The bounds on the higher derivatives are needed when computing the limit of the orbital integrals. This is done by establishing uniform bounds over compact subsets on the kernels and their derivatives, and by proving that heat kernels converge in a weak sense. Ultimately, we get pointwise convergence, which combined with the uniform bounds on the heat kernels, gives the convergence of the orbital integrals.

0.10 The heat kernel for large b, Toponogov, and local index

In chapter 15, the hypoelliptic heat kernel is studied as $b \to +\infty$. Note that by (0.3), for large $b > 0$, after rescaling $Y \in TX \oplus N$,

$$
\mathcal{L}_b^X \simeq \frac{1}{2} b^4 \left| [Y^{TX}, Y^N] \right|^2 + \frac{1}{2} |Y|^2 + \nabla_{Y^{TX}}^{C^\infty \left(TX \oplus N, \pi^* (\Lambda^{\cdot}(T^*X \oplus N^*) \widehat{\otimes} F) \right)}
$$
$$
+ \widehat{c} \left(\mathrm{ad} \left(Y^{TX} \right) \right) - c \left(\mathrm{ad} \left(Y^{TX} \right) + i\theta \mathrm{ad} \left(Y^N \right) \right) - i\rho^E \left(Y^N \right) + \mathcal{O} \left(1/b^2 \right).
\tag{0.1}
$$

Equation (0.1) indicates that the diffusion associated with the scalar part of $\mathcal{L}_b^X$ tends to propagate along the geodesic flow. Still, because we have to control the corresponding heat kernel, we ultimately need to obtain a quantitative estimate on how much this diffusion differs from the geodesic flow.

This question is dealt with in two steps.

1. Let $\varphi_t|_{t \in \mathbf{R}}$ be the geodesic flow in $\mathcal{X}$. Then $\gamma \varphi_1^{-1}$ is a symplectic transformation of $\mathcal{X}$, whose fixed point set $\mathcal{F}_\gamma$ is simply related to $X(\gamma)$, the critical set of d_γ. A purely geometric question, which is dealt with in the end of chapter 3, is to find how much the return map $\varphi_1^{-1} \gamma$ differs from the identity away from $\mathcal{F}_\gamma$. The corresponding quantitative estimates are obtained by using Toponogov's theorem repeatedly.

2. In chapter 15, these estimates are combined with the rough bounds on the heat kernel $q_{b,t}^X$ to obtain the proper uniform estimates for b large.

   Local index theoretic methods are used to control local cancellations in the supertrace of the heat kernel near $\mathcal{F}_\gamma$ as $b \to +\infty$.

0.11 The hypoelliptic Laplacian and the wave equation

A crucial observation is made in sections 12.3 and 14.2, which relates the heat equation for the scalar version of $\mathcal{L}_b^X$ to the classical wave equation on X. It is shown that after averaging in the fibre variables, the heat equation on $\mathcal{X}$ or $\widehat{\mathcal{X}}$ descends to a nonlinear version of the wave equation on X. This observation is at the heart of some of the key probabilistic arguments used in chapters 12 and 14 to establish the uniform Gaussian decay in (0.1). More fundamentally, it is connected with the fact that the Hamiltonian differential equation of order 1 for the Hamiltonian flow on $\mathcal{X}$ descends to a differential equation of order 2 on X for the geodesics. In some sense, this descent argument propagates to the heat equation for the hypoelliptic Laplacian.

0.12 The organization of the book

The book is divided into two parts. A first part, which includes chapters 1–9, contains the construction of the objects which are considered in the book, the geometric results which are needed and their proof, the statement of the main results and their proofs. The analytic results which are needed in the proofs are themselves stated without proof.

The detailed proof of the analytic results is deferred to a second part, which includes chapters 10–15.

This book is organized as follows. In chapter 1, we recall general results on Clifford and Heisenberg algebras.

In chapter 2, we construct the hypoelliptic Laplacian $\mathcal{L}_b^X$ over $\widehat{\mathcal{X}}$.

In chapter 3, we establish various geometric results. If $\gamma \in G$ is semisimple, we introduce the displacement function d_γ, its critical set $X(\gamma)$, and the coordinate system on X, which one obtains from the normals to $X(\gamma)$. Also we study the return map $\varphi_1^{-1}\gamma$ using Toponogov's theorem.

In chapter 4, we define the elliptic and hypoelliptic orbital integrals, and we establish identity (0.1).

In chapter 5, if $Y_0^{\mathfrak{k}} \in \mathfrak{k}(\gamma)$, we evaluate the heat kernel for a hypoelliptic operator $\mathcal{P}_{a,Y_0^{\mathfrak{k}}}$ acting on $\mathfrak{p} \oplus \mathfrak{g}$, and we compute the supertrace $J_\gamma\left(Y_0^{\mathfrak{k}}\right)$ of this heat kernel. In chapter 9, the operator $\mathcal{P}_{a,Y_0^{\mathfrak{k}}}$ will appear as a rescaled limit of $\mathcal{L}_b^X$ when $b \to +\infty$. This chapter can be read independently.

In chapter 6, we state without proof our main result, which expresses the elliptic orbital integrals associated with the heat kernel as a Gaussian integral on $\mathfrak{k}(\gamma)$. Also we show how to derive from this formula a corresponding formula for other kernels, which include the wave kernel.

In chapter 7, we show the compatibility of our formula for the orbital integrals to the Atiyah-Singer index theorem [AS68a, AS68b] and to the Lefschetz fixed point formulas of Atiyah-Bott [ABo67, ABo68]. Also we recover results of Moscovici-Stanton [MoSt91] that are related to the evaluation of the Ray-Singer analytic torsion of locally symmetric spaces.

In chapter 8, we evaluate explicitly the integrals over $\mathfrak{k}(\gamma)$ when γ verifies a simple commutation relation, and we recover Selberg's trace formula when $G = \mathrm{SL}_2(\mathbf{R})$.

In chapter 9, we prove the formula that was stated in chapter 6. The proof relies on estimates established in the second part of the book.

In chapter 10, we establish detailed results on the variational problem associated with the action $H_{b,t}$ in (0.1) when X is an Euclidean vector space E, and we state various versions of Mehler's formula. Also we establish key estimates on integrals involving the heat kernel of the harmonic oscillator. This chapter can be read independently.

In chapter 11, given $b > 0$, we adapt the functional analytic methods of [BL08] to construct the resolvent and the heat kernel for the hypoelliptic Laplacian.

In chapter 12, we obtain rough estimates for the heat kernel that is associated with a scalar hypoelliptic operator on $\mathcal{X}$ and on $\widehat{\mathcal{X}}$. Such estimates are given for b bounded, and for b large. Also we study the limit of the heat kernel as $b \to 0$.

In chapter 13, we obtain the analogue of the uniform estimates in (0.1) for the scalar heat kernels on $\mathcal{X}$ and on $\widehat{\mathcal{X}}$.

In chapter 14, we establish estimates for the heat kernel $q_{b,t}^X$ for bounded b, which include (0.1).

Finally, in chapter 15, we establish the required estimates on $q_{b,t}^X$ for large b.

As explained before, many ideas and techniques used in the present book have already been tested in our previous work with Lebeau [BL08]. Still the present book is largely self-contained, except for the analytic results of [BL08, chapter 15], whose use can in part be avoided. Some familiarity with [B08b] could be useful. Also we have tried to make the reading easier, by separating the results from their proofs, and geometric arguments from analytic arguments. Also we tried to justify the role of probabilistic arguments as much as possible. The length of the book can be partly explained by the fact that we rederive many of the technical results of [BL08] by different methods.

If $\mathcal{A}$ is a $\mathbf{Z}_2$-graded algebra, if $a, b \in \mathcal{A}$, $[a, b]$ will be our notation for the supercommutator of a, b, so that

$$[a, b] = ab - (-1)^{\deg(a)\deg(b)} ba. \tag{0.1}$$

Also, in most of the book, we will use Einstein summation conventions.

Moreover, in the whole text, constants C will always be positive. If a constant depends on a parameter ϵ, it will often be written as C_ϵ. Also positive constants will often be denoted using the same notation, although they may well be different.

The results contained in this book have been announced in [B09b].

Chapter One

Clifford and Heisenberg algebras

The purpose of this chapter is to recall various results on Clifford algebras and Heisenberg algebras. The results of this chapter will be used in our construction of the hypoelliptic Laplacian over a symmetric space.

This chapter is organized as follows. In section 1.1, we introduce the Clifford algebra of a vector space V equipped with a symmetric bilinear form B.

In section 1.2, we specialize the construction of the Clifford algebra to the case of $V \oplus V^*$.

In section 1.3, if (V, ω) is a symplectic vector space, we construct the associated Heisenberg algebra.

In section 1.4, we specialize the construction of the Heisenberg algebra to the case of $V \oplus V^*$.

In section 1.5, we consider the combination of the Clifford and Heisenberg algebras for $V \oplus V^*$, and we construct the complex $\left(\Lambda^{\cdot} (V^*) \otimes S^{\cdot} (V^*), \overline{d} \right)$, which is the subcomplex of polynomial forms in the de Rham complex.

Finally, in section 1.6, when V is equipped with a scalar product, this complex is related to a Witten complex over V.

1.1 The Clifford algebra of a real vector space

Let V be a finite dimensional real vector space of dimension n, let B be a real-valued symmetric bilinear form on V.

Let $c(V)$ be the Clifford algebra associated to (V, B). Namely, $c(V)$ is the algebra generated over $\mathbf{R}$ by $1, a \in V$ and the commutation relations for $a, b \in V$,

$$ab + ba = -2B(a, b). \qquad (1.1.1)$$

We denote by $\widehat{c}(V)$ the Clifford algebra associated to $-B$. Then $c(V), \widehat{c}(V)$ are filtered by length, and their corresponding Gr is just $\Lambda^{\cdot}(V)$. Also they are $\mathbf{Z}_2$-graded algebras. If B is zero, then $c(V), \widehat{c}(V)$ coincide with $\Lambda^{\cdot}(V)$.

In the sequel, we assume that B is nondegenerate, so that V and V^* can be identified by the form B. Let $\varphi : V \to V^*$ denote the corresponding isomorphism, so that if $a, b \in B$,

$$\langle \varphi a, b \rangle = B(a, b). \qquad (1.1.2)$$

Let B^* be the corresponding bilinear form on V^*. Then B^* induces a nondegenerate symmetric bilinear form on $\Lambda^{\cdot}(V^*)$, which we still denote B^*.

If $a \in V$, let $c(a), \widehat{c}(a) \in \operatorname{End}(\Lambda^{\cdot}(V^{*}))$ be given by

$$c(a) = \varphi a \wedge - i_{a}, \qquad\qquad \widehat{c}(a) = \varphi a \wedge + i_{a}. \qquad (1.1.3)$$

Then $c(a)$ and $\widehat{c}(a)$ are odd operators, which are respectively antisymmetric and symmetric with respect to B^{*}.

If $a, b \in V$, then

$$[c(a), c(b)] = -2B(a, b), \quad [\widehat{c}(a), \widehat{c}(b)] = 2B(a, b), \qquad [c(a), \widehat{c}(b)] = 0. \qquad (1.1.4)$$

By (1.1.4), we find that $\Lambda^{\cdot}(V^{*})$ is a $c(V)$ and a $\widehat{c}(V)$ Clifford module. Also if $a \in V$, $\widehat{c}(a)$ is the negative of $c(a)$ associated with $-B$. This is a discrepancy that is unavoidable, given the fact that the definition of the operator $\widehat{c}(a)$ is well established by tradition. If A lies in $c(V)$ or $\widehat{c}(V)$, we denote by $c(A), \widehat{c}(A)$ the corresponding endomorphisms of $\Lambda^{\cdot}(V^{*})$ associated with the representation of the Clifford algebras $c(V), \widehat{c}(V)$ defined in (1.1.3).

The maps $A \in c(V) \to c(A) 1 \in \Lambda^{\cdot}(V^{*})$, $A' \in \widehat{c}(V) \to (-1)^{\deg(A')} \widehat{c}(A')$ $1 \in \Lambda^{\cdot}(V^{*})$ identify $c(V), \widehat{c}(V)$ as vector spaces to $\Lambda^{\cdot}(V^{*})$. If $\alpha \in \Lambda^{\cdot}(V^{*})$, we will denote by $c(\alpha), \widehat{c}(\alpha)$ the corresponding elements in $c(V), \widehat{c}(V)$. There will be no risk of confusion with the above definition of $c(A), \widehat{c}(A)$.

Let $e_{1}, \dots, e_{n}$ be a basis of V, and let $e_{1}^{*}, \dots, e_{n}^{*}$ be the dual basis of V with respect to B, so that $B(e_{i}, e_{j}^{*}) = \delta_{ij}$. Note that if $a \in V$,

$$a = \sum_{i=1}^{n} B(e_{i}^{*}, a) e_{i}. \qquad (1.1.5)$$

Let $e^{1}, \dots, e^{n}$ be the basis of V^{*} which is dual to the basis $e_{1}, \dots, e_{n}$. Of course for $1 \leq i \leq n, e^{i} = \varphi e_{i}^{*}$.

Take $\alpha \in \Lambda^{p}(V^{*})$, so that

$$\alpha = \sum_{1 \leq i_{1} < \dots < i_{p} \leq n} \alpha(e_{i_{1}}, \dots, e_{i_{p}}) e^{i_{1}} \wedge \dots \wedge e^{i_{p}}. \qquad (1.1.6)$$

Then

$$c(\alpha) = \frac{1}{p!} \sum_{1 \leq i_{1}, \dots, i_{p} \leq n} \alpha(e_{i_{1}}^{*}, \dots, e_{i_{p}}^{*}) c(e_{i_{1}}) \dots c(e_{i_{p}}), \qquad (1.1.7)$$

$$\widehat{c}(\alpha) = \frac{1}{p!} \sum_{1 \leq i_{1}, \dots, i_{p} \leq n} \alpha(e_{i_{1}}^{*}, \dots, e_{i_{p}}^{*}) \widehat{c}(e_{i_{1}}) \dots \widehat{c}(e_{i_{p}}).$$

Note that when replacing B by $-B$, e_{i}^{*} is changed to $-e_{i}^{*}$ and $c(e_{i})$ into $-\widehat{c}(e_{i})$, which explains why there is no sign in the second formula in (1.1.7).

Moreover, the map $f \to if$ induces the canonical isomorphism

$$c(V) \otimes_{\mathbf{R}} \mathbf{C} \simeq \widehat{c}(V) \otimes_{\mathbf{R}} \mathbf{C}. \qquad (1.1.8)$$

It follows from (1.1.8) that the complexified Clifford algebra $c(V) \otimes_{\mathbf{R}} \mathbf{C}$ does not depend on the choice of B, which can as well be taken to be a scalar product. However, the Clifford representation which to $a \in V$ associates the corresponding element in $c(V)$ depends on B.

Let $\mathcal{A}(V)$ be the Lie algebra of the endomorphisms of V that are anti-symmetric with respect to B. Then $\mathcal{A}(V)$ embeds as a Lie algebra in $c(V)$. If $A \in \mathcal{A}(V)$, the image $c(A)$ of A in $c(V)$ is given by

$$c(A) = \frac{1}{4}B\left(Ae_i^*, e_j^*\right) c(e_i) c(e_j). \tag{1.1.9}$$

Moreover, if $a \in V$,

$$[c(A), c(a)] = c(Aa). \tag{1.1.10}$$

When replacing B by $-B$, we denote by $\widehat{c}(A)$ the corresponding element in $\widehat{c}(V)$. By (1.1.10), we obtain

$$\widehat{c}(A) = -\frac{1}{4}B\left(Ae_i^*, e_j^*\right) \widehat{c}(e_i) \widehat{c}(e_j). \tag{1.1.11}$$

Instead of (1.1.10), we now have

$$[\widehat{c}(A), \widehat{c}(a)] = \widehat{c}(Aa). \tag{1.1.12}$$

When acting on $\Lambda^{\cdot}(V^*)$ as indicated in (1.1.3), $c(A), \widehat{c}(A)$ are antisymmetric operators with respect to B^*.

Recall that if $A \in \mathrm{End}(V)$, A acts on $\Lambda^{\cdot}(V^*)$, and this action is given by

$$A|_{\Lambda^{\cdot}(V^*)} = -\left\langle Ae_i, e^j \right\rangle e^i i_{e_j}. \tag{1.1.13}$$

Then one verifies easily that if $A \in \mathcal{A}(V)$,

$$A|_{\Lambda^{\cdot}(V^*)} = c(A) + \widehat{c}(A). \tag{1.1.14}$$

Let $N^{\Lambda^{\cdot}(V^*)}$ be the number operator acting on $\Lambda^{\cdot}(V^*)$, i.e., $N^{\Lambda^{\cdot}(V^*)}$ acts on $\Lambda^p(V^*)$ by multiplication by p. Then one verifies easily that

$$N^{\Lambda^{\cdot}(V^*)} - \frac{n}{2} = \frac{1}{2}\sum_{i=1}^{n} c(e_i^*) \widehat{c}(e_i). \tag{1.1.15}$$

Let (V', B') be another pair like (V, B). Then $(V \oplus V', B \oplus B')$ is still another such a pair, and moreover

$$c(V \oplus V') = c(V) \widehat{\otimes} c(V'), \qquad \widehat{c}(V \oplus V') = \widehat{c}(V) \widehat{\otimes} \widehat{c}(V'). \tag{1.1.16}$$

In (1.1.16), $\widehat{\otimes}$ denotes the $\mathbf{Z}_2$-graded tensor product.

Let V, V' be real Euclidean vector spaces, let $c(V), \widehat{c}(V), c(V'), \widehat{c}(V')$ be the corresponding Euclidean Clifford algebras. We denote by V'_- the vector space V' equipped with the negative of its Euclidean product. By (1.1.16), we find that if $W = V \oplus V'_-$, then

$$c(W) = c(V) \widehat{\otimes} \widehat{c}(V'), \qquad \widehat{c}(W) = \widehat{c}(V) \widehat{\otimes} c(V'). \tag{1.1.17}$$

1.2 The Clifford algebra of $V \oplus V^*$

Let V be a real vector space, and let V^* be its dual. Put $W = V \oplus V^*$. Then W is canonically equipped with minus half of the symmetric bilinear pairing between V and V^*. If $a \in V, b \in V^*$, put

$$c(a) = i_a, \qquad\qquad c(b) = b \wedge. \tag{1.2.1}$$

Then (1.2.1) turns $\Lambda^{\cdot}(V^*)$ into a $c(W)$ Clifford module. Moreover, inspection of (1.1.3) and (1.2.1) shows that once a nondegenerate symmetric bilinear form B is fixed on V,

$$c(V \oplus V^*) \simeq c(V) \widehat{\otimes} \widehat{c}(V). \tag{1.2.2}$$

Let $I \in \mathrm{End}(V \oplus V^*)$ be the map that is -1 on V and 1 on V^*. Then $I \in \mathcal{A}(V \oplus V^*)$. Let $e_1, \ldots, e_n$ be a basis of V, and let $e^1, \ldots, e^n$ be the corresponding dual basis of V^*. Then

$$c(I) = \sum_{i=1}^{n} c\left(e^i\right) c(e_i) - \frac{n}{2}. \tag{1.2.3}$$

Recall that $N^{\Lambda^{\cdot}(V^*)}$ is the number operator of $\Lambda^{\cdot}(V^*)$. By (1.2.3), we have the identity of operators acting on $\Lambda^{\cdot}(V^*)$,

$$c(I) = N^{\Lambda^{\cdot}(V^*)} - n/2. \tag{1.2.4}$$

Assume now that B is a scalar product on V. Let S^V be the Hermitian vector space of V spinors. Then S^V is a $c(V)$ Clifford module. Moreover, if $a \in V$, $c(a)$ acts as a skew-adjoint automorphism of S^V. Let V_- denote the vector space V equipped with the bilinear form $-B$. If V is replaced by V_-, S^V is of course a $c(V_-)$ Clifford module. However, if $a \in V$, the associated Clifford action $c_-(a)$ is given by

$$c_-(a) = ic(a), \tag{1.2.5}$$

so that $c_-(a)$ acts like a self-adjoint automorphism of S^V. If $A \in \mathrm{End}(V)$ is antisymmetric, $c(A)$ and $c_-(A)$ coincide.

1.3 The Heisenberg algebra

Let V be a finite dimensional real vector space of dimension n. Let $S^{\cdot}(V^*) = \bigoplus_{i=0}^{+\infty} S^i(V^*)$ be the symmetric algebra of V^*, i.e., the polynomial algebra of V.

Let ω be an antisymmetric bilinear form on V. The Heisenberg algebra $b(V)$ is the algebra generated by 1, $a \in V$ and the commutation relations for $e, f \in V$,

$$ef - fe = 2\omega(e, f). \tag{1.3.1}$$

Similarly we denote by $\widehat{b}(V)$ the Heisenberg algebra associated with the form $-\omega$. Again $b(V), \widehat{b}(V)$ are filtered by length, and the corresponding $\mathrm{Gr}^{\cdot}$ is just $S^{\cdot}(V)$. If ω vanishes, $b(V), \widehat{b}(V)$ coincide with $S^{\cdot}(V)$. In the sequel, we assume that ω is nondegenerate, i.e., ω is a symplectic form on V, so that n is even.

If $e \in V$, let $e^* \in V^*$ correspond to e by the symplectic form, so that if $f \in V$,

$$\langle e^*, f \rangle = \omega(e, f). \tag{1.3.2}$$

If $e \in V, f \in V^*$, then e and f act on $S^{\cdot}(V^*)$ as annihilation and creation operators. More precisely e acts on $S^{\cdot}(V^*)$ by the derivation ∇_e, and f by multiplication. If $e \in V$, set

$$b(e) = e^* - \nabla_e, \qquad\qquad \widehat{b}(e) = e^* + \nabla_e. \qquad (1.3.3)$$

The operators $b(e), \widehat{b}(e)$ act on $S^{\cdot}(V^*)$.

In the sequel, the algebras $b(V), \widehat{b}(V)$ will be considered as trivially $\mathbf{Z}_2$-graded, so that supercommutators in this algebra are just ordinary commutators. Similarly the vector space $S^{\cdot}(V^*)$ will be considered as even, or as trivially $\mathbf{Z}_2$-graded. Then

$$[b(e), b(f)] = 2\omega(e,f), \quad \left[\widehat{b}(e,)\widehat{b}(f)\right] = -2\omega(e,f), \quad \left[b(e),\widehat{b}(f)\right] = 0. \qquad (1.3.4)$$

Equation (1.3.4) indicates that $S^{\cdot}(V^*)$ is a $b(V)$ and a $\widehat{b}(V)$ Heisenberg module. Moreover, $\widehat{b}(e)$ is the negative of the $b(e)$ associated with $-\omega$. The maps $A \in b(V) \to b(A)1 \in S^{\cdot}(V^*), A' \in \widehat{b}(V) \to (-1)^{\deg(A')}\widehat{b}(A')1 \in S^{\cdot}(V^*)$ are isomorphisms of vector spaces. If $\alpha \in S^{\cdot}(V^*)$, we denote by $b(\alpha), \widehat{b}(\alpha)$ the corresponding elements in $b(V), \widehat{b}(V)$. Then we have the obvious analogue of (1.1.7).

If $e_1, \ldots, e_n$ is a basis of V, let $e_1^*, \ldots, e_n^*$ be the corresponding symplectically dual basis of V, so that $\omega(e_i^*, e_j) = \delta_{i,j}$. If $e \in V$,

$$e = \sum_{i=1}^n \omega(e_i^*, e)e_i = -\sum_{i=1}^n \omega(e_i, e)e_i^*. \qquad (1.3.5)$$

Let $\mathcal{B}(V)$ be the algebra of real endomorphisms of V that are antisymmetric with respect to ω. This is exactly the Lie algebra of the symplectic group $\mathrm{Sp}(V)$. Then $\mathcal{B}(V)$ embeds as a Lie algebra in $b(V)$. If $A \in \mathcal{B}(V)$, the image $b(A)$ of A in $b(V)$ is given by

$$b(A) = \frac{1}{4}\omega(Ae_i^*, e_j^*)b(e_i)b(e_j). \qquad (1.3.6)$$

Moreover, we have the analogue of (1.1.10), i.e., if $A \in \mathcal{B}(V), a \in V$,

$$[b(A), b(a)] = b(Aa). \qquad (1.3.7)$$

When replacing ω by $-\omega$, we denote by $\widehat{b}(A)$ the corresponding element of $\widehat{b}(V)$. Then

$$\widehat{b}(A) = -\frac{1}{4}\omega(Ae_i^*, e_j^*)\widehat{b}(e_i)\widehat{b}(e_j). \qquad (1.3.8)$$

Moreover, if $a \in V$,

$$\left[\widehat{b}(A), \widehat{b}(a)\right] = \widehat{b}(Aa). \qquad (1.3.9)$$

1.4 The Heisenberg algebra of $V \oplus V^*$

If V is a real vector space of dimension n, and if V^* is its dual, set $W = V \oplus V^*$. Then W is canonically equipped with the symplectic form

$$\omega \left((e, f), (e', f') \right) = \langle f, e' \rangle - \langle f', e \rangle . \qquad (1.4.1)$$

If $e \in V, f \in V^*$, set

$$b(e) = \nabla_e, \qquad\qquad b(f) = f. \qquad (1.4.2)$$

Then (1.4.2) turns $S^{\cdot}(V^*)$ into a $b(W)$ Heisenberg module that is associated with minus half of the symplectic form ω.

Moreover, inspection of (1.3.3) and (1.4.2) shows that if n is even and if ω is a symplectic form on V,

$$b(V \oplus V^*) \simeq b(V) \otimes \widehat{b}(V). \qquad (1.4.3)$$

Let $I \in \mathrm{End}(V \oplus V^*)$ be defined as in section 1.2. Then $I \in \mathcal{B}(V \oplus V^*)$. Moreover,

$$b(I) = \sum_{i=1}^{n} b\left(e^i\right) b\left(e_i\right) + \frac{n}{2}. \qquad (1.4.4)$$

Let $N^{S^{\cdot}(V^*)}$ be the number operator of $S^{\cdot}(V^*)$. Then when acting on $S^{\cdot}(V^*)$, we get

$$b(I) = N^{S^{\cdot}(V^*)} + \frac{n}{2}. \qquad (1.4.5)$$

Let us now assume that V is equipped with a scalar product. Then each $S^i(V^*)$ inherits an associated scalar product. We normalize this scalar product so that if $V = \mathbf{R}$, then

$$\left\| 1^{\otimes i} \right\|^2 = i!, \qquad (1.4.6)$$

and we equip $S^{\cdot}(V^*)$ with the direct sum of the scalar products on the $S^i(V^*)$. Let $\overline{S^{\cdot}(V^*)}$ denote the Hilbert completion of the prehilbertian vector space $S^{\cdot}(V^*)$.

For $a > 0$, let δ_a be the dilation

$$\delta_a P(Y) = P(aY). \qquad (1.4.7)$$

Let Δ^V denote the Laplacian on V. Let $L_2(V)$ be the corresponding Hilbert space of square integrable real functions on V. Recall that $S^{\cdot}(V^*)$ has been identified with the algebra of polynomials on V.

Definition 1.4.1. Let $T : S^{\cdot}(V^*) \to L_2(V)$ be the map

$$TP = \pi^{-n/4} \delta_{\sqrt{2}} \exp\left(- |Y|^2 / 4 \right) e^{-\Delta^V / 2} P. \qquad (1.4.8)$$

Note that since P is a polynomial, $e^{-\Delta^V / 2} P$ is defined by taking the obvious formal expansion of $\exp\left(-\Delta^V / 2 \right)$. Its inverse, the Bargmann kernel, is given by

$$Bf = \pi^{n/4} e^{\Delta^V / 2} \exp\left(|Y|^2 / 4 \right) \delta_{1/\sqrt{2}} f. \qquad (1.4.9)$$

Note that in (1.4.9), $e^{\Delta^V / 2}$ is defined via the standard heat kernel of V.

Then T extends to an isometry from $\overline{S^{\cdot}(V^*)} \to L_2(V)$, and B to the inverse isometry from $L_2(V)$ into $\overline{S^{\cdot}(V^*)}$.

If $e \in V$, let $e^* \in V^*$ correspond to e by the scalar product of V. If $e \in V$, set

$$b\left(e\right) = \frac{1}{\sqrt{2}}\left(\nabla_e + \langle e, Y \rangle\right), \qquad b\left(e^*\right) = \frac{1}{\sqrt{2}}\left(-\nabla_e + \langle e, Y \rangle\right). \qquad (1.4.10)$$

One verifies easily that T, B intertwine the operators $b\left(e\right), b\left(e^*\right)$ in (1.4.2) and (1.4.10). In the above representation of the Heisenberg algebra,

$$b\left(I\right) - \frac{n}{2} = \frac{1}{2}\left(-\Delta^V + |Y|^2 - n\right), \qquad (1.4.11)$$

i.e., $b\left(I\right) - \frac{n}{2}$ is the standard harmonic oscillator acting on $L_2\left(V\right)$ as an unbounded operator.

Let $\mathrm{Aut}\left(V\right)$ be the group of automorphisms of V. Then $\mathrm{Aut}\left(V\right)$ acts isometrically on $L_2\left(V\right)$. There is a corresponding isometric action of $\mathrm{Aut}\left(V\right)$ on $\overline{S^{\cdot}\left(V^*\right)}$. If $h \in \mathrm{Aut}\left(V\right)$ is an isometry, the action of h on $\overline{S^{\cdot}\left(V^*\right)}$ is just the action of h induced by the natural action on $S^{\cdot}\left(V^*\right)$. This is not the case in general. The above considerations are directly related to the metaplectic representation for $V \oplus V^*$.

1.5 The Clifford-Heisenberg algebra of $V \oplus V^*$

Let V be a finite dimensional real vector space of dimension n. As we saw in sections 1.2 and 1.4, we can define the canonical Clifford algebra $c\left(V \oplus V^*\right)$ associated to minus half of the canonical symmetric pairing on $V \oplus V^*$ and the canonical Heisenberg algebra $b\left(V \oplus V^*\right)$ associated to minus half of the symplectic form of $V \oplus V^*$. In the sequel, we will often not distinguish between the elements of the algebra $c\left(V \oplus V^*\right) \otimes b\left(V \oplus V^*\right)$, and their natural action on $\Lambda^{\cdot}\left(V^*\right) \otimes S^{\cdot}\left(V^*\right)$, which was defined in (1.2.1) and (1.4.2).

Moreover, the Clifford algebras are $\mathbf{Z}_2$-graded algebras, while the Heisenberg algebras are trivially $\mathbf{Z}_2$-graded. We equip $c\left(V \oplus V^*\right) \otimes b\left(V \oplus V^*\right)$ with the corresponding $\mathbf{Z}_2$-grading, which just comes from the grading of $c\left(V \oplus V^*\right)$. Similarly we equip $\Lambda^{\cdot}\left(V^*\right) \otimes S^{\cdot}\left(V^*\right)$ with the $\mathbf{Z}_2$-grading coming from the $\mathbf{Z}_2$-grading of $\Lambda^{\cdot}\left(V^*\right)$.

Let $e_1, \ldots, e_n$ be a basis of V, let $e^1, \ldots, e^n$ be the corresponding dual basis of V^*. Put

$$\overline{d} = \sum_{i=1}^{n} c\left(e^i\right) b\left(e_i\right), \qquad \overline{d}^* = \sum_{i=1}^{m} c\left(e_i\right) b\left(e^i\right). \qquad (1.5.1)$$

Of course $\overline{d}, \overline{d}^*$ do not depend on the choice of the basis $e_1, \ldots, e_n$. Clearly

$$\overline{d}^2 = 0, \qquad\qquad \overline{d}^{*,2} = 0. \qquad (1.5.2)$$

Moreover, using the notation in (1.2.3), (1.4.4),

$$\left[\overline{d}, \overline{d}^*\right] = c\left(I\right) + b\left(I\right). \qquad (1.5.3)$$

Using (1.2.4), (1.4.5), and (1.5.3), we obtain

$$\left[\overline{d}, \overline{d}^*\right] = N^{S^{\cdot}\left(V^*\right)} + N^{\Lambda^{\cdot}\left(V^*\right)}. \qquad (1.5.4)$$

Let us now give a differential geometric interpretation of (1.5.4). Let Y be the tautological section of V over V. Then Y can be identified with the corresponding radial vector field.

Let $(\Omega^{\cdot}(V), d^V)$ be the de Rham complex of V. Let L_Y be the Lie derivative associated with Y. By Cartan's formula, we have the identity of operators acting on $\Omega^{\cdot}(V)$,

$$L_Y = \left[d^V, i_Y \right]. \tag{1.5.5}$$

Note that $\Lambda^{\cdot}(V^*) \otimes S^{\cdot}(V^*)$ is the vector subspace of $\Omega^{\cdot}(V)$ of polynomial forms on V, and is $\mathbf{Z}$-graded by the operator $N^{\Lambda^{\cdot}(V^*)}$. Then d^V, i_Y, L_Y act on $\Lambda^{\cdot}(V^*) \otimes S^{\cdot}(V^*)$. One then verifies easily that

$$\overline{d} = d^V |_{\Lambda^{\cdot}(V^*) \otimes S^{\cdot}(V^*)}, \qquad \overline{d}^* = i_Y |_{\Lambda^{\cdot}(V^*) \otimes S^{\cdot}(V^*)}, \tag{1.5.6}$$
$$N^{S^{\cdot}(V^*)} + N^{\Lambda^{\cdot}(V^*)} = L_Y |_{\Lambda^{\cdot}(V^*) \otimes S^{\cdot}(V^*)}.$$

Identity (1.5.4) is just a reflection of (1.5.5), (1.5.6).

1.6 The Clifford-Heisenberg algebra of $V \oplus V^*$ when V is Euclidean

We still consider the Clifford algebra $c(V \oplus V^*)$ and the Heisenberg algebra $b(V \oplus V^*)$ as in section 1.5.

Assume that V is equipped with a scalar product $\langle \rangle$. Let $e_1, \ldots, e_n$ be an orthonormal basis of V, and let $e^1, \ldots, e^n$ be the corresponding dual basis of V^*. Put

$$\mathcal{D} = \frac{1}{\sqrt{2}} \sum_{i=1}^n c\left(e^i - e_i\right) b\left(e_i - e^i\right), \quad \mathcal{E} = \frac{1}{\sqrt{2}} \sum_{i=1}^n c\left(e^i + e_i\right) b\left(e_i + e^i\right). \tag{1.6.1}$$

Then $\mathcal{D}, \mathcal{E}$ do not depend on the choice of the orthonormal basis $e_1, \ldots, e_n$. Moreover,

$$\overline{d} + \overline{d}^* = \frac{1}{\sqrt{2}} \left(\mathcal{D} + \mathcal{E}\right), \tag{1.6.2}$$

so that

$$\left[\overline{d}, \overline{d}^*\right] = \frac{1}{2} \left(\mathcal{D} + \mathcal{E}\right)^2. \tag{1.6.3}$$

Similarly, set

$$\mathcal{D}' = \frac{1}{\sqrt{2}} \sum_{i=1}^n c\left(e^i + e_i\right) b\left(e_i - e^i\right), \quad \mathcal{E}' = \frac{1}{\sqrt{2}} \sum_{i=1}^n c\left(e^i - e_i\right) b\left(e_i + e^i\right). \tag{1.6.4}$$

Then instead of (1.6.2), we now have

$$\overline{d} - \overline{d}^* = \frac{1}{\sqrt{2}} \left(\mathcal{D}' + \mathcal{E}'\right), \tag{1.6.5}$$

and instead of (1.6.3), we get

$$\left[\overline{d},\overline{d}^*\right] = -\frac{1}{2}\left(\mathcal{D}'+\mathcal{E}'\right)^2. \tag{1.6.6}$$

Recall that d^V is the de Rham operator on V, and that Y is the tautological section of V. Let us identify V and V^* by the scalar product. In particular Y^* will denote the section of V^* that corresponds to Y by the scalar product. Then using the actions of $c\left(V\oplus V^*\right), b\left(V\oplus V^*\right)$ on $\Lambda^\cdot\left(V^*\right), L_2\left(V\right)$ given in (1.2.1), (1.4.10), we have the identities of operators acting as unbounded operators on $\Lambda^\cdot\left(V^*\right)\otimes L_2\left(V\right)$,

$$
\begin{aligned}
\overline{d} &= \frac{1}{\sqrt{2}}\left(d^V + Y^*\wedge\right), & \overline{d}^* &= \frac{1}{\sqrt{2}}\left(d^{V*}+i_Y\right), \\
\mathcal{D} &= d^V + d^{V*}, & \mathcal{E} &= Y^*\wedge + i_Y, \\
\mathcal{D}' &= d^V - d^{V*}, & \mathcal{E}' &= Y^* - i_Y.
\end{aligned}
\tag{1.6.7}
$$

Note that

$$\overline{d} = \frac{1}{\sqrt{2}}\exp\left(-|Y|^2/2\right) d^V \exp\left(|Y|^2/2\right), \tag{1.6.8}$$

i.e., $\sqrt{2}\,\overline{d}$ is the Witten twist [Wi82] of d^V associated with the function $|Y|^2/2$, and $\sqrt{2}\,\overline{d}^*$ is its adjoint.

Now we use the conventions in (1.1.3) instead of the ones in (1.2.1), that is we consider instead the Clifford algebras $c\left(V\right), \widehat{c}\left(V\right)$ associated with $(V,\langle\rangle)$, and their action on $\Lambda^\cdot\left(V^*\right)$. By (1.6.7), we get

$$
\begin{aligned}
\mathcal{D} &= \sum_{i=1}^{n} c\left(e_i\right)\nabla_{e_i}, & \mathcal{E} &= \widehat{c}\left(Y\right), \\
\mathcal{D}' &= \sum_{i=1}^{n} \widehat{c}\left(e_i\right)\nabla_{e_i}, & \mathcal{E}' &= c\left(Y\right).
\end{aligned}
\tag{1.6.9}
$$

By (1.6.9), $\mathcal{D}$ is a classical Dirac operator. Moreover, by (1.6.3) and (1.6.9), we get

$$\left[\overline{d},\overline{d}^*\right] = \frac{1}{2}\left(-\Delta^V + |Y|^2 - n\right) + N^{\Lambda^\cdot\left(V^*\right)}, \tag{1.6.10}$$

which is just a version of (1.5.4).

By (1.5.4), the kernel of the operator $\left[\overline{d},\overline{d}^*\right]$ in $\Lambda^\cdot\left(V^*\right)\otimes S^\cdot\left(V^*\right)$ is generated by 1 and so it is 1-dimensional and concentrated in total degree 0. Equivalently the kernel of the unbounded operator $\left[\overline{d},\overline{d}^*\right]$ acting on $\Lambda^\cdot\left(V^*\right)\otimes L_2\left(V\right)$ is 1-dimensional and generated by the function $\exp\left(-|Y|^2/2\right)/\pi^{n/4}$. The above kernels coincide with $\ker\overline{d}\cap\ker\overline{d}^*$.

Now we consider the Clifford algebras associated with $(V, -\langle,\rangle)$. Also $e_1,\ldots,e_n$ still denotes an orthonormal basis of V with respect to $\langle\rangle$. Instead of (1.6.9), we get

$$\mathcal{D} = -\sum_{i=1}^{n} \widehat{c}\left(e_i\right) \nabla_{e_i}, \qquad\qquad \mathcal{E} = -c\left(Y\right), \qquad\qquad (1.6.11)$$

$$\mathcal{D}' = -\sum_{i=1}^{n} c\left(e_i\right) \nabla_{e_i}, \qquad\qquad \mathcal{E}' = -\widehat{c}\left(Y\right).$$

Chapter Two

The hypoelliptic Laplacian on $X = G/K$

The purpose of this chapter is to construct the hypoelliptic Laplacian $\mathcal{L}_b^X, b > 0$ acting on the total space of a vector bundle $TX \oplus N \simeq \mathfrak{g}$ over $X = G/K$. The operator $\mathcal{L}_b^X$ will be obtained using general constructions involving Clifford algebras and Heisenberg algebras, and also the Dirac operator of Kostant [Ko97].

This chapter is organized as follows. In section 2.1, we introduce a pair (G, K), the symmetric space $X = G/K$, and the vector bundles TX, N on X.

In section 2.2, we construct the canonical flat connection on $TX \oplus N \simeq \mathfrak{g}$.

In section 2.3, we define Clifford algebras associated with $\mathfrak{g}$.

In section 2.4, we obtain flat connections on $\Lambda^{\cdot}(T^*X \oplus N^*)$.

In section 2.5, we construct the Casimir operator for G.

In section 2.6, we introduce the canonical 3-form $\kappa^{\mathfrak{g}}$, and its image in the Clifford algebras of $\mathfrak{g}$.

In section 2.7, we construct the Dirac operator of Kostant.

In section 2.8, we describe the Clifford-Heisenberg algebra of $\mathfrak{g} \oplus \mathfrak{g}^*$, and the operators $\overline{d}, \overline{d}^*$ of chapter 1 associated with $\mathfrak{p}, \mathfrak{k}$.

In section 2.9, we construct the Dirac operator $\mathfrak{D}_b$. This operator acts on $C^{\infty}(G \times \mathfrak{g}, \Lambda^{\cdot}(\mathfrak{g}^*))$.

In section 2.10, we prove a simple identity relating a part of $\mathfrak{D}_b$ to the Kostant Dirac operator.

In section 2.11, we give a key formula for $\mathfrak{D}_b^2$.

In section 2.12, we consider the action of $\mathfrak{D}_b$ on functions over G that are K-equivariant with respect to a given representation of K. If E is a finite dimensional representation of K, we obtain this way an operator $\mathfrak{D}_b^X$, which is a differential operator acting on the total space $\widehat{\mathcal{X}}$ of $TX \oplus N$.

In section 2.13, we construct the elliptic Laplacian $\mathcal{L}^X$ on X, and also the hypoelliptic Laplacian $\mathcal{L}_b^X$, which is a second order hypoelliptic operator acting on $\widehat{\mathcal{X}}$.

In section 2.14, we consider the effect of the scaling of the form B on the above constructions.

In section 2.15, we prove the relevant Bianchi identity.

In section 2.16, we give a key formula relating $\mathcal{L}_b^X$ to $\mathcal{L}^X$.

In section 2.17, if $a \in \mathfrak{g}$, we consider the associated vector field a^{TX} on X and the corresponding Lie derivative L_a.

Finally, in section 2.18, we give various formulas involving the operator

$\mathcal{L}_b^X + L_a$.

This chapter is closely related to what we did in [B08b] in the case where G is a compact Lie group.

2.1 A pair (G, K)

Let G be a real reductive Lie group, and let K be a compact maximal subgroup of G. We will assume that G and K are connected. Let $\theta \in \mathrm{Aut}\,(G)$ be the Cartan involution, so that K is the fixed point set of θ.

Let $\mathfrak{g}$ be the Lie algebra of G, and let $\mathfrak{k} \subset \mathfrak{g}$ be the Lie algebra of K. If $g \in G$, we denote by $C\,(g)$ the corresponding inner automorphism of G, so that if $g' \in G$,

$$C\,(g)\,g' = gg'g^{-1}. \tag{2.1.1}$$

If $g \in G, a \in \mathfrak{g}$, let $\mathrm{Ad}\,(g)\,a \in \mathfrak{g}$ denote the action of g on a via the adjoint representation. Finally, if $a, b \in \mathfrak{g}$, set

$$\mathrm{ad}\,(a)\,b = [a, b]. \tag{2.1.2}$$

Then ad is the derivative of the map $g \in G \to \mathrm{Ad}\,(g) \in \mathrm{Aut}\,(\mathfrak{g})$.

The Cartan involution θ acts naturally as a Lie algebra automorphism of $\mathfrak{g}$. Then $\mathfrak{k}$ is the eigenspace of θ associated with the eigenvalue 1. Let $\mathfrak{p}$ be the eigenspace of θ associated with the eigenvalue -1, so that

$$\mathfrak{g} = \mathfrak{p} \oplus \mathfrak{k}. \tag{2.1.3}$$

Then we have the obvious

$$[\mathfrak{k}, \mathfrak{k}] \subset \mathfrak{k}, \qquad [\mathfrak{k}, \mathfrak{p}] \subset \mathfrak{p}, \qquad [\mathfrak{p}, \mathfrak{p}] \subset \mathfrak{k}. \tag{2.1.4}$$

Set

$$m = \dim \mathfrak{p}, \qquad n = \dim \mathfrak{k}, \tag{2.1.5}$$

so that

$$\dim \mathfrak{g} = m + n. \tag{2.1.6}$$

Let B be a real-valued nondegenerate bilinear symmetric form on $\mathfrak{g}$ which is invariant under the adjoint action of G on $\mathfrak{g}$, and also under θ. Then (2.1.3) is an orthogonal splitting of $\mathfrak{g}$ with respect to B. We assume B to be positive on $\mathfrak{p}$, and negative on $\mathfrak{k}$.

Let $\langle\ \rangle$ be the obvious scalar product on $\mathfrak{g}$ such that the splitting (2.1.3) is still orthogonal, and which coincides up to sign with B on $\mathfrak{k}$ and $\mathfrak{p}$, so that $\langle\,,\rangle = -B\,(\cdot, \theta\cdot)$. We will denote by $|\ |$ the corresponding norm.

In the sequel, given a vector subspace $\mathfrak{e}$ of $\mathfrak{g}$, we will often consider its orthogonal in $\mathfrak{g}$ with respect to B or to $\langle\ \rangle$. Most of the time, the splitting $\mathfrak{g} = \mathfrak{p} \oplus \mathfrak{k}$ induces a corresponding splitting of $\mathfrak{e}$, so that the orthogonal spaces to $\mathfrak{e}$ with respect to B or $\langle\ \rangle$ coincide.

Let $\omega^{\mathfrak{g}}$ be the canonical left-invariant 1-form on G with values in $\mathfrak{g}$, and let $\omega^{\mathfrak{p}}, \omega^{\mathfrak{k}}$ be the $\mathfrak{p}, \mathfrak{k}$ components of $\omega^{\mathfrak{g}}$, so that

$$\omega^{\mathfrak{g}} = \omega^{\mathfrak{p}} + \omega^{\mathfrak{k}}. \tag{2.1.7}$$

Then

$$d\omega^{\mathfrak{g}} = -\frac{1}{2}\left[\omega^{\mathfrak{g}}, \omega^{\mathfrak{g}}\right]. \tag{2.1.8}$$

By (2.1.4), equation (2.1.8) splits as

$$d\omega^{\mathfrak{p}} = -\left[\omega^{\mathfrak{k}}, \omega^{\mathfrak{p}}\right], \qquad d\omega^{\mathfrak{k}} = -\frac{1}{2}\left[\omega^{\mathfrak{k}}, \omega^{\mathfrak{k}}\right] - \frac{1}{2}\left[\omega^{\mathfrak{p}}, \omega^{\mathfrak{p}}\right]. \tag{2.1.9}$$

Let $p : G \to G/K$ be the obvious K-principal bundle. The connection form associated with the splitting (2.1.3) is just $\omega^{\mathfrak{k}}$. Let Ω be the curvature of this connection. By (2.1.9),

$$\Omega = -\frac{1}{2}\left[\omega^{\mathfrak{p}}, \omega^{\mathfrak{p}}\right]. \tag{2.1.10}$$

It follows from the above that $X = G/K$ is a symmetric space. Also the tangent bundle TX is given by

$$TX = G \times_K \mathfrak{p}. \tag{2.1.11}$$

The scalar product of $\mathfrak{p}$ descends to a Riemannian metric g^{TX} on $TX = G \times_K \mathfrak{p}$, and G acts isometrically on the left on X. Also the connection on TX induced by the splitting (2.1.3) is precisely the Levi-Civita connection ∇^{TX}.

More generally if V is a real Euclidean vector space, and if $\rho^V : K \to \mathrm{Aut}\,(V)$ is a representation of K by isometries of V, then $W = G \times_K V$ is a Euclidean vector bundle on X, and the left action of G on X lifts to an isometric action on W. Moreover, W is canonically equipped with a metric preserving connection ∇^W, which is G-invariant. These considerations also extend to the case where V is a complex Hermitian vector space.

Set

$$N = G \times_K \mathfrak{k}. \tag{2.1.12}$$

Let g^N be the metric of N. Then

$$TX \oplus N = G \times_K \mathfrak{g}. \tag{2.1.13}$$

As we just saw, the vector bundles TX, N are equipped with canonical metric preserving connections ∇^{TX}, ∇^N. We denote by $\nabla^{TX \oplus N}$ the corresponding connection on $TX \oplus N$. The Cartan involution θ descends to the involution of $TX \oplus N$ that is -1 on TX, 1 on N. Finally, B descends to a symmetric form on $TX \oplus N$, which is such that TX and N are orthogonal with respect to B, B is positive on TX and negative on N.

Also $\Lambda^{\cdot}(\mathfrak{g}^*), \Lambda^{\cdot}(\mathfrak{p}^*), \Lambda^{\cdot}(\mathfrak{k}^*)$ and $S^{\cdot}(\mathfrak{g}^*), S^{\cdot}(\mathfrak{p}^*), S^{\cdot}(\mathfrak{k}^*)$ are Euclidean representations of K. The corresponding Euclidean vector bundles are given by $\Lambda^{\cdot}(T^*X \oplus N^*), \Lambda^{\cdot}(T^*X), \Lambda^{\cdot}(N^*)$ and $S^{\cdot}(T^*X \oplus N^*), S^{\cdot}(T^*X), S^{\cdot}(N^*)$. As was explained in section 1.4, we identify the Hilbert completions of the symmetric algebras to the corresponding L_2 spaces, so that they coincide with the Hilbert bundles $L_2(TX \oplus N), L_2(TX), L_2(N)$.

2.2 The flat connection on $TX \oplus N$

Since G acts on $\mathfrak{g}$, the map $(g,a) \in G \times \mathfrak{g} \to \mathrm{Ad}\,(g)\,a \in \mathfrak{g}$ induces the canonical identification of $G \times_K \mathfrak{g} \simeq G \times \mathfrak{g}$, so that the bundle $G \times_K \mathfrak{g}$ is canonically trivial on X. Therefore,

$$TX \oplus N \simeq \mathfrak{g}. \tag{2.2.1}$$

Note that (2.2.1) identifies the form B. Therefore $TX \oplus N \simeq \mathfrak{g}$ is equipped with a flat connection $\nabla^{TX \oplus N, f}$, and this trivial connection projects on the given connections on TX, N. In fact one finds easily that

$$\nabla^{TX \oplus N, f} = \nabla^{TX \oplus N} + \mathrm{ad}\,(\cdot). \tag{2.2.2}$$

The term $\mathrm{ad}\,(\cdot)$ interchanges TX and N. When lifted to G, (2.2.2) just corresponds to the identity (2.1.7). Also note that the connections $\nabla^{TX \oplus N}$ and $\nabla^{TX \oplus N, f}$ preserve the form B.

A similar construction can be done on $T^*X \oplus N^*$. By identifying $TX \oplus N$ and $T^*X \oplus N^*$ by the form B, we obtain exactly the same connections as before.

If we identify $TX \oplus N$ with $T^*X \oplus N^*$ via the metric $g^{TX \oplus N} = g^{TX} \oplus g^N$, we obtain another flat connection $\nabla^{TX \oplus N, f*}$ on $TX \oplus N$, which is given by

$$\nabla^{TX \oplus N, f*} = \nabla^{TX \oplus N} - \mathrm{ad}\,(\cdot). \tag{2.2.3}$$

Equivalently,

$$\nabla^{TX \oplus N, f*} = \theta^{-1} \nabla^{TX \oplus N, f} \theta. \tag{2.2.4}$$

2.3 The Clifford algebras of $\mathfrak{g}$

Recall that $\mathfrak{g}$ is equipped with the symmetric nondegenerate bilinear form B. Let $c(\mathfrak{g}), \widehat{c}(\mathfrak{g})$ denote the Clifford algebras associated with $(\mathfrak{g}, B), (\mathfrak{g}, -B)$ as in section 1.1. Since G acts on $(\mathfrak{g}, B)$, G acts by automorphisms of $c(\mathfrak{g}), \widehat{c}(\mathfrak{g})$.

By restricting B to $\mathfrak{p}, \mathfrak{k}$, we get the Clifford algebras $c(\mathfrak{p}), \widehat{c}(\mathfrak{p}), c(\mathfrak{k}), \widehat{c}(\mathfrak{k})$. Since the splitting $\mathfrak{g} = \mathfrak{p} \oplus \mathfrak{k}$ is orthogonal with respect to B, we get

$$c(\mathfrak{g}) = c(\mathfrak{p}) \widehat{\otimes} c(\mathfrak{k}), \qquad \widehat{c}(\mathfrak{g}) = \widehat{c}(\mathfrak{p}) \widehat{\otimes} \widehat{c}(\mathfrak{k}). \tag{2.3.1}$$

Similarly the vector bundle $TX \oplus N$ is also equipped with the bilinear form B. Let $c(TX \oplus N), \widehat{c}(TX \oplus N)$ be the associated bundles of Clifford algebras. Also we equip TX, N with the restriction of B to TX, N. Let $c(TX), c(N), \widehat{c}(TX), \widehat{c}(N)$ denote the corresponding bundles of algebras. As in (2.3.1), we get

$$c(TX \oplus N) = c(TX) \widehat{\otimes} c(N), \qquad \widehat{c}(TX \oplus N) = \widehat{c}(TX) \widehat{\otimes} \widehat{c}(N). \tag{2.3.2}$$

2.4 The flat connections on $\Lambda^{\cdot}(T^*X \oplus N^*)$

From $\nabla^{TX \oplus N, f}$, we get a flat connection $\nabla^{\Lambda(T^*X \oplus N^*), f}$ on $\Lambda(T^*X \oplus N^*)$. By (1.1.14), (2.2.2), we get

$$\nabla^{\Lambda(T^*X \oplus N^*), f} = \nabla^{\Lambda(T^*X \oplus N^*)} + c(\mathrm{ad}\,(\cdot)) + \widehat{c}(\mathrm{ad}\,(\cdot)). \tag{2.4.1}$$

By (1.1.14), (2.2.3), the flat connection $\nabla^{\Lambda(T^*X \oplus N^*), f^*}$ on $\Lambda(T^*X \oplus N^*)$ that is associated with $\nabla^{TX \oplus N, f*}$ is given by

$$\nabla^{\Lambda(T^*X \oplus N^*), f*} = \nabla^{\Lambda(T^*X \oplus N^*)} - c\left(\mathrm{ad}\left(\cdot\right)\right) - \widehat{c}\left(\mathrm{ad}\left(\cdot\right)\right). \tag{2.4.2}$$

Note that θ acts on $\Lambda(T^*X \oplus N^*)$. Recall that $N^{\Lambda^{\cdot}(T^*X)}$ denotes the number operator of $\Lambda^{\cdot}(T^*X)$. Then $N^{\Lambda^{\cdot}(T^*X)}$ also acts on $\Lambda^{\cdot}(T^*X \oplus N^*)$. We have the obvious identity

$$\theta = (-1)^{N^{\Lambda^{\cdot}(T^*X)}}. \tag{2.4.3}$$

Then

$$\nabla^{\Lambda(T^*X \oplus N^*), f*} = \theta \nabla^{\Lambda(T^*X \oplus N^*), f} \theta^{-1}. \tag{2.4.4}$$

The connections $\nabla^{\Lambda^{\cdot}(T^*X \oplus N^*), f}$ and $\nabla^{\Lambda^{\cdot}(T^*X \oplus N^*), f*}$ preserve the total degree of the forms.

Definition 2.4.1. In the sequel, $\nabla^{\Lambda^{\cdot}(T^*X \oplus N^*), f*, \widehat{f}}$ denotes the connection on $\Lambda^{\cdot}(T^*X \oplus N^*)$ that is given by

$$\nabla_{\cdot}^{\Lambda^{\cdot}(T^*X \oplus N^*), f*, \widehat{f}} = \nabla_{\cdot}^{\Lambda^{\cdot}(T^*X \oplus N^*)} - c\left(\mathrm{ad}\left(\cdot\right)\right) + \widehat{c}\left(\mathrm{ad}\left(\cdot\right)\right). \tag{2.4.5}$$

Since K is connected, K acts by oriented isometries on $\mathfrak{g} = \mathfrak{p} \oplus \mathfrak{k}$. Therefore the vector bundles TX, N can be oriented, and G acts on these vector bundles by oriented isometries. Also since G is connected, the identification $TX \oplus N \simeq \mathfrak{g}$ preserves the orientation. Finally, G acts on TX, N and preserves their orientation.

Let $e_1, \ldots, e_m$ be an oriented orthonormal basis of TX. Set

$$\tau = c\left(e_1\right) \ldots c\left(e_m\right), \qquad \widehat{\tau} = \widehat{c}\left(e_1\right) \ldots \widehat{c}\left(e_m\right). \tag{2.4.6}$$

Then $\tau, \widehat{\tau}$ do not depend on the choice of the basis. A trivial computation shows that

$$\tau^2 = (-1)^{m(m+1)/2}, \qquad \widehat{\tau}^2 = (-1)^{m(m-1)/2}, \tag{2.4.7}$$

$$(-1)^{N^{\Lambda^{\cdot}(T^*X)}} = (-1)^{m(m+1)/2} \tau \widehat{\tau}.$$

Let $*^{TX}$ be the star operator acting on $\Lambda^{\cdot}(T^*X)$, which is associated with the given metric on TX and with the orientation of TX. Then $*^{TX}$ is an even operator if m is even, an odd operator if m is odd. Classically, if m is even,

$$\tau = *^{TX} i^{N^{\Lambda^{\cdot}(T^*X)}\left(N^{\Lambda^{\cdot}(T^*X)}-1\right)}, \tag{2.4.8}$$

and if m is odd,

$$\tau = *^{TX} i^{N^{\Lambda^{\cdot}(T^*X)}\left(N^{\Lambda^{\cdot}(T^*X)}+1\right)}. \tag{2.4.9}$$

From the above, it follows that

$$\tau N^{\Lambda^{\cdot}(T^*X)} \tau^{-1} = m - N^{\Lambda^{\cdot}(T^*X)}. \tag{2.4.10}$$

Also observe that τ acts naturally on $\Lambda^{\cdot}(T^*X \oplus N^*)$.

Proposition 2.4.2. *The connection* $\nabla^{\Lambda^{\cdot}(T^*X \oplus N^*), f*, \hat{f}}$ *is flat. Moreover,*

$$\nabla^{\Lambda^{\cdot}(T^*X \oplus N^*), f*, \hat{f}} = \tau \nabla^{\Lambda^{\cdot}(T^*X \oplus N^*), f} \tau^{-1}. \tag{2.4.11}$$

Proof. The first part of the proposition follows from the flatness of the connections $\nabla^{TX \oplus N, f}$ and $\nabla^{TX \oplus N, f*}$ over $TX \oplus N$. Also if $e \in TX, f \in N$, one verifies easily that

$$\tau c(e) c(f) \tau^{-1} = -c(e) c(f). \tag{2.4.12}$$

By (2.4.12), we get

$$\tau c(\text{ad}(\cdot)) \tau^{-1} = -c(\text{ad}(\cdot)). \tag{2.4.13}$$

Equation (2.4.11) follows from (2.4.1), (2.4.5) and (2.4.13). The proof of our proposition is completed. $\qquad\square$

Remark 2.4.3. Note that in general, the connection $\nabla^{\Lambda^{\cdot}(T^*X \oplus N^*), f*, \hat{f}}$ does not preserve the total degree of the form.

Let $\widehat{K}$ be the universal cover of K. Let $S^{\mathfrak{p}}$ be the spinors associated with the Euclidean space $\mathfrak{p}$. Then $c(\mathfrak{p})$ acts naturally on $S^{\mathfrak{p}}$.

If m is even, then $S^{\mathfrak{p}}$ is $\mathbf{Z}_2$-graded, it splits as $S^{\mathfrak{p}} = S^{\mathfrak{p}}_+ \oplus S^{\mathfrak{p}}_-$, and $S^{\mathfrak{p}}_{\pm}$ is of dimension $2^{m/2-1}$. Then $i^{m/2}\tau$ is just the involution defining the grading of $S^{\mathfrak{p}}$. Also,

$$\Lambda^{\cdot}(\mathfrak{p}^*) \otimes_{\mathbf{R}} \mathbf{C} = S^{\mathfrak{p}} \widehat{\otimes} S^{\mathfrak{p}*}. \tag{2.4.14}$$

If m is odd, then $S^{\mathfrak{p}}$ is of dimension $2^{(m-1)/2}$, and moreover,

$$\Lambda^{\cdot}(\mathfrak{p}^*) \otimes_{\mathbf{R}} \mathbf{C} = S^{\mathfrak{p}} \widehat{\otimes} S^{\mathfrak{p}*} \oplus S^{\mathfrak{p}} \widehat{\otimes} S^{\mathfrak{p}*}. \tag{2.4.15}$$

Moreover, $i^{(m+1)/2}\tau$ acts on $S^{\mathfrak{p}}$ like 1.

2.5 The Casimir operator

In the sequel, we identify $\mathfrak{g}$ to the vector space of left-invariant vector fields on G. The enveloping algebra $U(\mathfrak{g})$ will be identified with the algebra of left-invariant differential operators on G.

Let $C^{\mathfrak{g}} \in U(\mathfrak{g})$ be the Casimir element of G. If $e_1, \ldots, e_{m+n}$ is a basis of $\mathfrak{g}$ and if $e_1^*, \ldots, e_{m+n}^*$ is the dual basis of $\mathfrak{g}$ with respect to B, then

$$C^{\mathfrak{g}} = -\sum_{i=1}^{m+n} e_i^* e_i. \tag{2.5.1}$$

Moreover, $C^{\mathfrak{g}}$ lies in the center of $U(\mathfrak{g})$.

If $e_1, \ldots, e_m$ is an orthonormal basis of $\mathfrak{p}$, and if $e_{m+1}, \ldots, e_{m+n}$ is an orthonormal basis of $\mathfrak{k}$, then

$$e_i^* = e_i \text{ for } 1 \le i \le m, \tag{2.5.2}$$
$$- e_i \text{ for } m + 1 \le i \le m + n.$$

In particular,

$$C^{\mathfrak{g}} = -\sum_{i=1}^{m} e_i^2 + \sum_{i=m+1}^{m+n} e_i^2. \tag{2.5.3}$$

The Casimir operator $C^{\mathfrak{k}}$ of K will be calculated with respect to the bilinear form induced by B on $\mathfrak{k}$ so that

$$C^{\mathfrak{k}} = \sum_{i=m+1}^{m+n} e_i^2. \tag{2.5.4}$$

Put

$$C^{\mathfrak{g},H} = -\sum_{i=1}^{m} e_i^2. \tag{2.5.5}$$

By (2.5.3)–(2.5.5), we get

$$C^{\mathfrak{g}} = C^{\mathfrak{g},H} + C^{\mathfrak{k}}. \tag{2.5.6}$$

Moreover,

$$\left[C^{\mathfrak{g},H}, C^{\mathfrak{k}} \right] = 0. \tag{2.5.7}$$

Let $\rho^V : K \to \mathrm{Aut}\,(V)$ be an orthogonal or unitary representation of K on a finite dimensional Euclidean or Hermitian vector space V. We denote by $C^{\mathfrak{k},V} \in \mathrm{End}\,(V)$ the corresponding Casimir operator acting on V, so that

$$C^{\mathfrak{k},V} = \sum_{i=m+1}^{m+n} \rho^{V,2}\,(e_i). \tag{2.5.8}$$

In particular $C^{\mathfrak{k},\mathfrak{k}} \in \mathrm{End}\,(\mathfrak{k})\,, C^{\mathfrak{k},\mathfrak{p}} \in \mathrm{End}\,(\mathfrak{p})$ denote the Casimir operators associated with the actions of K on $\mathfrak{k}, \mathfrak{p}$.

The Casimir operator $C^{\mathfrak{g}}$ will act as a differential operator on $X = G/K$, and on sections of vector bundles like $W = G \times_K V$.

2.6 The form $\kappa^{\mathfrak{g}}$

Let $\kappa^{\mathfrak{g}} \in \Lambda^3\,(\mathfrak{g}^*)$ be such that if $a, b, c \in \mathfrak{g}$,

$$\kappa^{\mathfrak{g}}\,(a, b, c) = B\,([a, b]\,, c)\,. \tag{2.6.1}$$

Then $\kappa^{\mathfrak{g}}$ is G-invariant. It induces a closed left and right invariant 3-form on G. If $e_1, \ldots, e_{n+m}$ is a basis of $\mathfrak{g}$ and if $e^1, \ldots, e^{m+n}$ is the corresponding dual basis of $\mathfrak{g}^*$, then

$$\kappa^{\mathfrak{g}} = \frac{1}{6} \sum \kappa^{\mathfrak{g}}\,(e_i, e_j, e_k)\, e^i \wedge e^j \wedge e^k. \tag{2.6.2}$$

Recall that the symmetric bilinear form B induces a corresponding bilinear form B^* on $\Lambda^{\cdot}\,(\mathfrak{g}^*)$. By (2.6.2), we find easily that

$$B^*\,(\kappa^{\mathfrak{g}}, \kappa^{\mathfrak{g}}) = \frac{1}{6} \sum \kappa^{\mathfrak{g}}\,(e_i, e_j, e_k)\, \kappa^{\mathfrak{g}}\,(e_i^*, e_j^*, e_k^*)\,. \tag{2.6.3}$$

Equivalently,

$$B^* \left(\kappa^{\mathfrak{g}}, \kappa^{\mathfrak{g}} \right) = \frac{1}{6} \sum B \left([e_i, e_j], [e_i^*, e_j^*] \right). \tag{2.6.4}$$

From now on, except when explicitly indicated, the basis $e_1, \ldots, e_{n+m}$ is taken as in (2.5.2), (2.5.3).

Let $\kappa^{\mathfrak{k}}$ be the analogue of $\kappa^{\mathfrak{g}}$ associated to the restriction of B to $\mathfrak{k}$. By (2.6.4), we get

$$B^* \left(\kappa^{\mathfrak{k}}, \kappa^{\mathfrak{k}} \right) = -\frac{1}{6} \sum_{m+1 \leq i,j \leq m+n} \|[e_i, e_j]\|^2. \tag{2.6.5}$$

Of course,

$$\left| \kappa^{\mathfrak{k}} \right|^2 = -B^* \left(\kappa^{\mathfrak{k}}, \kappa^{\mathfrak{k}} \right). \tag{2.6.6}$$

Note that we can also rewrite (2.6.5) in the form

$$B^* \left(\kappa^{\mathfrak{k}}, \kappa^{\mathfrak{k}} \right) = \frac{1}{6} \mathrm{Tr}^{\mathfrak{k}} \left[C^{\mathfrak{k}, \mathfrak{k}} \right]. \tag{2.6.7}$$

Let R^X be the Ricci tensor of $X = G/K$, let S^X be its scalar curvature. Then one finds easily that

$$R^X = C^{\mathfrak{k}, \mathfrak{p}}, \tag{2.6.8}$$

$$S^X = - \sum_{1 \leq i,j \leq m} \|[e_i, e_j]\|^2 = \mathrm{Tr}^{\mathfrak{p}} \left[C^{\mathfrak{k}, \mathfrak{p}} \right].$$

Let S^K be the scalar curvature of K. Then one verifies easily that

$$S^K = \frac{1}{4} \sum_{m+1 \leq i,j \leq m+n} \|[e_i, e_j]\|^2 = -\frac{1}{4} \mathrm{Tr}^{\mathfrak{k}} \left[C^{\mathfrak{k}, \mathfrak{k}} \right]. \tag{2.6.9}$$

Proposition 2.6.1. *The following identity holds:*

$$B^* \left(\kappa^{\mathfrak{g}}, \kappa^{\mathfrak{g}} \right) = -\frac{1}{2} \sum_{1 \leq i,j \leq m} \|[e_i, e_j]\|^2 - \frac{1}{6} \sum_{m+1 \leq i,j \leq m+n} \|[e_i, e_j]\|^2. \tag{2.6.10}$$

Equivalently,

$$B^* \left(\kappa^{\mathfrak{g}}, \kappa^{\mathfrak{g}} \right) = \frac{1}{2} S^X - \frac{2}{3} S^K = \frac{1}{2} \mathrm{Tr}^{\mathfrak{p}} \left[C^{\mathfrak{k}, \mathfrak{p}} \right] + \frac{1}{6} \mathrm{Tr}^{\mathfrak{k}} \left[C^{\mathfrak{k}, \mathfrak{k}} \right]. \tag{2.6.11}$$

Proof. Clearly,

$$\kappa^{\mathfrak{g}} = \frac{1}{2} \sum_{\substack{1 \leq i,j \leq m \\ m+1 \leq k \leq m+n}} \kappa^{\mathfrak{g}} \left(e_i, e_j, e_k \right) e^i \wedge e^j \wedge e^k + \kappa^{\mathfrak{k}}. \tag{2.6.12}$$

By (2.6.5) and (2.6.12), we get (2.6.10). Also (2.6.11) follows from (2.6.5)–(2.6.9) and from (2.6.10). □

Let $c(\kappa^{\mathfrak{g}}) \in c(\mathfrak{g}), \widehat{c}(\kappa^{\mathfrak{g}}) \in \widehat{c}(\mathfrak{g})$ correspond to $\kappa^{\mathfrak{g}} \in \Lambda^{\cdot}(\mathfrak{g}^*)$ as in section 1.1. When replacing B by $-B$, $\kappa^{\mathfrak{g}}$ is replaced by $-\kappa^{\mathfrak{g}}$. By (1.1.7), we get

$$c \left(\kappa^{\mathfrak{g}} \right) = \frac{1}{6} \kappa^{\mathfrak{g}} \left(e_i^*, e_j^*, e_k^* \right) c(e_i) c(e_j) c(e_k), \tag{2.6.13}$$

$$\widehat{c} \left(-\kappa^{\mathfrak{g}} \right) = -\frac{1}{6} \kappa^{\mathfrak{g}} \left(e_i^*, e_j^*, e_k^* \right) \widehat{c}(e_i) \widehat{c}(e_j) \widehat{c}(e_k).$$

2.7 The Dirac operator of Kostant

Recall that G acts by automorphisms on the Clifford algebras $c\left(\mathfrak{g}\right), \widehat{c}\left(\mathfrak{g}\right)$, so that if $e \in \mathfrak{g}$, the action of g on $c\left(e\right), \widehat{c}\left(e\right)$ is just $c\left(e\right) \to c\left(\operatorname{Ad}\left(g\right) e\right), \widehat{c}\left(e\right) \to \widehat{c}\left(\operatorname{Ad}\left(g\right) e\right)$. Similarly the adjoint action of G on $\mathfrak{g}$ induces a corresponding action on $U\left(\mathfrak{g}\right)$. Set

$$\mathcal{A}^{\mathfrak{g}} = c\left(\mathfrak{g}\right) \otimes U\left(\mathfrak{g}\right), \qquad\qquad \widehat{\mathcal{A}}^{\mathfrak{g}} = \widehat{c}\left(\mathfrak{g}\right) \otimes U\left(\mathfrak{g}\right). \qquad (2.7.1)$$

Then $\mathcal{A}^{\mathfrak{g}}, \widehat{\mathcal{A}}^{\mathfrak{g}}$ are $\mathbf{Z}_2$-graded algebras. Moreover, G acts on $\mathcal{A}^{\mathfrak{g}}, \widehat{\mathcal{A}}^{\mathfrak{g}}$.

Definition 2.7.1. Let $D^{\mathfrak{g}} \in \mathcal{A}^{\mathfrak{g}}, \widehat{D}^{\mathfrak{g}} \in \widehat{\mathcal{A}}^{\mathfrak{g}}$ be the Dirac operators,

$$D^{\mathfrak{g}} = \sum_{i=1}^{m+n} c\left(e_i^*\right) e_i + \frac{1}{2} c\left(\kappa^{\mathfrak{g}}\right), \quad \widehat{D}^{\mathfrak{g}} = \sum_{i=1}^{m+n} \widehat{c}\left(e_i^*\right) e_i + \frac{1}{2} \widehat{c}\left(-\kappa^{\mathfrak{g}}\right). \quad (2.7.2)$$

Note that $\widehat{D}^{\mathfrak{g}}$ is the analogue of $D^{\mathfrak{g}}$, which is associated to $-B$. Moreover, $D^{\mathfrak{g}}, \widehat{D}^{\mathfrak{g}}$ are G-invariant.

In the sequel, we will mostly work with $\widehat{D}^{\mathfrak{g}}$. However, corresponding objects can be as well attached to $D^{\mathfrak{g}}$.

Assume that the basis $e_1, \ldots, e_{m+n}$ is taken as in (2.5.2), so that for $1 \le i \le m$, $B\left(e_i, e_i\right) = 1$, and for $m + 1 \le i \le m + n, B\left(e_i, e_i\right) = -1$.

By (2.6.13), we get

$$\widehat{c}\left(-\kappa^{\mathfrak{k}}\right) = \frac{1}{6} \sum_{m+1 \le i,j,k \le m+n} \kappa^{\mathfrak{k}}\left(e_i, e_j, e_k\right) \widehat{c}\left(e_i\right) \widehat{c}\left(e_j\right) \widehat{c}\left(e_k\right). \qquad (2.7.3)$$

If $e \in \mathfrak{k}$, let $\operatorname{ad}\left(e\right)|_{\mathfrak{p}}$ be the restriction of $\operatorname{ad}\left(e\right)$ to $\mathfrak{p}$. Then $\widehat{c}\left(\operatorname{ad}\left(e\right)\right) \in \widehat{c}\left(\mathfrak{p}\right)$. Using (1.1.11) and proceeding as in (2.6.12), we get

$$\widehat{c}\left(-\kappa^{\mathfrak{g}}\right) = -2 \sum_{i=m+1}^{m+n} \widehat{c}\left(e_i\right) \widehat{c}\left(\operatorname{ad}\left(e_i\right)|_{\mathfrak{p}}\right) + \widehat{c}\left(-\kappa^{\mathfrak{k}}\right). \qquad (2.7.4)$$

Put

$$\widehat{D}_H^{\mathfrak{g}} = \sum_{i=1}^{m} \widehat{c}\left(e_i\right) e_i, \qquad\qquad\qquad\qquad\qquad (2.7.5)$$

$$\widehat{D}_V^{\mathfrak{g}} = -\sum_{i=m+1}^{m+n} \widehat{c}\left(e_i\right)\left(e_i + \widehat{c}\left(\operatorname{ad}\left(e_i\right)|_{\mathfrak{p}}\right)\right) + \frac{1}{2} \widehat{c}\left(-\kappa^{\mathfrak{k}}\right).$$

By (2.7.2), (2.7.4), and (2.7.5), we get

$$\widehat{D}^{\mathfrak{g}} = \widehat{D}_H^{\mathfrak{g}} + \widehat{D}_V^{\mathfrak{g}}. \qquad (2.7.6)$$

Now we recall a fundamental result of Kostant [Ko76, Ko97].

Theorem 2.7.2. *The following identities hold:*

$$\left[\widehat{D}_H^{\mathfrak{g}}, \widehat{D}_V^{\mathfrak{g}}\right] = 0, \qquad \widehat{D}^{\mathfrak{g},2} = -C^{\mathfrak{g}} - \frac{1}{4} B^*\left(\kappa^{\mathfrak{g}}, \kappa^{\mathfrak{g}}\right). \qquad (2.7.7)$$

Proof. First, we establish the second identity in (2.7.7). We take $e_1, \ldots, e_{m+n}$ as in (2.5.2) and (2.7.5). In particular for $1 \leq i \leq m + n$,

$$\widehat{c}(e_i)\,\widehat{c}(e_i^*) = 1. \tag{2.7.8}$$

By (2.7.2), we get

$$\widehat{D}^{\mathfrak{g},2} = B\,(e_i^*, e_i^*)\,e_i^2 + \frac{1}{2}\widehat{c}(e_i^*)\,\widehat{c}(e_j^*)\,[e_i, e_j]$$
$$+ \frac{1}{2}\,[\widehat{c}(e_i^*), \widehat{c}(-\kappa^{\mathfrak{g}})]\,e_i + \frac{1}{4}\widehat{c}(-\kappa^{\mathfrak{g}})^2. \tag{2.7.9}$$

Moreover,

$$B\,(e_i^*, e_i^*)\,e_i^2 = -C^{\mathfrak{g}}. \tag{2.7.10}$$

Also one verifies easily that

$$\frac{1}{2}\widehat{c}(e_i^*)\,\widehat{c}(e_j^*)\,[e_i, e_j] + \frac{1}{2}\,[\widehat{c}(e_i^*), \widehat{c}(-\kappa^{\mathfrak{g}})]\,e_i = 0. \tag{2.7.11}$$

We will now show that

$$\widehat{c}(-\kappa^{\mathfrak{g}})^2 = -B^*\,(\kappa^{\mathfrak{g}}, \kappa^{\mathfrak{g}}). \tag{2.7.12}$$

Indeed by (2.6.13),

$$\widehat{c}(-\kappa^{\mathfrak{g}})^2 = \frac{1}{36}\kappa^{\mathfrak{g}}\,(e_i, e_j, e_k)\,\kappa^{\mathfrak{g}}\,(e_{i'}^*, e_{j'}^*, e_{k'}^*)$$
$$\widehat{c}(e_i^*)\,\widehat{c}(e_j^*)\,\widehat{c}(e_k^*)\,\widehat{c}(e_{i'})\,\widehat{c}(e_{j'})\,\widehat{c}(e_{k'}). \tag{2.7.13}$$

In (2.7.13), the indices (i, j, k) and (i', j', k') are all distinct. By antisymmetry the sum over disjoint triples of (i, j, k), (i', j', k') vanishes. By using in particular (2.7.8), we get

$$\widehat{c}(-\kappa^{\mathfrak{g}})^2 = \frac{1}{12}B\,([e_i, e_j], [e_i^*, e_{j'}^*])\,\widehat{c}(e_i^*)\,\widehat{c}(e_j^*)\,\widehat{c}(e_{i'})\,\widehat{c}(e_{j'}), \tag{2.7.14}$$

which can also be written as

$$\widehat{c}(-\kappa^{\mathfrak{g}})^2 = \frac{1}{12}B\,([e_i, e_j], [e_{i'}, e_{j'}^*])\,\widehat{c}(e_i^*)\,\widehat{c}(e_j^*)\,\widehat{c}(e_{i'}^*)\,\widehat{c}(e_{j'}). \tag{2.7.15}$$

Using the G-invariance of B and the Jacobi identity, we find that given j', the sum in (2.7.15) over the distinct i, j, i' vanishes identically. By (2.5.2), the same considerations apply to the sum in (2.7.14). Using (2.7.8) and this last identity, we finally obtain

$$\widehat{c}(-\kappa^{\mathfrak{g}})^2 = -\frac{1}{6}B\,([e_i, e_j], [e_i^*, e_j^*]), \tag{2.7.16}$$

which by (2.6.4) is just (2.7.12). From (2.7.9)–(2.7.12), we get the second identity in (2.7.7).

Let us now establish the first identity in (2.7.7). Indeed,

$$\widehat{D}^{\mathfrak{g},2} = \widehat{D}_H^{\mathfrak{g},2} + \widehat{D}_V^{\mathfrak{g},2} + \left[\widehat{D}_H^{\mathfrak{g}}, \widehat{D}_V^{\mathfrak{g}}\right]. \tag{2.7.17}$$

Now $\left[\widehat{D}_H^{\mathfrak{g}}, \widehat{D}_V^{\mathfrak{g}}\right]$ is the only component in (2.7.17) that is odd in $\widehat{c}(\mathfrak{k})$. From the second identity in (2.7.7), we find that this component vanishes. Of course a direct proof can also be given. This completes the proof of our theorem. $\qquad\square$

Remark 2.7.3. By changing B into $-B$, we get similar identities for $D^{\mathfrak{g}}$. In particular,

$$D^{\mathfrak{g},2} = C^{\mathfrak{g}} + \frac{1}{4} B^* \left(\kappa^{\mathfrak{g}}, \kappa^{\mathfrak{g}} \right). \tag{2.7.18}$$

2.8 The Clifford-Heisenberg algebra of $\mathfrak{g} \oplus \mathfrak{g}^*$

Now we use temporarily the conventions of section 1.5 in the case where $V = \mathfrak{g}$. In particular we will consider the Clifford algebra $c\,(\mathfrak{g} \oplus \mathfrak{g}^*)$ and the Heisenberg algebra $b\,(\mathfrak{g} \oplus \mathfrak{g}^*)$, and we disregard for the moment the bilinear form B. Similar considerations will be applied to $\mathfrak{p}, \mathfrak{k}$. Of course,

$$\mathfrak{g} \oplus \mathfrak{g}^* = (\mathfrak{p} \oplus \mathfrak{p}^*) \oplus (\mathfrak{k} \oplus \mathfrak{k}^*), \tag{2.8.1}$$

so that

$$c\,(\mathfrak{g} \oplus \mathfrak{g}^*) = c\,(\mathfrak{p} \oplus \mathfrak{p}^*)\, \widehat{\otimes}\, c\,(\mathfrak{k} \oplus \mathfrak{k}^*), \quad b\,(\mathfrak{g} \oplus \mathfrak{g}^*) = b\,(\mathfrak{p} \oplus \mathfrak{p}^*) \otimes b\,(\mathfrak{k} \oplus \mathfrak{k}^*). \tag{2.8.2}$$

We define the operators $\overline{d}^{\mathfrak{g}}, \overline{d}^{\mathfrak{g}*} \in c\,(\mathfrak{g} \oplus \mathfrak{g}^*) \otimes b\,(\mathfrak{g} \oplus \mathfrak{g}^*)$, $\overline{d}^{\mathfrak{p}}, \overline{d}^{\mathfrak{p}*} \in c\,(\mathfrak{p} \oplus \mathfrak{p}^*) \otimes b\,(\mathfrak{p} \oplus \mathfrak{p}^*)$ and $\overline{d}^{\mathfrak{k}}, \overline{d}^{\mathfrak{k}*} \in c\,(\mathfrak{k} \oplus \mathfrak{k}^*) \otimes b\,(\mathfrak{k} \oplus \mathfrak{k}^*)$ as in (1.5.1). Of course,

$$\overline{d}^{\mathfrak{g}} = \overline{d}^{\mathfrak{p}} + \overline{d}^{\mathfrak{k}}, \qquad\qquad \overline{d}^{\mathfrak{g}*} = \overline{d}^{\mathfrak{p}*} + \overline{d}^{\mathfrak{k}*}. \tag{2.8.3}$$

Also, the operators $\overline{d}^{\mathfrak{g}}, \overline{d}^{\mathfrak{g}*}$ are G-invariant.

If we consider $\mathfrak{p}, \mathfrak{k}$ as Euclidean vector spaces, using (1.6.7), we may view the operators $\overline{d}^{\mathfrak{p}}, \overline{d}^{\mathfrak{p}*}$ as acting on $\Lambda^{\cdot} (\mathfrak{p}^*) \otimes L_2\,(\mathfrak{p})$, and $\overline{d}^{\mathfrak{k}}, \overline{d}^{\mathfrak{k}*}$ as acting on $\Lambda^{\cdot} (\mathfrak{k}^*) \otimes L_2\,(\mathfrak{k})$. This is what we will do in the sequel, because it is the most suitable representation in a geometric context. Still, it is very easy to go back to the original representation.

If $Y \in \mathfrak{g}$, we split Y in the form

$$Y = Y^{\mathfrak{p}} + Y^{\mathfrak{k}}, \tag{2.8.4}$$

with $Y^{\mathfrak{p}} \in \mathfrak{p}, Y^{\mathfrak{k}} \in \mathfrak{k}$.

From now on, the Clifford operators will be associated with $(\mathfrak{g}, B)$.

First, we consider the Euclidean vector space $\mathfrak{p}$, and the associated Clifford algebras $c\,(\mathfrak{p}), \widehat{c}\,(\mathfrak{p})$. Recall that the operators $\mathcal{D}^{\mathfrak{p}}, \mathcal{E}^{\mathfrak{p}}$ associated to $\mathfrak{p}$ were defined in (1.6.1), (1.6.9). We will not distinguish these objects from their action on $\Lambda^{\cdot} (\mathfrak{p}^*) \otimes S^{\cdot} (\mathfrak{p}^*)$, which was described in (1.6.7)–(1.6.9). This will also be done in the sequel for other operators.

Let $e_1, \ldots, e_m$ be an orthonormal basis of $\mathfrak{p}$. By (1.6.9), we get

$$\mathcal{D}^{\mathfrak{p}} = \sum_{i=1}^{m} c\,(e_i)\, \nabla_{e_i}, \qquad\qquad \mathcal{E}^{\mathfrak{p}} = \widehat{c}\,(Y^{\mathfrak{p}}). \tag{2.8.5}$$

By (1.6.2),

$$\overline{d}^{\mathfrak{p}} + \overline{d}^{\mathfrak{p}*} = \frac{1}{\sqrt{2}} \left(\mathcal{D}^{\mathfrak{p}} + \mathcal{E}^{\mathfrak{p}} \right). \tag{2.8.6}$$

When acting on $\Lambda^{\cdot}(\mathfrak{p}^*) \otimes L_2(\mathfrak{p})$, the operator in (2.8.6) is self-adjoint.

By (1.5.4),

$$\left[\overline{d}^{\mathfrak{p}}, \overline{d}^{\mathfrak{p}*} \right] = N^{S^{\cdot}(\mathfrak{p}^*)} + N^{\Lambda^{\cdot}(\mathfrak{p}^*)}. \tag{2.8.7}$$

By (1.6.10), we can rewrite (2.8.7) in the form

$$\frac{1}{2}\left(\mathcal{D}^{\mathfrak{p}} + \mathcal{E}^{\mathfrak{p}} \right)^2 = \frac{1}{2}\left(-\Delta^{\mathfrak{p}} + |Y^{\mathfrak{p}}|^2 - m \right) + N^{\Lambda^{\cdot}(\mathfrak{p}^*)}. \tag{2.8.8}$$

The above operators are K-invariant.

Now we consider the vector space $\mathfrak{k}$. It is here crucial to consider again the Clifford algebras associated with $(\mathfrak{k}, B|_{\mathfrak{k}})$. Let $e_{m+1}, \ldots, e_{m+n}$ be an orthonormal basis of $\mathfrak{k}$. We define the operators $\mathcal{D}^{\mathfrak{k}}, \mathcal{E}^{\mathfrak{k}}$ by the formulas

$$\mathcal{D}^{\mathfrak{k}} = \sum_{i=m+1}^{m+n} c(e_i^*) \nabla_{e_i}, \qquad \mathcal{E}^{\mathfrak{k}} = \widehat{c}\left(Y^{\mathfrak{k}} \right). \tag{2.8.9}$$

Let $\mathcal{D}^{\mathfrak{k}\prime}, \mathcal{E}^{\mathfrak{k}\prime}$ be the operators defined in (1.6.4), which are associated with the Euclidean vector space $(\mathfrak{k}, -B|_{\mathfrak{k}})$. Comparing the second line of (1.6.11) with (2.8.9), we get

$$\mathcal{D}^{\mathfrak{k}\prime} = \mathcal{D}^{\mathfrak{k}}, \qquad \mathcal{E}^{\mathfrak{k}\prime} = -\mathcal{E}^{\mathfrak{k}}. \tag{2.8.10}$$

By (1.6.5) and (2.8.10), we obtain

$$\overline{d}^{\mathfrak{k}*} - \overline{d}^{\mathfrak{k}} = \frac{1}{\sqrt{2}}\left(-\mathcal{D}^{\mathfrak{k}} + \mathcal{E}^{\mathfrak{k}} \right). \tag{2.8.11}$$

In particular when acting on $\Lambda^{\cdot}(\mathfrak{k}^*) \otimes L_2(\mathfrak{k})$, the operator in (2.8.11) is skew-adjoint. Again the above constructions are K-equivariant.

By (1.5.4),

$$\left[\overline{d}^{\mathfrak{k}}, \overline{d}^{\mathfrak{k}*} \right] = N^{S^{\cdot}(\mathfrak{k}^*)} + N^{\Lambda^{\cdot}(\mathfrak{k}^*)}. \tag{2.8.12}$$

Let $\Delta^{\mathfrak{k}}$ denote the classical Euclidean Laplacian of $\mathfrak{k}$, and let $|Y^{\mathfrak{k}}|$ denote the Euclidean norm of $Y^{\mathfrak{k}}$. By (1.6.10) and (2.8.11), we can rewrite (2.8.12) in the form

$$\frac{1}{2}\left(-i\mathcal{D}^{\mathfrak{k}} + i\mathcal{E}^{\mathfrak{k}} \right)^2 = \frac{1}{2}\left(-\Delta^{\mathfrak{k}} + |Y^{\mathfrak{k}}|^2 - n \right) + N^{\Lambda^{\cdot}(\mathfrak{k}^*)}. \tag{2.8.13}$$

As we saw before, the above constructions are K-equivariant. However, because we used the splitting $\mathfrak{g}$ into $\mathfrak{p} \oplus \mathfrak{k}$, we lost the G-equivariance of the operators $\overline{d}^{\mathfrak{g}}, \overline{d}^{\mathfrak{g}*}$. Finally, the above operators commute with the obvious action of K on $\Lambda^{\cdot}(\mathfrak{p}^*) \otimes S^{\cdot}(\mathfrak{p}^*)$ and $\Lambda^{\cdot}(\mathfrak{k}^*) \otimes S^{\cdot}(\mathfrak{k}^*)$, and so with the action $\rho^{\Lambda^{\cdot}(\mathfrak{g}^*) \otimes S^{\cdot}(\mathfrak{g}^*)}$ of K on $\Lambda^{\cdot}(\mathfrak{g}^*) \otimes S^{\cdot}(\mathfrak{g}^*)$.

2.9 The operator $\mathfrak{D}_b$

If $k \in K$, the action of k on $C^{\infty}(G, \Lambda^{\cdot}(\mathfrak{g}^*) \otimes S^{\cdot}(\mathfrak{g}^*))$ is given by

$$ks(g) = \rho^{\Lambda^{\cdot}(\mathfrak{g}^*) \otimes S^{\cdot}(\mathfrak{g}^*)}(k) s(gk). \tag{2.9.1}$$

Also we have the identification

$$C^{\infty}\left(G, \Lambda^{\cdot}\left(\mathfrak{g}^{*}\right) \otimes C^{\infty}\left(\mathfrak{g}\right)\right) = C^{\infty}\left(G \times \mathfrak{g}, \Lambda^{\cdot}\left(\mathfrak{g}^{*}\right)\right). \tag{2.9.2}$$

The action of K on (2.9.2) that corresponds to (2.9.1) is given by

$$ks\left(g, Y\right) = \rho^{\Lambda^{\cdot}\left(\mathfrak{g}^{*}\right)}\left(k\right) s\left(gk, \operatorname{Ad}\left(k^{-1}\right) Y\right). \tag{2.9.3}$$

Definition 2.9.1. For $b > 0$, let $\mathfrak{D}_b \in \operatorname{End}\left(C^{\infty}\left(G \times \mathfrak{g}, \Lambda^{\cdot}\left(\mathfrak{g}^{*}\right)\right)\right)$ be given by

$$\mathfrak{D}_b = \widehat{D}^{\mathfrak{g}} + ic\left(\left[Y^{\mathfrak{k}}, Y^{\mathfrak{p}}\right]\right) + \frac{1}{b}\left(\mathcal{D}^{\mathfrak{p}} + \mathcal{E}^{\mathfrak{p}} - i\mathcal{D}^{\mathfrak{k}} + i\mathcal{E}^{\mathfrak{k}}\right). \tag{2.9.4}$$

The operator $\mathfrak{D}_b$ commutes with K. Note that by (2.8.6), (2.8.11), and (2.9.4),

$$\mathfrak{D}_b = \widehat{D}^{\mathfrak{g}} + ic\left(\left[Y^{\mathfrak{k}}, Y^{\mathfrak{p}}\right]\right) + \frac{\sqrt{2}}{b}\left(\overline{\partial}^{\mathfrak{p}} - i\overline{\partial}^{\mathfrak{k}} + \overline{\partial}^{\mathfrak{p}*} + i\overline{\partial}^{\mathfrak{k}*}\right). \tag{2.9.5}$$

2.10 The compression of the operator $\mathfrak{D}_b$

If $Y = Y^{\mathfrak{p}} + Y^{\mathfrak{k}} \in \mathfrak{g}$, we have the obvious identity

$$|Y|^2 = |Y^{\mathfrak{p}}|^2 + |Y^{\mathfrak{k}}|^2. \tag{2.10.1}$$

As we saw in section 1.6, the kernel $H \subset \Lambda^{\cdot}\left(\mathfrak{g}^{*}\right) \otimes L_2\left(\mathfrak{g}\right)$ of the operator $\mathcal{D}^{\mathfrak{p}} + \mathcal{E}^{\mathfrak{p}} - i\mathcal{D}^{\mathfrak{k}} + i\mathcal{E}^{\mathfrak{k}}$ is 1-dimensional and spanned by $\exp\left(-|Y|^2/2\right)$. Let P be the orthogonal projection operator on H. Of course P acts on $C^{\infty}\left(G, \Lambda^{\cdot}\left(\mathfrak{g}^{*}\right) \otimes L_2\left(\mathfrak{g}\right)\right)$.

Proposition 2.10.1. *The following identity holds:*

$$P\left(\widehat{D}^{\mathfrak{g}} + ic\left(\left[Y^{\mathfrak{k}}, Y^{\mathfrak{p}}\right]\right)\right) P = 0. \tag{2.10.2}$$

Proof. Since H is of degree 0 in $\Lambda^{\cdot}\left(\mathfrak{g}^{*}\right)$, and since $\widehat{D}^{\mathfrak{g}} + ic\left(\left[Y^{\mathfrak{k}}, Y^{\mathfrak{p}}\right]\right)$ is of odd degree, (2.10.2) follows. $\qquad\square$

2.11 A formula for $\mathfrak{D}_b^2$

We denote by $\Delta^{\mathfrak{p} \oplus \mathfrak{k}}$ the standard Euclidean Laplacian on the Euclidean vector space $\mathfrak{g} = \mathfrak{p} \oplus \mathfrak{k}$.

We use the notation in (1.1.11), so that if $U \in \mathfrak{g}$,

$$\widehat{c}\left(\operatorname{ad}\left(U\right)\right) = -\frac{1}{4} B\left(\left[U, e_i^{*}\right], e_j^{*}\right) \widehat{c}\left(e_i\right) \widehat{c}\left(e_j\right). \tag{2.11.1}$$

Also if $V \in \mathfrak{k}$, $\operatorname{ad}\left(V\right)|_{\mathfrak{p}}$ acts as an antisymmetric endomorphism of $\mathfrak{p}$, so that by (1.1.9),

$$c\left(\operatorname{ad}\left(V\right)|_{\mathfrak{p}}\right) = \frac{1}{4} \sum_{1 \le i,j \le m} \langle\left[V, e_i\right], e_j\rangle c\left(e_i\right) c\left(e_j\right). \tag{2.11.2}$$

Finally, if $W \in \mathfrak{p}$, $\mathrm{ad}\,(W)$ exchanges $\mathfrak{k}$ and $\mathfrak{p}$ and is antisymmetric with respect to B, i.e., it is symmetric with respect to the scalar product on $\mathfrak{g}$. Moreover, by (1.1.9),

$$c\,(\mathrm{ad}\,(W)) = \frac{1}{2} \sum_{\substack{m+1 \leq i \leq m+n \\ 1 \leq j \leq m}} \langle [W, e_i^*], e_j \rangle\, c\,(e_i)\, c\,(e_j). \tag{2.11.3}$$

If $a \in \mathfrak{g}$, we denote by ∇_a^V the corresponding differentiation operator along $\mathfrak{g}$. In particular $\nabla_{[Y^{\mathfrak{p}}, Y^{\mathfrak{k}}]}^V$ denotes the differentiation operator in the direction $[Y^{\mathfrak{p}}, Y^{\mathfrak{k}}] \in \mathfrak{p}$. If $Y \in \mathfrak{g}$, we denote by $\underline{Y}^{\mathfrak{p}} + i\underline{Y}^{\mathfrak{k}}$ the section of $U\,(\mathfrak{g}) \otimes_{\mathbf{R}} \mathbf{C}$ associated to $Y^{\mathfrak{p}} + iY^{\mathfrak{k}} \in \mathfrak{g} \otimes_{\mathbf{R}} \mathbf{C}$. Recall that $N^{\Lambda^{\cdot}(\mathfrak{g}^*)}$ is the number operator of $\Lambda^{\cdot}(\mathfrak{g}^*)$.

Theorem 2.11.1. *The following identity holds:*

$$\frac{\mathfrak{D}_b^2}{2} = \frac{\widehat{D}^{\mathfrak{g},2}}{2} + \frac{1}{2}\left|[Y^{\mathfrak{k}}, Y^{\mathfrak{p}}]\right|^2 + \frac{1}{2b^2}\left(-\Delta^{\mathfrak{p} \oplus \mathfrak{k}} + |Y|^2 - m - n\right) + \frac{N^{\Lambda^{\cdot}(\mathfrak{g}^*)}}{b^2}$$

$$+ \frac{1}{b}\left(\underline{Y}^{\mathfrak{p}} + i\underline{Y}^{\mathfrak{k}} - i\nabla_{[Y^{\mathfrak{k}}, Y^{\mathfrak{p}}]}^V + \widehat{c}\left(\mathrm{ad}\left(Y^{\mathfrak{p}} + iY^{\mathfrak{k}}\right)\right)\right.$$

$$\left. + 2ic\left(\mathrm{ad}\left(Y^{\mathfrak{k}}\right)|_{\mathfrak{p}}\right) - c\left(\mathrm{ad}\left(Y^{\mathfrak{p}}\right)\right)\right). \tag{2.11.4}$$

Proof. By equations (1.6.10), (2.6.13), (2.7.2), (2.8.6), (2.8.11), (2.9.4), and (2.11.1)–(2.11.3), we get (2.11.4) easily. $\qquad\qquad\square$

2.12 The action of $\mathfrak{D}_b$ on quotients by K

In the sequel, if V is a real vector space and if E is a complex vector space, we will denote by $V \otimes E$ the complex vector space $V \otimes_{\mathbf{R}} E$. We will use such a convention in the whole book.

Let E be a finite dimensional Hermitian vector space, let $\rho^E : K \to \mathrm{Aut}\,(E)$ be a unitary representation of K. Let F be the corresponding vector bundle over $X = G/K$, i.e.,

$$F = G \times_K E. \tag{2.12.1}$$

Let g^F be the natural Hermitian metric on F, let ∇^F be the canonical Hermitian connection of F, and let R^F be its curvature. Then K acts unitarily on $\Lambda^{\cdot}(\mathfrak{g}^*) \otimes S^{\cdot}(\mathfrak{g}^*) \otimes E$. Equivalently K acts unitarily on $\Lambda^{\cdot}(\mathfrak{g}^*) \otimes L_2(\mathfrak{g}) \otimes E$. If $k \in K$, and $s \in \Lambda^{\cdot}(\mathfrak{g}^*) \otimes L_2(\mathfrak{g}) \otimes E$, ks is given by

$$ks\,(Y) = \rho^{\Lambda^{\cdot}(\mathfrak{g}^*) \otimes E}\,(k)\, s\left(\mathrm{Ad}\left(k^{-1}\right) Y\right). \tag{2.12.2}$$

Equation (2.12.2) is to be compared with (2.9.3).

If $e \in \mathfrak{k}$, then $[e, Y]$ is a Killing vector field on $\mathfrak{g} = \mathfrak{p} \oplus \mathfrak{k}$, and the corresponding Lie derivative $L_{[e,Y]}^V$ acting on $C^\infty\,(\mathfrak{g}, \Lambda^{\cdot}(\mathfrak{g}^*))$ is given by

$$L_{[e,Y]}^V = \nabla_{[e,Y]}^V + \sum_{1 \leq k \leq m+n} e^k i_{[e,e_k]}. \tag{2.12.3}$$

Equivalently,

$$L^V_{[e,Y]} = \nabla^V_{[e,Y]} - (c + \widehat{c}) \, (\mathrm{ad}\,(e)).\tag{2.12.4}$$

Recall that the tangent bundle TX is given by (2.1.11), and that the vector bundle N was defined in (2.1.12). We have the identity of vector bundles over X,

$$G \times_K (\Lambda^{\cdot}\,(\mathfrak{g}^*) \otimes S^{\cdot}\,(\mathfrak{g}^*) \otimes E) = \Lambda^{\cdot}\,(T^*X \oplus N^*) \otimes S^{\cdot}\,(T^*X \oplus N^*) \otimes F.\tag{2.12.5}$$

Of course, (2.12.5) is equivalent to

$$G \times_K (\Lambda^{\cdot}\,(\mathfrak{g}^*) \otimes L_2\,(\mathfrak{g}) \otimes E) = \Lambda^{\cdot}\,(T^*X \oplus N^*) \otimes L_2\,(T^*X \oplus N^*) \otimes F.\tag{2.12.6}$$

Let $\widehat{\mathcal{X}}$ be the total space of $TX \oplus N$ over X, and let $\widehat{\pi} : \widehat{\mathcal{X}} \to X$ be the obvious projection. As we saw in (2.2.1),

$$\widehat{\mathcal{X}} = X \times \mathfrak{g},\tag{2.12.7}$$

and $\widehat{\pi} : \widehat{\mathcal{X}} \to X$ is the obvious projection $X \times \mathfrak{g} \to X$. Let $Y = Y^{TX} + Y^N, Y^{TX} \in TX, Y^N \in N$ be the canonical section of $\widehat{\pi}^*\,(TX \oplus N)$ over $\widehat{\mathcal{X}}$. Then

$$|Y|^2 = \left|Y^{TX}\right|^2 + \left|Y^N\right|^2.\tag{2.12.8}$$

Definition 2.12.1. Let $\mathcal{H}$ be the vector space of smooth sections over X of the vector bundle $C^\infty\,(TX \oplus N, \widehat{\pi}^*\,(\Lambda^{\cdot}\,(T^*X \oplus N^*) \otimes F))$.

The vector space $\mathcal{H}$ can be identified with the space of smooth sections s of $\Lambda^{\cdot}\,(\mathfrak{g}^*) \otimes E$ over $G \times \mathfrak{g}$, which are such that if $k \in K$,

$$s\,(gk, Y) = \rho^{\Lambda^{\cdot}\,(\mathfrak{g}^*) \otimes E}\,(k^{-1})\,s\,(g, \mathrm{Ad}\,(k)\,Y).\tag{2.12.9}$$

Equivalently, an action of K on $C^\infty\,(G \times \mathfrak{g}, \Lambda^{\cdot}\,(\mathfrak{g}^*) \otimes E)$ can be defined as in (2.9.3). Equation (2.12.9) just asserts that s is fixed by the action of K.

If s is taken as before, if $(g, Y) \in G \times \mathfrak{g}$, set

$$\sigma\,(g, Y) = \rho^{\Lambda^{\cdot}\,(\mathfrak{g}^*)}\,(g)\,s\,(g, \mathrm{Ad}\,(g^{-1})\,Y).\tag{2.12.10}$$

Then (2.12.10) is equivalent to the fact that if $k \in K$,

$$\sigma\,(gk, Y) = \rho^E\,(k^{-1})\,\sigma\,(g, Y).\tag{2.12.11}$$

Equation (2.12.11) says that σ is a section of $\widehat{\pi}^* F$ over $G \times \mathfrak{g}$. From the above we deduce that we have the canonical isomorphism of vector bundles over X,

$$C^\infty\,(TX \oplus N, \widehat{\pi}^*\,(\Lambda^{\cdot}\,(T^*X \oplus N^*) \otimes F)) \simeq C^\infty\,(\mathfrak{g}, \Lambda^{\cdot}\,(\mathfrak{g}^*) \otimes \widehat{\pi}^* F).\tag{2.12.12}$$

If $\gamma \in G$, the action of γ on $\mathcal{H}$ is given by the map $s\,(g, Y) \to \gamma s\,(g, Y) = s\,(\gamma^{-1}g, Y)$. Equivalently, it is given by the map $\sigma\,(g, Y) \to \gamma\sigma\,(g, Y) = \rho^{\Lambda^{\cdot}\,(\mathfrak{g}^*)}\,(\gamma)\,\sigma\,(\gamma^{-1}g, \mathrm{Ad}\,(\gamma^{-1})\,Y)$.

If $e \in \mathfrak{k}$, let $\nabla_{e,\ell}$ be the differentiation operator with respect to the left-invariant vector field e. By (2.12.9), we deduce that if $e \in \mathfrak{k}, s \in \mathcal{H}$,

$$\nabla_{e,\ell} s = \left(L^V_{[e,Y]} - \rho^E(e) \right) s. \tag{2.12.13}$$

As we saw in section 2.1, the splitting $\mathfrak{g} = \mathfrak{p} \oplus \mathfrak{k}$ induces a connection on the K-principal bundle $p : G \to X = G/K$. Let $\nabla^{C^\infty(TX \oplus N, \widehat{\pi}^*(\Lambda^{\cdot}(T^*X \oplus N^*) \otimes F))}$ be the corresponding connection on $C^\infty(TX \oplus N, \widehat{\pi}^*(\Lambda^{\cdot}(T^*X \oplus N^*) \otimes F))$.

The vector bundle $TX \oplus N$ is equipped with the canonical connection $\nabla^{TX \oplus N}$. If $e \in TX$, we still denote by $e \in T\widehat{\mathcal{X}}$ the horizontal lift of e with respect to this connection.

The operators $\widehat{D}^{\mathfrak{g}}, \widehat{D}^{\mathfrak{g}}_H, \widehat{D}^{\mathfrak{g}}_V$ were defined in (2.7.2), (2.7.5). Since they are K-invariant, they descend to operators $\widehat{D}^{\mathfrak{g},X}, \widehat{D}^{\mathfrak{g},X}_H, \widehat{D}^{\mathfrak{g},X}_V$ acting on $\mathcal{H}$.

We still denote by $C^{\mathfrak{g}}$ the action of the Casimir operator on $\mathcal{H}$. We will use the same conventions for $C^{\mathfrak{g},H}, C^{\mathfrak{k}}$. Let $\Delta^{H,X}$ be the Bochner Laplacian acting on $\mathcal{H}$. The operator $\Delta^{H,X}$ is given by

$$\Delta^{H,X} = \sum_{i=1}^{m} \nabla^{C^\infty(TX \oplus N, \widehat{\pi}^*(\Lambda^{\cdot}(T^*X \oplus N^*) \otimes F)),2}_{e_i}. \tag{2.12.14}$$

By comparing (2.5.5) and (2.12.14), we find that

$$C^{\mathfrak{g},H} = -\Delta^{H,X}. \tag{2.12.15}$$

We still equip the Lie algebra $\mathfrak{k}$ with the bilinear form induced by B. As we saw in (2.12.2), the group K acts on the left $C^\infty(\mathfrak{g}, \Lambda^{\cdot}(\mathfrak{g}^*) \otimes E)$. We still denote by $C^{\mathfrak{k}}$ the action of the Casimir operator of K on this vector space. Then

$$C^{\mathfrak{k}} = \sum_{i=m+1}^{m+n} \left(L^V_{[e_i,Y]} - \rho^E(e_i) \right)^2. \tag{2.12.16}$$

The operator $C^{\mathfrak{k}}$ is nonpositive. It descends to an operator acting along the fibres of $C^\infty(TX \oplus N, \widehat{\pi}^*(\Lambda^{\cdot}(T^*X \oplus N^*) \otimes F))$, which we still denote by $C^{\mathfrak{k}}$.

Theorem 2.12.2. *The following identities hold:*

$$\widehat{D}^{\mathfrak{g},X} = \widehat{D}^{\mathfrak{g},X}_H + \widehat{D}^{\mathfrak{g},X}_V, \qquad \left[\widehat{D}^{\mathfrak{g},X}_H, \widehat{D}^{\mathfrak{g},X}_V \right] = 0,$$

$$\widehat{D}^{\mathfrak{g},X}_H = \sum_{i=1}^{m} \widehat{c}(e_i) \nabla^{C^\infty(TX \oplus N, \widehat{\pi}^*(\Lambda^{\cdot}(T^*X \oplus N^*) \otimes F))}_{e_i},$$

$$\widehat{D}^{\mathfrak{g},X}_V = - \sum_{i=m+1}^{m+n} \widehat{c}(e_i) \left(L^V_{[e_i,Y]} + \widehat{c}(\mathrm{ad}(e_i)|_{\mathfrak{p}}) - \rho^E(e_i) \right) + \frac{1}{2} \widehat{c}(-\kappa^{\mathfrak{k}}),$$

$$\tag{2.12.17}$$

$$\widehat{D}^{\mathfrak{g},X,2} = -C^{\mathfrak{g}} - \frac{1}{4} B^*(\kappa^{\mathfrak{g}}, \kappa^{\mathfrak{g}}),$$

$$C^{\mathfrak{g}} = C^{\mathfrak{g},H} + C^{\mathfrak{k}}.$$

Proof. This follows from (2.5.6), (2.7.5)–(2.7.7), and (2.12.13). □

Remark 2.12.3. The operator $\widehat{D}_H^{\mathfrak{g},X}$ is just a standard Dirac operator over X.

Since the operator $\mathfrak{D}_b$ is K-invariant, it also descends to an operator acting on $\mathcal{H}$, which we denote by $\mathfrak{D}_b^X$. The operators $\mathcal{D}^{\mathfrak{p}}, \mathcal{E}^{\mathfrak{p}}, \mathcal{D}^{\mathfrak{k}}, \mathcal{E}^{\mathfrak{k}}$, descend to operators $\mathcal{D}^{TX}, \mathcal{E}^{TX}, \mathcal{D}^N, \mathcal{E}^N$. Let $\Delta^{TX \oplus N}$ be the standard Euclidean Laplacian acting along the fibres of $TX \oplus N$. Finally, $Y = Y^{TX} + Y^N$ denotes the tautological section of $\widehat{\pi}^* (TX \oplus N)$ over $\widehat{\mathcal{X}}$.

The vector bundle $TX \oplus N$ is equipped with the nondegenerate symmetric bilinear form induced by B. Let $dv_{\widehat{\mathcal{X}}}$ be the induced volume form on $\widehat{\mathcal{X}}$. This is just the volume form coming from the Riemannian metric on X, and from the Euclidean scalar product on $TX \oplus N$.

Combining B^* with the Hermitian metric g^F on F produces a Hermitian form ( ) on $\Lambda^{\cdot} (T^*X \oplus N^*) \otimes F$.

The Cartan involution θ acts on $\widehat{\mathcal{X}}$, so that $\theta \left(Y^{TX} + Y^N \right) = -Y^{TX} + Y^N$.

Definition 2.12.4. Let η be the Hermitian form on the space of smooth compactly supported sections of $\widehat{\pi}^* (\Lambda^{\cdot} (T^*X \oplus N^*) \otimes F)$ over $\widehat{\mathcal{X}}$,

$$\eta(s, s') = \int_{\widehat{\mathcal{X}}} (s \circ \theta, s') \, dv_{\widehat{\mathcal{X}}}. \tag{2.12.18}$$

Note that $\theta\mathrm{ad}\left(Y^N \right)$ is an endomorphism of $TX \oplus N$ that preserves TX and N, and which is antisymmetric with respect to B (and also with respect to the scalar product of $\mathfrak{g}$), so that $c\left(\theta\mathrm{ad}\left(Y^N \right) \right)$ is well defined. Similarly $\rho^E \left(Y^N \right)$ descends to a section of $\mathrm{End}\left(F \right)$.

Observe that Y^{TX} is the canonical section of $\widehat{\pi}^* TX$ over $\widehat{\mathcal{X}}$. Then

$$\nabla_{Y^{TX}}^{C^\infty (TX \oplus N, \widehat{\pi}^* (\Lambda^{\cdot} (T^*X \oplus N^*) \otimes F))}$$

will denote covariant differentiation with respect to the given canonical connection in the horizontal direction corresponding to Y^{TX}.

Theorem 2.12.5. *The following identities hold:*

$$\mathfrak{D}_b^X = \widehat{D}^{\mathfrak{g},X} + ic \left([Y^N, Y^{TX}] \right) + \frac{1}{b} \left(\mathcal{D}^{TX} + \mathcal{E}^{TX} - i\mathcal{D}^N + i\mathcal{E}^N \right),$$

$$\frac{1}{2}\mathfrak{D}_b^{X,2} = \frac{1}{2}\widehat{D}^{\mathfrak{g},X,2} + \frac{1}{2} \left| [Y^N, Y^{TX}] \right|^2 + \frac{1}{2b^2} \left(-\Delta^{TX \oplus N} + |Y|^2 - m - n \right)$$

$$\tag{2.12.19}$$

$$+ \frac{N^{\Lambda^{\cdot} (T^*X \oplus N^*)}}{b^2} + \frac{1}{b} \left(\nabla_{Y^{TX}}^{C^\infty (TX \oplus N, \widehat{\pi}^* (\Lambda^{\cdot} (T^*X \oplus N^*) \otimes F))} + \widehat{c} \left(\mathrm{ad}\left(Y^{TX} \right) \right) \right.$$

$$\left. - c \left(\mathrm{ad}\left(Y^{TX} \right) + i\theta\mathrm{ad}\left(Y^N \right) \right) - i\rho^E \left(Y^N \right) \right).$$

The operator $\mathfrak{D}_b^X$ is formally skew-adjoint with respect to η, and the operator $\mathfrak{D}_b^{X,2}$ is formally self-adjoint with respect to η.

Proof. The first identity in (2.12.19) follows from (2.9.4), and the second identity from (2.11.4), (2.12.4), and (2.12.13). By using the considerations we made after (1.1.3), and also the first identity in (2.12.19), one verifies easily that $\mathfrak{D}_b^X$ is skew-adjoint with respect to η. It is now obvious that $\mathfrak{D}_b^{X,2}$ is self-adjoint with respect to η. $\qquad\square$

Remark 2.12.6. Equation (2.12.19) for $\mathfrak{D}_b^{X,2}$ will play a critical role in the sequel. Observe in particular that the term $-i\nabla^V_{[Y^{\mathfrak{k}}, Y^{\mathfrak{p}}]}$ in (2.11.4) has now disappeared, and besides that the quadratic and quartic expressions in $Y^{\mathfrak{k}}, Y^{\mathfrak{p}}$ have the proper positive sign.

The connection $\nabla^{\Lambda^{\cdot}(T^*X \oplus N^*), f*, \widehat{f}}$ was defined in Definition 2.4.1. Let

$$\nabla^{C^\infty(TX \oplus N, \widehat{\pi}^*(\Lambda^{\cdot}(T^*X \oplus N^*) \otimes F)), f*, \widehat{f}}_{Y^{TX}}$$

be the connection on $C^\infty(TX \oplus N, \widehat{\pi}^*(\Lambda^{\cdot}(T^*X \oplus N^*) \otimes F))$ that is induced by $\nabla^{\Lambda^{\cdot}(T^*X \oplus N^*), f*, \widehat{f}}, \nabla^F$. Then we can rewrite the last identity in (2.12.19) in the form

$$\frac{1}{2}\mathfrak{D}_b^{X,2} = \frac{1}{2}\widehat{D}^{\mathfrak{g},X,2} + \frac{1}{2}\left|[Y^N, Y^{TX}]\right|^2 + \frac{1}{2b^2}\left(-\Delta^{TX \oplus N} + |Y|^2 - m - n\right)$$
$$+ \frac{N^{\Lambda^{\cdot}(T^*X \oplus N^*)}}{b^2} + \frac{1}{b}\left(\nabla^{C^\infty(TX \oplus N, \widehat{\pi}^*(\Lambda^{\cdot}(T^*X \oplus N^*) \otimes F)), f*, \widehat{f}}_{Y^{TX}}\right.$$
$$\left. - c\left(i\theta \operatorname{ad}\left(Y^N\right)\right) - i\rho^E\left(Y^N\right)\right). \quad (2.12.20)$$

The fact that the connection $\nabla^{\Lambda^{\cdot}(T^*X \oplus N^*), f*, \widehat{f}}$ is flat makes the operator in (2.12.20) almost look like a scalar operator. This will be of utmost importance in the sequel.

2.13 The operators $\mathcal{L}^X$ and $\mathcal{L}_b^X$

The Casimir operator $C^{\mathfrak{g}}$ descends to an operator acting on $C^\infty(X, F)$, which will still be denoted $C^{\mathfrak{g}}$.

Let $\Delta^{H,X}$ be the Bochner Laplacian acting on $C^\infty(X, F)$. As in (2.12.15), we get

$$C^{\mathfrak{g},H} = -\Delta^{H,X}. \quad (2.13.1)$$

Also the Casimir operator $C^{\mathfrak{k},E}$ descends to $C^{\mathfrak{k},F} \in \operatorname{End}(F)$. By (2.5.6) and (2.13.1), we get

$$C^{\mathfrak{g}} = -\Delta^{H,X} + C^{\mathfrak{k},F}. \quad (2.13.2)$$

Definition 2.13.1. Let $\mathcal{L}^X$ act on $C^\infty(X, F)$ by the formula

$$\mathcal{L}^X = \frac{1}{2}C^{\mathfrak{g}} + \frac{1}{8}B^*\left(\kappa^{\mathfrak{g}}, \kappa^{\mathfrak{g}}\right). \quad (2.13.3)$$

For $b > 0$, let $\mathcal{L}_b^X$ act on $C^\infty\left(\widehat{\mathcal{X}}, \widehat{\pi}^*\left(\Lambda^{\cdot}\left(T^*X \oplus N^*\right) \otimes F\right)\right)$ by the formula

$$\mathcal{L}_b^X = -\frac{1}{2}\widehat{D}^{\mathfrak{g},X,2} + \frac{1}{2}\mathfrak{D}_b^{X,2}. \tag{2.13.4}$$

By (2.12.19), we get

$$\mathcal{L}_b^X = \frac{1}{2}\left|\left[Y^N, Y^{TX}\right]\right|^2 + \frac{1}{2b^2}\left(-\Delta^{TX \oplus N} + |Y|^2 - m - n\right) + \frac{N^{\Lambda^{\cdot}(T^*X \oplus N^*)}}{b^2}$$

$$+ \frac{1}{b}\left(\nabla_{Y^{TX}}^{C^\infty(TX \oplus N, \widehat{\pi}^*(\Lambda^{\cdot}(T^*X \oplus N^*) \otimes F))} + \widehat{c}\left(\mathrm{ad}\left(Y^{TX}\right)\right)\right.$$

$$\left. - c\left(\mathrm{ad}\left(Y^{TX}\right) + i\theta\mathrm{ad}\left(Y^N\right)\right) - i\rho^E\left(Y^N\right)\right). \tag{2.13.5}$$

The fibres of $\Lambda^{\cdot}\left(T^*X \oplus N^*\right) \otimes F$ are naturally equipped with a Hermitian product, which is denoted $\langle\rangle$. Let $\langle\rangle$ be the Hermitian product on the space of smooth compactly supported sections of $\widehat{\pi}^*\left(\Lambda^{\cdot}\left(T^*X \oplus N^*\right) \otimes F\right)$ over $\widehat{\mathcal{X}}$,

$$\langle s, s'\rangle = \int_{\widehat{\mathcal{X}}} \langle s, s'\rangle \, dv_{\widehat{\mathcal{X}}}. \tag{2.13.6}$$

Theorem 2.13.2. *The operator $\mathcal{L}_b^X$ is formally self-adjoint with respect to η. Moreover, $\frac{\partial}{\partial t} + \mathcal{L}_b^X$ is hypoelliptic.*

Also $\frac{1}{b}\nabla_{Y^{TX}}^{C^\infty(TX \oplus N, \widehat{\pi}^*(\Lambda^{\cdot}(T^*X \oplus N^*) \otimes F))}$ is formally skew-adjoint with respect to $\langle\rangle$, and $\mathcal{L}_b^X - \frac{1}{b}\nabla_{Y^{TX}}^{C^\infty(TX \oplus N, \widehat{\pi}^*(\Lambda^{\cdot}(T^*X \oplus N^*) \otimes F))}$ is formally self-adjoint with respect to $\langle\rangle$.

Proof. The first part of our theorem follows from Theorem 2.12.5 and from (2.13.4). The second part is a consequence of a general result by Hörmander [Hö67] on second order differential operators.

Clearly $\frac{1}{b}\nabla_{Y^{TX}}^{C^\infty(TX \oplus N, \widehat{\pi}^*(\Lambda^{\cdot}(T^*X \oplus N^*) \otimes F))}$ is formally skew-adjoint with respect to $\langle\rangle$. The first line in the right-hand side of (2.13.5) is formally self-adjoint with respect to $\langle\rangle$. Except for $\frac{1}{b}\nabla_{Y^{TX}}^{C^\infty(TX \oplus N, \widehat{\pi}^*(\Lambda^{\cdot}(T^*X \oplus N^*) \otimes F))}$, this is also the case for all terms in the right-hand side of (2.13.5). Indeed since $\mathrm{ad}\left(Y^{TX}\right)$ exchanges TX and N, this is clear for $\widehat{c}\left(\mathrm{ad}\left(Y^{TX}\right)\right)$ and $c\left(\mathrm{ad}\left(Y^{TX}\right)\right)$. Also since $\theta\mathrm{ad}\left(Y^N\right)$ is antisymmetric with respect to the scalar product of $\Lambda^{\cdot}\left(T^*X \oplus N^*\right)$ and preserves TX and N, $c\left(\theta\mathrm{ad}\left(Y^N\right)\right)$ is antisymmetric, and so $ic\left(\theta\mathrm{ad}\left(Y^N\right)\right)$ is self-adjoint. Finally, $i\rho^E\left(Y^N\right)$ is clearly self-adjoint. The proof of our theorem is completed. □

The operator $\mathcal{L}_b^X$ will be called a hypoelliptic Laplacian.

By proceeding as in (2.12.20), we can rewrite (2.13.5) in the form

$$\mathcal{L}_b^X = \frac{1}{2}\left|\left[Y^N, Y^{TX}\right]\right|^2 + \frac{1}{2b^2}\left(-\Delta^{TX \oplus N} + |Y|^2 - m - n\right) + \frac{N^{\Lambda^{\cdot}(T^*X \oplus N^*)}}{b^2}$$

$$+ \frac{1}{b}\left(\nabla_{Y^{TX}}^{C^\infty(TX \oplus N, \widehat{\pi}^*(\Lambda^{\cdot}(T^*X \oplus N^*) \otimes F)), f*, \widehat{f}} - c\left(i\theta\mathrm{ad}\left(Y^N\right)\right) - i\rho^E\left(Y^N\right)\right).$$

$$\tag{2.13.7}$$

2.14 The scaling of the form B

For $t > 0$, we denote with an extra subscript t the above operators associated with the form B/t.

By (2.12.19), it is clear that

$$t^{N^{\Lambda^{\cdot}(T^*X \oplus N^*)}/2} \mathfrak{D}_{b,t}^X t^{-N^{\Lambda^{\cdot}(T^*X \oplus N^*)}/2} = \sqrt{t} \widehat{D}^{\mathfrak{g},X} + \frac{i}{\sqrt{t}} c\left([Y^N, Y^{TX}]\right)$$

$$+ \frac{1}{b}\left(\sqrt{t} D^{TX} + \frac{1}{\sqrt{t}} \mathcal{E}^{TX} - i\sqrt{t} D^N + \frac{i}{\sqrt{t}} \mathcal{E}^N\right). \quad (2.14.1)$$

For $a > 0$, set

$$K_a s\left(x, Y\right) = s\left(x, aY\right). \quad (2.14.2)$$

By (2.14.1), we get

$$K_{\sqrt{t}} t^{N^{\Lambda^{\cdot}(T^*X \oplus N^*)}/2} \mathfrak{D}_{b,t}^X t^{-N^{\Lambda^{\cdot}(T^*X \oplus N^*)}/2} K_{\sqrt{t}}^{-1} = \sqrt{t} \mathfrak{D}_{\sqrt{t}b}^X. \quad (2.14.3)$$

By (2.13.4), (2.14.3), we obtain

$$K_{\sqrt{t}} t^{N^{\Lambda^{\cdot}(T^*X \oplus N^*)}/2} \mathcal{L}_{b,t}^X t^{-N^{\Lambda^{\cdot}(T^*X \oplus N^*)}/2} K_{\sqrt{t}}^{-1} = t \mathcal{L}_{\sqrt{t}b}^X. \quad (2.14.4)$$

2.15 The Bianchi identity

The classical Bianchi identity says that

$$\left[\mathfrak{D}_b^X, \mathfrak{D}_b^{X,2}\right] = 0. \quad (2.15.1)$$

We will establish a refinement of the Bianchi identity along the lines of [B08b, Proposition 3.12].

Proposition 2.15.1. *For any $b > 0$, the following identity holds:*

$$\left[\mathfrak{D}_b^X, \mathcal{L}_b^X\right] = 0. \quad (2.15.2)$$

Proof. We start from the Bianchi identity (2.15.1). By (2.12.17), $\widehat{D}^{\mathfrak{g},X,2}$ coincides with $-C^{\mathfrak{g}} - \frac{1}{4} B^*\left(\kappa^{\mathfrak{g}}, \kappa^{\mathfrak{g}}\right)$, so that

$$\left[\mathfrak{D}_b^X, \widehat{D}^{\mathfrak{g},X,2}\right] = 0. \quad (2.15.3)$$

By (2.13.4), (2.15.1), and (2.15.3), we get (2.15.2). $\qquad \square$

2.16 A fundamental identity

Put

$$\alpha = \frac{1}{2}\left(-\Delta^{TX \oplus N} + |Y|^2 - m - n\right) + N^{\Lambda^{\cdot}(T^*X \oplus N^*)},$$

$$\beta = \nabla_{Y^{TX}}^{C^{\infty}(TX \oplus N, \widehat{\pi}^*(\Lambda^{\cdot}(T^*X \oplus N^*) \otimes F))} + \widehat{c}\left(\mathrm{ad}\left(Y^{TX}\right)\right) \quad (2.16.1)$$

$$- c\left(\mathrm{ad}\left(Y^{TX}\right) + i\theta \mathrm{ad}\left(Y^N\right)\right) - i\rho^E\left(Y^N\right),$$

$$\gamma = \frac{1}{2}\left|[Y^N, Y^{TX}]\right|^2.$$

By (2.13.5), we get

$$\mathcal{L}_b^X = \frac{\alpha}{b^2} + \frac{\beta}{b} + \gamma. \tag{2.16.2}$$

We denote by H the fibrewise kernel of α, so that

$$H = \left\{ \exp\left(-|Y|^2/2\right) \right\} \otimes F. \tag{2.16.3}$$

Let $H^\perp$ be the orthogonal to H in $L_2\left(\widehat{\mathcal{X}}, \widehat{\pi}^*\left(\Lambda^{\cdot}\left(T^*X \oplus N^*\right) \otimes F\right)\right)$.

Note that β maps H into $H^\perp$. Let α^{-1} be the inverse of α restricted to $H^\perp$. Let $P, P^\perp$ be the orthogonal projections on $H, H^\perp$. We embed $L_2\left(X, F\right)$ into $L_2\left(\widehat{\mathcal{X}}, \widehat{\pi}^*\left(\Lambda^{\cdot}\left(T^*X \oplus N^*\right) \otimes F\right)\right)$ via the isometric embedding $s \to \widehat{\pi}^* s \exp\left(-|Y|^2/2\right)/\pi^{(m+n)/4}$.

Now we establish an analogue of [B05, Theorem 3.14], [B08a, Theorem 1.14], and [B08b, Theorem 3.11], this last reference being especially relevant.

Theorem 2.16.1. *The following identity holds:*

$$P\left(\gamma - \beta\alpha^{-1}\beta\right)P = \mathcal{L}^X. \tag{2.16.4}$$

Proof. We can rewrite the first equation in (2.12.19) in the form

$$\frac{1}{\sqrt{2}}\mathfrak{D}_b^X = E + \frac{F}{b}. \tag{2.16.5}$$

Using (2.13.4), and comparing (2.16.2) and (2.16.5), we get

$$\alpha = F^2, \qquad \beta = [E, F], \qquad \gamma = E^2 - \frac{1}{2}\widehat{D}^{\mathfrak{g},X,2}. \tag{2.16.6}$$

Moreover, as we saw in section 1.5, H is just the kernel of F, so that $PF = 0, FP = 0$. From (2.16.6), we obtain

$$P\left(\gamma - \beta\alpha^{-1}\beta\right)P = P\left(E^2 - EP^\perp E - \frac{1}{2}\widehat{D}^{X,2}\right)P. \tag{2.16.7}$$

Equivalently,

$$P\left(\gamma - \beta\alpha^{-1}\beta\right)P = (PEP)^2 - \frac{1}{2}P\widehat{D}^{X,2}P. \tag{2.16.8}$$

By equation (2.10.2) in Proposition 2.10.1, we get

$$PEP = 0. \tag{2.16.9}$$

Using (2.7.7), (2.13.3), (2.16.8), and (2.16.9), we get (2.16.4). $\qquad\square$

Another proof for our theorem is to use the explicit formulas for α, β, γ in (2.16.1), and to get (2.16.4) by an explicit computation. A similar explicit computation has been done in the proof of [B08b, Theorem 3.11], to which the reader is referred. Since such a computation will be needed in chapter 14, we will give the main steps of this computation.

Set

$$R(Y) = \widehat{c}\left(\mathrm{ad}\left(Y^{TX}\right)\right) - c\left(\mathrm{ad}\left(Y^{TX}\right) + i\theta\mathrm{ad}\left(Y^N\right)\right). \qquad (2.16.10)$$

Let $e_1, \ldots, e_m$ be an orthonormal basis of TX, let $e_{m+1}, \ldots, e_{m+n}$ be an orthonormal basis of N. By (1.1.9), (1.1.11), we get

$$\widehat{c}\left(\mathrm{ad}\left(Y^{TX}\right)\right) - c\left(\mathrm{ad}\left(Y^{TX}\right)\right) = \sum_{\substack{1 \le i \le m \\ m+1 \le j \le m+n}} \left\langle\left[Y^{TX}, e_i\right], e_j\right\rangle$$

$$\left(e^i \wedge e^j - i_{e_i} i_{e_j}\right), \qquad (2.16.11)$$

$$c\left(i\theta\mathrm{ad}\left(Y^N\right)\right) = -\frac{i}{4}\sum_{1 \le i,j \le m+n}\left\langle\left[Y^N, e_i\right], e_j\right\rangle c\left(e_i\right) c\left(e_j\right).$$

Also by (1.1.3), we get

$$c\left(e_i\right) = e^i - i_{e_i}, 1 \le i \le m, \qquad (2.16.12)$$
$$-e^i - i_{e_i}, m + 1 \le i \le m + n.$$

Let $\mathbf{P}$ be the projection from $\Lambda^{\cdot}\left(T^*X \oplus N^*\right)$ on $\Lambda^0\left(T^*X \oplus N^*\right) = \mathbf{R}$. Set

$$\mathbf{P}^\perp = 1 - \mathbf{P}. \qquad (2.16.13)$$

Note that by (2.16.10)–(2.16.12),

$$\mathbf{P}R(Y)\mathbf{P} = 0. \qquad (2.16.14)$$

Definition 2.16.2. For $Y \in TX \oplus N$, set

$$S(Y) = \mathbf{P}R(Y)\left(1 + N^{\Lambda^{\cdot}\left(T^*X \oplus N^*\right)}\right)^{-1} R(Y)\mathbf{P}. \qquad (2.16.15)$$

Then $S(Y) \in \mathbf{R}$.

Note that $\mathrm{ad}^2\left(Y^{TX}\right)$ maps N into itself, and moreover,

$$\mathrm{Tr}^N\left[\mathrm{ad}^2\left(Y^{TX}\right)\right] = -\left\langle C^{\mathfrak{t},\mathfrak{p}}Y^{TX}, Y^{TX}\right\rangle. \qquad (2.16.16)$$

Proposition 2.16.3. *The following identity holds:*

$$S(Y) = \left(\frac{1}{3}\mathrm{Tr}^N\left[\mathrm{ad}^2\left(Y^{TX}\right)\right] + \frac{1}{24}\mathrm{Tr}^{TX \oplus N}\left[-\mathrm{ad}^2\left(Y^N\right)\right]\right)\mathbf{P}. \qquad (2.16.17)$$

Proof. Using (2.16.10), (2.16.11), we get

$$\mathbf{P}R\left(Y^{TX}\right)\left(1 + N^{\Lambda^{\cdot}\left(T^*X \oplus N^*\right)}\right)^{-1} R\left(Y^{TX}\right)\mathbf{P} = \frac{1}{3}\sum_{i=1}^{m}\left|\left[Y^{TX}, e_i\right]\right|^2 \mathbf{P}$$

$$= \frac{1}{3}\sum_{i=m+1}^{m+n}\left|\left[Y^{TX}, e_i\right]\right|^2 \mathbf{P} = \frac{1}{3}\mathrm{Tr}^N\left[\mathrm{ad}^2\left(Y^{TX}\right)\right]\mathbf{P}. \qquad (2.16.18)$$

Also using (2.16.10), (2.16.11), and (2.16.12), we obtain

$$\mathbf{P}R\left(Y^N\right)\left(1+N^{\Lambda^\cdot\left(T^*X\oplus N^*\right)}\right)^{-1}R\left(Y^N\right)\mathbf{P} = \frac{1}{24}\sum_{i=1}^{m+n}\left|\left[Y^N,e_i\right]\right|^2\mathbf{P}$$

$$= \frac{1}{24}\mathrm{Tr}^{TX\oplus N}\left[-\mathrm{ad}^2\left(Y^N\right)\right]\mathbf{P}. \quad (2.16.19)$$

Finally, because in the first equation in (2.16.11), $1 \leq i \leq m, m+1 \leq j \leq m+n$, and in the second equation, either $1 \leq i,j \leq m$ or $m+1 \leq i,j \leq m+n$, we get easily

$$\mathbf{P}R\left(Y^{TX}\right)\left(1+N^{\Lambda^\cdot\left(T^*X\oplus N^*\right)}\right)^{-1}R\left(Y^N\right)\mathbf{P}$$

$$= \mathbf{P}R\left(Y^N\right)\left(1+N^{\Lambda^\cdot\left(T^*X\oplus N^*\right)}\right)^{-1}R\left(Y^{TX}\right)\mathbf{P} = 0. \quad (2.16.20)$$

By (2.16.18)–(2.16.20), we get (2.16.17). $\qquad\square$

Definition 2.16.4. Put

$$\delta = -\frac{3}{16}\mathrm{Tr}\left[C^{\mathfrak{k},\mathfrak{p}}\right] - \frac{1}{48}\mathrm{Tr}\left[C^{\mathfrak{k},\mathfrak{k}}\right]. \quad (2.16.21)$$

Proposition 2.16.5. *The following identity holds:*

$$PS\left(Y\right)P = \delta. \quad (2.16.22)$$

Proof. If $u \in \mathfrak{g}$, we have the easy identity

$$P\left\langle u,Y\right\rangle^2 P = \frac{1}{2}|u|^2. \quad (2.16.23)$$

By (2.16.16), (2.16.17) and (2.16.23), we get (2.16.22). $\qquad\square$

Proposition 2.16.6. *The following identity holds:*

$$P\left(\frac{1}{2}\left|\left[Y^{TX},Y^N\right]\right|^2 - R\left(Y\right)\left(1+N^{\Lambda^\cdot\left(T^*X\oplus N^*\right)}\right)^{-1}R\left(Y\right)\right)P$$

$$= \frac{1}{8}B^*\left(\kappa^{\mathfrak{g}},\kappa^{\mathfrak{g}}\right)P. \quad (2.16.24)$$

Proof. Using (2.16.23), we get

$$P\frac{1}{2}\left|\left[Y^{TX},Y^N\right]\right|^2 P = -\frac{1}{8}\mathrm{Tr}\left[C^{\mathfrak{k},\mathfrak{p}}\right]. \quad (2.16.25)$$

By (2.6.11) and (2.16.21), we obtain

$$\delta + \frac{1}{8}\mathrm{Tr}\left[C^{\mathfrak{k},\mathfrak{p}}\right] = -\frac{1}{8}B^*\left(\kappa^{\mathfrak{g}},\kappa^{\mathfrak{g}}\right). \quad (2.16.26)$$

From (2.16.15), (2.16.22), (2.16.25), and (2.16.26), we get (2.16.24). $\qquad\square$

Using again (2.16.23), we get

$$-P\nabla^{C^\infty(TX\oplus N,\widehat{\pi}^*(\Lambda^\cdot(T^*X\oplus N^*)\otimes F)),2}_{Y^{TX}}P = -\frac{1}{2}\Delta^{H,X}, \qquad (2.16.27)$$

$$P\rho^E\left(Y^N\right)^2 P = \frac{1}{2}C^{\mathfrak{k},F}.$$

Finally, note that the functions $\langle u, Y\rangle \exp\left(-\left|Y\right|^2/2\right)$ span the eigenspace of the harmonic oscillator $\frac{1}{2}\left(-\Delta^{TX\oplus N} + \left|Y\right|^2 - m - n\right)$ that is associated with the eigenvalue 1. This explains why in (2.16.4), we may as well replace α by $1 + N^{\Lambda^\cdot(T^*X\oplus N^*)}$. By combining (2.13.2), (2.13.3), (2.16.1), (2.16.11), (2.16.24), and (2.16.27), we get

$$P\left(\gamma - \beta\alpha^{-1}\beta\right)P = -P\nabla^{C^\infty(TX\oplus N,\widehat{\pi}^*(\Lambda^\cdot(T^*X\oplus N^*)\otimes F)),2}_{Y^{TX}}P$$

$$+ P\left(\frac{1}{2}\left|\left[Y^{TX},Y^N\right]\right|^2 - R(Y)\left(1 + N^{\Lambda^\cdot(T^*X\oplus N^*)}\right)^{-1}R(Y)\right)P$$

$$+ P\rho^E\left(Y^N\right)^2 P = -\frac{1}{2}\Delta^{H,X} + \frac{1}{2}C^{\mathfrak{k},F} + \frac{1}{8}B^*\left(\kappa^{\mathfrak{g}}, \kappa^{\mathfrak{g}}\right) = \mathcal{L}^X. \quad (2.16.28)$$

This completes the computational proof of Theorem 2.16.1.

2.17 The canonical vector fields on X

Recall that the curvature Ω of the canonical connection on the K-principal bundle $p : G \to G/K$ is given by (2.1.10). Let R^{TX} be the curvature of the connection ∇^{TX}. If $a, b, c \in \mathfrak{p}$, R^{TX} is just the equivariant representation of the map $a, b, c \in \mathfrak{p} \to -\left[[a,b],c\right] \in \mathfrak{p}$.

Observe that if $a, b, c, d \in \mathfrak{g}$, then

$$B\left(-\left[[a,b],c\right],d\right) = -B\left(\left[a,b\right],\left[c,d\right]\right). \qquad (2.17.1)$$

In particular,

$$B\left(-\left[[a,b],b\right],a\right) = B\left(\left[a,b\right],\left[a,b\right]\right). \qquad (2.17.2)$$

Since B is negative on $\mathfrak{k}$, from (2.17.2), by taking $a, b \in \mathfrak{p}$, we deduce that X has nonpositive curvature.

Since X has nonpositive curvature, given a base point, the exponential map $T_x X \to X$ is a covering. Since X is simply connected, this map is one to one.

If $a \in \mathfrak{g}$, let $\nabla_{a,r}$ denote differentiation on G with respect to the right-invariant vector field associated with a. The operator $\nabla_{a,r}$ descends to an operator L_a acting on $C^\infty(X, F)$. The operator L_a is just the infinitesimal version of the action of G on $C^\infty(X, F)$.

Let a^{TX} be the vector field on X associated with a. Then L_a is a Lie derivative operator acting on $C^\infty(X, F)$ that is associated with a^{TX}. We can write L_a in the form

$$L_a = \nabla^F_{a^{TX}} - \mu^F(a). \qquad (2.17.3)$$

In (2.17.3), $\mu^F(a)$ is a skew-adjoint section of $\mathrm{End}\,(F)$.

Let us now give a more precise version of (2.17.3). If $g \in G$, we can split $\mathrm{ad}\,(g^{-1})\,a \in \mathfrak{g}$ according to the splitting $\mathfrak{g} = \mathfrak{p} \oplus \mathfrak{k}$, in the form

$$\mathrm{Ad}\,(g^{-1})\,a = (\mathrm{Ad}\,(g^{-1})\,a)^{\mathfrak{p}} + (\mathrm{Ad}\,(g^{-1})\,a)^{\mathfrak{k}}. \tag{2.17.4}$$

Tautologically, the functions $(\mathrm{Ad}\,(g^{-1})\,a)^{\mathfrak{p}}, (\mathrm{Ad}\,(g^{-1})\,a)^{\mathfrak{k}}$ define sections a^{TX}, a^N of TX, N over X, the section a^{TX} being just the above vector field. Under the identification $TX \oplus N \simeq \mathfrak{g}$ given in (2.2.1), a^{TX}, a^N are precisely the components of $a \in \mathfrak{g}$ with respect to this splitting, so that

$$a = a^{TX} + a^N. \tag{2.17.5}$$

Also $\rho^E\left((\mathrm{Ad}\,(g^{-1})\,a)^{\mathfrak{k}}\right)$ descends to a skew-adjoint section of $\mathrm{End}\,(F)$, which will be denoted $\rho^E(a^N)$.

From the above considerations, we get

$$\mu^F(a) = \rho^E(a^N). \tag{2.17.6}$$

By (2.17.3) and (2.17.6), we obtain

$$L_a = \nabla^F_{a^{TX}} - \rho^E(a^N). \tag{2.17.7}$$

Moreover, it is well-known that

$$\nabla^F \mu^F(a) + i_{a^{TX}} R^F = 0. \tag{2.17.8}$$

Note that (2.17.8) follows easily from the above considerations on $\mu^F(a)$.

On the other hand, a is a flat section of $TX \oplus N \simeq \mathfrak{g}$, i.e., if $A \in TX$,

$$\nabla^{TX \oplus N, f}_A a = 0. \tag{2.17.9}$$

From (2.2.2) and (2.17.9), we get

$$\nabla^{TX}_A a^{TX} + [A, a^N] = 0, \qquad \nabla^N_A a^N + [A, a^{TX}] = 0. \tag{2.17.10}$$

When L_a acts on sections on TX, it is just the Lie derivative operator $L_{a^{TX}}$. Classically,

$$L_{a^{TX}} = \nabla^{TX}_{a^{TX}} - \nabla^{TX}_. a^{TX}. \tag{2.17.11}$$

Comparing with (2.17.3), (2.17.6), we get

$$\mu^{TX}(a) = \rho^{TX}(a^N) = \nabla^{TX}_. a^{TX}. \tag{2.17.12}$$

Of course the last equality in (2.17.12) also follows from (2.17.10).

2.18 Lie derivatives and the operator $\mathcal{L}^X_b$

If $e \in TX \oplus N$, let ∇^V_e denote the corresponding differentiation operator along the fibre $TX \oplus N$, and $L^V_{[e,Y]}$ denotes the fibrewise Lie derivative operator associated with the fibrewise vector field $[e, Y]$.

Take $a \in \mathfrak{g}$. The action of the Lie derivative operator L_a on $C^\infty(X, F)$ is given by (2.17.7). When replacing E by $C^\infty(\mathfrak{g}, \Lambda^{\cdot}(\mathfrak{g}^*) \otimes E)$, by (2.12.13) and (2.17.7), the action of L_a on

$$C^\infty(X, C^\infty(TX \oplus N, \pi^*(\Lambda^{\cdot}(T^*X \oplus N^*) \otimes F)))$$

is given by

$$L_a = \nabla^{C^\infty(TX \oplus N, \pi^*(\Lambda^{\cdot}(T^*X \oplus N^*) \otimes F))}_{a^{TX}} + L^V_{[a^N, Y]} - \rho^E(a^N). \qquad (2.18.1)$$

In (2.18.1), the fibrewise Lie derivative operator $L^V_{[a^N, Y]}$ is given by (2.12.4), i.e.,

$$L^V_{[a^N, Y]} = \nabla^V_{[a^N, Y]} - (c + \widehat{c})(\mathrm{ad}(a^N)). \qquad (2.18.2)$$

Definition 2.18.1. Set

$$\mathcal{L}^X_{a,b} = \mathcal{L}^X_b + L_a. \qquad (2.18.3)$$

In the sequel, we use the notation in (2.17.5).

Theorem 2.18.2. *The following identity holds:*

$$\mathcal{L}^X_{a,b} = \frac{1}{2} \left| [Y^N, Y^{TX}] \right|^2 + \frac{1}{2b^2} \left(-\Delta^{TX \oplus N} + |Y|^2 - m - n \right) + \frac{N^{\Lambda^{\cdot}(T^*X \oplus N^*)}}{b^2}$$

$$+ \frac{1}{b} \left(\nabla^{C^\infty(TX \oplus N, \pi^*(\Lambda^{\cdot}(T^*X \oplus N^*) \otimes F)), f_*, \widehat{f}}_{Y^{TX} + ba^{TX}} - \widehat{c}(\mathrm{bad}(a)) \right.$$

$$\left. - c \left(i\theta \mathrm{ad}(Y^N) + \mathrm{bad}(a^N - a^{TX}) \right) - \rho^E(iY^N + ba^N) \right) + \nabla^V_{[a^N, Y]}.$$

$$(2.18.4)$$

Moreover, the operator $\mathcal{L}^X_{a,b}$ commutes with $Z(a)$.

Proof. Equation (2.18.4) follows from (2.4.5), (2.13.7), (2.18.1), and (2.18.2). Since $\mathcal{L}^X_b$ commutes with G, and L_a commutes with $Z(a)$, $\mathcal{L}^X_{a,b}$ commutes with $Z(a)$. $\qquad \square$

Chapter Three

The displacement function and the return map

The purpose of this chapter is to study the displacement function d_γ on X that is associated with a semisimple element $\gamma \in G$. If $\varphi_t, t \in \mathbf{R}$ denotes the geodesic flow on the total space $\mathcal{X}$ of the tangent bundle of X, the critical set $X(\gamma) \subset X$ of d_γ can be easily related to the fixed point set $\mathcal{F}_\gamma \subset \mathcal{X}$ of the symplectic transformation $\gamma^{-1}\varphi_1$ of $\mathcal{X}$. We study the nondegeneracy of $\gamma^{-1}\varphi_1 - 1$ along $\mathcal{F}_\gamma$. More fundamentally, we give important quantitative estimates on how much $\varphi_{1/2}$ differs from $\varphi_{-1/2}\gamma$ away from $\mathcal{F}_\gamma$. These quantitative estimates are based on Toponogov's theorem [BaGSc85]. They will play an essential role in chapter 15, where we establish the results needed to study the hypoelliptic orbital integrals as $b \to +\infty$, which are shown to concentrate in the proper way near a natural lift $\widehat{\mathcal{F}}_\gamma$ of $\mathcal{F}_\gamma$ in $\widehat{\mathcal{X}}$.

This chapter is organized as follows. In section 3.1, if $\gamma \in G$ is semisimple, we recall key properties of the displacement function $d_\gamma(x) = d(x, \gamma x)$ on X along the lines of Ballmann-Gromov-Schroeder [BaGSc85]. In particular d_γ is a convex function, its critical set $X(\gamma)$ is introduced, and the factorization of γ as a product of commuting hyperbolic and elliptic elements is obtained. By conjugation, we will write γ in the form $\gamma = e^a k^{-1}, a \in \mathfrak{p}, k \in K, \mathrm{Ad}(k) a = a$.

In section 3.2, when $a \in \mathfrak{p}$, we show that the function $\psi_a = \left|a^{TX}\right|^2 / 2$ is a Morse-Bott function, whose critical set is the submanifold $X(e^a)$.

In section 3.3, we show that $X(\gamma)$ is the symmetric space associated with the centralizer $Z(\gamma)$ of γ.

In section 3.4, we show that d_γ^2 is a Morse-Bott function. Also we describe the normal bundle $N_{X(\gamma)/X}$, and the corresponding normal coordinate system that identifies X to the total space of $N_{X(\gamma)/X}$.

In section 3.5, we describe the obvious analogue of the return map along the geodesics in $X(\gamma)$ that connect $x, \gamma x, x \in X(\gamma)$. We also interpret $X(\gamma)$ as the fixed point set of a symplectic transformation of the cotangent bundle of X associated with the geodesic flow.

In section 3.6, we consider the obvious lift of the return map to the total space $\widehat{\mathcal{X}}$ of $TX \oplus N$.

In section 3.7, we evaluate the connection form on the principal bundle $p : G \to X = G/K$ using the parallel transport trivialization.

In section 3.8, we define a family of distances and pseudodistances on $\mathcal{X}$ and $\widehat{\mathcal{X}}$.

In section 3.9, we establish important estimates involving the flow $\varphi_t|_{t \in \mathbf{R}}$, which are based on Toponogov's theorem.

Finally, in section 3.10, we construct a flat subbundle $(TX \oplus N)(\gamma)$ of $TX \oplus N$. This subbundle will be used in chapter 9 when applying local index methods to hypoelliptic orbital integrals as $b \to +\infty$.

Reading sections 3.6–3.10 can be deferred to the point where the corresponding results are used in the text.

3.1 Convexity, the displacement function, and its critical set

If $\gamma \in G$, we denote by $Z(\gamma) \subset G$ the centralizer of γ, and by $\mathfrak{z}(\gamma)$ its Lie algebra. Then

$$\mathfrak{z}(\gamma) = \{f \in \mathfrak{g}, \operatorname{Ad}(\gamma) f = f\}. \tag{3.1.1}$$

Let $Z^0(\gamma)$ be the connected component of the identity in $Z(\gamma)$.

If $a \in \mathfrak{g}$, let $Z(a) \subset G$ be the stabilizer of a, and let $\mathfrak{z}(a)$ be its Lie algebra. Then

$$\mathfrak{z}(a) = \ker \operatorname{ad}(a). \tag{3.1.2}$$

Definition 3.1.1. A function $h : X \to \mathbf{R}$ is said to be convex if for any geodesic $t \in \mathbf{R} \to x_t \in X$ with constant speed, the function $t \in \mathbf{R} \to h(x_t) \in \mathbf{R}$ is convex. A subset $Y \subset X$ is said to be convex if when $y, y' \in Y$, the geodesic connecting y and y' is included in Y.

Let d be the Riemannian distance in X. If $\gamma \in G$, the displacement function d_γ is given by

$$d_\gamma(x) = d(x, \gamma x). \tag{3.1.3}$$

By [BaGSc85, §6.1], the function d_γ is convex on X. Also if $g \in G$,

$$d_{C(g)\gamma}(gx) = d_\gamma(x). \tag{3.1.4}$$

By (3.1.4), the function d_γ is $Z(\gamma)$-invariant. Moreover,

$$d_{\theta\gamma}(\theta x) = d_\gamma(x). \tag{3.1.5}$$

By [E96, 2.19.21], $\gamma \in G$ is said to be semisimple if the function d_γ has a minimum value in X. Let $X(\gamma) \subset X$ be the subset of X where d_γ takes its minimum value m_γ. By [BaGSc85, p. 78], $X(\gamma)$ is a closed convex subset of X. Since d_γ is convex and is smooth on $X \setminus X(\gamma)$, by [BaGSc85, 1.2], d_γ has no critical points on $X \setminus X(\gamma)$. Moreover, the action of $Z(\gamma)$ on X preserves $X(\gamma)$. By (3.1.5), if γ is semisimple, $\theta\gamma$ is semisimple, and moreover,

$$X(\theta\gamma) = \theta X(\gamma), \qquad m_{\theta\gamma} = m_\gamma. \tag{3.1.6}$$

The element $\gamma \in G$ is said to be elliptic if γ is semisimple and $m_\gamma = 0$, i.e., if γ fixes some point in X. In this case, $X(\gamma)$ is just the set of fixed points of γ. Equivalently γ is elliptic if and only if it is conjugate in G to an element of K. Finally, γ is said to be hyperbolic if it is conjugated in G to $e^a, a \in \mathfrak{p}$.

Elliptic or hyperbolic elements are semisimple. By [E96, Theorem 2.19.23], γ is semisimple if and only if there exist $h, e \in G$, with h hyperbolic and e elliptic such that $\gamma = he = eh$. Also e, h are uniquely determined. Finally,

$$Z(\gamma) = Z(h) \cap Z(e). \qquad (3.1.7)$$

If $g \in G$ is such that $pg = x$, then $p\gamma g = \gamma x$. If $t \in [0,1] \to x_t \in X$ is the geodesic connecting x and γx, let g. be the horizontal lift of x. in G such that $g_0 = g$. We define the parallel transport element $k \in K$ by the formula

$$g_1 = \gamma g k. \qquad (3.1.8)$$

Theorem 3.1.2. *Let $\gamma \in G$ be semisimple. If $g \in G, x = pg \in X$, then $x \in X(\gamma)$ if and only if there exist $a \in \mathfrak{p}, k \in K$ such that $\mathrm{Ad}(k) a = a$, and moreover,*

$$\gamma = C(g)\left(e^a k^{-1}\right). \qquad (3.1.9)$$

Also $C(g) e^a \in G, C(g) k \in G$ are uniquely determined by γ. If $g_t = g e^{ta}$, then $t \in [0,1] \to x_t = pg_t$ is the unique geodesic connecting x and γx. Moreover

$$m_\gamma = |a|. \qquad (3.1.10)$$

Finally, $k \in K$ is the parallel transport along the above geodesic.

Proof. If $x \in X(\gamma)$, the unique geodesic connecting x and γx is of minimal length among the geodesics connecting y and γy, for $y \in X$. We claim that $x \in X(\gamma)$ if and only if x is represented by $g \in G$ such that there is $a \in \mathfrak{p}, k \in K$ with

$$\gamma = C(g)\left(e^a k^{-1}\right), \mathrm{Ad}(k) a = a. \qquad (3.1.11)$$

Indeed the geodesic connecting x and γx is represented in G by $t \in [0,1] \to g_t = g e^{ta}$, and its length is $|a|$. There is $k \in K$ such that

$$g e^a = \gamma g k, \qquad (3.1.12)$$

and (3.1.8), (3.1.12) show that k is precisely the parallel transport along x.. The condition $\mathrm{Ad}(k) a = a$ is exactly the one that guarantees that x minimizes d_γ. As we saw before Theorem 3.1.2, in (3.1.9), $C(g) e^a$ and $C(g) k$ are uniquely determined. The proof of our theorem is completed. □

3.2 The norm of the canonical vector fields

Take $a \in \mathfrak{g}$. Since the bilinear form B is G-invariant, by (2.17.4), we get

$$\left|a^{TX}\right|^2 - \left|a^N\right|^2 = B(a, a). \qquad (3.2.1)$$

Set

$$\psi_a = \frac{1}{2}\left|a^N\right|^2 = \frac{1}{2}\left|a^{TX}\right|^2 - \frac{1}{2}B(a, a). \qquad (3.2.2)$$

We denote by $\nabla \psi_a$ the gradient of ψ_a, and by $\nabla^{TX} \nabla \psi_a$ its Hessian, i.e., the covariant derivative of $\nabla \psi_a$ with respect to ∇^{TX}. If $e \in TX$, we identify e with the corresponding element in $\mathfrak{g}$ via the identification $TX \oplus N \simeq \mathfrak{g}$.

Proposition 3.2.1. *The following identities hold:*

$$\nabla \psi_a = \left[a^{TX}, a^N \right], \qquad \nabla^{TX} \nabla \psi_a = \operatorname{ad}^2 \left(a^{TX} \right) - \operatorname{ad}^2 \left(a^N \right). \qquad (3.2.3)$$

If $e \in TX$,

$$\left\langle \nabla_e^{TX} \nabla \psi_a, e \right\rangle = |[a, e]|^2. \qquad (3.2.4)$$

In particular the function ψ_a is convex.

Proof. Clearly,

$$\nabla \psi_a = \left\langle a^N, \nabla^N a^N \right\rangle. \qquad (3.2.5)$$

Using (2.17.10) and (3.2.5), and the fact that $\operatorname{ad} \left(a^{TX} \right)$ acts as a symmetric operator on $TX \oplus N$, we get the first identity in (3.2.3). By (2.17.10) and the first identity in (3.2.3), we get the second identity. Since $\operatorname{ad} \left(a^{TX} \right)$ acts on $TX \oplus N$ as a symmetric operator, and $\operatorname{ad} \left(a^N \right)$ as an antisymmetric operator, equation (3.2.4) follows from (3.2.3). Since (3.2.4) is nonnegative, the function ψ_a is convex. The proof of our proposition is completed. $\square$

Remark 3.2.2. Using Proposition 3.2.1, the functions $\left| a^{TX} \right|, \left| a^N \right|$ can be shown to be convex. Since along geodesics, a^{TX} is a Jacobi field, and since X has nonpositive curvature, the convexity of $\left| a^{TX} \right|$ is also a consequence of [BaGSc85, Remark p.5].

Let $a^{TX\prime}$ be the 1-form dual to a^{TX} by the metric.

Proposition 3.2.3. *If $U, V \in TX$, then*

$$da^{TX\prime} (U, V) = 2 \left\langle \left[a^N, U \right], V \right\rangle. \qquad (3.2.6)$$

Proof. Since $\nabla^{TX} a^{TX}$ is antisymmetric,

$$da^{TX\prime} (U, V) = 2 \left\langle \nabla_U^{TX} a^{TX}, V \right\rangle. \qquad (3.2.7)$$

By (2.17.10) and (3.2.7), we get (3.2.6). $\square$

In the sequel, we take $a \in \mathfrak{p}$. Set $x = p1$. The geodesic coordinate system centered at x allows us to identify $\mathfrak{p}$ with X. More precisely, the identification is given by $Y^{\mathfrak{p}} \in \mathfrak{p} \to \exp (Y^{\mathfrak{p}}) x \in X$. Along the geodesics centered at x, we trivialize TX, N by parallel transport with respect to the connections ∇^{TX}, ∇^N.

Proposition 3.2.4. *The following identities hold:*

$$a^{TX} (Y^{\mathfrak{p}}) = \cosh (\operatorname{ad} (Y^{\mathfrak{p}})) a, \qquad a^N (Y^{\mathfrak{p}}) = - \sinh (\operatorname{ad} (Y^{\mathfrak{p}})) a. \qquad (3.2.8)$$

so that

$$\left| a^{TX} \right|^2 (Y^{\mathfrak{p}}) = |\cosh (\operatorname{ad} (Y^{\mathfrak{p}})) a|^2, \qquad \left| a^N \right|^2 = |\sinh (\operatorname{ad} (Y^{\mathfrak{p}})) a|^2. \qquad (3.2.9)$$

In particular,

$$\left| a^{TX} \right|^2 (Y^{\mathfrak{p}}) \geq |a|^2 + |[a, Y^{\mathfrak{p}}]|^2. \qquad (3.2.10)$$

Proof. Recall that a^N vanishes at x. Since $Y^{\mathfrak{p}}$ is a parallel section of TX along the geodesic $t \in \mathbf{R} \to \exp(tY^{\mathfrak{p}})x \in \mathfrak{p}$, equation (3.2.8) follows from (2.17.10). Equation (3.2.9) follows from (3.2.8).

By (3.2.9), we get

$$\left|a^{TX}\right|^2 (Y^{\mathfrak{p}}) = \frac{1}{2}|a|^2 + \frac{1}{2} \langle \cosh(2\mathrm{ad}(Y^{\mathfrak{p}})) a, a \rangle. \qquad (3.2.11)$$

Since $\mathrm{ad}^2(Y^{\mathfrak{p}})$ is a symmetric nonnegative endomorphism of $\mathfrak{p}$, we can diagonalize it on an orthonormal basis of $\mathfrak{p}$, with nonnegative eigenvalues. The inequality (3.2.10) is now a consequence of the fact that for $t \in \mathbf{R}$, $\cosh(t)$ dominates its asymptotic expansion. $\qquad\square$

Remark 3.2.5. In the trivialization used in Proposition 3.2.4, we get

$$a^{TX} + a^N = \exp(-\mathrm{ad}(Y^{\mathfrak{p}})) a. \qquad (3.2.12)$$

On the other hand, $\exp(\mathrm{ad}(Y^{\mathfrak{p}}))\left(a^{TX} + a^N\right) \in \mathfrak{g}$ corresponds to $a^{TX} + a^N$ in the identification $TX \oplus N \simeq \mathfrak{g}$. By (3.2.12), we find that this is precisely a, which has indeed to be the case.

Since ψ_a is convex, its critical set coincides with the set where it reaches its minimum value. Also this set is convex.

Theorem 3.2.6. *In the geodesic coordinate system centered at $x = p1$,*

$$X(e^a) = \mathfrak{z}(a) \cap \mathfrak{p}. \qquad (3.2.13)$$

If $a \in \mathfrak{p}$, the minimum value of ψ_a is 0, and the minimum set of ψ_a is just $X(e^a)$. Equivalently,

$$X(e^a) = \left\{x \in X, a^N = 0\right\}. \qquad (3.2.14)$$

The section a^N is nondegenerate along the manifold $X(e^a)$, and the function ψ_a is a Morse-Bott function.

Proof. Clearly the length of the path $s \in [0, 1] \to e^{sa}x$ that connects x and $e^a x$ is just $\left|a_x^{TX}\right|$. By Theorem 3.1.2, $p1 \in X(e^a)$, and so by equation (3.1.10) in Theorem 3.1.2, $m_{e^a} = |a|$. Therefore,

$$|a| \le d_{e^a} \le \left|a^{TX}\right|. \qquad (3.2.15)$$

By (3.2.2), the minimum value of ψ_a is 0, and the minimum value of $\left|a^{TX}\right|$ is $|a|$.

As we just saw, $p1 \in X(e^a)$. Take $f \in \mathfrak{p}$, and assume that $x = pe^f \in X(e^a)$. Since $X(e^a)$ is convex, $t \in [0, 1] \to x_t = pe^{tf} \in X$ is a geodesic included in $X(e^a)$. Set

$$\varphi_f(t) = d_{e^a}(x_t). \qquad (3.2.16)$$

Then

$$\phi_f(t) = |a|. \qquad (3.2.17)$$

Let $c_t(s), 0 \leq s \leq 1$ be the unique geodesic connecting x_t and $e^a x_t$, and let $E_f(t)$ be its energy. By (3.2.17),

$$E_f(t) = \frac{1}{2}|a|^2. \tag{3.2.18}$$

By (3.2.18), we get

$$E_f''(0) = 0. \tag{3.2.19}$$

By Theorem 3.1.2, $c_0(s) = pe^{sa}$. Let $J_{f,s}, 0 \leq s \leq 1$ be the Jacobi field $J_{f,s} = \frac{\partial}{\partial t} c_t(s)|_{t=0}$. In the trivialization given by parallel transport,

$$\ddot{J}_f - \mathrm{ad}^2(a) J_f = 0, \tag{3.2.20}$$
$$J_{f,0} = f, \quad J_{f,1} = f.$$

Classically,

$$E_f''(0) = \int_0^1 \left(\left| \dot{J}_f \right|^2 + |[a, J_f]|^2 \right) ds. \tag{3.2.21}$$

By (3.2.19), (3.2.21), we get

$$[a, f] = 0, \tag{3.2.22}$$

so that $f \in \mathfrak{z}(a) \cap \mathfrak{p}$. Conversely, if $f \in \mathfrak{z}(a) \cap \mathfrak{p}$, then $e^f \in Z(e^a)$. By Theorem 3.1.2, $pe^f \in X(e^a)$, which gives (3.2.13). By Proposition 3.2.4 and by (3.2.13), a^N vanishes on $X(e^a)$. Combining this with (3.2.15), we get (3.2.14).

On $X(e^a)$, $a^{TX} = a$. By (2.17.10), we find that over $X(e^a)$, if $A \in TX$,

$$\nabla_A^N a^N = [a, A]. \tag{3.2.23}$$

Since $\mathrm{ad}(a)$ exchanges $\mathfrak{p}$ and $\mathfrak{k}$, $\mathfrak{z}(a)$ is the direct sum of its $\mathfrak{p}$ and its $\mathfrak{k}$ parts. Let $\mathfrak{z}^\perp(a) \subset \mathfrak{g}$ be the orthogonal to $\mathfrak{z}(a)$ in $\mathfrak{g}$ with respect to B, which is also the orthogonal to $\mathfrak{z}(a)$ with respect to the scalar product of $\mathfrak{g}$. Since $\mathrm{ad}(a)$ is symmetric, it acts as an invertible endomorphism of $\mathfrak{z}^\perp(a)$. By (3.2.23), we conclude that a^N is nondegenerate along its zero set.

By (3.2.3), we get

$$\nabla^{TX} \nabla \psi_a|_{X(e^a)} = \mathrm{ad}^2(a)|_{\mathfrak{p}}. \tag{3.2.24}$$

By the same argument as before, (3.2.24) is a nondegenerate quadratic form on $\mathfrak{z}^\perp(a) \cap \mathfrak{p}$, i.e., ψ_a is a Morse-Bott function. The proof of our theorem is completed. □

Remark 3.2.7. There is a corresponding statement when replacing $a \in \mathfrak{p}$ by $b \in \mathfrak{k}$, and exchanging the roles of TX and N. However, since $X(e^{sb})$ may change when $s \in \mathbf{R}^*$, some care has to be given to the precise formulation.

Proposition 3.2.8. The following identity holds:

$$Z(e^a) = Z(a). \tag{3.2.25}$$

Proof. Clearly $Z(a) \subset Z(e^a)$. Conversely take $g \in Z(e^a)$. By Theorem 3.1.2, $pg \in X(e^a)$. We can write uniquely g in the form

$$g = e^b k', \ b \in \mathfrak{p}, k' \in K. \tag{3.2.26}$$

For $t \in \mathbf{R}$, set

$$x_t = e^{tb} p1. \tag{3.2.27}$$

Then $x.$ is a geodesic in X connecting $p1$ and pg. Since $X(e^a)$ is totally geodesic, for any $t \in \mathbf{R}$, $x_t \in X(e^a)$. Using equation (3.2.13) in Theorem 3.2.6, we conclude that $b \in \mathfrak{z}(a)$, and so $e^b \in Z(a)$.

Using (3.2.26), we find that $k' \in Z(e^a)$. Moreover,

$$C(k') e^a = e^{\mathrm{Ad}(k')a}, \tag{3.2.28}$$

and so

$$e^a = e^{\mathrm{Ad}(k')a}. \tag{3.2.29}$$

Moreover, $a \in \mathfrak{p}$, $\mathrm{Ad}(k') a \in \mathfrak{p}$, and so by (3.2.29), we get

$$\mathrm{Ad}(k') a = a, \tag{3.2.30}$$

i.e., $k' \in Z(a)$.

Since $e^b \in Z(a), k' \in Z(a)$, we find that $g \in Z(a)$. This completes the proof of our proposition. $\qquad\square$

3.3 The subset $X(\gamma)$ as a symmetric space

Let $\gamma \in G$ be semisimple. We fix one $g_0 \in G$ such that (3.1.9) holds with $g = g_0$. Set

$$\mathfrak{p}_{\gamma,g_0} = \mathrm{ad}(g_0) \mathfrak{p}, \mathfrak{k}_{\gamma,g_0} = \mathrm{ad}(g_0) \mathfrak{k}. \tag{3.3.1}$$

Then $\mathfrak{g} = \mathfrak{p}_{\gamma,g_0} \oplus \mathfrak{k}_{\gamma,g_0}$ is a Cartan decomposition of $\mathfrak{g}$.

Because of the above, we may as well assume that $g_0 = 1$, so that

$$\gamma = e^a k^{-1}, \qquad\qquad a \in \mathfrak{p}, \tag{3.3.2}$$
$$k \in K, \qquad\qquad \mathrm{Ad}(k) a = a.$$

By (3.1.7) and (3.3.2), we get

$$Z(\gamma) = Z(e^a) \cap Z(k). \tag{3.3.3}$$

Using Proposition 3.2.8, we can rewrite (3.3.3) in the form

$$Z(\gamma) = Z(a) \cap Z(k). \tag{3.3.4}$$

The Lie algebra $\mathfrak{z}(k)$ of $Z(k)$ is given by

$$\mathfrak{z}(k) = \{f \in \mathfrak{g}, \mathrm{Ad}(k) f = f\}. \tag{3.3.5}$$

By (3.3.4), we obtain

$$\mathfrak{z}(\gamma) = \mathfrak{z}(a) \cap \mathfrak{z}(k). \tag{3.3.6}$$

Put

$$\mathfrak{p}(\gamma) = \mathfrak{z}(\gamma) \cap \mathfrak{p}, \qquad\qquad \mathfrak{k}(\gamma) = \mathfrak{z}(\gamma) \cap \mathfrak{k}. \qquad (3.3.7)$$

From (3.3.6), we find that

$$\mathfrak{z}(\gamma) = \mathfrak{p}(\gamma) \oplus \mathfrak{k}(\gamma). \qquad (3.3.8)$$

By (3.3.8), the restriction of B to $\mathfrak{z}(\gamma)$ is nondegenerate.

Set

$$K(\gamma) = K \cap Z(\gamma). \qquad (3.3.9)$$

By (3.3.6), a lies in the center of $\mathfrak{z}(\gamma)$. Let $\mathfrak{z}^{a,\perp}(\gamma)$ be the orthogonal to a in $\mathfrak{z}(\gamma)$, let $\mathfrak{p}^{a,\perp}(\gamma)$ be the orthogonal to a in $\mathfrak{p}(\gamma)$. By (3.3.8), we get

$$\mathfrak{z}^{a,\perp}(\gamma) = \mathfrak{p}^{a,\perp}(\gamma) \oplus \mathfrak{k}(\gamma). \qquad (3.3.10)$$

Moreover, $\mathfrak{z}^{a,\perp}(\gamma)$ is a Lie algebra.

Let $Z^{a,\perp,0}(\gamma)$ be the connected Lie subgroup of $Z^0(\gamma)$ that is associated with the Lie algebra $\mathfrak{z}^{a,\perp}(\gamma)$. Note that if $a \neq 0$,

$$Z^0(\gamma) \simeq Z^{a,\perp,0}(\gamma) \times \mathbf{R}, \qquad (3.3.11)$$

so that e^{ta} maps into $t|a|$.

Let $\mathfrak{z}^{\perp}(\gamma)$ be the orthogonal space to $\mathfrak{z}(\gamma)$ in $\mathfrak{g}$ with respect to B. Then $\mathfrak{z}^{\perp}(\gamma)$ splits as

$$\mathfrak{z}^{\perp}(\gamma) = \mathfrak{p}^{\perp}(\gamma) \oplus \mathfrak{k}^{\perp}(\gamma), \qquad (3.3.12)$$

where $\mathfrak{p}^{\perp}(\gamma) \subset \mathfrak{p}, \mathfrak{k}^{\perp}(\gamma) \subset \mathfrak{k}$ are the orthogonal spaces to $\mathfrak{p}(\gamma), \mathfrak{k}(\gamma)$ with respect to the scalar products induced by B. Moreover,

$$\mathfrak{g} = \mathfrak{z}(\gamma) \oplus \mathfrak{z}^{\perp}(\gamma). \qquad (3.3.13)$$

Theorem 3.3.1. *The set $X(\gamma)$ is preserved by θ. Moreover,*

$$X(\gamma) = X(e^a) \cap X(k). \qquad (3.3.14)$$

The set $X(\gamma)$ is a submanifold of X. In the geodesic coordinate system centered at p1, then

$$X(\gamma) = \mathfrak{p}(\gamma). \qquad (3.3.15)$$

The action of $Z^0(\gamma)$ on $X(\gamma)$ is transitive. More precisely the map $g \in Z^0(\gamma) \to pg \in X$ induces the identification of $Z^0(\gamma)$-manifolds,

$$X(\gamma) \simeq Z^0(\gamma)/K \cap Z^0(\gamma). \qquad (3.3.16)$$

Also $K \cap Z^0(\gamma)$ coincides with the connected component of the identity $K^0(\gamma)$ in $K(\gamma)$.

Similarly, the action of $Z(\gamma)$ on $X(\gamma)$ is transitive, and we have the identification of $Z(\gamma)$-manifolds,

$$X(\gamma) \simeq Z(\gamma)/K(\gamma). \qquad (3.3.17)$$

The embedding $K(\gamma) \to Z(\gamma)$ induces the isomorphism of finite groups,

$$K^0(\gamma) \backslash K(\gamma) \simeq Z^0(\gamma) \backslash Z(\gamma). \qquad (3.3.18)$$

The map $g \in Z^{a,\perp,0}(\gamma) \to pg \in X$ surjects on a convex submanifold $X^{a,\perp}(\gamma)$ of $X(\gamma)$, and moreover,

$$X^{a,\perp}(\gamma) = Z^{a,\perp,0}(\gamma)/K^0(\gamma). \tag{3.3.19}$$

If $a \neq 0$, we have the identification of Riemannian $Z^0(\gamma)$ manifolds,

$$X(\gamma) \simeq X^{a,\perp}(\gamma) \times \mathbf{R} \tag{3.3.20}$$

so that the action of e^{ta} on $X(\gamma)$ is just the translation by $t|a|$ on $\mathbf{R}$. In particular, γ acts on $X(\gamma)$ by translation by $|a|$.

Proof. By Theorem 3.1.2 and by (3.3.2), we know that $p1 \in X(\gamma)$. More generally, the uniqueness of the decomposition of γ into the product of commuting hyperbolic and elliptic elements shows that $g \in G$ represents $x \in X(\gamma)$ if and only if there exist $a' \in \mathfrak{p}, k' \in K$ such that

$$C\left(g^{-1}\right) e^a = e^{a'}, \qquad C\left(g^{-1}\right) k = k', \qquad \mathrm{Ad}\left(k'\right) a' = a'. \tag{3.3.21}$$

By (3.3.21), $X(\gamma) \subset X(e^a) \cap X(k)$. If $x \in X(e^a) \cap X(k)$, x is represented by $g \in G$ such that the first two conditions in (3.3.21) are verified. Since $k \in Z(e^a)$, then $k' \in Z(e^{a'})$. By Proposition 3.2.8, $k' \in Z(a')$, so that the last condition in (3.3.21) is also verified, i.e., $x \in X(\gamma)$. This completes the proof of (3.3.14).

Also $X(k) \subset X$ is the fixed point set of k. In the geodesic coordinate system centered at $p1$, $X(k)$ is represented by $\mathfrak{p}(k) \subset \mathfrak{p}$. By Theorem 3.2.6, in the same coordinate system, $X(e^a)$ is represented by $\mathfrak{z}(a) \cap \mathfrak{p}$. Therefore $X(\gamma)$ is represented by $\mathfrak{p}(\gamma)$, which is just (3.3.15). Then $X(\gamma)$ is preserved by θ.

It follows that the map $g \in Z(\gamma) \to pg \in X(\gamma)$ is surjective. Since $X(\gamma)$ is connected, p induces a surjection $Z^0(\gamma) \to X(\gamma)$. Then $Z^0(\gamma)$ acts transitively on $X(\gamma)$, and the stabilizer of $p1 \in X(\gamma)$ in $Z^0(\gamma)$ is just $K \cap Z^0(\gamma)$, so that (3.3.16) holds. Since $Z^0(\gamma)$ is connected and $X(\gamma)$ is contractible, $K \cap Z^0(\gamma)$ is also connected. Therefore $K \cap Z^0(\gamma)$ is the connected component of the identity in $K(\gamma)$.

By the above, the action of $Z(\gamma)$ on $X(\gamma)$ is transitive, so that (3.3.17) holds. By (3.3.16), (3.3.17), we get (3.3.18). Since K is compact, the group in (3.3.18) is finite.

Equation (3.3.20) follows from (3.3.11). The fact that the action of e^{ta} on $X(\gamma)$ is just translation by $t|a|$ also follows from (3.3.11). Since $X(\gamma) \subset X(k)$, k acts trivially on $X(\gamma)$, so that the actions of γ on $X(\gamma)$ coincides with the action of e^a. The proof of our theorem is completed. $\square$

Remark 3.3.2. Identity (3.3.20) is proved in [BaGSc85, Lemma 6.5] in a more general context. The identity in [BaGSc85] can be rewritten in the form

$$Z^0(\gamma) \times_{K^0(\gamma)} \mathfrak{p}^\perp(\gamma) = \left(Z^{a,\perp,0}(\gamma) \times_{K^0(\gamma)} \mathfrak{p}^\perp(\gamma)\right) \times \mathbf{R}. \tag{3.3.22}$$

The action of γ on the right-hand side of (3.3.22) is given by

$$(g, f, s) \to \left(g, \mathrm{Ad}\left(k^{-1}\right) f, s + |a|\right). \tag{3.3.23}$$

3.4 The normal coordinate system on X based at $X(\gamma)$

On $X(e^a)$, the vector field a^{TX} restricts to the vector field $a^{TX(e^a)}$, and the restriction of a^{TX} to $X(\gamma)$ is equal to $a^{TX(\gamma)}$. By (2.17.10) and by Theorem 3.2.6, the vector field $a^{TX(e^a)}$ is of constant norm $|a|$, its integral curves are geodesics, and it is a parallel section of $TX(e^a)$. A similar result holds for $a^{TX(\gamma)}$.

Let $N_{X(\gamma)/X}$ be the orthogonal bundle to $TX(\gamma)$ in TX. Under the identification $X(\gamma) \simeq Z^0(\gamma)/K^0(\gamma)$, one verifies easily that

$$N_{X(\gamma)/X} = Z^0(\gamma) \times_{K^0(\gamma)} \mathfrak{p}^\perp(\gamma). \qquad (3.4.1)$$

Let $\mathcal{N}_{X(\gamma)/X}$ be the total space of $N_{X(\gamma)/X}$. Using the geodesics normal to $X(\gamma)$, we obtain an identification of a neighborhood of $X(\gamma)$ in $\mathcal{N}_{X(\gamma)/X}$ to a tubular neighborhood of $X(\gamma)$ in X. As we shall see, this diffeomorphism identifies $\mathcal{N}_{X(\gamma)/X}$ to X.

Let $p_\gamma : X \to X(\gamma)$ be the projection defined in [BaGSc85, p. 8], i.e., if $x \in X$, $p_\gamma x \in X(\gamma)$ is the unique element in $X(\gamma)$ that minimizes the distance $d(x,y)$, $y \in X(\gamma)$. Let $|\nabla d_\gamma|$ be the norm of ∇d_γ with respect to the metric of TX.

Theorem 3.4.1. *The map*

$$\rho_\gamma : (g, f) \in Z^0(\gamma) \times_{K^0(\gamma)} \mathfrak{p}^\perp(\gamma) \to pge^f = gpe^f \in X \qquad (3.4.2)$$

is a diffeomorphism of $Z^0(\gamma)$-spaces. The map $(g, f) \to (e^a g, \mathrm{Ad}(k^{-1}) f)$ corresponds to the action of γ on X, and the map $Z^0(\gamma) \times_{K^0(\gamma)} \mathfrak{p}^\perp(\gamma) \to Z^0(\gamma)/K^0(\gamma)$ corresponds to $p_\gamma : X \to X(\gamma)$.
If $(g, f) \in Z^0(\gamma) \times_{K^0(\gamma)} \mathfrak{p}^\perp(\gamma)$, then

$$d_\gamma(\rho_\gamma(g, f)) = d_\gamma(\rho_\gamma(1, f)). \qquad (3.4.3)$$

There exists $C_\gamma > 0$ such that if $f \in \mathfrak{p}^\perp(\gamma), |f| \geq 1$, then

$$d_\gamma(\rho_\gamma(1, f)) \geq |a| + C_\gamma |f|. \qquad (3.4.4)$$

There exist $C'_\gamma > 0, C''_\gamma > 0$ such that if $f \in \mathfrak{p}^\perp(\gamma), |f| \geq 1$,

$$|\nabla d_\gamma(\rho_\gamma(1, f))| \geq C'_\gamma, \qquad (3.4.5)$$

and for $|f| \leq 1$,

$$|\nabla d_\gamma^2(\rho_\gamma(1, f))/2| \geq C''_\gamma |f|. \qquad (3.4.6)$$

In particular the function $d_\gamma^2/2$ is a Morse-Bott function, whose critical set is $X(\gamma)$, and its Hessian on $X(\gamma)$ is given by the symmetric positive endomorphism of $\mathfrak{p}^\perp(\gamma)$,

$$\nabla^{TX}\nabla d_\gamma^2/2|_{X(\gamma)} = \frac{\mathrm{ad}(a)}{\sinh(\mathrm{ad}(a))} \left(2\mathrm{ch}(\mathrm{ad}(a)) - (\mathrm{Ad}(k) + \mathrm{Ad}(k^{-1}))\right). \qquad (3.4.7)$$

Proof. If g, f are taken as in our theorem, $t \in \mathbf{R} \to pge^{tf} \in X$ is a geodesic normal to $X(\gamma)$. Let J_t be the Jacobi field corresponding to the variation of such a geodesic. Since $X(\gamma)$ is totally geodesic in X, we know that $J_0 \in TX(\gamma), \frac{D}{Dt}J_0 \in N_{X(\gamma)/X}$, so that

$$\frac{d}{dt}|J|^2_{t=0} = 2\left\langle J_0, \frac{D}{Dt}J_0 \right\rangle = 0. \tag{3.4.8}$$

Since X has nonpositive curvature, by [BaGSc85, Remark p.5], $|J_t|$ is a convex function of t. We deduce from (3.4.8) that $|J_t| \geq |J_0|$. If $J_0 \neq 0$, for any $t > 0$, $J_t \neq 0$. If $J_0 = 0$, if for a given $t > 0$, $J_t = 0$, by convexity, $J_s = 0$ for any $s \in [0, t]$, so that $\frac{D}{Dt}J_0 = 0$, and J. vanishes identically. We have thus shown that the map ρ_γ is nonsingular.

The unique geodesic that connects $p_\gamma x$ to x is normal to $X(\gamma)$, so that ρ_γ is surjective. Therefore ρ_γ is a covering map. Since X is contractible, ρ_γ is a diffeomorphism.

If $g \in Z(\gamma)$, g preserves $X(\gamma)$ and maps geodesics normal to $X(\gamma)$ into geodesics of the same type. One verifies trivially that the obvious left actions of $Z(\gamma)$ correspond via ρ_γ.

Clearly, if $(g, f) \in Z^0(\gamma) \times_{K^0(\gamma)} \mathfrak{p}^\perp(\gamma)$, using (3.3.3), we get

$$\gamma\rho_\gamma(g, f) = p\gamma ge^f = pe^a k^{-1} ge^f = pe^a ge^{\mathrm{Ad}(k^{-1})f}. \tag{3.4.9}$$

By (3.4.9), we find that the action of γ on X corresponds to the diffeomorphism $(g, f) \to (e^a g, \mathrm{Ad}(k^{-1})f)$ of $Z^0(\gamma) \times_{K^0(\gamma)} \mathfrak{p}^\perp(\gamma)$. Since d_γ is $Z(\gamma)$-invariant, we get (3.4.3).

We use the same notation as in the proof of Theorem 3.2.6. For $t \in \mathbf{R}$, set

$$\varphi_f(t) = d_\gamma\left(pe^{tf}\right). \tag{3.4.10}$$

Since the function d_γ is convex, the function $t \in \mathbf{R} \to \varphi_f(t) \in \mathbf{R}$ is convex.

Assume first that γ is elliptic, so that $a = 0$, and $\gamma = k^{-1}$. In this case, by Toponogov's theorem [BaGSc85, section 1.4], we get

$$d\left(pe^f, pe^{\mathrm{Ad}(k^{-1})f}\right) \geq \left|\left(\mathrm{Ad}(k^{-1}) - 1\right)f\right|. \tag{3.4.11}$$

Since no eigenvalue of $\mathrm{Ad}(k^{-1})$ is equal to 1 on $\mathfrak{p}^\perp(\gamma)$, by (3.4.11), there is $C_\gamma > 0$ such that for any $f \in \mathfrak{p}^\perp(\gamma)$,

$$d_\gamma\left(pe^f\right) \geq C_\gamma|f|, \tag{3.4.12}$$

from which (3.4.4) follows. By (3.4.12), since the function φ_f is convex, for $f \in \mathfrak{p}^\perp(\gamma), |f| = 1$,

$$\varphi'_f(0)_+ \geq C_\gamma. \tag{3.4.13}$$

Using (3.4.13) and the convexity of $\varphi_f(t)$, for any $t > 0$, we get

$$\varphi'_f(t) \geq C_\gamma. \tag{3.4.14}$$

By (3.4.12) and (3.4.14), we get (3.4.5) and (3.4.6). Moreover, by (3.4.13), the function d^2_γ is a Morse-Bott function. Also from the above, we get easily

$$\nabla^{TX}\nabla d^2_\gamma/2|_{X(\gamma)} = \left(1 - \mathrm{Ad}(k^{-1})\right)\left(1 - \mathrm{Ad}(k)\right), \tag{3.4.15}$$

which is just (3.4.7) when $a = 0$.

Let us now assume that γ is nonelliptic, i.e., $a \neq 0$. Take $f \in \mathfrak{p}^{\perp}(\gamma), |f| = 1$. The minimum of $\varphi_f(t)$ is at $t = 0$, and $\varphi_f(t) \geq \varphi_f(0) = |a|$. Since $a \neq 0$, φ_f is smooth. We claim that there is $c_\gamma > 0$ such that for any f taken as before,

$$\varphi_f''(0) \geq c_\gamma. \tag{3.4.16}$$

Indeed let $c_t(s), 0 \leq s \leq 1$ be the unique geodesic in X connecting pe^{tf} and γpe^{tf}. Let $E_f(t)$ be the energy of c_t, so that

$$E_f(t) = \frac{1}{2}\varphi_f^2(t). \tag{3.4.17}$$

By (3.4.17), we get

$$E_f''(0) = |a|\,\varphi_f''(0). \tag{3.4.18}$$

Let $J_{f,s}, 0 \leq s \leq 1$ be the Jacobi field $J_{f,s} = \frac{\partial}{\partial t}c_t(s)|_{t=0}$. In the trivialization given by parallel transport,

$$\ddot{J}_f - \operatorname{ad}^2(a) J_f = 0, \tag{3.4.19}$$
$$J_{f,0} = f, \; J_{f,1} = \operatorname{Ad}(k^{-1}) f.$$

As in (3.2.21), we get

$$E_f''(0) = \int_0^1 \left(\left|\dot{J}_f\right|^2 + |[a, J_f]|^2 \right) ds. \tag{3.4.20}$$

If $E_f''(0) = 0$, by (3.4.19), (3.4.20), we would get $[a, f] = 0, \operatorname{Ad}(k^{-1}) f = f$, which contradicts the fact that $f \in \mathfrak{p}^{\perp}(\gamma), f \neq 0$. Therefore there is $C_\gamma > 0$ such that

$$E_f''(0) \geq C_\gamma. \tag{3.4.21}$$

By (3.4.18) and (3.4.21), we get (3.4.16).

Since $\varphi_f'(0) = 0$, using (3.4.16) and Taylor's formula, there exist $c_\gamma' > 0, \epsilon_\gamma \in\,]0, 1/2]$ such that for f taken as before,

$$\varphi_f'(\epsilon_\gamma) \geq c_\gamma'. \tag{3.4.22}$$

Since φ_f is convex and $\varphi_f \geq |a|$, by (3.4.22), for $t \in \mathbf{R}$,

$$\varphi_f(t) \geq |a| + c_\gamma'(t - \epsilon_\gamma). \tag{3.4.23}$$

Since $\epsilon_\gamma \in\,]0, 1/2]$, for $|f| \geq 1$,

$$|f| - \epsilon_\gamma \geq \frac{1}{2}|f|. \tag{3.4.24}$$

By (3.4.23), (3.4.24), we deduce that if $f \in \mathfrak{p}^{\perp}(\gamma), |f| \geq 1$, then

$$d_\gamma(pe^f) \geq |a| + \frac{c_\gamma'}{2}|f|, \tag{3.4.25}$$

which is just (3.4.4). If $f \in \mathfrak{p}^{\perp}(\gamma), |f| = 1$, since φ_f is convex, by (3.4.22), for $t \geq \epsilon_\gamma$,

$$\varphi_f'(t) \geq c_\gamma', \tag{3.4.26}$$

which gives (3.4.5). Moreover,

$$\varphi_f'(t) = \int_0^t \varphi_f''(s)\,ds. \tag{3.4.27}$$

Since $\varphi_f''(t) \geq 0$, and $\varphi_f''(0) > 0$, there exists $C_\gamma > 0$ such that for $0 \leq t \leq 1$,

$$\varphi_f'(t) \geq C_\gamma t. \tag{3.4.28}$$

This gives (3.4.6). Also by (3.4.16), the function d_γ is a Morse-Bott function with critical set $X(\gamma)$, and so is the function $d_\gamma^2/2$. Let us now compute its Hessian on $X(\gamma)$. By (3.4.19), (3.4.20), we get

$$\nabla\nabla d_\gamma^2/2|_{X(\gamma)}(f,f) = \left\langle \dot{J}_{f,1}, \dot{J}_{f,1} \right\rangle - \left\langle \dot{J}_{f,0}, \dot{J}_{f,0} \right\rangle. \tag{3.4.29}$$

Moreover, the unique solution of equation (3.4.19) is given by

$$J_{f,s} = \cosh(\mathrm{sad}(a))\,f + \frac{\sinh(\mathrm{sad}(a))}{\sinh(\mathrm{ad}(a))}\left(\mathrm{Ad}(k^{-1}) - \cosh(a)\right)f. \tag{3.4.30}$$

By (3.4.30), we get

$$\dot{J}_{f,0} = \frac{\mathrm{ad}(a)}{\sinh(\mathrm{ad}(a))}\left(\mathrm{Ad}(k^{-1}) - \cosh(\mathrm{ad}(a))\right)f,$$

$$\dot{J}_{f,1} = \mathrm{ad}(a)\sinh(\mathrm{ad}(a))\,f + \frac{\mathrm{ad}(a)}{\sinh(\mathrm{ad}(a))}\cosh(\mathrm{ad}(a)) \tag{3.4.31}$$
$$\left(\mathrm{Ad}(k^{-1}) - \cosh(\mathrm{ad}(a))\right)f.$$

By (3.4.29), (3.4.31), we get (3.4.7).

The proof of our theorem is completed. $\qquad\square$

Remark 3.4.2. Take $x, x' \in X, x' \neq x$, and let $u_x, v_{x'}$ be the unit tangent vectors along the geodesic connecting x to x'. Take γ as in Theorem 3.4.1, and $x \in X$ such that $x' = \gamma x \neq x$. Then

$$\nabla d_\gamma(x) = (\gamma^* v - u)(x). \tag{3.4.32}$$

The inequalities in (3.4.5) and (3.4.6) can be reformulated using (3.4.32).

By [BaGSc85, p. 10], the distance function to $X(\gamma)$, denoted $d(\cdot, X(\gamma))$, is a convex function. Tautologically,

$$d(\rho_\gamma(1,f), X(\gamma)) = |f|. \tag{3.4.33}$$

The inequalities in (3.4.4)–(3.4.6) can be rewritten using (3.4.33).

Let dx be the volume element on X, let dy be the volume element on $X(\gamma)$, let df be the volume element on $\mathfrak{p}^\perp(\gamma)$. Then $dydf$ is a volume element on $Z^0(\gamma) \times_{K^0(\gamma)} \mathfrak{p}^\perp(\gamma)$ that is $Z^0(\gamma)$-invariant. There is a smooth positive function $r(f)$ on $\mathfrak{p}^\perp(\gamma)$ that is $K^0(\gamma)$-invariant, such that we have the identity of volume elements on X,

$$dx = r(f)\,dydf, \tag{3.4.34}$$

and moreover,

$$r(0) = 1. \tag{3.4.35}$$

Using the equation of Jacobi fields, one finds easily that there exist $C > 0, C' > 0$ such that for any $f \in \mathfrak{p}^{\perp}(\gamma)$,

$$r(f) \leq C \exp\left(C' |f|\right). \tag{3.4.36}$$

The group $K^0(\gamma)$ acts on the left on K and on $\mathfrak{p}^{\perp}(\gamma)$, and so it also acts on the left on $\mathfrak{p}^{\perp}(\gamma) \times K$. Over $K^0(\gamma) \setminus K$, one obtains the vector bundle $\mathfrak{p}^{\perp}(\gamma)_{K^0(\gamma)} \times K$.

Similarly K acts on the right on K, and this action extends to an action on the right on $\mathfrak{p}^{\perp}(\gamma) \times K$.

Theorem 3.4.3. *The map*

$$\sigma_{\gamma} : (g, f, k') \in Z^0(\gamma) \times_{K^0(\gamma)} \left(\mathfrak{p}^{\perp}(\gamma) \times K\right) \to ge^f k' \in G \tag{3.4.37}$$

is a diffeomorphism of left $Z^0(\gamma)$-spaces, and of right K-spaces. The map $(g, f, k') \to (g, f)$ corresponds to the projection $p : G \to X = G/K$. Moreover, under this diffeomorphism, we have the identity of right K-spaces,

$$\mathfrak{p}^{\perp}(\gamma)_{K^0(\gamma)} \times K = Z^0(\gamma) \setminus G. \tag{3.4.38}$$

Proof. The first two statements in our theorem follow from Theorem 3.4.1. The remainder of our theorem is now obvious. $\qquad\qquad\square$

Remark 3.4.4. Equation (3.4.38) just says that $Z^0(\gamma) \setminus G$ fibres over $K^0(\gamma) \setminus K$ with fibre $\mathfrak{p}^{\perp}(\gamma)$. Also note that in Theorem 3.4.3, we may as well replace $Z^0(\gamma), K^0(\gamma)$ by $Z(\gamma), K(\gamma)$.

Let dk be the Haar measure on K that gives volume 1 to K. Let dg be the measure on G,

$$dg = dx dk. \tag{3.4.39}$$

Then dg is a left-invariant Haar measure on G. Since G is unimodular, it is also a right-invariant Haar measure.

By (3.4.34) and (3.4.39), we get

$$dg = r(f) dy df dk. \tag{3.4.40}$$

Let dk'^0 be the Haar measure on $K^0(\gamma)$ that gives volume 1 to $K^0(\gamma)$, and let du^0 be a K-invariant volume form on $K^0(\gamma) \setminus K$, so that

$$dk = dk'^0 du^0. \tag{3.4.41}$$

By (3.4.40), (3.4.41), we get

$$dg = r(f) dy dk'^0 df du^0. \tag{3.4.42}$$

Set

$$dz^0 = dy dk'^0. \tag{3.4.43}$$

Then dz^0 is a left and right Haar measure on $Z^0(\gamma)$, so that (3.4.42) can be written in the form

$$dg = r(f) dz^0 df du^0. \tag{3.4.44}$$

Also by (3.4.38), $r\,(f)\,dfdu^0$ can be viewed as a measure on $Z^0\,(\gamma)\setminus G$ such that

$$dg = dz^0 r\,(f)\,dfdu^0. \tag{3.4.45}$$

Let dv^0 be the measure on $Z^0\,(\gamma)\setminus G$,

$$dv^0 = r\,(f)\,dfdu^0. \tag{3.4.46}$$

By (3.4.45), we see that dv^0 is exactly the canonical measure on $Z^0\,(\gamma)\setminus G$ that is canonically associated with dg and dz^0, so that

$$dg = dz^0 dv^0. \tag{3.4.47}$$

Another interpretation for the identity $dv^0 = r\,(f)\,dfdu^0$ can be given via equation (3.4.38), which simply expresses the obvious factorization of dv^0.

When replacing $Z^0\,(\gamma)\,, K^0\,(\gamma)$ by $Z\,(\gamma)\,, K\,(\gamma)$, one can define measures dk', du, dz, dv on $K\,(\gamma)\,, K\,(\gamma)\setminus K, Z\,(\gamma)\,, Z\,(\gamma)\setminus G$ such that the analogue of the above identities still holds.

3.5 The return map along the minimizing geodesics in $X\,(\gamma)$

Let $\pi : \mathcal{X} \to X$ be the total space of the tangent bundle TX to X, let $\mathcal{X}^*$ be the total space of the cotangent bundle T^*X. The manifold $\mathcal{X}^*$ is a symplectic manifold, whose symplectic form is denoted ω. The identification of the fibres TX and T^*X by the metric g^{TX} identifies the manifolds $\mathcal{X}$ and $\mathcal{X}^*$.

Let V be the vector field on $\mathcal{X}^*$ that generates the geodesic flow. The vector field V is the Hamiltonian vector field associated with the Hamiltonian $\mathcal{H}\,(x,p) = \frac{1}{2}\,|p|^2$. Let $\varphi_t|_{t\in\mathbf{R}}$ be the corresponding group of diffeomorphisms of $\mathcal{X}$. When identifying $\mathcal{X}$ and $\mathcal{X}^*$, we may consider φ_t as a flow of symplectic diffeomorphisms of $\mathcal{X}$. If $\left(x, Y^{TX}\right) \in \mathcal{X}$, if $\left(x_t, Y_t^{TX}\right) = \varphi_t\left(x, Y^{TX}\right)$, then $t \in \mathbf{R} \to x_t \in X$ is the unique geodesic in X such that $x_0 = x, \dot{x}_0 = Y^{TX}$.

Let $\gamma \in G$ be a semisimple element such that (3.3.2) holds. The action of γ on X lifts to $\mathcal{X}$ and $\mathcal{X}^*$. Since γ is an isometry, these actions correspond by the above identification. Also γ preserves the symplectic form of $\mathcal{X}$ or $\mathcal{X}^*$. It follows from the above that $\gamma^{-1}\varphi_1$ is a symplectic diffeomorphism of $\mathcal{X}$.

Recall that by (3.3.16), $X\,(\gamma) \simeq Z^0\,(\gamma)\,/K^0\,(\gamma)$. Let $i_a : X \to \mathcal{X}$ be the embedding $x \in X \to \left(x, a^{TX}\right) \in \mathcal{X}$. Set

$$\mathcal{F}_\gamma = \left\{z \in \mathcal{X}, \gamma^{-1}\varphi_1\,(z) = z\right\}. \tag{3.5.1}$$

Proposition 3.5.1. *The following identity of submanifolds of $\mathcal{X}$ holds:*

$$\mathcal{F}_\gamma = i_a X\,(\gamma). \tag{3.5.2}$$

Proof. By the considerations we made in the proof of Theorem 3.1.2, it is clear that $i_a X\,(\gamma) \subset \mathcal{F}_\gamma$. Conversely if $(x,y) \in \mathcal{F}_\gamma$, then $x \in X$ is a critical point for the displacement function d_γ. As we saw above, from the results of [BaGSc85], we find that x minimizes d_γ, i.e., $x \in X(\gamma)$. The geodesic $x_t = \pi\varphi_t(x, y)$ is the minimizing geodesic connecting x and γx, so that $\dot{x} = a^{TX}$. This completes the proof of our proposition. $\qquad\square$

If $x \in X(\gamma)$, then $(x, a^{TX}) \in \mathcal{F}_\gamma$, and the differential $d(\gamma^{-1}\varphi_1)$ is an automorphism of $T_{(x, a^{TX})}\mathcal{X}$.

Recall that $\mathfrak{z}(a) = \ker \operatorname{ad}(a)$ is the Lie algebra of $Z(a)$. From now on, we will use the notation

$$\mathfrak{z}(a) = \mathfrak{z}_0. \tag{3.5.3}$$

Put

$$\mathfrak{p}_0 = \ker \operatorname{ad}(a) \cap \mathfrak{p}, \qquad \mathfrak{k}_0 = \ker \operatorname{ad}(a) \cap \mathfrak{k}. \tag{3.5.4}$$

Then

$$\mathfrak{z}_0 = \mathfrak{p}_0 \oplus \mathfrak{k}_0. \tag{3.5.5}$$

Let $\mathfrak{z}_0^\perp, \mathfrak{p}_0^\perp, \mathfrak{k}_0^\perp$ be the orthogonal spaces to $\mathfrak{z}_0, \mathfrak{p}_0, \mathfrak{k}_0$ in $\mathfrak{g}, \mathfrak{p}, \mathfrak{k}$, so that

$$\mathfrak{z}_0^\perp = \mathfrak{p}_0^\perp \oplus \mathfrak{k}_0^\perp. \tag{3.5.6}$$

Moreover,

$$\mathfrak{p}_0^\perp = \operatorname{ad}(a)\,\mathfrak{k}, \qquad \mathfrak{k}_0^\perp = \operatorname{ad}(a)\,\mathfrak{p}. \tag{3.5.7}$$

The above vector spaces are preserved by $\operatorname{ad}(a), \operatorname{Ad}(k), \operatorname{Ad}(\gamma)$.

By (3.3.6), $\mathfrak{z}(\gamma)$ is a Lie subalgebra of $\mathfrak{z}_0$. We will view $\mathfrak{p}(\gamma)$ as a vector subspace of $\mathfrak{p}_0 \oplus \{0\} \subset \mathfrak{p}_0 \oplus \mathfrak{k}_0 = \mathfrak{z}_0$.

Using the Levi-Civita connection ∇^{TX} on TX, we have the splitting,

$$T\mathcal{X} \simeq \pi^*(TX \oplus TX). \tag{3.5.8}$$

In (3.5.8), the first copy of TX is identified with its horizontal lift, and the second copy is the tangent bundle along the fibre.

Observe that $\operatorname{ad}(a)$ is an odd invertible endomorphism of $\mathfrak{p}_0^\perp \oplus \mathfrak{k}_0^\perp$. Let ρ be the isomorphism from $\mathfrak{p}_0^\perp \oplus \mathfrak{k}_0^\perp \to \mathfrak{p}_0^\perp \oplus \mathfrak{p}_0^\perp$ given by

$$\rho(e, f) = (e, -\operatorname{ad}(a)\,f). \tag{3.5.9}$$

Our computations will not depend on the choice of $x \in X(\gamma)$. It will be convenient to take $x = p1$.

Theorem 3.5.2. *The following identities hold at* $(x, a^{TX}) \in \mathcal{F}_\gamma$:

$$d\gamma^{-1}\varphi_1\big|_{\mathfrak{p}_0 \oplus \mathfrak{p}_0} = \begin{pmatrix} \operatorname{Ad}(k)\big|_{\mathfrak{p}_0} & \operatorname{Ad}(k)\big|_{\mathfrak{p}_0} \\ 0 & \operatorname{Ad}(k)\big|_{\mathfrak{p}_0} \end{pmatrix}, \tag{3.5.10}$$

$$d\gamma^{-1}\varphi_1\big|_{\mathfrak{p}_0^\perp \oplus \mathfrak{p}_0^\perp} = \rho\operatorname{Ad}(\gamma^{-1})\big|_{\mathfrak{z}_0^\perp}\rho^{-1}.$$

The eigenspace of $d\gamma^{-1}\varphi_1$ *associated with the eigenvalue 1 is just* $T\mathcal{F}_\gamma \simeq \mathfrak{p}(\gamma) \oplus \{0\} \subset \mathfrak{p}_0 \oplus \mathfrak{p}_0$.

Proof. Let $t \in [0, 1] \to x_t \in X$ be the geodesic connecting x and γx. Consider the Jacobi field $J_t \in T_{x_t} X$, so that

$$\ddot{J} + R_{x_t}^{TX}(J, \dot{x}) \dot{x} = 0. \qquad (3.5.11)$$

In (3.5.11), the differentials $\dot{J}, \ddot{J}$ are taken with respect to the Levi-Civita connection along $x.$. With respect to the splitting (3.5.8) of $T\mathcal{X}$, the differential $d\varphi_1$ at $(x_0, \dot{x}_0)$ is given by the linear map $\left(J_0, \dot{J}_0\right) \to \left(J_1, \dot{J}_1\right)$.

Recall that R^{TX} was calculated at the beginning of section 3.1. In the parallel transport trivialization with respect to the Levi-Civita connection, $\dot{x}$ is the constant $a \in \mathfrak{p}$, and as in (3.4.19), equation (3.5.11) can be written in the form

$$\ddot{J} - \mathrm{ad}^2(a) J = 0. \qquad (3.5.12)$$

In (3.5.12), $J_t \in \mathfrak{p}$. As we saw before,

$$d\varphi_1\left(J_0, \dot{J}_0\right) = \left(J_1, \dot{J}_1\right). \qquad (3.5.13)$$

By (3.3.2) and (3.5.13), we deduce that

$$d\gamma^{-1}\varphi_1\left(J_0, \dot{J}_0\right) = \mathrm{Ad}(k)\left(J_1, \dot{J}_1\right). \qquad (3.5.14)$$

By (3.5.12), (3.5.14), we get the first identity in (3.5.10). Observe that ρ in (3.5.9) commutes with $\mathrm{Ad}(k)$. If $\left(J_0, \dot{J}_0\right) \in \mathfrak{p}_0^\perp \oplus \mathfrak{p}_0^\perp$, then $\left(J_t, \dot{J}_t\right) \in \mathfrak{p}_0^\perp \oplus \mathfrak{p}_0^\perp$. Let $H \in \mathfrak{p}_0^\perp \oplus \mathfrak{k}_0^\perp$ be given by

$$H = \rho^{-1}\left(J, \dot{J}\right). \qquad (3.5.15)$$

By (3.5.12), we obtain

$$\dot{H} + [a, H] = 0, \qquad (3.5.16)$$

so that

$$H_1 = \mathrm{Ad}\left(e^{-a}\right) H_0. \qquad (3.5.17)$$

By (3.5.14), (3.5.15), (3.5.17), we get the second identity in (3.5.10).

By (3.5.10), the kernel of $d\gamma^{-1}\varphi_1 - 1$ in $\mathfrak{p}_0 \oplus \mathfrak{p}_0$ is just the kernel of $\mathrm{Ad}(k) - 1$ in the first copy of $\mathfrak{p}_0 \oplus \mathfrak{p}_0$, which, by (3.3.6), coincides with $\mathfrak{p}(\gamma) \oplus \{0\}$. Equivalently, $\ker\left(d\gamma^{-1}\varphi_1 - 1\right)$ is just $T\mathcal{F}_\gamma$. The proof of our theorem is completed. □

Remark 3.5.3. We know that $d\gamma^{-1}\varphi_1$ is a symplectic transformation of $\mathfrak{p}_0 \oplus \mathfrak{p}_0$. The conditions in (3.3.2) on γ show directly that the right-hand side of (3.5.10) defines a symplectic transformation of $\mathfrak{p}_0 \oplus \mathfrak{p}_0$.

3.6 The return map on $\widehat{\mathcal{X}}$

Recall that $\widehat{\mathcal{X}}$ is the total space of $TX \oplus N$. Let $\tau : \widehat{\mathcal{X}} \to \mathcal{X}$ be the obvious projection. We lift the flow $\varphi_t|_{t \in \mathbf{R}}$ to a flow of diffeomorphisms of $\widehat{\mathcal{X}}$. The lifted flow acts on N by parallel transport with respect to the connection ∇^N. If $\left(x, Y^{TX}, Y^N\right) \in \widehat{\mathcal{X}}$, then $\varphi_t\left(x, Y^{TX}, Y^N\right) = \left(\underline{x}_t, \underline{Y}_t^{TX}, \underline{Y}_t^N\right)$, $\underline{x}_{\cdot}$ being the geodesic in X with speed $\underline{Y}_t^{TX}$, and $\underline{Y}_{\cdot}^N$ being the parallel transport of Y^N along $\underline{x}_{\cdot}$. Similarly γ acts on $\widehat{\mathcal{X}}$.

As in (3.5.1), set

$$\widehat{\mathcal{F}}_\gamma = \left\{z \in \widehat{\mathcal{X}}, \gamma^{-1}\varphi_1(z) = z\right\}. \tag{3.6.1}$$

Then

$$\tau\widehat{\mathcal{F}}_\gamma = \mathcal{F}_\gamma. \tag{3.6.2}$$

Recall that by (3.3.14), $X(\gamma) \subset X\left(k^{-1}\right)$. In particular, k^{-1} acts as a parallel isometry on the vector bundle $N|_{X(\gamma)}$. If $x = p1$, the action of k^{-1} on $N|_{X(\gamma)}$ is induced by the action of $\mathrm{Ad}\left(k^{-1}\right)$ on $\mathfrak{k}$. Set

$$N\left(k^{-1}\right) = \left\{Y^N \in N|_{X(\gamma)}, \mathrm{Ad}\left(k^{-1}\right)Y^N = Y^N\right\}. \tag{3.6.3}$$

Then $N\left(k^{-1}\right)$ is associated with the eigenspace of $\mathfrak{k}$ corresponding to the eigenvalue 1 of $\mathrm{Ad}\left(k^{-1}\right)$, i.e., with the Lie algebra $\mathfrak{k}\left(k^{-1}\right)$ of the centralizer of k^{-1} in K. Clearly $N\left(k^{-1}\right)$ is a vector bundle on $X(\gamma)$. Let $\mathcal{N}\left(k^{-1}\right)$ be the total space of $N\left(k^{-1}\right)$.

Let $\widehat{i}_a$ be the embedding $\left(x, Y^N\right) \in \mathcal{N}\left(k^{-1}\right) \to \left(x, a^{TX}, Y^N\right) \in \widehat{\mathcal{X}}$. Now we establish the obvious analogue of Proposition 3.5.1.

Proposition 3.6.1. *The following identity of submanifolds of $\widehat{\mathcal{X}}$ holds:*

$$\widehat{\mathcal{F}}_\gamma = \widehat{i}_a\mathcal{N}\left(k^{-1}\right). \tag{3.6.4}$$

Proof. Using Proposition 3.5.1 and (3.6.2), we get (3.6.4). $\qquad\square$

3.7 The connection form in the parallel transport trivialization

Put $x_0 = p1$. Let $a \in \mathfrak{p} \to x = pe^a$ be the geodesic coordinate system centered at x_0. This coordinate system induces a trivialization of the K-bundle $p : G \to X = G/K$ that is given by $(a, k) \in \mathfrak{p} \times K \to e^a k \in G$. Equivalently, this is the trivialization of G using the parallel transport with respect to the canonical connection on the K-bundle $p : G \to X = G/K$ along geodesics centered at x_0.

Let Γ be the $\mathfrak{k}$-valued connection form in the above trivialization. Let J be a Jacobi field along the geodesic $x_s = e^{sa}x_0$ such that $J_0 = 0, \dot{J}_0 = U \in \mathfrak{p}$. By proceeding as in (3.4.19), (3.4.30), we get

$$J_s = \frac{\sinh\left(\mathrm{sad}(a)\right)}{\mathrm{ad}(a)}U. \tag{3.7.1}$$

Let $|\,|$ denote the norm in $T_{x_0}X$, and $\|\,\|$ the norm in T_xX. By (3.7.1), we get

$$\|U\|_{T_xX} = \left|\frac{\sinh\left(\mathrm{ad}(a)\right)}{\mathrm{ad}(a)}U\right|. \tag{3.7.2}$$

Recall that Ω denotes the curvature of the canonical connection. One finds easily that

$$\Gamma_a (U) = - \int_0^1 \Omega (J, \dot{x}) \, ds. \qquad (3.7.3)$$

By (2.1.10), we can rewrite (3.7.3) in the form

$$\Gamma_a = \int_0^1 [J, a] \, ds. \qquad (3.7.4)$$

By (3.7.1), (3.7.4), we get

$$\Gamma_a (U) = - \frac{\cosh (\mathrm{ad} \, (a)) - 1}{\mathrm{ad} \, (a)} U. \qquad (3.7.5)$$

As it should be, the expression after ad in the right-hand side of (3.7.5) is in $\mathfrak{k}$. Put

$$V = J_1. \qquad (3.7.6)$$

By (3.7.1), we get

$$U = \frac{\mathrm{ad} \, (a)}{\sinh (\mathrm{ad} \, (a))} V. \qquad (3.7.7)$$

By (3.7.5), (3.7.7), we obtain

$$\Gamma_a (U) = - \frac{\cosh (\mathrm{ad} \, (a)) - 1}{\sinh (\mathrm{ad} \, (a))} V. \qquad (3.7.8)$$

The critical fact on $\Gamma_a (U)$ is that as a linear function of V, it is uniformly bounded.

The above trivialization of G induces a corresponding trivialization of $TX \oplus N$ with respect to the connection $\nabla^{TX \oplus N}$ along geodesics centered at x_0. Let $\Gamma^{TX \oplus N}$ be the associated connection form on $TX \oplus N \simeq \mathfrak{p} \oplus \mathfrak{k}$. By (3.7.5), we get

$$\Gamma_a^{TX \oplus N} (U) = - \mathrm{ad} \left(\frac{\cosh (\mathrm{ad} \, (a)) - 1}{\mathrm{ad} \, (a)} U \right). \qquad (3.7.9)$$

Remark 3.7.1. There is a more direct way to derive the above formulas. Indeed, consider the exponential map $f \in \mathfrak{g} \to \exp (f) \in G$. Recall that the left-invariant forms $\omega^{\mathfrak{g}}, \omega^{\mathfrak{p}}, \omega^{\mathfrak{k}}$ were defined in section 2.1. One verifies easily that if $f \in \mathfrak{g}$,

$$(\exp^* \omega^{\mathfrak{g}})_f = \frac{1 - \exp (-\mathrm{ad} \, (f))}{\mathrm{ad} \, (f)}. \qquad (3.7.10)$$

From (3.7.10), if $a \in \mathfrak{p}$, we get

$$(\exp^* \omega^{\mathfrak{p}})_a |_{\mathfrak{p}} = \frac{\sinh (\mathrm{ad} \, (a))}{\mathrm{ad} \, (a)} |_{\mathfrak{p}}, \qquad (\exp^* \omega^{\mathfrak{k}})_a |_{\mathfrak{p}} = - \frac{\cosh (\mathrm{ad} \, (a)) - 1}{\mathrm{ad} \, (a)} |_{\mathfrak{p}},$$
$$(3.7.11)$$

$$(\exp^* \omega^{\mathfrak{p}})_a |_{\mathfrak{k}} = - \frac{\cosh (\mathrm{ad} \, (a)) - 1}{\mathrm{ad} \, (a)} |_{\mathfrak{k}}, \qquad (\exp^* \omega^{\mathfrak{k}})_a |_{\mathfrak{k}} = \frac{\sinh (\mathrm{ad} \, (a))}{\mathrm{ad} \, (a)} |_{\mathfrak{k}}.$$

Then equations (3.7.1), (3.7.5) follow from the first line in (3.7.11).

3.8 Distances and pseudodistances on $\mathcal{X}$ and $\widehat{\mathcal{X}}$

If $x, x' \in X$, let $s \in [0,1] \to x_s \in X$ be the geodesic connecting x and x'. Let $\tau_x^{x'}$ be the parallel transport from $T_{x'}X$ into T_xX with respect to the Levi-Civita connection along this geodesic. If $(x, f), (x', f') \in \mathcal{X}$, set

$$\delta\left((x, f), (x', f')\right) = d(x, x') + \left|\tau_x^{x'} f' - f\right|. \tag{3.8.1}$$

Note that δ is symmetric, and only vanishes on the diagonal of $\mathcal{X}$. Also it is invariant under the left action of G on $\mathcal{X}$. However, δ is not a distance. Moreover,

$$\delta\left((x, f), (x', f')\right) \le d(x, x') + |f| + |f'|. \tag{3.8.2}$$

We will say that δ is a pseudodistance on $\mathcal{X}$.

Take $x_0 \in X$. If $(x, f), (x', f') \in \mathcal{X}$, set

$$d_{x_0}\left((x, f), (x', f')\right) = d(x, x') + \left|\tau_{x_0}^{x'} f' - \tau_{x_0}^x f\right|. \tag{3.8.3}$$

Then d_{x_0} is a distance on $\mathcal{X}$. Moreover,

$$\delta\left((x, f), (x', f')\right) = d_x\left((x, f), (x', f')\right) = d_{x'}\left((x, f), (x', f')\right). \tag{3.8.4}$$

In the sequel, we may and we will assume that $x_0 = p1$. We use the notation of section 3.7. As mentioned after equation (3.7.8), Γ^{TX} is a bounded section of $T^*X \otimes \mathrm{End}(TX)$, when T^*X and $\mathrm{End}(TX)$ are equipped with their natural metrics. Since the connection ∇^{TX} preserves the metric, we deduce easily that

$$\left|\tau_{x_0}^x \tau_x^{x'} \tau_{x'}^{x_0} - 1\right| \le Cd(x, x'). \tag{3.8.5}$$

If $(x, f), (x', f') \in \mathcal{X}$,

$$\left|\tau_{x_0}^x f - \tau_{x_0}^{x'} f'\right| \le \left|f - \tau_x^{x'} f'\right| + \left|\left(\tau_{x_0}^x \tau_x^{x'} \tau_{x'}^{x_0} - 1\right) \tau_{x_0}^{x'} f'\right|. \tag{3.8.6}$$

By (3.8.5), (3.8.6), we get

$$\left|\tau_{x_0}^x f - \tau_{x_0}^{x'} f'\right| \le \left|f - \tau_x^{x'} f'\right| + Cd(x, x')|f'|. \tag{3.8.7}$$

Similarly,

$$\left|\tau_x^{x'} f' - f\right| \le \left|\tau_{x_0}^x f - \tau_{x_0}^{x'} f'\right| + Cd(x, x')|f'|. \tag{3.8.8}$$

By (3.8.1), (3.8.3), and (3.8.7), we get

$$d_{x_0}\left((x, f), (x', f')\right) \le \delta\left((x, f), (x', f')\right) + Cd(x, x')|f'|. \tag{3.8.9}$$

Similarly, using (3.8.1), (3.8.3), and (3.8.8), we obtain

$$\delta\left((x, f), (x', f')\right) \le d_{x_0}\left((x, f), (x', f')\right) + Cd(x, x')|f'|. \tag{3.8.10}$$

The above definitions extend to $\widehat{\mathcal{X}}$. Indeed, if $x, x' \in X$, we still denote by $\tau_x^{x'}$ the parallel transport from $(TX \oplus N)_{x'}$ into $(TX \oplus N)_x$ with respect to the connection $\nabla^{TX \oplus N}$ along the geodesic connecting x and x'. If $(x, f), (x', f') \in \widehat{\mathcal{X}}$, we still define the pseudodistance $\delta\left((x, f), (x', f')\right)$ as in (3.8.1). Similarly, if $x_0 \in X$, we define the distance d_{x_0} on $\widehat{\mathcal{X}}$ as in (3.8.3). All the above inequalities established over $\mathcal{X}$ remain valid on $\widehat{\mathcal{X}}$. In particular equation (3.7.8) still implies inequality (3.8.5).

3.9 The pseudodistance and Toponogov's theorem

We use the notation of section 3.5. By Proposition 3.5.1, the fixed point set $\mathcal{F}_\gamma$ of the map $\gamma^{-1}\varphi_1$ in $\mathcal{X}$ is just $i_a X(\gamma)$. It follows that if $(x, Y^{TX}) \in \mathcal{X}, x \notin X(\gamma)$, then $\gamma^{-1}\varphi_1(x, Y^{TX}) \neq (x, Y^{TX})$.

For $a \in \mathbf{R}$, let δ_a be the dilation

$$\delta_a(x, Y^{TX}) = (x, aY^{TX}). \tag{3.9.1}$$

Since for $t \neq 0$,

$$\varphi_t = \delta_{1/t}\varphi_1\delta_t, \tag{3.9.2}$$

we deduce from the above that if $x \notin X(\gamma)$, for any $t \in \mathbf{R}$,

$$\gamma^{-1}\varphi_t(x, Y^{TX}) \neq (x, Y^{TX}). \tag{3.9.3}$$

Take $(x, Y^{TX}) \in \mathcal{X}, x \notin X(\gamma)$. In the sequel, we will assume that

$$\left|Y^{TX}\right| = 1. \tag{3.9.4}$$

Set

$$(x', Y^{TX\prime}) = \gamma(x, Y^{TX}). \tag{3.9.5}$$

From the above, it follows that for any $t \in \mathbf{R}$,

$$\varphi_{t/2}(x, Y^{TX}) \neq \varphi_{-t/2}(x', Y^{TX\prime}). \tag{3.9.6}$$

Let $s \in [0, 1] \to \mathbf{x}_s \in X$ be the geodesic connecting x to x'. Set

$$Y_0^{TX} = \dot{\mathbf{x}}_0, \qquad\qquad Y_1^{TX} = \dot{\mathbf{x}}_1. \tag{3.9.7}$$

Then

$$(x', Y_1^{TX}) = \varphi_1(x, Y_0^{TX}). \tag{3.9.8}$$

Moreover,

$$d_\gamma(x) = \left|Y_0^{TX}\right| = \left|Y_1^{TX}\right|. \tag{3.9.9}$$

By equation (3.4.32),

$$\nabla d_\gamma(x) = \frac{1}{d_\gamma(x)}\left(\gamma^* Y_1^{TX} - Y_0^{TX}\right). \tag{3.9.10}$$

Assume that $x = \rho_\gamma(1, f), f \in \mathfrak{p}^\perp(\gamma)$. By (3.4.4), (3.4.5), and (3.9.10), for $|f| \geq 1$,

$$\left|\gamma^* Y_1^{TX} - Y_0^{TX}\right| \geq C'_\gamma d_\gamma(x) \geq C'_\gamma(|a| + C_\gamma|f|). \tag{3.9.11}$$

Moreover, by (3.4.6) and (3.9.10), for $|f| \leq 1$,

$$\left|\gamma^* Y_1^{TX} - Y_0^{TX}\right| \geq C''_\gamma|f|. \tag{3.9.12}$$

Theorem 3.9.1. *Given $\beta > 0$, there exists $C_{\gamma,\beta} > 0$ such that if $x \in X$ is such that $d(x, X(\gamma)) \geq \beta$, if $Y^{TX} \in T_x X, \left|Y^{TX}\right| = 1$, for $t \geq 0$, then*

$$\delta\left(\varphi_t(x, Y^{TX}), \varphi_{-t}\gamma(x, Y^{TX})\right) \geq C_{\gamma,\beta}. \tag{3.9.13}$$

Proof. We may as well assume that $x = \rho_\gamma(1, f)$, $f \in \mathfrak{p}^\perp(\gamma)$.

Let $\theta, 0 \le \theta \le \pi$ be the angle between Y_0^{TX} and $\gamma^* Y_1^{TX}$. Because of (3.9.9), (3.9.11) and (3.9.12), this angle cannot be 0. By (3.9.9), (3.9.11), (3.9.12), we find that given $\beta > 0$, and $|f| \ge \beta$, there is $\theta_{\gamma,\beta}, 0 < \theta_{\gamma,\beta} \le \pi$ such that

$$\theta_{\gamma,\beta} \le \theta \le \pi. \tag{3.9.14}$$

We use the notation in (3.9.5). Set

$$\underline{x}_t = \pi \varphi_t\left(x, Y^{TX}\right), \qquad \underline{x}'_t = \pi \varphi_{-t}\left(x', Y^{TX'}\right). \tag{3.9.15}$$

Then $\underline{x}, \underline{x}'$ are geodesics in X. Moreover,

$$d\left(x, \underline{x}_t\right) = d\left(x', \underline{x}'_t\right) = t. \tag{3.9.16}$$

Set

$$d_t = d\left(\underline{x}_t, \underline{x}'_t\right). \tag{3.9.17}$$

Since d is a convex function, the function $t \in \mathbf{R} \to d_t \in \mathbf{R}_+$ is convex. Let $s \in [0, 1] \to \mathbf{x}_{s,t}$ be the geodesic connecting $\underline{x}_t$ and $\underline{x}'_t$. Of course, when $t = 0$, this is just the geodesic $s \in [0, 1] \to \mathbf{x}_s$.

By (3.4.32),

$$\dot{d}_0 = -\frac{1}{d_\gamma(x)} \left\langle \gamma^* Y_1^{TX} + Y_0^{TX}, Y^{TX} \right\rangle. \tag{3.9.18}$$

Assume first that

$$\left\langle \gamma^* Y_1^{TX} + Y_0^{TX}, Y^{TX} \right\rangle \le 0. \tag{3.9.19}$$

Since $t \to d_t$ is convex, by (3.9.18) and (3.9.19), for any $t \ge 0$,

$$d_t \ge d_\gamma(x). \tag{3.9.20}$$

Also by (3.4.4), there is $a_{\gamma,\beta} > 0$ such that for $x \in X, d(x, X(\gamma)) \ge \beta$,

$$d_\gamma(x) \ge a_{\gamma,\beta}. \tag{3.9.21}$$

Using (3.9.20), (3.9.21), we get

$$d_t \ge a_{\gamma,\beta}, \tag{3.9.22}$$

which is compatible with (3.9.13).

In the remainder of the proof, we will assume that

$$\left\langle \gamma^* Y_1^{TX} + Y_0^{TX}, Y^{TX} \right\rangle > 0, \tag{3.9.23}$$

so that

$$\dot{d}_0 < 0. \tag{3.9.24}$$

By (3.9.16), we get

$$d_t \ge d_\gamma(x) - 2t. \tag{3.9.25}$$

By (3.9.21), (3.9.25), we find that if $t \le a_{\gamma,\beta}/4$, then

$$d_t \ge a_{\gamma,\beta}/2. \tag{3.9.26}$$

By (3.9.26), we deduce that to establish (3.9.13), we may as well take $t > a_{\gamma,\beta}/4$.

Note that if u, v are unit vectors in $\mathfrak{p}$, their angle $\alpha(u, v)$, $0 \leq \alpha(u, v) \leq \pi$ is just the distance of u, v in the unit sphere $S^{\mathfrak{p}}$. It follows that if u, v, w are unit vectors,

$$\alpha(u, w) \leq \alpha(u, v) + \alpha(v, w). \tag{3.9.27}$$

Put

$$\theta_0 = \alpha\left(Y_0^{TX}, Y^{TX}\right), \qquad \theta_1 = \alpha\left(Y_1^{TX}, \gamma_* Y^{TX}\right). \tag{3.9.28}$$

Using (3.9.14), (3.9.23), and (3.9.27), we find that

$$\theta_{\gamma,\beta} \leq \theta \leq \theta_0 + \theta_1 < \pi. \tag{3.9.29}$$

Given $t > 0$, consider the quadrangle Γ_t in Figure 3.1 made of the geodesics $s \in [0,1] \to \mathbf{x}_s$ connecting x and x', $u \in [0,t] \to \underline{x}'_u$ connecting x' and $\underline{x}'_t$, $s \in [0,1] \to \mathbf{x}_{s,t}$ connecting $\underline{x}_t$ and $\underline{x}'_t$, and $u \in [0,t] \to \underline{x}_u$ connecting x to $\underline{x}_t$. The two lower angles of this quadrangle are just θ_0, θ_1. We denote by $\sigma_{0,t}, \sigma_{1,t}$ the two upper angles. Using the form of Toponogov's theorem given in [BaGSc85, p7, Exercice (i)], we get

$$\theta_0 + \theta_1 + \sigma_{0,t} + \sigma_{1,t} \leq 2\pi. \tag{3.9.30}$$

Set

$$\theta_{0,t} = \pi - \sigma_{0,t}, \qquad \theta_{1,t} = \pi - \sigma_{1,t}. \tag{3.9.31}$$

Then (3.9.30) can be written in the form

$$\theta_{0,t} + \theta_{1,t} \geq \theta_0 + \theta_1. \tag{3.9.32}$$

The same arguments as before show that $\theta_{0,t} + \theta_{1,t}$ is an increasing function of t. However, this fact will not be used in the sequel.

Set

$$\underline{z}_t = \mathbf{x}_{1/2,t}. \tag{3.9.33}$$

Equivalently $\underline{z}_t$ is the middle point between $\underline{x}_t$ and $\underline{x}'_t$.

Consider the geodesic triangle T_t in Figure 3.1 with vertices $x, x', \underline{z}_t$. Let $u \in [0,t] \to x_u^t, x_u'^t$ be the geodesics connecting x, x' to $\underline{z}_t$. Let $\alpha_{0,t}, \alpha_{1,t}, \gamma_t$ be the angles at $x, x', \underline{z}_t$. By Toponogov's theorem,

$$\alpha_{0,t} + \alpha_{1,t} + \gamma_t \leq \pi. \tag{3.9.34}$$

Set

$$f_t = \frac{\partial x_u^t}{\partial u}\Big|_{u=t}, \qquad g_t = \frac{\partial x_u'^t}{\partial u}\Big|_{u=t}. \tag{3.9.35}$$

Then f_t, g_t lie in $T_{\underline{z}_t} X$, and their angle is γ_t. Therefore,

$$|f_t + g_t|^2 = |f_t|^2 + |g_t|^2 + 2|f_t||g_t|\cos(\gamma_t). \tag{3.9.36}$$

By (3.9.34), (3.9.36), we obtain

$$|f_t + g_t|^2 \geq |f_t|^2 + |g_t|^2 - 2|f_t||g_t|\cos(\alpha_{0,t} + \alpha_{1,t}). \tag{3.9.37}$$

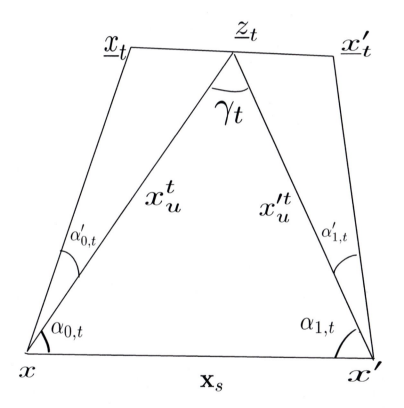

Figure 3.1

In the sequel, we will fix the value of a parameter $c \in]0,1]$. If

$$\frac{d_t}{2} > \inf(ca_{\gamma,\beta}/4, 1), \tag{3.9.38}$$

then (3.9.13) is verified. Therefore, we may as well assume that

$$\frac{d_t}{2} \le \inf(ca_{\gamma,\beta}/4, 1). \tag{3.9.39}$$

Since $t > a_{\gamma,\beta}/4$, from (3.9.39), we get

$$d_t \le 2ct. \tag{3.9.40}$$

Clearly,

$$t - \frac{d_t}{2} \le t|f_t| \le t + \frac{d_t}{2}, \qquad t - \frac{d_t}{2} \le t|g_t| \le t + \frac{d_t}{2}. \tag{3.9.41}$$

By (3.9.40), (3.9.41), we get

$$1 - c \le |f_t| \le 1 + c, \qquad 1 - c \le |g_t| \le 1 + c. \tag{3.9.42}$$

By (3.9.37), (3.9.42), we get

$$|f_t + g_t|^2 \ge 2\left((1-c)^2 - (1+c)^2 \cos(\alpha_{0,t} + \alpha_{1,t})\right). \tag{3.9.43}$$

Let $\tau_{\underline{x}_t}^{\underline{z}_t}, \tau_{\underline{x}'_t}^{\underline{z}_t}$ be the parallel transport from $T_{\underline{z}_t}X$ into $T_{\underline{x}_t}X, T_{\underline{x}'_t}X$ along the geodesic $\mathbf{x}_{.,t}$ with respect to the Levi-Civita connection. Using (3.9.16), (3.9.40), and equation (13.1.10) in Proposition 13.1.1, we get

$$\left|t\dot{\underline{x}}_t - \tau_{\underline{x}_t}^{\underline{z}_t} t f_t\right| \le (1 + Ct)\frac{d_t}{2}, \qquad \left|t\dot{\underline{x}}'_t - \tau_{\underline{x}'_t}^{\underline{z}_t} t g_t\right| \le (1 + Ct)\frac{d_t}{2}. \tag{3.9.44}$$

In (3.9.44), the norms are calculated with respect to the metric in $T_{\underline{x}_t}X$ or $T_{\underline{x}'_t}X$.

Recall that $\tau_{\underline{x}_t}^{\underline{x}'_t}$ is the parallel transport from $T_{\underline{x}'_t}X$ into $T_{\underline{x}_t}X$ along the geodesic $\mathbf{x}_{.,t}$ with respect to the Levi-Civita connection. By (3.9.43) and (3.9.44), we get

$$\left|\dot{\underline{x}}_t + \tau_{\underline{x}_t}^{\underline{x}'_t}\dot{\underline{x}}'_t\right| \ge \left(2(1-c)^2 - 2(1+c)^2 \cos(\alpha_{0,t} + \alpha_{1,t})\right)^{1/2} - (1 + Ct)\frac{d_t}{t}. \tag{3.9.45}$$

Let $\alpha'_{0,t}, \alpha'_{1,t}$ denote the angles at x, x' made by $\dot{x}_0, \frac{\partial}{\partial u}x_u^t|_{u=0}$ and by $\dot{x}'_0, \frac{\partial}{\partial u}x_u'^t|_{u=0}$. By Toponogov's theorem, these angles are smaller than the corresponding angles in the Euclidean triangles with similar lengths. Because of (3.9.39), there is $C' > 0$ such that

$$\alpha'_{0,t} \le \frac{C'}{2}\frac{d_t}{t}, \qquad \alpha'_{1,t} \le \frac{C'}{2}\frac{d_t}{t}. \tag{3.9.46}$$

Also by (3.9.27),

$$\alpha_{0,t} \ge \theta_0 - \alpha'_{0,t}, \qquad \alpha_{1,t} \ge \theta_1 - \alpha'_{1,t}. \tag{3.9.47}$$

By (3.9.46), (3.9.47), we get

$$\alpha_{0,t} + \alpha_{1,t} \ge \theta_0 + \theta_1 - C'\frac{d_t}{t}. \tag{3.9.48}$$

By (3.9.29), (3.9.40), and (3.9.48), we see that for $c \leq \theta_{\gamma,\beta}/4C'$,

$$\alpha_{0,t} + \alpha_{1,t} \geq \frac{\theta_{\gamma,\beta}}{2}. \tag{3.9.49}$$

By (3.9.45) and (3.9.49), we conclude that

$$\left| \dot{x}_t + \tau \frac{x_t'}{\dot{x}_t} \dot{x}_t' \right| \geq \left(2 \left(1 - c\right)^2 - 2 \left(1 + c\right)^2 \cos \left(\theta_{\gamma,\beta}/2\right) \right)^{1/2} - \left(1 + Ct\right) \frac{d_t}{t}. \tag{3.9.50}$$

Clearly there is $c_{\gamma,\beta} \in]0, 1]$ such that for $c \in]0, c_{\gamma,\beta}]$,

$$\left(2 \left(1 - c\right)^2 - 2 \left(1 + c\right)^2 \cos \left(\theta_{\gamma,\beta}/2\right) \right)^{1/2} \geq \sin \left(\theta_{\gamma,\beta}/4\right). \tag{3.9.51}$$

Recall that we assumed that $t > a_{\gamma,\beta}/4$. Then there is $\eta_{\gamma,\beta} \in]0, 1]$ such that if $d_t \leq \eta_{\gamma,\beta}$,

$$\left(1 + Ct\right) \frac{d_t}{t} \leq \frac{1}{2} \sin \left(\theta_{\gamma,\beta}/4\right). \tag{3.9.52}$$

By (3.9.50)–(3.9.52), we see that if $d_t \leq \eta_{\gamma,\beta}$,

$$\left| \dot{x}_t + \tau \frac{x_t'}{\dot{x}_t} \dot{x}_t' \right| \geq \frac{1}{2} \sin \left(\theta_{\gamma,\beta}/4\right), \tag{3.9.53}$$

which is compatible with (3.9.13). If $d_t > \eta_{\gamma,\beta}$ this is also the case. The proof of our theorem is completed. $\qquad\square$

Theorem 3.9.2. *Given* $\beta > 0, M > 0$, *there exists* $C_{\gamma,\beta,M} > 0$ *such that if* $x \in X$ *is such that* $d \left(x, X \left(\gamma\right)\right) \geq \beta$, *if* $Y^{TX} \in T_x X$, *for* $0 \leq t \leq M$,

$$\delta \left(\varphi_t \left(x, Y^{TX}\right), \varphi_{-t}\gamma \left(x, Y^{TX}\right) \right) \geq C_{\gamma,\beta,M}. \tag{3.9.54}$$

Proof. Clearly,

$$\varphi_t \left(x, 0\right) = \left(x, 0\right). \tag{3.9.55}$$

Therefore, by (3.9.21), (3.9.54) holds for $Y^{TX} = 0$. In the sequel, we may as well assume that $Y^{TX} \neq 0$. By (3.9.2),

$$\varphi_t \left(x, Y^{TX}\right) = \delta_{|Y^{TX}|} \varphi_{|Y^{TX}|t} \left(x, \frac{Y^{TX}}{|Y^{TX}|} \right). \tag{3.9.56}$$

Assume first that $\left| Y^{TX} \right| \leq \frac{d_\gamma(x)}{4M}$. For $0 \leq t \leq M$,

$$d \left(\pi\varphi_{|Y^{TX}|t} \left(x, \frac{Y^{TX}}{|Y^{TX}|} \right), \pi\varphi_{-|Y^{TX}|t}\gamma \left(x, \frac{Y^{TX}}{|Y^{TX}|} \right) \right) \geq d_\gamma \left(x\right)/2. \tag{3.9.57}$$

By (3.9.21), (3.9.57) is compatible with (3.9.54).

So we may as well assume that $\left| Y^{TX} \right| > \frac{d_\gamma(x)}{4M}$. By the same arguments as before, there exists $d_{\gamma,\beta,M} > 0$ such that $\left| Y^{TX} \right| \geq d_{\gamma,\beta,M}$. Then (3.9.54) follows from (3.9.13) in Theorem 3.9.1. The proof of our theorem is completed. $\qquad\square$

Theorem 3.9.3. *Given* $\beta > 0, \mu > 0$, *there exists* $C_{\gamma,\beta,\mu} > 0$ *such that if* $f \in \mathfrak{p}^\perp(\gamma), |f| \leq \beta, x = \rho_\gamma(1, f), |Y^{TX} - a^{TX}| \geq \mu$,

$$\delta\left(\varphi_{1/2}\left(x, Y^{TX}\right), \varphi_{-1/2}\gamma\left(x, Y^{TX}\right)\right) \geq C_{\gamma,\beta,\mu}. \tag{3.9.58}$$

Proof. We will use the notation of the proof of Theorem 3.9.1. By a compactness argument, we find that given $M > 0$, if $|f| \leq \beta, |Y^{TX}| \leq M$, if (x, Y^{TX}) stays away from $\mathcal{F}_\gamma = i_a X(\gamma)$, then $\delta\left(\varphi_{1/2}\left(x, Y^{TX}\right), \varphi_{-1/2}\gamma\left(x, Y^{TX}\right)\right)$ has a positive lower bound.

For $|f| \leq \beta$, $d_\gamma(x)$ remains uniformly bounded. We claim that there exists $M_{\gamma,\beta} > 0, C_{\gamma,\beta} > 0$ such that for $t \geq M_{\gamma,\beta}, |Y^{TX}| = 1$,

$$\delta\left(\varphi_t\left(x, Y^{TX}\right), \varphi_{-t}\gamma\left(x, Y^{TX}\right)\right) \geq C_{\gamma,\beta}. \tag{3.9.59}$$

We still denote by $\underline{z}_t$ the middle point of the geodesic segment connecting $\underline{x}_t$ to $\underline{x}'_t$. Consider the geodesic triangle T_t with vertices $x, x', \underline{z}_t$. By Toponogov's theorem, the angles of this triangle are smaller than the angles of the Euclidean triangle whose edges have the same lengths. In particular as $t \to +\infty$, the angle at $\underline{z}_t$ tends uniformly to 0. By proceeding as in the proof of Theorem 3.9.1, we get (3.9.59).

Now we establish (3.9.58). By the first part of our proof, we may assume that $|Y^{TX}| \geq 2M_{\gamma,\beta}$. Then we use equation (3.9.56) with $t = 1/2$, and also (3.9.59), and we get (3.9.58). The proof of our theorem is completed. $\square$

Theorem 3.9.4. *Given* $\nu > 0$, *there exists* $C_\nu > 0$ *such that if* $f \in \mathfrak{p}^\perp(\gamma), |f| \leq 1, x = \rho_\gamma(1, f), Y^{TX} \in T_x X, |Y^{TX}| \leq \nu$, *then*

$$\delta\left(\varphi_{1/2}\left(x, Y^{TX}\right), \varphi_{-1/2}\gamma\left(x, Y^{TX}\right)\right) \geq C_\nu\left(|f| + |Y^{TX} - a^{TX}|\right). \tag{3.9.60}$$

Proof. Here (x, Y^{TX}) varies in a compact set. If (x, Y^{TX}) stays away from $\mathcal{F}_\gamma = i_a X(\gamma)$, (3.9.60) is trivial. For $f = 0, Y^{TX} = a^{TX}$, both sides of (3.9.60) vanish. Moreover, by Theorem 3.5.2, the return map $\gamma^{-1}\varphi_1$ is nondegenerate along $\mathcal{F}_\gamma$. More precisely $T\mathcal{F}_\gamma$ is the eigenspace of $d\gamma^{-1}\varphi_1|_{\mathcal{F}_\gamma}$ that is associated with the eigenvalue 1. It follows that for $|f|$ and $|Y^{TX} - a^{TX}|$ small enough, (3.9.60) still holds. The proof of our theorem is completed. $\square$

If $x, x' \in X$, we still denote by $\tau_x^{x'}$ the parallel transport from $N_{x'}$ into N_x along the geodesic connecting x to x' with respect to the connection ∇^N.

Now we use the notation of section 3.6. If $(x, Y) \notin \widehat{i_a}\mathcal{N}\left(k^{-1}\right)$, this means that either $x \notin X(\gamma)$, or that if $x \in X(\gamma)$, either $Y^{TX} \neq a^{TX}$ or $Y^N \notin N\left(k^{-1}\right)$.

Take $(x, Y) \in \widehat{\mathcal{X}}$. As in (3.9.5), set

$$(x', Y') = \gamma(x, Y). \tag{3.9.61}$$

As in (3.9.15), set

$$\underline{x}_t = \widehat{\pi}\varphi_t(x, Y), \qquad \underline{x}'_t = \widehat{\pi}\varphi_{-t}(x', Y'). \tag{3.9.62}$$

Let $\underline{Y}^N, \underline{Y}^{N\prime}$ be the parallel transports of $Y^N, Y^{N\prime}$ along $\underline{x}_\cdot, \underline{x}'_\cdot$ with respect to ∇^N. Equivalently, $\underline{Y}^N, \underline{Y}^{N\prime}$ are the projections on N of the vertical components of $\varphi_\cdot(x, Y), \varphi_{-\cdot}(x', Y')$.

Assume that $x = \rho_\gamma(1, f)$. Given $\beta > 0$, if $|f| \leq \beta, \left|Y^{TX} - a^{TX}\right| \leq \beta$, then $\left(x, Y^{TX}\right)$ varies in a compact subset of $\mathcal{X}$. Moreover, $\underline{x}_{1/2}, \underline{x}'_{1/2}$ remain at finite distance from each other, this distance tending to 0 as $\beta \to 0$.

We trivialize N along the geodesics that are normal to $X(\gamma)$ with respect to the connection ∇^N. In particular $\mathrm{Ad}\left(k^{-1}\right)$ now acts on the vector bundle N.

Theorem 3.9.5. *There exists $c_\gamma > 0$ such that $f \in \mathfrak{p}^\perp(\gamma), |f| \leq 1, x = \rho_\gamma(1, f)$, if $\left|Y^{TX} - a^{TX}\right| \leq 1$, then*

$$\left| \tau^{\underline{x}'_{1/2}}_{\underline{x}_{1/2}} \underline{Y}^{N\prime}_{1/2} - \underline{Y}^N_{1/2} \right| \geq \left| \left(\mathrm{Ad}\left(k^{-1}\right) - 1\right) Y^N \right| - c_\gamma \left(|f| + \left|Y^{TX} - a^{TX}\right| \right) \left|Y^N\right| . \tag{3.9.63}$$

Proof. If we make $f = 0, Y^{TX} = a^{TX}, x' = \gamma x$, then

$$\tau^{x'}_x Y^{N\prime} - Y^N = \left(\mathrm{Ad}\left(k^{-1}\right) - 1\right) Y^N. \tag{3.9.64}$$

Since f, Y^{TX} vary in a compact set, using finite increments, we get (3.9.63). The proof of our theorem is completed. $\square$

3.10 The flat bundle $(TX \oplus N)(\gamma)$

Let $(TX \oplus N)(\gamma), (TX \oplus N)^\perp(\gamma)$ be the vector subbundles of $TX \oplus N$ associated with $\mathfrak{z}(\gamma), \mathfrak{z}^\perp(\gamma)$ via the identification $TX \oplus N = \mathfrak{g}$. Then

$$TX \oplus N = (TX \oplus N)(\gamma) \oplus (TX \oplus N)^\perp(\gamma), \tag{3.10.1}$$

and the splitting (3.10.1) is orthogonal with respect to B. Moreover, the vector bundles in the right-hand side of (3.10.1) are flat with respect to $\nabla^{TX \oplus N, f}$. Since the action of γ on $TX \oplus N$ just corresponds to the action of $\mathrm{Ad}(\gamma)$ on $\mathfrak{g}$, it follows that γ preserves $(TX \oplus N)(\gamma)$ and $(TX \oplus N)^\perp(\gamma)$.

Since $X(\gamma)$ is of the same type as X, over $X(\gamma)$, there is a vector bundle $N(\gamma)$ on $X(\gamma)$ such that $TX(\gamma) \oplus N(\gamma) \simeq \mathfrak{z}(\gamma)$. One verifies easily that the restriction of $(TX \oplus N)(\gamma)$ to $X(\gamma)$ coincides with $TX(\gamma) \oplus N(\gamma)$.

Also we have the dual splitting,

$$T^*X \oplus N^* = (T^*X \oplus N^*)(\gamma) \oplus (T^*X \oplus N^*)^\perp(\gamma). \tag{3.10.2}$$

Note that by (3.1.4), if $g \in G$, then g maps $X(\gamma)$ into $X(C(g)\gamma)$ isometrically, and it maps $(TX \oplus N)(\gamma)$ into $(TX \oplus N)(C(g)\gamma)$. In particular $Z(\gamma)$ acts on $(TX \oplus N)(\gamma)$.

The flat vector bundle $(TX \oplus N)(\gamma)$ will play an important role in chapter 9 in our construction of a Getzler rescaling.

Chapter Four

Elliptic and hypoelliptic orbital integrals

The purpose of this chapter is to construct semisimple orbital integrals associated with the heat kernel for the hypoelliptic Laplacian $\mathcal{L}_b^X$. By making $b \to 0$, we show that the corresponding supertrace coincides with the orbital integral associated with the standard elliptic heat kernel. In the next chapters, the evaluation of these elliptic orbital integrals will be obtained by making $b \to +\infty$.

This chapter is organized as follows. In section 4.1, we introduce an algebra $\mathcal{Q}$ of invariant kernels $q(x, x')$ acting on the vector space of bounded continuous sections of F.

In section 4.2, if $\gamma \in G$ is semisimple, the orbital integrals of such kernels are constructed, and they are shown to vanish on commutators.

In section 4.3, when replacing F by $\Lambda^\cdot (T^*X \oplus N^*) \otimes S^\cdot (TX \oplus N) \otimes F$, we construct an associated algebra $\mathfrak{Q}$ of continuous kernels acting on $C^b \left(\widehat{\mathcal{X}}, \widehat{\pi}^* (\Lambda^\cdot (T^*X \oplus N^*) \otimes F) \right)$. We obtain the corresponding orbital integrals, and we show that they vanish on supercommutators.

In section 4.4, if A is a constant section of $\mathrm{End}(F)$ and if $\mathcal{L}_A^X = \mathcal{L}^X + A$, we construct the orbital integrals associated with the heat kernel for $\exp \left(-t\mathcal{L}_A^X \right)$, which is known to lie in $\mathcal{Q}$.

In section 4.5, if $\mathcal{L}_{A,b}^X = \mathcal{L}_b^X + A$ is the hypoelliptic Laplacian, the orbital integrals associated with the heat kernel for $\exp \left(-\mathcal{L}_{A,b}^X \right)$ are obtained. The proof that such heat kernels lie in the algebra $\mathfrak{Q}$ is deferred to chapter 14.

In section 4.6, we prove the fundamental fact that the above elliptic and hypoelliptic orbital integrals coincide.

In section 4.7, we give a slightly different approach to the orbital integrals associated with a semisimple element $\gamma = e^a k^{-1}$, by incorporating the action of a in the considered elliptic or hypoelliptic Laplacians.

Finally, in section 4.8, if $\Gamma \subset G$ is a cocompact subgroup of G, and $Z = \Gamma \backslash X, \widehat{Z} = \Gamma \backslash \widehat{\mathcal{X}}$, the elliptic and hypoelliptic traces and supertraces are shown to be equal.

In the whole chapter, we make the same assumptions as in chapters 2 and 3, and we use the corresponding notation. Also if $V = V_+ \oplus V_-$ is a finite dimensional $\mathbf{Z}_2$-graded vector space, if $\tau = \pm 1$ is the involution of V that defines the $\mathbf{Z}_2$-grading, if $A \in \mathrm{End}(V)$, we define its supertrace $\mathrm{Tr}_s [A]$ by

$$\mathrm{Tr}_s [A] = \mathrm{Tr} [\tau A]. \tag{4.0.1}$$

The same notation will be used for trace class operators when V is infinite dimensional.

4.1 An algebra of invariant kernels on X

Let $C^b(X, F)$ be the Banach space of continuous bounded sections of F over X. Let $C^b_K(G, E)$ be the vector space of continuous bounded functions s on G with values in E such that if $k \in K$,

$$s(gk) = \rho^E(k^{-1}) s(g). \tag{4.1.1}$$

Of course we have the canonical isomorphism

$$C^b(X, F) = C^b_K(G, E). \tag{4.1.2}$$

Let Q be an operator acting on $C^b(X, F)$ with continuous kernel $q(x, x')$ with respect to the volume dx'. If $x, x' \in X$, then $q(x, x') \in \mathrm{Hom}(F_{x'}, F_x)$.

It will often be more convenient to view q as a continuous function $q(g, g')$ defined on $G \times G$ with values in $\mathrm{End}(E)$, which is such that if $k \in K$,

$$q(gk, g') = \rho^E(k^{-1}) q(g, g'), \qquad q(g, g'k) = q(g, g') \rho^E(k). \tag{4.1.3}$$

In the sequel, we will assume that Q commutes with the left action of G on $C^b(X, F)$. This is just to say that if $x, x' \in X, g \in G$,

$$q(gx, gx') = gq(x, x') g^{-1}, \tag{4.1.4}$$

where in the right-hand side of (4.1.4), g^{-1} maps $F_{gx'}$ onto $F_{x'}$, and g maps F_x into F_{gx}. Equivalently, if we consider instead the kernel $q(g, g')$, if $g'' \in G$, then

$$q(g''g, g''g') = q(g, g'). \tag{4.1.5}$$

Set

$$q(g) = q(1, g). \tag{4.1.6}$$

By (4.1.3), (4.1.5), we get

$$\begin{aligned} q(g, g') &= q(g^{-1}g'), \\ q(k^{-1}g) &= \rho^E(k^{-1}) q(g), \\ q(gk) &= q(g) \rho^E(k). \end{aligned} \tag{4.1.7}$$

From (4.1.7), we obtain

$$q(k^{-1}gk) = \rho^E(k^{-1}) q(g) \rho^E(k). \tag{4.1.8}$$

By (4.1.8), we find that $\mathrm{Tr}^E[q(g)]$ is invariant when replacing g by $k^{-1}gk$, with $k \in K$.

In the sequel, since there is no risk of confusion, we will use the same notation q for the various versions of the corresponding kernel Q. Also $|q(g)|$ will denote the norm of $q(g) \in \mathrm{End}(E)$.

Definition 4.1.1. Let $\mathcal{Q}$ be the vector space of continuous kernels $q(g)$ taken as above such that there exist $C > 0, C' > 0$ for which

$$|q(g)| \leq C \exp\left(-C'd^2(p1, pg)\right). \tag{4.1.9}$$

Equation (4.1.9) is equivalent to the fact that $x, x' \in X$,

$$|q(x, x')| \leq C \exp\left(-C'd^2(x, x')\right). \tag{4.1.10}$$

In (4.1.10), $|q(x, x')|$ denotes the norm of $q(x, x') \in \mathrm{Hom}(F_{x'}, F_x)$ with respect to the Hermitian product on F.

Let $dY^{\mathfrak{p}}$ be the volume form on $\mathfrak{p}$ with respect to its scalar product. If $Y^{\mathfrak{p}} \to x \in X$ is the geodesic coordinate system considered in section 3.2, the volume form dx can be written as

$$dx = \delta (Y^{\mathfrak{p}}) dY^{\mathfrak{p}}, \tag{4.1.11}$$

where δ is a positive smooth function such that $\delta (0) = 1$. As in (3.4.36), from the equation for Jacobi fields, we deduce that there exist $c > 0, C > 0$ such that

$$\delta (Y^{\mathfrak{p}}) \le c \exp (C |Y^{\mathfrak{p}}|). \tag{4.1.12}$$

By (4.1.10), if $q \in \mathcal{Q}$,

$$|q (Y^{\mathfrak{p}})| \le C \exp \left(-C' |Y^{\mathfrak{p}}|^2\right). \tag{4.1.13}$$

By (4.1.11), (4.1.13), we get

$$\int_X |q (p1, x)| \, dx < +\infty. \tag{4.1.14}$$

From (4.1.14), we deduce that there exists $C > 0$ such that for $x \in X$,

$$\int_X |q (x, x')| \, dx' \le C. \tag{4.1.15}$$

If $q \in \mathcal{Q}, s \in C^b (X, F)$, set

$$Qs (x) = \int_X q (x, x') s (x') \, dx'. \tag{4.1.16}$$

By (4.1.10) and (4.1.15), Q is a continuous operator from $C^b (X, F)$ into itself.

Proposition 4.1.2. *The vector space $\mathcal{Q}$ is an algebra with respect to the composition of operators acting on $C^b (X, F)$.*

Proof. If $q, q' \in \mathcal{Q}$ and if Q, Q' are the corresponding operators, the kernel of the operator QQ' should be given by

$$q * q' (x, x') = \int_X q (x, y) q' (y, x') \, dy. \tag{4.1.17}$$

Equations (4.1.10), (4.1.15) and dominated convergence guarantee that $q * q' (x, x')$ is continuous and uniformly bounded. If $y \in X$, either $d (x, y) \ge d (x, x') / 2$ or $d (x', y) \ge d (x, x') / 2$. It is now obvious that $q * q'$ verifies (4.1.10). The proof of our proposition is completed. $\square$

4.2 Orbital integrals

Let $\gamma \in G$ be semisimple. Because the properties of γ that will be used later are conjugation independent, we may as well assume that γ is as in (3.3.2), so that

$$\gamma = e^a k^{-1}, \qquad\qquad a \in \mathfrak{p}, \tag{4.2.1}$$
$$k \in K, \qquad\qquad \mathrm{Ad} (k) a = a.$$

Let $q \in Q$. If $x \in X$, $q(x, \gamma x) \in \text{Hom}(F_{\gamma x}, F_x)$. By still denoting by γ the obvious element of $\text{Hom}(F_x, F_{\gamma x})$, we find that $\gamma q(x, \gamma x) \in \text{End}(F_{\gamma x})$, so that $\text{Tr}^F[\gamma q(x, \gamma x)]$ is well-defined.

Let $h(y)$ be a compactly supported bounded measurable function on $X(\gamma)$.

Theorem 4.2.1. *The function* $\text{Tr}^F[\gamma q(x, \gamma x)] h(p_\gamma x)$ *is integrable on* X, *and moreover,*

$$\int_X \text{Tr}^F[\gamma q(x, \gamma x)] h(p_\gamma x)\, dx$$
$$= \int_{\mathfrak{p}^\perp(\gamma)} \text{Tr}^E[q(e^{-f} \gamma e^f)]\, r(f)\, df \int_{X(\gamma)} h(y)\, dy. \quad (4.2.2)$$

Proof. We temporarily assume the function $\text{Tr}^F[\gamma q(x, \gamma x)] h(p_\gamma x)$ to be integrable on X. Using (3.4.2) and (3.4.34), we get

$$\int_X \text{Tr}^F[\gamma q(x, \gamma x)] h(p_\gamma x)\, dx$$
$$= \int_{Z^0(\gamma) \times_{K^0(\gamma)} \mathfrak{p}^\perp(\gamma)} \text{Tr}^F[\gamma q(pye^f, \gamma pye^f)] h(y)\, r(f)\, df dy. \quad (4.2.3)$$

By (4.1.5),

$$\int_{Z^0(\gamma) \times_{K^0(\gamma)} \mathfrak{p}^\perp(\gamma)} \text{Tr}^F[\gamma q(pye^f, \gamma pye^f)] h(y)\, r(f)\, df dy$$
$$= \int_{\mathfrak{p}^\perp(\gamma)} \text{Tr}^E[q(e^{-f} \gamma e^f)]\, r(f)\, df \int_{X(\gamma)} h(y)\, dy. \quad (4.2.4)$$

By (4.2.3), (4.2.4), we get (4.2.2).

Moreover,

$$d(e^f p1, \gamma e^f p1) = d_\gamma(\rho_\gamma(1, f)). \quad (4.2.5)$$

By (3.4.4) and (4.2.5), we see that if q has compact support, the integral in the variable f in (4.2.4) is only over a bounded set of f, so that it is indeed well-defined.

More generally, if q is such that (4.1.9) or (4.1.10) hold, using (3.4.4), we get

$$\left| q(e^f p1, \gamma e^f p1) \right| \le C \exp\left(-C' |f|^2\right). \quad (4.2.6)$$

By (3.4.36) and (4.2.6), the integral in the variable f in (4.2.4) is still finite. The proof of our theorem is completed. $\square$

We will reformulate the above results in a different way. Indeed using (3.4.39), (4.1.5)–(4.1.8) and the fact that $\text{Vol}(K) = 1$, we get

$$\int_X \text{Tr}^F[\gamma q(x, \gamma x)] h(p_\gamma x)\, dx = \int_G \text{Tr}^E[q(g^{-1} \gamma g)] h(p_\gamma pg)\, dg. \quad (4.2.7)$$

By (3.4.37) and (3.4.47), we get

$$\int_G \mathrm{Tr}^E \left[q \left(g^{-1} \gamma g \right) \right] h \left(p_\gamma p g \right) dg$$

$$= \int_{Z^0(\gamma) \backslash G} \mathrm{Tr}^E \left[q \left(v^{-1} \gamma v \right) \right] dv \int_{Z^0(\gamma)} h \left(p z^0 \right) dz^0. \quad (4.2.8)$$

Using (3.4.46) and the fact that $\mathrm{Vol} \left(K^0 \left(\gamma \right) \backslash K \right) = 1$, we get

$$\int_{Z^0(\gamma) \backslash G} \mathrm{Tr}^E \left[q \left(v^{-1} \gamma v \right) \right] dv = \int_{\mathfrak{p}^\perp(\gamma)} \mathrm{Tr}^E \left[q \left(e^{-f} \gamma e^f \right) \right] r \left(f \right) df. \quad (4.2.9)$$

By (3.4.43) and using the fact that $\mathrm{Vol} \left(K^0 \left(\gamma \right) \right) = 1$, we have the identity

$$\int_{Z^0(\gamma)} h \left(p z^0 \right) dz^0 = \int_{X(\gamma)} h \left(y \right) dy. \quad (4.2.10)$$

Therefore (4.2.7)–(4.2.10) is equivalent to (4.2.2).

We will denote by $[\gamma]$ the conjugacy class of γ in G.

Definition 4.2.2. If $\gamma \in G$ is semisimple, put

$$\mathrm{Tr}^{[\gamma]} [Q] = \int_{Z^0(\gamma) \backslash G} \mathrm{Tr}^E \left[q \left(v^{-1} \gamma v \right) \right] dv = \int_{\mathfrak{p}^\perp(\gamma)} \mathrm{Tr}^E \left[q \left(e^{-f} \gamma e^f \right) \right] r \left(f \right) df.$$
$$(4.2.11)$$

By proceeding as in (4.2.3), (4.2.4), we get

$$\mathrm{Tr}^{[\gamma]} [Q] = \int_{\mathfrak{p}^\perp(\gamma)} \mathrm{Tr}^F \left[\gamma q \left(\gamma^{-1} e^f p1, e^f p1 \right) \right] r \left(f \right) df$$

$$= \int_{\mathfrak{p}^\perp(\gamma)} \mathrm{Tr}^F \left[\gamma q \left(e^f p1, \gamma e^f p1 \right) \right] r \left(f \right) df. \quad (4.2.12)$$

Integrals like (4.2.11), (4.2.12) are called orbital integrals.

If $h \in G$, the map $g \in G \to C \left(h \right) g \in G$ induces a map $Z \left(\gamma \right) \backslash G \to Z \left(C \left(h \right) \gamma \right) \backslash G$, which maps the volume dv of $Z^0 \left(\gamma \right) \backslash G$ into the corresponding volume on $Z^0 \left(C \left(h \right) \gamma \right) \backslash G$. Also dv is invariant under the right action of G on $Z^0 \left(\gamma \right) \backslash G$. Therefore, as the notation indicates, $\mathrm{Tr}^{[\gamma]} [Q]$ only depends on the conjugacy class $[\gamma]$ of γ in G.

By Proposition 4.1.2, $\mathcal{Q}$ is an algebra. If $q, q' \in \mathcal{Q}$, the function $q * q' \left(g \right)$ associated with QQ' is given by

$$q * q' \left(g \right) = \int_G q \left(h \right) q' \left(h^{-1} g \right) dh = \int_G q \left(g h^{-1} \right) q' \left(h \right) dh. \quad (4.2.13)$$

In the sequel, $[Q, Q']$ denotes the commutator of Q and Q'.

Theorem 4.2.3. *For any semisimple element $\gamma \in \Gamma$,*

$$\mathrm{Tr}^{[\gamma]} \left[[Q, Q'] \right] = 0. \quad (4.2.14)$$

Proof. If $\gamma = 1$, then

$$\mathrm{Tr}^{[\gamma]}[Q] = \mathrm{Tr}^E[q(1)]. \qquad (4.2.15)$$

In this case, (4.2.14) follows from (4.2.13) and (4.2.15).

Let δ_γ be the current on G so that

$$\int_G f\delta_\gamma = \int_{Z^0(\gamma)\backslash G} f\left(v^{-1}\gamma v\right) dv. \qquad (4.2.16)$$

Since dv is invariant under the right action of G on $Z^0(\gamma)\backslash G$, δ_γ is a current on G that is invariant by conjugation. By (4.2.11),

$$\mathrm{Tr}^{[\gamma]}[Q] = \int_G \mathrm{Tr}^E[q]\,\delta_\gamma = \mathrm{Tr}^E\left[q * \delta_{\gamma^{-1}}(1)\right]. \qquad (4.2.17)$$

Let $R_{\gamma^{-1}}$ be the operator

$$R_{\gamma^{-1}}f(g) = \int_{Z^0(\gamma)\backslash G} f\left(gv^{-1}\gamma^{-1}v\right) dv. \qquad (4.2.18)$$

Then $\delta_{\gamma^{-1}}$ can be thought of as the kernel associated with the operator $R_{\gamma^{-1}}$.

We can rewrite (4.2.17) in the form

$$\mathrm{Tr}^{[\gamma]}[Q] = \mathrm{Tr}^{[1]}\left[QR_{\gamma^{-1}}\right]. \qquad (4.2.19)$$

Since $\delta_{\gamma^{-1}}$ is invariant under conjugation, then

$$q * \delta_{\gamma^{-1}} = \delta_{\gamma^{-1}} * q, \qquad (4.2.20)$$

which just says that

$$QR_{\gamma^{-1}} = R_{\gamma^{-1}}Q. \qquad (4.2.21)$$

Then using (4.2.14) with $\gamma = 1$, (4.2.19) and (4.2.21), we get

$$\mathrm{Tr}^{[\gamma]}[[Q, Q']] = \mathrm{Tr}^{[1]}\left[[Q, Q']R_{\gamma^{-1}}\right] = \mathrm{Tr}^{[1]}\left[[Q, Q'R_{\gamma^{-1}}]\right] = 0, \qquad (4.2.22)$$

which is precisely (4.2.14). The proof of our theorem is completed. $\qquad \square$

4.3 Infinite dimensional orbital integrals

In the sequel, we will use the above formalism, when replacing the finite dimensional vector space E by the infinite dimensional vector space $\mathcal{E} = \Lambda^{\cdot}(\mathfrak{p}^* \oplus \mathfrak{k}^*) \otimes S^{\cdot}(\mathfrak{p}^* \oplus \mathfrak{k}^*) \otimes E$. Equivalently, we will replace the vector bundle F on X by the infinite dimensional vector bundle

$$\mathcal{F} = \Lambda^{\cdot}(T^*X \oplus N^*) \otimes S^{\cdot}(T^*X \oplus N^*) \otimes F.$$

Most of the time, $S^{\cdot}(\mathfrak{p}^* \oplus \mathfrak{k}^*)$ and $S^{\cdot}(T^*X \oplus N^*)$ will be replaced by their completion $\overline{S^{\cdot}(\mathfrak{p}^* \oplus \mathfrak{k}^*)} = L_2(\mathfrak{p} \oplus \mathfrak{k})$ and $\overline{S^{\cdot}(T^*X \oplus N^*)} = L_2(TX \oplus N)$. To avoid introducing extra notation, we still denote by $\mathcal{E}, \mathcal{F}$ the completions of the above vector spaces.

Of course the fact that $\mathcal{E}, \mathcal{F}$ are infinite dimensional creates new analytic difficulties, but the algebraic formalism that we described above remains

valid. A minor change with respect to what we did before is that traces will
be replaced by supertraces.

Let $C^b\left(\widehat{\mathcal{X}}, \widehat{\pi}^* \left(\Lambda^{\cdot}\left(T^*X \oplus N^*\right) \otimes F\right)\right)$ denote the vector space of continuous bounded sections of $\widehat{\pi}^* \left(\Lambda^{\cdot}\left(T^*X \oplus N^*\right)\right) \otimes F$ over $\widehat{\mathcal{X}}$.

Let $C^b_K\left(G \times \mathfrak{g}, \Lambda^{\cdot}\left(\mathfrak{p}^* \oplus \mathfrak{k}^*\right) \otimes E\right)$ denote the vector space of continuous
bounded sections of $\Lambda^{\cdot}\left(\mathfrak{p}^* \oplus \mathfrak{k}^*\right) \otimes E$ on $G \times \mathfrak{g}$ such that if $k \in K$,

$$s\left(gk, \mathrm{Ad}\left(k^{-1}\right) Y\right) = \rho^{\Lambda^{\cdot}\left(\mathfrak{p}^* \oplus \mathfrak{k}^*\right) \otimes E}\left(k^{-1}\right) s\left(g, Y\right). \tag{4.3.1}$$

Instead of (4.1.2), we now have

$$C^b\left(\widehat{\mathcal{X}}, \widehat{\pi}^* \left(\Lambda^{\cdot}\left(T^*X \oplus N^*\right) \otimes F\right)\right) = C^b_K\left(G \times \mathfrak{g}, \Lambda^{\cdot}\left(\mathfrak{p}^* \oplus \mathfrak{k}^*\right) \otimes E\right). \tag{4.3.2}$$

Recall that $dY^{\mathfrak{p}}$ is the volume element on $\mathfrak{p}$. Let $dY^{\mathfrak{k}}$ denote the volume
element of $\mathfrak{k}$ with respect to its scalar product. Let dY be the volume element
of $\mathfrak{g} = \mathfrak{p} \oplus \mathfrak{k}$,

$$dY = dY^{\mathfrak{p}} dY^{\mathfrak{k}}. \tag{4.3.3}$$

The above volume elements define volume elements dY^{TX}, dY^N, dY on the
fibres of $TX, N, TX \oplus N$.

Recall that K acts unitarily on the vector space $\mathcal{E}$. Our kernel $q\left(g\right)$ now
acts as an endomorphism of $\mathcal{E}$, and verifies (4.1.7). In what follows, the
operator $q\left(g\right)$ is given by continuous kernels $q\left(g, Y, Y'\right), Y, Y' \in \mathfrak{g}$, with
$q\left(g, Y, Y'\right) \in \mathrm{End}\left(\Lambda^{\cdot}\left(\mathfrak{g}^*\right) \otimes E\right)$. Let $q\left(\left(x, Y\right), \left(x', Y'\right)\right), \left(x, Y\right), \left(x', Y'\right) \in \widehat{\mathcal{X}}$
be the corresponding kernel on $\widehat{\mathcal{X}}$.

Definition 4.3.1. Let $\mathfrak{Q}$ denote the vector space of continuous kernels
$q\left(g, Y, Y'\right)$ taken as above such that there exist $C > 0, C' > 0$ for which

$$\left|q\left(g, Y, Y'\right)\right| \leq C \exp\left(-C'\left(d^2\left(p1, pg\right) + |Y|^2 + |Y'|^2\right)\right). \tag{4.3.4}$$

Equation (4.3.4) is equivalent to

$$\left|q\left(\left(x, Y\right), \left(x', Y'\right)\right)\right| \leq C \exp\left(-C'\left(d^2\left(x, x'\right) + |Y|^2 + |Y'|^2\right)\right). \tag{4.3.5}$$

Of course the norm in the right-hand side of (4.3.5) is calculated with respect
to the Hermitian product on $\Lambda^{\cdot}\left(T^*X \oplus N^*\right) \otimes F$.

If $q \in \mathfrak{Q}, s \in C^b\left(\widehat{\mathcal{X}}, \widehat{\pi}^* \left(\Lambda^{\cdot}\left(T^*X \oplus N^*\right) \otimes F\right)\right)$, set

$$Qs\left(x, Y\right) = \int_{\widehat{\mathcal{X}}} q\left(\left(x, Y\right), \left(x', Y'\right)\right) s\left(x', Y'\right) dx' dY'. \tag{4.3.6}$$

By proceeding as in section 4.1, by (4.3.5), (4.3.6), the operator Q acts on
$C^b\left(\widehat{\mathcal{X}}, \widehat{\pi}^* \left(\Lambda^{\cdot}\left(T^*X \oplus N^*\right) \otimes F\right)\right)$.

Proposition 4.3.2. *The vector space $\mathfrak{Q}$ is an algebra with respect to the
composition of operators acting on* $C^b\left(\widehat{\mathcal{X}}, \widehat{\pi}^* \left(\Lambda^{\cdot}\left(T^*X \oplus N^*\right) \otimes F\right)\right)$.

Proof. If $q, q' \in \mathfrak{Q}$, if Q, Q' are the corresponding operators, the kernel of the operator QQ' should be given by

$$q * q' \left((x, Y), (x', Y') \right) = \int_{\widehat{\mathcal{X}}} q \left((x, Y), (z, Z) \right) q' \left((z, Z), (x', Y') \right) dz dZ. \tag{4.3.7}$$

By proceeding as in the proof of Proposition 4.1.2, one finds easily that $q * q' \in \mathfrak{Q}$, which completes the proof of our proposition. $\qquad\square$

If $q \in \mathfrak{Q}$, by (4.3.4), if $g \in G$, the operator $q(g) \in \operatorname{End}(L_2(\mathfrak{p} \oplus \mathfrak{k}) \otimes E)$ is trace class. Moreover, the supertrace $\operatorname{Tr_s}^{\mathcal{E}} [q(g)]$ is given by

$$\operatorname{Tr_s}^{\mathcal{E}} [q(g)] = \int_{\mathfrak{g}} \operatorname{Tr_s}^{\Lambda^{\cdot}(\mathfrak{p}^* \oplus \mathfrak{k}^*) \otimes E} [q(g, Y, Y)] \, dY. \tag{4.3.8}$$

As before, $\gamma \in G$ is a semisimple element of G, such that (4.2.1) holds.

Definition 4.3.3. We define $\operatorname{Tr_s}^{[\gamma]} [Q]$ as in (4.2.11), i.e.,

$$\operatorname{Tr_s}^{[\gamma]} [Q] = \int_{(Z^0(\gamma) \backslash G) \times \mathfrak{g}} \operatorname{Tr_s}^{\Lambda^{\cdot}(\mathfrak{p}^* \oplus \mathfrak{k}^*) \otimes E} \left[q \left(v^{-1} \gamma v, Y, Y \right) \right] dv dY$$

$$= \int_{\mathfrak{p}^{\perp}(\gamma) \times \mathfrak{g}} \operatorname{Tr_s}^{\Lambda^{\cdot}(\mathfrak{p}^* \oplus \mathfrak{k}^*) \otimes E} \left[q \left(e^{-f} \gamma e^{f}, Y, Y \right) \right] r(f) \, df dY. \tag{4.3.9}$$

By Theorem 3.4.1 and by (4.3.5), the integral in (4.3.9) is well defined. As in (4.2.12), we can rewrite (4.3.9) in the form

$$\operatorname{Tr_s}^{[\gamma]} [Q]$$
$$= \int_{\mathfrak{p}^{\perp}(\gamma)} \left[\int_{TX \oplus N} \operatorname{Tr_s}^{\Lambda^{\cdot}(T^*X \oplus N^*) \otimes F} \left[\gamma q \left(\gamma^{-1} \left(e^{f} p 1, Y \right), \left(e^{f} p 1, Y \right) \right) \right] dY \right]$$
$$r(f) \, df$$
$$= \int_{\mathfrak{p}^{\perp}(\gamma)} \left[\int_{TX \oplus N} \operatorname{Tr_s}^{\Lambda^{\cdot}(T^*X \oplus N^*) \otimes F} \left[\gamma q \left(\left(e^{f} p 1, Y \right), \gamma \left(e^{f} p 1, Y \right) \right) \right] dY \right]$$
$$r(f) \, df. \tag{4.3.10}$$

Expressions such as (4.3.9), (4.3.10) will be called orbital supertraces.

We now use the notation $[\ ]$ for the supercommutator. Let $Q, Q' \in \mathfrak{Q}$ be two operators taken as before.

Theorem 4.3.4. *For any semisimple element $\gamma \in G$,*

$$\operatorname{Tr_s}^{[\gamma]} [[Q, Q']] = 0. \tag{4.3.11}$$

Proof. Recall that if $V = V_+ \oplus V_-$ is a finite dimensional vector space, the supertrace vanishes on supercommutators. The proof of our theorem is then an easy modification of the proof of Theorem 4.2.3. $\qquad\square$

4.4 The orbital integrals for the elliptic heat kernel of X

Let A be a self-adjoint element of $\mathrm{End}\,(E)$ that is K-invariant. Then A descends to a self-adjoint parallel section of $\mathrm{End}\,(F)$.

Recall that the operator $\mathcal{L}^X$ acting on $C^\infty\,(X, F)$ was defined in (2.13.3).

Definition 4.4.1. Let $\mathcal{L}_A^X$ be the operator acting on $C^\infty\,(X, F)$,

$$\mathcal{L}_A^X = \mathcal{L}^X + A. \tag{4.4.1}$$

Let $C^{\infty,c}\,(X, F)$ be the vector space of smooth sections of F over X that have compact support. Since X is complete, $\mathcal{L}_A^X$ is an essentially self-adjoint operator on $C^{\infty,c}\,(X, F)$.

Let $C^{\infty,b}\,(X, F)$ be the vector space of smooth sections of F, which are bounded together with their derivatives of arbitrary order. For $t > 0$, the heat operator $\exp\left(-t\mathcal{L}_A^X\right)$ acts on $C^{\infty,b}\,(X, F)$. This operator commutes with the left action of G on $C^{\infty,b}\,(X, F)$. Let $p_t^X\,(x, x')$ be the smooth kernel for the operator $\exp\left(-t\mathcal{L}_A^X\right)$ with respect to the volume element dx' on X.

Proposition 4.4.2. For any $t > 0$, $p_t^X \in \mathcal{Q}$.

Proof. The fact that $p_t^X\,(x, x')$ verifies (4.1.10) is well-known, and can be established using finite propagation speed for the wave equation [CP81, section 7.8], [T81, section 4.4]. □

Let $\gamma \in G$ be a semisimple element. It follows from the results of section 4.2 and from Proposition 4.4.2 that for $t > 0$, the orbital integral $\mathrm{Tr}^{[\gamma]}\left[\exp\left(-t\mathcal{L}_A^X\right)\right]$ is well-defined.

4.5 The orbital supertraces for the hypoelliptic heat kernel

Recall that the operator $\mathcal{L}_b^X$ was defined in equation (2.13.4), and acts on

$$C^\infty\left(\widehat{\mathcal{X}}, \widehat{\pi}^*\left(\Lambda^\cdot\,(T^*X \oplus N^*) \otimes F\right)\right).$$

Let $\mathcal{L}_{A,b}^X$ be the operator acting on the same vector space,

$$\mathcal{L}_{A,b}^X = \mathcal{L}_b^X + A. \tag{4.5.1}$$

By Theorem 2.13.2, the operator $\frac{\partial}{\partial t} + \mathcal{L}_{A,b}^X$ is a hypoelliptic operator, and $\mathcal{L}_{A,b}^X$ is formally self-adjoint with respect to η.

By (2.13.5), with the exception of the quartic term $\frac{1}{2}\left|\left[Y^N, Y^{TX}\right]\right|^2$, the operator $\mathcal{L}_{A,b}^X$ has exactly the same structure as the hypoelliptic Laplacian constructed in [B05], whose analytic study was done in detail in Bismut-Lebeau [BL08]. Moreover, the above quartic term is nonnegative, and it commutes with

$$\nabla_{Y^{TX}}^{C^\infty(TX \oplus N, \widehat{\pi}^*(\Lambda^\cdot(T^*X \oplus N^*) \otimes F))}.$$

We will see in section 11.8 that for $t > 0$, the heat operator $\exp\left(-t\mathcal{L}_{A,b}^X\right)$ is well-defined. Also, there is an associated smooth kernel $q_{b,t}^X\left((x, Y), (x', Y')\right)$.

Definition 4.5.1. Let $\mathbf{P}$ be the projection from $\Lambda^{\cdot}(T^*X \oplus N^*) \otimes F$ on $\Lambda^0(T^*X \oplus N^*) \otimes F = F$. For $t > 0, (x, Y), (x', Y') \in \widehat{\mathcal{X}}$, put

$$q_{0,t}^X((x, Y), (x', Y')) = \mathbf{P}p_t^X(x, x') \pi^{-(m+n)/2} \exp\left(-\frac{1}{2}\left(|Y|^2 + |Y'|^2\right)\right) \mathbf{P}. \tag{4.5.2}$$

Theorem 4.5.2. Given $\epsilon > 0, M > 0, \epsilon \leq M$, there exist $C > 0, C' > 0$ such that for $0 < b \leq M, \epsilon \leq t \leq M, (x, Y), (x', Y') \in \widehat{\mathcal{X}}$,

$$\left|q_{b,t}^X((x, Y), (x', Y'))\right| \leq C \exp\left(-C'\left(d^2(x, x') + |Y|^2 + |Y'|^2\right)\right). \tag{4.5.3}$$

In particular, for any $b > 0, t > 0, q_{b,t} \in \mathfrak{Q}$.
   Moreover, as $b \to 0$,

$$q_{b,t}^X((x, Y), (x', Y')) \to q_{0,t}^X((x, Y), (x', Y')). \tag{4.5.4}$$

Proof. The proof of our theorem will be given in chapter 14. $\qquad\square$

4.6 A fundamental equality

Let $\gamma \in G$ be semisimple as in (4.2.1). By the results of section 4.3 and by Theorem 4.5.2, for $t > 0$, $\mathrm{Tr_s}^{[\gamma]}\left[\exp\left(-t\mathcal{L}_{A,b}^X\right)\right]$ is well-defined.
   Now we establish the main result of this chapter, which should be viewed as an analogue of [B08b, Theorem 4.1], where a corresponding result was established in the case where G is a compact Lie group.

Theorem 4.6.1. *For any $b > 0, t > 0$, the following identity holds:*

$$\mathrm{Tr_s}^{[\gamma]}\left[\exp\left(-t\mathcal{L}_{A,b}^X\right)\right] = \mathrm{Tr}^{[\gamma]}\left[\exp\left(-t\mathcal{L}_A^X\right)\right]. \tag{4.6.1}$$

Proof. A first step in the proof will be to show that

$$\frac{\partial}{\partial b}\mathrm{Tr_s}^{[\gamma]}\left[\exp\left(-t\mathcal{L}_{A,b}^X\right)\right] = 0. \tag{4.6.2}$$

Indeed using Theorem 4.3.4, we find easily that

$$\frac{\partial}{\partial b}\mathrm{Tr_s}^{[\gamma]}\left[\exp\left(-t\mathcal{L}_{A,b}^X\right)\right] = -t\mathrm{Tr_s}^{[\gamma]}\left[\frac{\partial}{\partial b}\mathcal{L}_{A,b}^X \exp\left(-t\mathcal{L}_{A,b}^X\right)\right]. \tag{4.6.3}$$

By (2.13.4) and (4.5.1), we get

$$\frac{\partial}{\partial b}\mathcal{L}_{A,b}^X = \frac{1}{2}\left[\mathfrak{D}_b^X, \frac{\partial}{\partial b}\mathfrak{D}_b^X\right]. \tag{4.6.4}$$

By Proposition 2.15.1,

$$\left[\mathfrak{D}_b^X, \mathcal{L}_b^X\right] = 0. \tag{4.6.5}$$

Also, we have the trivial

$$\left[\mathfrak{D}_b^X, A\right] = 0. \tag{4.6.6}$$

By (4.5.1), (4.6.5), and (4.6.6), we obtain

$$\left[\mathfrak{D}_b^X, \mathcal{L}_{A,b}^X\right] = 0. \tag{4.6.7}$$

By Theorem 4.3.4 and by (4.6.3), (4.6.4) and (4.6.7), we get

$$\frac{\partial}{\partial b}\mathrm{Tr_s}^{[\gamma]}\left[\exp\left(-t\mathcal{L}_{A,b}^X\right)\right] = -\frac{t}{2}\mathrm{Tr_s}^{[\gamma]}\left[\left[\mathfrak{D}_b^X, \frac{\partial}{\partial b}\mathfrak{D}_b^X \exp\left(-t\mathcal{L}_{A,b}^X\right)\right]\right] = 0, \tag{4.6.8}$$

which is just (4.6.2).

We claim that as $b \to 0$,

$$\mathrm{Tr_s}^{[\gamma]}\left[\exp\left(-\mathcal{L}_{A,b}^X\right)\right] \to \mathrm{Tr}^{[\gamma]}\left[\exp\left(-\mathcal{L}_A^X\right)\right]. \tag{4.6.9}$$

Indeed using (3.4.4) and (4.5.3), we find that given $t > 0$, there exist $C > 0, C' > 0$ such that for $0 < b \le 1, f \in \mathfrak{p}^\perp(\gamma), Y \in (TX \oplus N)_{e^f p1}$,

$$\left|q_{b,t}^X\left(\left(e^f p1, Y\right), \gamma\left(e^f p1, Y\right)\right)\right| \le C \exp\left(-C'\left(|f|^2 + |Y|^2\right)\right). \tag{4.6.10}$$

By (3.4.36), (4.2.12), (4.3.10), (4.5.4), and (4.6.10), using dominated convergence, we get (4.6.9).

Our theorem now follows from (4.6.2) and (4.6.9). □

4.7 Another approach to the orbital integrals

We still take γ as in (4.2.1). We have the identity of operators acting on $C^\infty(G)$,

$$\gamma = e^{-\nabla_{a,r}}k^{-1}. \tag{4.7.1}$$

As explained in section 2.17, equation (4.7.1) descends to an identity of operators acting on $C^\infty(X, F)$,

$$\gamma = e^{-L_a}k^{-1}. \tag{4.7.2}$$

Take $t > 0$. Recall that $p_t^X(x, x'), x, x' \in X$ is the smooth kernel associated with $\exp\left(-t\mathcal{L}_A^X\right)$. Then $\gamma p_t^X\left(\gamma^{-1}x, x'\right)$ is the smooth kernel associated with $\gamma \exp\left(-t\mathcal{L}_A^X\right)$. By (4.7.2), we get

$$\gamma\exp\left(-t\mathcal{L}_A^X\right) = k^{-1}\exp\left(-t\mathcal{L}_A^X - L_a\right). \tag{4.7.3}$$

We will denote by $p_{a,t}^X(x, x'), x, x' \in X$ the smooth kernel associated with $\exp\left(-t\mathcal{L}_A^X - L_a\right)$. By (4.7.3), we get

$$\gamma p_t^X\left(\gamma^{-1}x, x'\right) = k^{-1}p_{a,t}^X(kx, x'). \tag{4.7.4}$$

Equations (4.7.3), (4.7.4) can be read off on G instead of on X. We already know that $p_t^X(x, x')$ lifts to a function $p_t^X(g)$ on G that verifies (4.1.6) and (4.1.7). Moreover, $p_{a,t}^X(x, x')$ lifts to a smooth function $p_{a,t}^X(g, g')$ on $G \times G$ that verifies (4.1.3). By (3.3.4), $\exp\left(-t\mathcal{L}_A^X - L_a\right)$ commutes with $Z(\gamma)$. Therefore (4.1.5) holds only when $g \in Z(\gamma)$. In particular the function

$p_{a,t}^X(kg, g)$ is well-defined on $Z^0(\gamma) \backslash G$. It follows from (4.7.4) that we have the equality of smooth functions on $Z^0(\gamma) \backslash G$ with values in $\text{End}(E)$,

$$p_t^X\left(g^{-1}\gamma g\right) = p_{a,t}^X(kg, g). \tag{4.7.5}$$

By (4.2.11) and (4.7.5), we get

$$\text{Tr}^{[\gamma]}\left[\exp\left(-t\mathcal{L}_A^X\right)\right] = \int_{Z^0(\gamma)\backslash G} \text{Tr}^E\left[p_t^X\left(v^{-1}\gamma v\right)\right] dv$$

$$= \int_{Z^0(\gamma)\backslash G} \text{Tr}^E\left[p_{a,t}^X(kv, v)\right] dv. \tag{4.7.6}$$

Of course a similar equation holds tautologically when replacing the operator $\mathcal{L}_A^X$ by the operator $\mathcal{L}_{A,b}^X$. If $q_{a,b,t}^X\left((x, Y), (x', Y')\right)$ denotes the smooth kernel associated with $\exp\left(-t\mathcal{L}_{A,b}^X - L_a\right)$, then we have the obvious analogue of (4.7.4), i.e.,

$$\gamma q_{b,t}^X\left(\gamma^{-1}(x, Y), (x, Y)\right) = k^{-1}q_{a,b,t}^X\left(k(x, Y), (x, Y)\right). \tag{4.7.7}$$

From (4.7.7), the orbital integral $\text{Tr}_s^{[\gamma]}\left(\exp\left(-t\mathcal{L}_{A,b}^X\right)\right)$, which is given by the integral (4.3.9), with q replaced by $q_{b,t}^X$, can be reexpressed using the kernel $q_{a,b,t}^X$ as in equation (4.7.6).

4.8 The locally symmetric space Z

Let $\Gamma \subset G$ be a cocompact subgroup of G. Then the elements of Γ are semisimple. Also Γ acts isometrically on X, and this action lifts to all the considered Euclidean or Hermitian vector bundles, and preserves the corresponding connections. Set

$$Z = \Gamma \backslash G / K. \tag{4.8.1}$$

Equivalently,

$$Z = \Gamma \backslash X. \tag{4.8.2}$$

Then Z is a compact locally symmetric space. If Γ is torsion free, or equivalently if Γ does not contain elliptic elements, then Z is a smooth manifold. Otherwise, in Γ there are only a finite number of elliptic conjugacy classes in Γ, and Z is an orbifold.

If Γ is torsion free, since X is contractible,

$$\Gamma = \pi_1(Z), \tag{4.8.3}$$

and X is the universal cover of Z.

A vector bundle like F descends to a vector bundle on Z, which we still denote by F. Except for the tangent bundle TX of X, which descends to the orbifold tangent bundle TZ of Z, we will use the same notation for the other vector bundles over X when descended to Z.

By definition, $C\left(Z,F\right)$ is the part of $C^{b}\left(X,F\right)$ that is fixed under the left action of Γ, i.e.,

$$C\left(Z,F\right) = C^{b}\left(X,F\right)^{\Gamma}. \qquad (4.8.4)$$

By (4.1.2) and (4.8.4), we obtain

$$C\left(Z,F\right) = C_{K}^{b}\left(G,E\right)^{\Gamma}. \qquad (4.8.5)$$

We use the notation of section 4.1. Let $Q \in \mathcal{Q}$. Since the operator Q commutes with the left action of G, Q descends to an operator Q^{Z} acting on $C\left(Z,F\right)$. Let $q^{Z}\left(z,z'\right), z, z' \in Z$ be the continuous kernel of Q^{Z} over Z. If z, z' are identified with their lifts in X, then

$$q^{Z}\left(z,z'\right) = \sum_{\gamma \in \Gamma} \gamma q\left(\gamma^{-1}z,z'\right) = \sum_{\gamma \in \Gamma} q\left(z,\gamma z'\right)\gamma. \qquad (4.8.6)$$

If we use the representation (4.8.5) of $C\left(Z,F\right)$, the kernel $q^{Z}\left(z,z'\right)$ can be identified with the function over $G \times G$,

$$q^{Z}\left(g,g'\right) = \sum_{\gamma \in \Gamma} q\left(g^{-1}\gamma g'\right). \qquad (4.8.7)$$

Let C be the set of conjugacy classes in Γ. If $[\gamma] \in C$, set

$$q^{X,[\gamma]}\left(g,g'\right) = \sum_{\gamma \in [\gamma]} q\left(g^{-1}\gamma g'\right). \qquad (4.8.8)$$

Then

$$q^{Z}\left(z,z'\right) = \sum_{[\gamma] \in C} q^{X,[\gamma]}\left(g,g'\right). \qquad (4.8.9)$$

The trace $\operatorname{Tr}\left[Q^{Z}\right]$ of Q^{Z} is given by

$$\operatorname{Tr}\left[Q^{Z}\right] = \int_{Z} \operatorname{Tr}\left[q^{Z}\left(z,z\right)\right] dz. \qquad (4.8.10)$$

Set

$$\operatorname{Tr}\left[Q^{Z,[\gamma]}\right] = \int_{Z} \operatorname{Tr}\left[q^{X,[\gamma]}\left(z,z\right)\right] dz. \qquad (4.8.11)$$

Then

$$\operatorname{Tr}\left[Q^{Z}\right] = \sum_{[\gamma] \in C} \operatorname{Tr}\left[Q^{Z,[\gamma]}\right]. \qquad (4.8.12)$$

On the other hand, one finds easily that

$$\operatorname{Tr}\left[Q^{Z,[\gamma]}\right] = \int_{\Gamma \cap Z(\gamma) \backslash G} \operatorname{Tr}\left[q\left(g^{-1}\gamma g\right)\right] dg. \qquad (4.8.13)$$

We use the notation of subsection 3.4. By (4.8.13), we get

$$\operatorname{Tr}\left[Q^{Z,[\gamma]}\right] = \operatorname{Vol}\left(\Gamma \cap Z\left(\gamma\right) \backslash Z\left(\gamma\right)\right) \int_{Z(\gamma) \backslash G} \operatorname{Tr}\left[q\left(g^{-1}\gamma g\right)\right] dv. \qquad (4.8.14)$$

Using Theorems 3.4.1, 3.4.3, and 4.2.1, if $\Gamma \cap Z(\gamma)$ is torsion free, so that $\Gamma \cap Z(\gamma)$ acts freely on $X(\gamma)$, we obtain

$$\mathrm{Tr}\left[Q^{Z,[\gamma]}\right] = \mathrm{Vol}\left(\Gamma \cap Z(\gamma) \setminus X(\gamma)\right) \int_{\mathfrak{p}^{\perp}(\gamma)} \mathrm{Tr}\left[q\left(e^{-f}\gamma e^{f}\right)\right] r(f)\, df. \tag{4.8.15}$$

By (4.2.11), we can rewrite (4.8.15) in the form

$$\mathrm{Tr}\left[Q^{Z,[\gamma]}\right] = \mathrm{Vol}\left(\Gamma \cap Z(\gamma) \setminus X(\gamma)\right) \mathrm{Tr}^{[\gamma]}[Q]. \tag{4.8.16}$$

Since $K(\gamma)$ acts freely on $\Gamma \cap Z(\gamma) \setminus Z(\gamma)$, we get

$$\mathrm{Vol}\left(\Gamma \cap Z(\gamma) \setminus Z(\gamma)\right) = \mathrm{Vol}\left(K(\gamma)\right) \mathrm{Vol}\left(\Gamma \cap Z(\gamma) \setminus X(\gamma)\right). \tag{4.8.17}$$

Also by (3.4.46) and its analogue for $Z(\gamma), K(\gamma)$, we obtain

$$\mathrm{Tr}^{[\gamma]}[Q] = \frac{\mathrm{Vol}\left(K(\gamma)\right)}{\mathrm{Vol}(K)} \int_{Z(\gamma)\setminus G} \mathrm{Tr}\left[q\left(g^{-1}\gamma g\right)\right] dv. \tag{4.8.18}$$

Since $\mathrm{Vol}(K) = 1$, from (4.8.16)–(4.8.18), we recover (4.8.14).

If $\Gamma \cap Z(\gamma)$ is not torsion free, let $\delta(\gamma)$ be the subgroup of the elements in $K(\gamma)$ that act on the right like the identity on $\Gamma \cap Z(\gamma) \setminus Z(\gamma)$. Then $\delta(\gamma)$ is a subgroup of the finite group $\Gamma \cap K(\gamma)$. Instead of (4.8.17), we get

$$\mathrm{Vol}\left(\Gamma \cap Z(\gamma) \setminus Z(\gamma)\right) = \frac{\mathrm{Vol}\left(K(\gamma)\right)}{|\delta(\gamma)|} \mathrm{Vol}\left(\Gamma \cap Z(\gamma) \setminus X(\gamma)\right). \tag{4.8.19}$$

Instead of (4.8.16), we obtain

$$\mathrm{Tr}\left[Q^{Z,[\gamma]}\right] = \frac{\mathrm{Vol}\left(\Gamma \cap Z(\gamma) \setminus X(\gamma)\right)}{|\delta(\gamma)|} \mathrm{Tr}^{[\gamma]}[Q]. \tag{4.8.20}$$

If Γ is not torsion free, then K acts locally freely on $\Gamma \setminus G$, and $Z = \Gamma \setminus G / K$ is an orbifold. Let $E \subset C$ be the finite set of elliptic conjugacy classes in Γ. If $e \in E$, let $[e]$ be the set of conjugacy classes in E that are conjugate to e in G. Let $k \in K$ represent the classes in $[e]$. By (4.8.14), we get

$$\sum_{[\gamma]\in[e]} \mathrm{Tr}\left[Q^{Z,[\gamma]}\right] = \sum_{[\gamma]\in[e]} \mathrm{Vol}\left(\Gamma \cap Z(\gamma) \setminus Z(\gamma)\right) \int_{Z(k)\setminus G} \mathrm{Tr}\left[q\left(g^{-1}kg\right)\right] dv. \tag{4.8.21}$$

Now we give a geometric interpretation of the right-hand side of (4.8.21). Let $(\Gamma \setminus G)_k \subset \Gamma \setminus G$ be the fixed point set of the right action of k. For each $[\gamma] \in [e]$, take one $\gamma \in [\gamma]$, and let $g_\gamma \in G$ be such that $g_\gamma k g_\gamma^{-1} = \gamma$. Then one verifies easily that

$$(\Gamma \setminus G)_k = \bigcup_{[\gamma]\in[e]} (Z(\gamma) \cap \Gamma \setminus Z(\gamma)) g_\gamma, \tag{4.8.22}$$

and the components in (4.8.22) do not intersect. Moreover, $K(k)$ acts on the right on $(\Gamma \setminus G)_k$. By (4.8.22), we can rewrite (4.8.21) in the form

$$\sum_{[\gamma]\in[e]} \mathrm{Tr}\left[Q^{Z,[\gamma]}\right] = \mathrm{Vol}\left((\Gamma \setminus G)_k\right) \int_{Z(k)\setminus G} \mathrm{Tr}\left[q\left(g^{-1}kg\right)\right] dv. \tag{4.8.23}$$

Let $\Delta(k)$ be the subgroup of $K(k)$ of the elements that act like the identity on $(\Gamma \setminus G)_k$. Then $\Delta(k)$ is a subgroup of the finite group $\Gamma \cap K(k)$. Instead of (4.8.17), we get

$$\mathrm{Vol}\left((\Gamma \setminus G)_k\right) = \frac{\mathrm{Vol}\left(K(k)\right)}{|\Delta(k)|} \mathrm{Vol}\left((\Gamma \setminus G)_k / K(k)\right). \tag{4.8.24}$$

Still using (3.4.46), we obtain

$$\mathrm{Tr}^{[k]}[Q] = \frac{\mathrm{Vol}\left(K(k)\right)}{\mathrm{Vol}(K)} \int_{Z(k) \setminus G} \mathrm{Tr}\left[q\left(g^{-1}kg\right)\right] dv. \tag{4.8.25}$$

By (4.8.23)–(4.8.25), we get

$$\sum_{[\gamma] \in [e]} \mathrm{Tr}\left[Q^{Z,[\gamma]}\right] = \frac{\mathrm{Vol}\left((\Gamma \setminus G)_k / K(k)\right)}{|\Delta(k)|} \mathrm{Tr}^{[k]}[Q]. \tag{4.8.26}$$

Equation (4.8.26) is of special interest in connection with the Kawasaki formula [Ka79], an analogue for orbifolds of the Atiyah-Singer index theorem, that will be briefly considered in section 7.7.

The vector bundle $TX \oplus N$ descends to $TZ \oplus N$. The total space $\widehat{\mathcal{X}}$ of the vector bundle $TX \oplus N$ over X descends to the total space $\widehat{\mathcal{Z}}$ of the vector bundle $TZ \oplus N$ over Z. We still denote by $\widehat{\pi}$ the projection $\widehat{\mathcal{Z}} \to Z$. The same arguments as before apply to the kernels $Q \in \mathfrak{Q}$.

The operator $\mathcal{L}_A^X$ descends to an operator $\mathcal{L}_A^Z$, and the operator $\mathcal{L}_{A,b}^X$ descends to an operator $\mathcal{L}_{A,b}^Z$. For $t > 0$, the operator $\exp\left(-t\mathcal{L}_A^X\right)$ is obviously trace class. Similarly, by proceeding as in [BL08, chapter 3], we find that the operator $\exp\left(-t\mathcal{L}_{A,b}^Z\right)$ is also trace class. Since the base Z is compact, the results of [BL08] can be imported without any change.

Theorem 4.8.1. *For any $t > 0, b > 0$,*
$$\mathrm{Tr}_s\left[\exp\left(-t\mathcal{L}_{A,b}^Z\right)\right] = \mathrm{Tr}\left[\exp\left(-t\mathcal{L}_A^Z\right)\right]. \tag{4.8.27}$$

Proof. We give two proofs. A first method is to use Theorem 4.6.1, and (4.8.12), (4.8.16).

Another method is first to establish the analogue of (4.6.2),

$$\frac{\partial}{\partial b} \mathrm{Tr}_s\left[\exp\left(-t\mathcal{L}_{A,b}^Z\right)\right] = 0. \tag{4.8.28}$$

This can be done by proceeding as in (4.6.4)–(4.6.8), while invoking at the last stage the fact that standard supertraces vanish on supercommutators. The next step is to show that as $b \to 0$,

$$\mathrm{Tr}_s\left[\exp\left(-t\mathcal{L}_{A,b}^Z\right)\right] \to \mathrm{Tr}\left[\exp\left(-t\mathcal{L}_A^Z\right)\right]. \tag{4.8.29}$$

To prove (4.8.29), we can use Theorem 4.5.2. Since the base Z is compact, we can also use the results of [BL08, section 3.4], and obtain (4.8.29). This is because by (2.16.1), (2.16.2), the structure of the operator $\mathcal{L}_b^X$ is essentially the same as the structure of the operator considered in [BL08]. The extra quartic term $\frac{1}{2}\left|\left[Y^N, Y^{TX}\right]\right|^2$ can be dealt with by the methods of sections 11.7 and 11.8. Finally, by equation (2.16.4) in Theorem 2.16.1, we identify the limit in (4.8.29).

The proof of our theorem is completed.   □

Remark 4.8.2. Theorem 4.8.1 will not be used in the sequel, i.e., we will ignore the spectral side of the trace formula, to concentrate on the orbital integrals.

Chapter Five

Evaluation of supertraces for a model operator

In this chapter, given a semisimple element $\gamma \in G$, we evaluate the supertrace of the heat kernel of a hypoelliptic operator acting over $\mathfrak{p} \times \mathfrak{g}$. In section 9.10, this operator will appear in the asymptotics as $b \to +\infty$ of a rescaled version of the operator $\mathcal{L}_b^X$.

In particular we obtain a function $J_\gamma\left(Y_0^{\mathfrak{k}}\right), Y_0^{\mathfrak{k}} \in \mathfrak{k}(\gamma)$, which will play a fundamental role in our final formula for the orbital integrals.

This chapter is organized as follows. In section 5.1, if $Y_0^{\mathfrak{k}} \in \mathfrak{k}(\gamma)$, we introduce the hypoelliptic operator $\mathcal{P}_{a,Y_0^{\mathfrak{k}}}$, its heat kernel, and a corresponding supertrace $J_\gamma\left(Y_0^{\mathfrak{k}}\right)$.

In section 5.2, by a conjugation of $\mathcal{P}_{a,Y_0^{\mathfrak{k}}}$, we obtain a simpler operator $\mathcal{Q}_{a,Y_0^{\mathfrak{k}}}$, where $\mathfrak{p} \times \mathfrak{p}$ and $\mathfrak{k}$ have been decoupled. The operator $\mathcal{Q}_{a,Y_0^{\mathfrak{k}}}$ splits naturally into a scalar part and a matrix part.

In section 5.3, we evaluate the trace of the heat kernel of the scalar part of $\mathcal{Q}_{a,Y_0^{\mathfrak{k}}}$.

In section 5.4, we compute the supertrace of the matrix part of the heat kernel.

Finally, in section 5.5, we give an explicit formula for $J_\gamma\left(Y_0^{\mathfrak{k}}\right)$.

5.1 The operator $\mathcal{P}_{a,Y_0^{\mathfrak{k}}}$ and the function $J_\gamma\left(Y_0^{\mathfrak{k}}\right)$

Let γ be a semisimple element of G, which is written in the form given in (4.2.1).

Definition 5.1.1. Let $\underline{\mathfrak{z}}(\gamma)$ denote another copy of $\mathfrak{z}(\gamma)$, and let $\underline{\mathfrak{z}}(\gamma)^*$ denote the corresponding copy of the dual of $\mathfrak{z}(\gamma)$. If $\alpha \in \Lambda^{\cdot}\left(\mathfrak{z}(\gamma)^*\right)$, let $\underline{\alpha}$ be the corresponding element in $\Lambda^{\cdot}\left(\underline{\mathfrak{z}}(\gamma)^*\right)$. Also $c(\mathfrak{z}(\gamma))$ denotes the Clifford algebra of $\left(\mathfrak{z}(\gamma), B|_{\mathfrak{z}(\gamma)}\right)$.

We will continue using the notation of chapter 2. In particular, E is taken as in section 2.12. Our operators act on $C^\infty\left(\mathfrak{p} \times \mathfrak{g}, \Lambda^{\cdot}(\mathfrak{g}^*) \widehat{\otimes} \Lambda^{\cdot}\left(\underline{\mathfrak{z}}^*(\gamma)\right) \widehat{\otimes} E\right)$. Of course,

$$\mathfrak{p} \times \mathfrak{g} = \mathfrak{p} \times (\mathfrak{p} \oplus \mathfrak{k}). \tag{5.1.1}$$

We denote by y the tautological section of the first copy of $\mathfrak{p}$ in the right-hand side of (5.1.1), and by $Y^{\mathfrak{g}} = Y^{\mathfrak{p}} + Y^{\mathfrak{k}}$ the tautological section of $\mathfrak{g} = \mathfrak{p} \oplus \mathfrak{k}$. Let dy be the volume form on $\mathfrak{p}$, let $dY^{\mathfrak{g}}$ be the volume form on $\mathfrak{g} = \mathfrak{p} \oplus \mathfrak{k}$.

Set

$$p = \dim \mathfrak{p}(\gamma), \qquad q = \dim \mathfrak{k}(\gamma), \qquad r = \dim \mathfrak{z}(\gamma), \qquad (5.1.2)$$

so that

$$r = p + q. \qquad (5.1.3)$$

Let $e_1, \ldots, e_p$ be an orthonormal basis of $\mathfrak{p}(\gamma)$, let $e_{p+1}, \ldots, e_r$ be an orthonormal basis of $\mathfrak{k}(\gamma)$. We denote with upper scripts the corresponding dual bases. Then $\underline{e}^1, \ldots, \underline{e}^r$ is a basis of $\mathfrak{z}(\gamma)^*$.

Put

$$\alpha = \sum_{i=1}^{r} c(e_i) \underline{e}^i. \qquad (5.1.4)$$

Recall that $\Delta^{\mathfrak{p} \oplus \mathfrak{k}}$ is the standard Laplacian on $\mathfrak{p} \oplus \mathfrak{k}$. Let $\Delta^{\mathfrak{p}}, \Delta^{\mathfrak{k}}$ be the Laplacians on $\mathfrak{p}, \mathfrak{k} \subset \mathfrak{g} = \mathfrak{p} \oplus \mathfrak{k}$. These operators act along $\mathfrak{p} \oplus \mathfrak{k}$, i.e., on the second factor in the right-hand side of (5.1.1).

Let ∇^H denote differentiation in the variable $y \in \mathfrak{p}$, and let ∇^V denote differentiation in the variable $Y^{\mathfrak{g}} \in \mathfrak{g}$.

Definition 5.1.2. If $Y_0^{\mathfrak{k}} \in \mathfrak{k}(\gamma)$, set

$$\mathcal{P}_{a,Y_0^{\mathfrak{k}}} = \frac{1}{2} \left| [Y^{\mathfrak{k}}, a] + [Y_0^{\mathfrak{k}}, Y^{\mathfrak{p}}] \right|^2 - \frac{1}{2} \Delta^{\mathfrak{p} \oplus \mathfrak{k}} + \alpha - \nabla^H_{Y^{\mathfrak{p}}}$$
$$- \nabla^V_{[a+Y_0^{\mathfrak{k}}, [a,y]]} - \widehat{c}(\mathrm{ad}(a)) + c\left(\mathrm{ad}(a) + i\theta \mathrm{ad}(Y_0^{\mathfrak{k}})\right). \qquad (5.1.5)$$

By Hörmander [Hö67], the operator $\frac{\partial}{\partial t} + \mathcal{P}_{a,Y_0^{\mathfrak{k}}}$ is hypoelliptic.

Proposition 5.1.3. *The following identity holds:*

$$\left[\frac{1}{2} \left| [Y^{\mathfrak{k}}, a] + [Y_0^{\mathfrak{k}}, Y^{\mathfrak{p}}] \right|^2, \nabla^H_{Y^{\mathfrak{p}}} + \nabla^V_{[a+Y_0^{\mathfrak{k}}, [a,y]]} \right] = 0. \qquad (5.1.6)$$

Proof. The term $\nabla^H_{Y^{\mathfrak{p}}}$ does not contribute to the commutator. As to the second term, its contribution will be

$$\left\langle [Y^{\mathfrak{k}}, a] + [Y_0^{\mathfrak{k}}, Y^{\mathfrak{p}}], [a, [[y,a], Y_0^{\mathfrak{k}}]] + [[a, [a,y]], Y_0^{\mathfrak{k}}] \right\rangle. \qquad (5.1.7)$$

Using Jacobi's identity and the fact that $[a, Y_0^{\mathfrak{k}}] = 0$, we get

$$[a, [[y,a], Y_0^{\mathfrak{k}}]] + [[a, [a,y]], Y_0^{\mathfrak{k}}] = [[y,a], [a, Y_0^{\mathfrak{k}}]] = 0. \qquad (5.1.8)$$

By (5.1.7), (5.1.8), we get (5.1.6). $\qquad \square$

Let Op be the set of scalar differential operators over $\mathfrak{p} \times \mathfrak{g}$. Clearly,

$$\mathcal{P}_{a,Y_0^{\mathfrak{k}}} \in \mathrm{Op} \otimes \mathrm{End}\left(\Lambda^{\cdot}(\mathfrak{z}^{\perp}(\gamma)^*)\right) \widehat{\otimes} c(\mathfrak{z}(\gamma)) \widehat{\otimes} \Lambda^{\cdot}(\mathfrak{z}^*(\gamma)). \qquad (5.1.9)$$

Let $R_{Y_0^{\mathfrak{k}}}((y, Y^{\mathfrak{g}}), (y', Y^{\mathfrak{g}'}))$ be the smooth kernel of $\exp\left(-\mathcal{P}_{a,Y_0^{\mathfrak{k}}}\right)$ with respect to the volume $dy dY^{\mathfrak{g}}$ on $\mathfrak{p} \times \mathfrak{g}$. From (5.1.9), we deduce that

$$R_{Y_0^{\mathfrak{k}}}((y, Y^{\mathfrak{g}}), (y', Y^{\mathfrak{g}'})) \in \mathrm{End}\left(\Lambda^{\cdot}(\mathfrak{z}^{\perp}(\gamma)^*)\right) \widehat{\otimes} c(\mathfrak{z}(\gamma)) \widehat{\otimes} \Lambda^{\cdot}(\mathfrak{z}^*(\gamma)).$$
$$(5.1.10)$$

If $A \in \mathrm{End}\left(\Lambda^{\cdot}\left(\mathfrak{z}^{\perp}\left(\gamma\right)^{*}\right)\right)$, let $\mathrm{Tr}_{\mathrm{s}}\left[A\right]$ be its supertrace. Let $\widehat{\mathrm{Tr}}_{\mathrm{s}}$ be the functional defined on $c\left(\mathfrak{z}\left(\gamma\right)\right)\widehat{\otimes}\Lambda^{\cdot}\left(\underline{\mathfrak{z}}^{*}\left(\gamma\right)\right)$ that vanishes on monomials of nonmaximal length, and gives the value $\left(-1\right)^{r}$ to $c\left(e_{1}\right)\underline{e}^{1}\ldots c\left(e_{r}\right)\underline{e}^{r}$. We combine these two functionals, and we get a functional $\widehat{\mathrm{Tr}}_{\mathrm{s}}$ that maps

$$\mathrm{End}\left(\Lambda^{\cdot}\left(\mathfrak{z}^{\perp}\left(\gamma\right)^{*}\right)\right)\widehat{\otimes}c\left(\mathfrak{z}\left(\gamma\right)\right)\widehat{\otimes}\Lambda^{\cdot}\left(\underline{\mathfrak{z}}^{*}\left(\gamma\right)\right)$$

into $\mathbf{R}$.

Observe that $\mathrm{Ad}\left(k^{-1}\right)$ induces an automorphism of $\Lambda^{\cdot}\left(\mathfrak{z}^{\perp}\left(\gamma\right)^{*}\right)$. Set

$$J_{\gamma}\left(Y_{0}^{\mathfrak{k}}\right)=\left(2\pi\right)^{r/2}\int_{\mathfrak{p}^{\perp}\left(\gamma\right)\times\left(\mathfrak{p}\oplus\mathfrak{k}^{\perp}\left(\gamma\right)\right)}\widehat{\mathrm{Tr}}_{\mathrm{s}}\left[\mathrm{Ad}\left(k^{-1}\right)\right.$$

$$\left. R_{Y_{0}^{\mathfrak{k}}}\left(\left(y,Y^{\mathfrak{g}}\right),\mathrm{Ad}\left(k^{-1}\right)\left(y,Y^{\mathfrak{g}}\right)\right)\right]dydY^{\mathfrak{g}}. \quad (5.1.11)$$

The main purpose of the present chapter is to compute $J_{\gamma}\left(Y_{0}^{\mathfrak{k}}\right)$.

5.2 A conjugate operator

Definition 5.2.1. Put

$$\mathcal{Q}_{a,Y_{0}^{\mathfrak{k}}}=\exp\left(\left\langle\left[\left[y,a\right],Y_{0}^{\mathfrak{k}}\right],Y^{\mathfrak{k}}\right\rangle\right)\mathcal{P}_{a,Y_{0}^{\mathfrak{k}}}\exp\left(-\left\langle\left[\left[y,a\right],Y_{0}^{\mathfrak{k}}\right],Y^{\mathfrak{k}}\right\rangle\right),$$

$$\underline{\mathcal{Q}}_{a,Y_{0}^{\mathfrak{k}}}=\exp\left(\left\langle\left[\left[y,a\right],Y_{0}^{\mathfrak{k}}\right],Y^{\mathfrak{k}}\right\rangle+\left\langle\left[a,\left[a,y\right]\right],Y^{\mathfrak{p}}\right\rangle\right)\mathcal{P}_{a,Y_{0}^{\mathfrak{k}}} \quad (5.2.1)$$

$$\exp\left(-\left\langle\left[\left[y,a\right],Y_{0}^{\mathfrak{k}}\right],Y^{\mathfrak{k}}\right\rangle-\left\langle\left[a,\left[a,y\right]\right],Y^{\mathfrak{p}}\right\rangle\right).$$

Proposition 5.2.2. *The following identities hold:*

$$\mathcal{Q}_{a,Y_{0}^{\mathfrak{k}}}=\frac{1}{2}\left(-\Delta^{\mathfrak{p}\oplus\mathfrak{k}}+\left|\left[a,Y^{\mathfrak{k}}\right]\right|^{2}+\left|\left[Y_{0}^{\mathfrak{k}},Y^{\mathfrak{p}}\right]\right|^{2}+\left|\left[\left[y,a\right],Y_{0}^{\mathfrak{k}}\right]\right|^{2}\right)$$

$$-\nabla_{Y^{\mathfrak{p}}}^{H}-\nabla_{\left[a,\left[a,y\right]\right]}^{V}+\alpha-\widehat{c}\left(\mathrm{ad}\left(a\right)\right)+c\left(\mathrm{ad}\left(a\right)+i\theta\mathrm{ad}\left(Y_{0}^{\mathfrak{k}}\right)\right), \quad (5.2.2)$$

$$\underline{\mathcal{Q}}_{a,Y_{0}^{\mathfrak{k}}}=\frac{1}{2}\left(-\Delta^{\mathfrak{p}\oplus\mathfrak{k}}+\left|\left[a,Y^{\mathfrak{k}}\right]\right|^{2}+\left|\left[Y_{0}^{\mathfrak{k}},Y^{\mathfrak{p}}\right]\right|^{2}+\left|\left[\left[y,a\right],Y_{0}^{\mathfrak{k}}\right]\right|^{2}\right)+\left|\left[a,Y^{\mathfrak{p}}\right]\right|^{2}$$

$$+\frac{1}{2}\left|\left[a,\left[a,y\right]\right]\right|^{2}-\nabla_{Y^{\mathfrak{p}}}^{H}+\alpha-\widehat{c}\left(\mathrm{ad}\left(a\right)\right)+c\left(\mathrm{ad}\left(a\right)+i\theta\mathrm{ad}\left(Y_{0}^{\mathfrak{k}}\right)\right).$$

Proof. Let $e_{1},\ldots,e_{m}$ be an orthonormal basis of $\mathfrak{p}$ and let $e_{m+1},\ldots,e_{m+n}$ be an orthonormal basis of $\mathfrak{k}$. Then

$$-\frac{1}{2}\Delta^{\mathfrak{k}}-\nabla_{\left[y,a\right],Y_{0}^{\mathfrak{k}}\right]}^{V}=-\frac{1}{2}\sum_{i=m+1}^{m+n}\left(\nabla_{e_{i}}^{V}+\left\langle\left[\left[y,a\right],Y_{0}^{\mathfrak{k}}\right],e_{i}\right\rangle\right)^{2}$$

$$+\frac{1}{2}\left|\left[\left[y,a\right],Y_{0}^{\mathfrak{k}}\right]\right|^{2}. \quad (5.2.3)$$

By (5.2.3), we get

$$\exp\left(\left\langle\left[[y,a],Y_0^{\mathfrak{k}}\right],Y^{\mathfrak{k}}\right\rangle\right)\left(-\frac{1}{2}\Delta^{\mathfrak{p}\oplus\mathfrak{k}}-\nabla_{Y^{\mathfrak{p}}}^H-\nabla_{[[y,a],Y_0^{\mathfrak{k}}]}^V\right)$$
$$\exp\left(-\left\langle\left[[y,a],Y_0^{\mathfrak{k}}\right],Y^{\mathfrak{k}}\right\rangle\right)$$
$$=-\frac{1}{2}\Delta^{\mathfrak{p}\oplus\mathfrak{k}}-\nabla_{Y^{\mathfrak{p}}}^H+\frac{1}{2}\left|\left[[y,a],Y_0^{\mathfrak{k}}\right]\right|^2+\left\langle\left[[Y^{\mathfrak{p}},a],Y_0^{\mathfrak{k}}\right],Y^{\mathfrak{k}}\right\rangle. \quad (5.2.4)$$

Moreover, using the fact that $\left[Y_0^{\mathfrak{k}},a\right]=0$, the G-invariance of B and the Jacobi identity, we get

$$\left\langle\left[[Y^{\mathfrak{p}},a],Y_0^{\mathfrak{k}}\right],Y^{\mathfrak{k}}\right\rangle=\left\langle\left[Y_0^{\mathfrak{k}},Y^{\mathfrak{p}}\right],[a,Y^{\mathfrak{k}}]\right\rangle. \quad (5.2.5)$$

By (5.1.5), (5.2.4), (5.2.5), we get the first identity in (5.2.2).

Moreover,

$$-\frac{1}{2}\Delta^{\mathfrak{p}}-\nabla_{[a,[a,y]]}^V=-\frac{1}{2}\sum_{1\leq i\leq m}\left(\nabla_{e_i}^V+\langle[a,[a,y]],e_i\rangle\right)^2+\frac{1}{2}\left|[a,[a,y]]\right|^2.$$
$$(5.2.6)$$

By (5.2.6), we get

$$\exp\left(\langle[a,[a,y]],Y^{\mathfrak{p}}\rangle\right)\left(-\frac{1}{2}\Delta^{\mathfrak{p}}-\nabla_{[a,[a,y]]}^V\right)\exp\left(-\langle[a,[a,y]],Y^{\mathfrak{p}}\rangle\right)$$
$$=-\frac{1}{2}\Delta^{\mathfrak{p}\oplus\mathfrak{k}}+\frac{1}{2}\left|[a,[a,y]]\right|^2. \quad (5.2.7)$$

Moreover,

$$\exp\left(\langle[a,[a,y]],Y^{\mathfrak{p}}\rangle\right)\nabla_{Y^{\mathfrak{p}}}^H\exp\left(-\langle[a,[a,y]],Y^{\mathfrak{p}}\rangle\right)=\nabla_{Y^{\mathfrak{p}}}^H-|[a,Y^{\mathfrak{p}}]|^2. \quad (5.2.8)$$

By the first identity in (5.2.2), by (5.2.7), and (5.2.8), we get the second identity in (5.2.2). The proof of our proposition is completed. $\square$

Remark 5.2.3. It is remarkable that in equation (5.2.2) for $\mathcal{Q}_{a,Y_0^{\mathfrak{k}}}$ and $\underline{\mathcal{Q}}_{a,Y_0^{\mathfrak{k}}}$, $(y,Y^{\mathfrak{p}})$ and $Y^{\mathfrak{k}}$ have become independent variables.

In the sequel, we will mostly use the operator $\mathcal{Q}_{a,Y_0^{\mathfrak{k}}}$, but the operator $\underline{\mathcal{Q}}_{a,Y_0^{\mathfrak{k}}}$ could be used as well.

We denote by $R'_{Y_0^{\mathfrak{k}}}\left((y,Y),(y',Y')\right)$ the smooth kernel associated with $\exp\left(-\mathcal{Q}_{a,Y_0^{\mathfrak{k}}}\right)$. Since $\mathrm{Ad}\,(k)\,a=a$, $\mathrm{Ad}\,(k)\,Y_0^{\mathfrak{k}}=Y_0^{\mathfrak{k}}$, we get

$$\left\langle\left[[\mathrm{Ad}\,(k^{-1})\,y,a],Y_0^{\mathfrak{k}}\right],\mathrm{Ad}\,(k^{-1})\,Y^{\mathfrak{k}}\right\rangle=\left\langle\left[[y,a],Y_0^{\mathfrak{k}}\right],Y^{\mathfrak{k}}\right\rangle. \quad (5.2.9)$$

It follows that in our evaluation of $J_\gamma\left(Y_0^{\mathfrak{k}}\right)$ in (5.1.11), we may and we will replace $R_{Y_0^{\mathfrak{k}}}$ by $R'_{Y_0^{\mathfrak{k}}}$.

5.3 An evaluation of certain infinite dimensional traces

Let $S_{Y_0^{\mathfrak{k}}}$ be the scalar part of $\mathcal{Q}_{a,Y_0^{\mathfrak{k}}}$, so that

$$S_{Y_0^{\mathfrak{k}}}=\frac{1}{2}\left(-\Delta^{\mathfrak{p}\oplus\mathfrak{k}}+\left|[Y^{\mathfrak{k}},a]\right|^2+\left|[Y_0^{\mathfrak{k}},Y^{\mathfrak{p}}]\right|^2+\left|[[y,a],Y_0^{\mathfrak{k}}]\right|^2\right)$$
$$-\nabla_{Y^{\mathfrak{p}}}^H-\nabla_{[a,[a,y]]}^V. \quad (5.3.1)$$

Let $T_{Y_0^{\mathfrak{k}}}\left((y, Y^{\mathfrak{g}}), (y', Y^{\mathfrak{g}'})\right)$ be the smooth kernel of $\exp\left(-S_{Y_0^{\mathfrak{k}}}\right)$ with respect to the volume $dy' dY^{\mathfrak{g}'}$ on $\mathfrak{p} \times \mathfrak{g}$.

It will also be convenient to consider the operator U acting on $\mathfrak{p} \times \mathfrak{p}$,

$$U = -\frac{1}{2}\Delta^{\mathfrak{p}} - \nabla^H_{Y^{\mathfrak{p}}} - \nabla^V_{[[y,a],a]}, \tag{5.3.2}$$

and the associated heat kernel $V\left((y, Y^{\mathfrak{p}}), (y', Y^{\mathfrak{p}'})\right)$.

Recall that $d\gamma^{-1}\varphi_1$ is given by equation (3.5.10) in Theorem 3.5.2. Then $d\gamma^{-1}\varphi_1$ preserves $\mathfrak{p}(\gamma) \oplus \mathfrak{p}(\gamma)$ and $\mathfrak{p}^\perp(\gamma) \oplus \mathfrak{p}^\perp(\gamma)$. By Theorem 3.5.2, $\mathfrak{p}(\gamma) \oplus \{0\} \subset \mathfrak{p}_0 \oplus \mathfrak{p}_0$ is the eigenspace of $d\gamma^{-1}\varphi_1$ associated with the eigenvalue 1. We denote by $d^\perp\gamma^{-1}\varphi_1$ the restriction of $d\gamma^{-1}\varphi_1$ to $\mathfrak{p}^\perp(\gamma) \oplus \mathfrak{p}^\perp(\gamma)$. Then $1 - d^\perp\gamma^{-1}\varphi_1$ is invertible.

Let $\mathfrak{p}_0^\perp(\gamma)$ be the orthogonal to $\mathfrak{p}(\gamma)$ in $\mathfrak{p}_0$. By (4.2.1), on $\mathfrak{p}_0$, $\mathrm{Ad}(\gamma)$ and $\mathrm{Ad}(k^{-1})$ coincide. By equation (3.5.10) in Theorem 3.5.2, we get

$$\det\left(1 - d^\perp\gamma^{-1}\varphi_1\right) = \det\left(1 - \mathrm{Ad}(k)\right)^2|_{\mathfrak{p}_0^\perp(\gamma)} \det\left(1 - \mathrm{Ad}(\gamma^{-1})\right)|_{\mathfrak{z}_0^\perp}. \tag{5.3.3}$$

Moreover, $\mathfrak{z}_0^\perp$ is even dimensional, the linear map $\mathrm{ad}(a)|_{\mathfrak{z}_0^\perp}$ is invertible, it exchanges $\mathfrak{p}_0^\perp$ and $\mathfrak{k}_0^\perp$, and it intertwines the action of $\mathrm{Ad}(k^{-1})$ on $\mathfrak{p}_0^\perp$ and $\mathfrak{k}_0^\perp$. Therefore (5.3.3) can be rewritten in the form

$$\det\left(1 - d^\perp\gamma^{-1}\varphi_1\right) = \det\left(1 - \mathrm{Ad}(k^{-1})\right)^2|_{\mathfrak{p}_0^\perp(\gamma)} \det\left(1 - \mathrm{Ad}(\gamma)\right)|_{\mathfrak{z}_0^\perp}. \tag{5.3.4}$$

Let $\mathfrak{k}_0^\perp(\gamma)$ be the orthogonal to $\mathfrak{k}(\gamma)$ in $\mathfrak{k}_0$. If $\mathfrak{z}_0^\perp(\gamma)$ denotes the orthogonal to $\mathfrak{z}(\gamma)$ in $\mathfrak{z}_0$, then

$$\mathfrak{z}_0^\perp(\gamma) = \mathfrak{p}_0^\perp(\gamma) \oplus \mathfrak{k}_0^\perp(\gamma). \tag{5.3.5}$$

Recall that the function $\widehat{A}(x)$ is given by

$$\widehat{A}(x) = \frac{x/2}{\sinh(x/2)}. \tag{5.3.6}$$

If V is a finite dimensional Hermitian vector space and if $B \in \mathrm{End}(V)$ is self-adjoint, note that $\frac{B/2}{\sinh(B/2)}$ is a self-adjoint positive endomorphism. Set

$$\widehat{A}(B) = \det{}^{1/2}\left[\frac{B/2}{\sinh(B/2)}\right]. \tag{5.3.7}$$

In (5.3.7), the square root is taken to be the positive square root.

Put

$$F = \mathrm{ad}^2(a), \qquad \underline{F} = \mathrm{ad}^2(a) - \mathrm{ad}^2\left(Y_0^{\mathfrak{k}}\right). \tag{5.3.8}$$

Theorem 5.3.1. *The following identity holds:*

$$\int_{\mathfrak{p}^\perp(\gamma) \times (\mathfrak{p} \oplus \mathfrak{k}^\perp(\gamma))} T_{Y_0^{\mathfrak{k}}}\left((y, Y^{\mathfrak{g}}), \mathrm{Ad}(k^{-1})(y, Y^{\mathfrak{g}})\right) dy dY^{\mathfrak{g}}$$

$$= \frac{1}{\left|\det\left(1 - \mathrm{Ad}(\gamma)\right)|_{\mathfrak{z}^\perp(\gamma)}\right|} (2\pi)^{-r/2} \widehat{A}^2\left(i\,\mathrm{ad}\left(Y_0^{\mathfrak{k}}\right)|_{\mathfrak{p}(\gamma)}\right)$$

$$\left[\det\left(e^{\sqrt{\underline{F}}/2} - e^{-\sqrt{\underline{F}}/2}\mathrm{Ad}(k^{-1})\right)|_{\mathfrak{p}^\perp(\gamma)}\right]^{-1}. \tag{5.3.9}$$

Proof. As explained in Remark 5.2.3, the operator $S_{Y_0^{\mathfrak{k}}}$ splits naturally as a sum of two commuting operators acting over $\mathfrak{p} \oplus \mathfrak{p}$ and over $\mathfrak{k}$. We will evaluate the trace of the heat kernels of the various components of $S_{Y_0^{\mathfrak{k}}}$, and take their products.

First, we evaluate the contribution of $\mathfrak{p}(\gamma) \times \mathfrak{z}(\gamma)$, while using the splitting $\mathfrak{z}(\gamma) = \mathfrak{p}(\gamma) \oplus \mathfrak{k}(\gamma)$. Let $\Delta^{\mathfrak{z}(\gamma)}$ be the Laplacian along $\mathfrak{z}(\gamma)$ with respect to its natural metric. Let $Y^{\mathfrak{p}(\gamma)}$ be the canonical section of $\mathfrak{p}(\gamma)$. Set

$$S_{Y_0^{\mathfrak{k}}}^{\mathfrak{p}(\gamma) \times \mathfrak{z}(\gamma)} = \frac{1}{2} \left| \left[Y_0^{\mathfrak{k}}, Y^{\mathfrak{p}(\gamma)} \right] \right|^2 - \frac{1}{2} \Delta^{\mathfrak{z}(\gamma)} - \nabla_{Y^{\mathfrak{p}(\gamma)}}^{H, \mathfrak{p}(\gamma)}. \tag{5.3.10}$$

Let $T_{Y_0^{\mathfrak{k}}}^{\mathfrak{p}(\gamma) \times \mathfrak{z}(\gamma)} \left((y, Y), (y', Y') \right)$ be the smooth kernel for $\exp \left(-S_{Y_0^{\mathfrak{k}}}^{\mathfrak{p}(\gamma) \times \mathfrak{z}(\gamma)} \right)$. We claim that

$$\int_{\mathfrak{p}(\gamma)} T_{Y_0^{\mathfrak{k}}}^{\mathfrak{p}(\gamma) \times \mathfrak{z}(\gamma)} \left(\left(0, Y^{\mathfrak{p}(\gamma)} \right), \left(0, Y^{\mathfrak{p}(\gamma)} \right) \right) dY^{\mathfrak{p}(\gamma)}$$

$$= (2\pi)^{-r/2} \widehat{A}^2 \left(iad \left(Y_0^{\mathfrak{k}} \right) |_{\mathfrak{p}(\gamma)} \right). \tag{5.3.11}$$

Indeed $\mathfrak{k}(\gamma)$ obviously contributes by $(2\pi)^{-q/2}$. If $\lambda \in \mathbf{R}_+$, let B_λ be the operator on $\mathbf{R}^2$,

$$B_\lambda = -\frac{1}{2} \frac{\partial^2}{\partial Y^2} + \frac{1}{2} \lambda^2 Y^2 - Y \frac{\partial}{\partial y}. \tag{5.3.12}$$

Let $\exp(-B_\lambda) ((y, Y), (y', Y'))$ be the smooth kernel associated with the operator $\exp(-B_\lambda)$. We will establish the identity

$$\int_{\mathbf{R}} \exp(-B_\lambda) ((0, Y), (0, Y)) dY = (2\pi)^{-1/2} \widehat{A}(\lambda), \tag{5.3.13}$$

from which (5.3.11) follows.

First, we will assume $\lambda > 0$. Then

$$B_\lambda = -\frac{1}{2} \frac{\partial^2}{\partial Y^2} + \frac{1}{2} \lambda^2 \left(Y - \frac{1}{\lambda^2} \frac{\partial}{\partial y} \right)^2 - \frac{1}{2\lambda^2} \frac{\partial^2}{\partial y^2}. \tag{5.3.14}$$

By making the translation $Y \to Y + \frac{1}{\lambda^2} \frac{\partial}{\partial y}$ as in [B05, section 3.10], or by using a simple Fourier transform argument in the variable y as in [BL08, Proposition 4.11.2], we find easily that B_λ can be replaced by

$$C_\lambda = \frac{\lambda}{2} \left(-\frac{\partial^2}{\partial Y^2} + Y^2 \right) - \frac{1}{2\lambda^2} \frac{\partial^2}{\partial y^2}. \tag{5.3.15}$$

Now

$$\mathrm{Tr} \left[\exp \left(-\frac{\lambda}{2} \left(-\frac{\partial^2}{\partial Y^2} + Y^2 \right) \right) \right] = \frac{1}{2 \sinh(\lambda/2)}, \tag{5.3.16}$$

from which (5.3.13) follows for $\lambda > 0$. By making $\lambda \to 0$, we also obtain (5.3.13) when $\lambda = 0$. This completes the proof of (5.3.11).

Now we study the contribution of $\mathfrak{p}^\perp(\gamma) \times \mathfrak{p}^\perp(\gamma)$ to the integral in (5.3.9). We denote by $U^{\mathfrak{p}^\perp(\gamma) \times \mathfrak{p}^\perp(\gamma)}$ the obvious analogue of the operator U in (5.3.2) that is associated with this vector space. Namely,

$$U^{\mathfrak{p}^\perp(\gamma) \times \mathfrak{p}^\perp(\gamma)} = -\frac{1}{2} \Delta^{\mathfrak{p}^\perp(\gamma)} - \nabla_{Y^{\mathfrak{p}^\perp(\gamma)}}^{H, \mathfrak{p}^\perp(\gamma)} - \nabla_{[a, [a, y]]}^{V, \mathfrak{p}^\perp(\gamma)}. \tag{5.3.17}$$

Let $V^{\mathfrak{p}^\perp(\gamma)\times\mathfrak{p}^\perp(\gamma)}\left((y,Y),(y',Y')\right)$ be the smooth kernel associated with the operator $\exp\left(-U^{\mathfrak{p}^\perp(\gamma)\oplus\mathfrak{p}^\perp(\gamma)}\right)$.

Let $t\in[0,1]\to w_t^{\mathfrak{p}^\perp(\gamma)}\in\mathfrak{p}^\perp(\gamma)$ be a standard Brownian motion with $w_0^{\mathfrak{p}^\perp(\gamma)}=0$. Consider the differential equation

$$\dot{y}=Y^{\mathfrak{p}^\perp(\gamma)},\qquad \dot{Y}^{\mathfrak{p}^\perp(\gamma)}=[a,[a,y]]+\dot{w}^{\mathfrak{p}^\perp(\gamma)},\qquad (5.3.18)$$

with given $\left(y_0,Y_0^{\mathfrak{p}^\perp(\gamma)}\right)\in\mathfrak{p}^\perp(\gamma)\times\mathfrak{p}^\perp(\gamma)$. By (5.3.18), we get

$$\ddot{y}-\mathrm{ad}^2(a)\,y=\dot{w}^{\mathfrak{p}^\perp(\gamma)}.\qquad (5.3.19)$$

Let h be a bounded measurable function defined on $\mathfrak{p}^\perp(\gamma)\times\mathfrak{p}^\perp(\gamma)$ with values in $\mathbf{R}$. If E denotes the expectation with respect to the Brownian motion $w^{\mathfrak{p}^\perp(\gamma)}$, by Itô's formula, we get

$$\exp\left(-U_{Y_0^t}^{\mathfrak{p}^\perp(\gamma)\times\mathfrak{p}^\perp(\gamma)}\right)h\left(y_0,Y_0^{\mathfrak{p}^\perp}\right)=E\left[h\left(y_1,Y_1^{\mathfrak{p}^\perp(\gamma)}\right)\right].\qquad (5.3.20)$$

Set

$$Z.=\left(y.,Y_.^{\mathfrak{p}^\perp(\gamma)}\right).\qquad (5.3.21)$$

Let M_a be the endomorphism of $\mathfrak{p}\times\mathfrak{p}$, which is given in matrix form by

$$M_a=\begin{bmatrix}0&1\\\mathrm{ad}^2(a)&0\end{bmatrix}.\qquad (5.3.22)$$

Of course M_a preserves $\mathfrak{p}^\perp(\gamma)\times\mathfrak{p}^\perp(\gamma)$ and commutes with $\mathrm{Ad}(k)$.

Here $w^{\mathfrak{p}^\perp(\gamma)}$ will be considered as taking its values in $\{0\}\times\mathfrak{p}^\perp(\gamma)$. Put $Z_0=\left(y_0,Y_0^{\mathfrak{p}^\perp}\right)$. By (5.3.18), we get

$$\dot{Z}=M_aZ+\dot{w}^{\mathfrak{p}^\perp(\gamma)},\qquad Z|_{t=0}=Z_0.\qquad (5.3.23)$$

Then we can rewrite equation (5.3.20) in the form

$$\exp\left(-U_{Y_0^t}^{\mathfrak{p}^\perp(\gamma)\times\mathfrak{p}^\perp(\gamma)}\right)h(Z_0)=E[h(Z_1)].\qquad (5.3.24)$$

In particular $V^{\mathfrak{p}^\perp(\gamma)\times\mathfrak{p}^\perp(\gamma)}(Z_0,Z_0')\,dZ_0'$ is the probability law of Z_1'. Let $R_{Z_0,Z_0'}$ be the probability law of the process $Z.$ in (5.3.23) conditional on $Z_1=Z_0'$.

Let $J^{\mathfrak{p}^\perp(\gamma)}$ be the antisymmetric operator $\frac{d}{dt}$ acting on smooth functions on $[0,1]$ with values in $\mathfrak{p}^\perp(\gamma)$, with the boundary condition $y_1=\mathrm{Ad}(k^{-1})\,y_0$. The corresponding operator $J^{\mathfrak{p}^\perp(\gamma),2}$ is symmetric and nonpositive. The crucial fact is that the operator $J^{\mathfrak{p}^\perp(\gamma),2}-\mathrm{ad}^2(a)|_{\mathfrak{p}^\perp(\gamma)}$ is symmetric and negative. Among the solutions of (5.3.19) there is the solution

$$\underline{y}=\left(J^{\mathfrak{p}^\perp(\gamma),2}-\mathrm{ad}^2(a)|_{\mathfrak{p}^\perp(\gamma)}\right)^{-1}\dot{w}^{\mathfrak{p}^\perp(\gamma)},\qquad (5.3.25)$$

which is the unique solution of (5.3.18) such that

$$\underline{y}_1=\mathrm{Ad}(k^{-1})\,\underline{y}_0,\qquad \underline{\dot{y}}_1=\mathrm{Ad}(k^{-1})\,\underline{\dot{y}}_0.\qquad (5.3.26)$$

Then $\underline{y}$ is a Gaussian process with covariance $\left(-J^{\mathfrak{p}^{\perp}(\gamma),2} + \mathrm{ad}^2(a)\right)^{-2}$.

Note that if $\underline{Y}^{\mathfrak{p}^{\perp}(\gamma)} = \dot{y}$, (5.3.26) is equivalent to

$$\left(\underline{y}_1, \underline{Y}_1^{\mathfrak{p}^{\perp}(\gamma)}\right) = \mathrm{Ad}\left(k^{-1}\right)\left(\underline{y}_0, \underline{Y}_0^{\mathfrak{p}^{\perp}(\gamma)}\right). \tag{5.3.27}$$

Let $J^{\mathfrak{p}^{\perp}(\gamma) \times \mathfrak{p}^{\perp}(\gamma)}$ be the antisymmetric operator $\frac{d}{dt}$ acting on smooth functions $Z = \left(y, Y^{\mathfrak{p}^{\perp}(\gamma)}\right)$ on $[0,1]$ with values in $\mathfrak{p}^{\perp}(\gamma) \times \mathfrak{p}^{\perp}(\gamma)$, and the boundary conditions $Z_1 = \mathrm{Ad}\left(k^{-1}\right) Z_0$. Then the operator $J^{\mathfrak{p}^{\perp}(\gamma) \times \mathfrak{p}^{\perp}(\gamma)} - M_a$ is invertible. Moreover, $\underline{Z} = (\underline{y}, \dot{\underline{y}})$ is given by

$$\underline{Z} = \left(J^{\mathfrak{p}^{\perp}(\gamma) \times \mathfrak{p}^{\perp}(\gamma)} - M_a\right)^{-1} \dot{w}^{\mathfrak{p}^{\perp}(\gamma)}. \tag{5.3.28}$$

Let Q be the probability law of $\underline{Z}$ on the set $C_k\left([0,1], \mathfrak{p}^{\perp}(\gamma) \oplus \mathfrak{p}^{\perp}(\gamma)\right)$ of continuous functions $\underline{Z}_t$ from $[0,1]$ into $\mathfrak{p}^{\perp}(\gamma) \oplus \mathfrak{p}^{\perp}(\gamma)$ such that $\underline{Z}_1 = \mathrm{Ad}\left(k^{-1}\right) \underline{Z}_0$. It is the probability law of a Gaussian process with covariance C given by

$$C = \left(J^{\mathfrak{p}^{\perp}(\gamma) \times \mathfrak{p}^{\perp}(\gamma)} - M_a\right)^{-1} \left(-J^{\mathfrak{p}^{\perp}(\gamma) \times \mathfrak{p}^{\perp}(\gamma)} - M_a^*\right)^{-1}. \tag{5.3.29}$$

By stationarity, the probability law of $\underline{Z}_t$ on $\mathfrak{p}^{\perp}(\gamma) \oplus \mathfrak{p}^{\perp}(\gamma)$ does not depend on $t \in S^1$. Also it is a centered Gaussian. We claim that its covariance is nondegenerate. Indeed this is an easy consequence of the fact that the covariance C in (5.3.29) is invertible. A more direct proof is to show that the covariance of $\underline{Z}_t$ is given by

$$\left(\mathrm{Ad}\left(k^{-1}\right) - e^{M_a}\right)^{-1} \int_0^1 e^{sM_a} e^{sM_a^*} ds \left(\mathrm{Ad}(k) - e^{M_a^*}\right)^{-1}, \tag{5.3.30}$$

which is obviously invertible. Equation (5.3.30) follows easily from (5.3.28).

We conclude that the probability law of $\underline{Z}_0$ is a centered Gaussian of the form $r(\underline{Z}) d\underline{Z}$.

If $Z_. = \left(y_., Y_.^{\mathfrak{p}^{\perp}(\gamma)}\right)$ is the solution of (5.3.23) with $Z_0 = \left(y_0, Y_0^{\mathfrak{p}^{\perp}(\gamma)}\right) \in \mathfrak{p}^{\perp}(\gamma) \times \mathfrak{p}^{\perp}(\gamma)$, we get

$$Z_t = e^{tM_a}\left(Z_0 - \underline{Z}_0\right) + \underline{Z}_t. \tag{5.3.31}$$

By (3.5.12) and (3.5.14),

$$d\gamma^{-1}\varphi_1 = \mathrm{Ad}(k) e^{M_a} = e^{M_a} \mathrm{Ad}(k). \tag{5.3.32}$$

Since $\underline{Z}_1 = \mathrm{Ad}\left(k^{-1}\right) \underline{Z}_0$, by (5.3.31), (5.3.32), we get

$$Z_1 = e^{M_a} Z_0 + \mathrm{Ad}\left(k^{-1}\right)\left(1 - d^{\perp}\gamma^{-1}\varphi_1\right) \underline{Z}_0. \tag{5.3.33}$$

Put

$$u_{Z_0}(Z') = \mathrm{Ad}(k)\left(1 - d^{\perp}\gamma^{-1}\varphi_1\right)^{-1}\left(Z' - e^{M_a} Z_0\right). \tag{5.3.34}$$

For $Z \in \mathfrak{p}^{\perp}(\gamma) \oplus \mathfrak{p}^{\perp}(\gamma)$, let Q_Z be the probability law of $\underline{Z}_.$ conditional on $\underline{Z}_0 = Z$. Let f be a bounded measurable function defined on

$C_k\left([0,1],\mathfrak{p}^\perp(\gamma)\times\mathfrak{p}^\perp(\gamma)\right)$ with values in $\mathbf{R}$, let f' be a bounded measurable function defined on $\mathfrak{p}^\perp(\gamma)\times\mathfrak{p}^\perp(\gamma)$ with values in $\mathbf{R}$. By (5.3.33), we obtain

$$\int_{C_k([0,1],\mathfrak{p}^\perp(\gamma)\oplus\mathfrak{p}^\perp(\gamma))} f\left(\underline{Z}\right) f'\left(Z_1\right) dQ\left(\underline{Z}\right)$$

$$= \int_{p^\perp(\gamma)\oplus\mathfrak{p}^\perp(\gamma)} E^{Q_{u_{Z_0}}(Z')}\left[f\left(\underline{Z}.\right)\right] f'\left(Z'\right) r\left(u_{Z_0}\left(Z'\right)\right) \frac{dZ'}{\left|\det\left(1-d^\perp\gamma^{-1}\varphi_1\right)\right|}. \tag{5.3.35}$$

By (5.3.35), the probability law of Z_1 is just $r\left(u_{Z_0}\left(Z'\right)\right)\frac{dZ'}{\left|\det(1-d^\perp\gamma^{-1}\varphi_1)\right|}$.

Now observe that by (5.3.32)–(5.3.34),

$$u_{Z_0}\left(\mathrm{Ad}\left(k^{-1}\right)Z_0\right) = Z_0. \tag{5.3.36}$$

By (5.3.35), (5.3.36), we get

$$V^{\mathfrak{p}^\perp(\gamma)\times\mathfrak{p}^\perp(\gamma)}\left(Z_0,\mathrm{Ad}\left(k^{-1}\right)Z_0\right) = \frac{r\left(Z_0\right)}{\left|\det\left(1-d^\perp\gamma^{-1}\varphi_1\right)\right|}. \tag{5.3.37}$$

Moreover, if we make $Z_1 = \mathrm{Ad}\left(k^{-1}\right)Z_0$ in (5.3.33), we get $Z_0 = \underline{Z}_0$, and in (5.3.31), $Z. = \underline{Z}.$. It follows from the above that we have the identity of positive measures on $C_k\left([0,1],\mathfrak{p}^\perp(\gamma)\oplus\mathfrak{p}^\perp(\gamma)\right)$,

$$R_{Z_0,\mathrm{Ad}(k^{-1})Z_0}V^{\mathfrak{p}^\perp(\gamma)\times\mathfrak{p}^\perp(\gamma)}\left(Z_0,\mathrm{Ad}\left(k^{-1}\right)Z_0\right)dZ_0 = \frac{Q}{\left|\det\left(1-d^\perp\gamma^{-1}\varphi_1\right)\right|}. \tag{5.3.38}$$

Let $S_{Y_0^{\mathfrak{e}}}^{\mathfrak{p}^\perp(\gamma)\times\mathfrak{p}^\perp(\gamma)}$ be the analogue of $S_{Y_0^{\mathfrak{e}}}$ for $\mathfrak{p}^\perp(\gamma)\times\mathfrak{p}^\perp(\gamma)$, i.e.,

$$S_{Y_0^{\mathfrak{e}}}^{\mathfrak{p}^\perp(\gamma)\times\mathfrak{p}^\perp(\gamma)} = \frac{1}{2}\left(-\Delta^{V,\mathfrak{p}^\perp(\gamma)} + \left|\left[Y_0^{\mathfrak{e}},Y^{\mathfrak{p}^\perp(\gamma)}\right]\right|^2 + \left|\left[\left[y,a\right],Y_0^{\mathfrak{e}}\right]\right|^2\right)$$
$$- \nabla_{Y^{\mathfrak{p}^\perp(\gamma)}}^{H,\mathfrak{p}^\perp(\gamma)} - \nabla_{[a,[a,y]]}^{V,\mathfrak{p}^\perp(\gamma)}. \tag{5.3.39}$$

Let $T_{Y_0^{\mathfrak{e}}}^{\mathfrak{p}^\perp(\gamma)\times\mathfrak{p}^\perp(\gamma)}\left(\left(y,Y^{\mathfrak{p}^\perp(\gamma)}\right),\left(y',Y^{\mathfrak{p}^\perp(\gamma)'}\right)\right)$ be the heat kernel associated with $\exp\left(-S_{Y_0^{\mathfrak{e}}}^{\mathfrak{p}^\perp(\gamma)\times\mathfrak{p}^\perp(\gamma)}\right)$. Using (5.3.17), (5.3.39) and the Feynman-Kac formula, we get

$$T_{Y_0^{\mathfrak{e}}}^{\mathfrak{p}^\perp(\gamma)\times\mathfrak{p}^\perp(\gamma)}\left(Z_0,\mathrm{Ad}\left(k^{-1}\right)Z_0\right) = E^{R_{Z_0,\mathrm{Ad}(k^{-1})Z_0}}$$
$$\left[\exp\left(-\frac{1}{2}\int_0^1\left(\left|\left[Y_0^{\mathfrak{e}},Y^{\mathfrak{p}^\perp(\gamma)}\right]\right|^2 + \left|\left[[y,a],Y_0^{\mathfrak{e}}\right]\right|^2\right)ds\right)\right]$$
$$V^{\mathfrak{p}^\perp(\gamma)\times\mathfrak{p}^\perp(\gamma)}\left(Z_0,\mathrm{Ad}\left(k^{-1}\right)Z_0\right). \tag{5.3.40}$$

By combining (5.3.38) and (5.3.40), we obtain

$$\int_{\mathfrak{p}^\perp(\gamma)\times\mathfrak{p}^\perp(\gamma)} T^{\mathfrak{p}^\perp(\gamma)\times\mathfrak{p}^\perp(\gamma)}\left(Z_0,\mathrm{Ad}\left(k^{-1}\right)Z_0\right)dZ_0$$

$$= \frac{1}{\left|\det\left(1-d^\perp\gamma^{-1}\varphi_1\right)\right|}\int_{C_k(\mathfrak{p}^\perp(\gamma)\times\mathfrak{p}^\perp(\gamma))}$$
$$E^Q\left[\exp\left(-\frac{1}{2}\int_0^1\left(\left|\left[Y_0^{\mathfrak{e}},Y^{\mathfrak{p}^\perp(\gamma)}\right]\right|^2 + \left|\left[[y,a],Y_0^{\mathfrak{e}}\right]\right|^2\right)ds\right)\right]. \tag{5.3.41}$$

Since Q is the probability law associated with the Gaussian process y. with covariance $\left(-J^{\mathfrak{p}^\perp(\gamma),2} + \mathrm{ad}^2(a)\big|_{\mathfrak{p}^\perp(\gamma)}\right)^{-2}$, we deduce from (5.3.41) the identity

$$\int_{\mathfrak{p}^\perp(\gamma)\times\mathfrak{p}^\perp(\gamma)} T^{\mathfrak{p}^\perp(\gamma)\times\mathfrak{p}^\perp(\gamma)}\left(Z_0, \mathrm{Ad}\left(k^{-1}\right)Z_0\right)dZ_0 = \frac{1}{\left|\det\left(1 - d^\perp\gamma^{-1}\varphi_1\right)\right|}$$

$$\frac{\det^{1/2}\left[\left(-J^{\mathfrak{p}^\perp(\gamma),2} + \mathrm{ad}^2(a)\right)^2\right]}{\det^{1/2}\left[\left(-J^{\mathfrak{p}^\perp(\gamma),2} + \mathrm{ad}^2(a)\right)\left(-J^{\mathfrak{p}^\perp(\gamma),2} + \mathrm{ad}^2(a) - \mathrm{ad}^2\left(Y_0^{\mathfrak{k}}\right)\right)\right]}. \quad (5.3.42)$$

The notation in (5.3.42) should be self-explanatory. Indeed each of the determinants in the right-hand side of (5.3.42) is given by a diverging infinite product of eigenvalues. However, the eigenvalues, which can be explicitly calculated, are of the order $4k^2\pi^2, k \in \mathbf{Z}$. It is then a trivial matter that the ratio of the infinite products in (5.3.42) can be made to converge. Similar considerations will apply later.

Using the notation in (5.3.8), we can rewrite (5.3.42) in the form

$$\int_{\mathfrak{p}^\perp(\gamma)\times\mathfrak{p}^\perp(\gamma)} T^{\mathfrak{p}^\perp(\gamma)\times\mathfrak{p}^\perp(\gamma)}\left(Z_0, \mathrm{Ad}\left(k^{-1}\right)Z_0\right)dZ_0$$

$$= \frac{1}{\left|\det\left(1 - d^\perp\gamma^{-1}\varphi_1\right)\right|}\frac{\det^{1/2}\left[-J^{\mathfrak{p}^\perp(\gamma),2} + F\big|_{\mathfrak{p}^\perp(\gamma)}\right]}{\det^{1/2}\left[-J^{\mathfrak{p}^\perp(\gamma),2} + \underline{F}\big|_{\mathfrak{p}^\perp(\gamma)}\right]}. \quad (5.3.43)$$

We will give a more explicit formula for (5.3.43). By [BL08, eq. (7.8.19)], if no eigenvalue of $\mathrm{Ad}(k)$ on $\mathfrak{p}^\perp(\gamma)$ is equal to 1, the operator $-J^{\mathfrak{p}^\perp(\gamma),2}$ is invertible, and we have the identity

$$\det\left(e^{\sqrt{F}/2} - e^{-\sqrt{F}/2}\mathrm{Ad}\left(k^{-1}\right)\right)\big|_{\mathfrak{p}^\perp(\gamma)} = \det\left(1 - \mathrm{Ad}\left(k^{-1}\right)\right)\big|_{\mathfrak{p}^\perp(\gamma)}$$

$$\det{}^{1/2}\left(1 - J^{\mathfrak{p}^\perp(\gamma),-2}F\big|_{\mathfrak{p}^\perp(\gamma)}\right). \quad (5.3.44)$$

By (5.3.44), if no eigenvalue of $\mathrm{Ad}(k)$ on $\mathfrak{p}^\perp(\gamma)$ is equal to 1,

$$\frac{\det^{1/2}\left[-J^{\mathfrak{p}^\perp(\gamma),2} + F\big|_{\mathfrak{p}^\perp(\gamma)}\right]}{\det^{1/2}\left[-J^{\mathfrak{p}^\perp(\gamma),2} + \underline{F}\big|_{\mathfrak{p}^\perp(\gamma)}\right]} = \frac{\det\left(e^{\sqrt{F}/2} - e^{-\sqrt{F}/2}\mathrm{Ad}\left(k^{-1}\right)\right)\big|_{\mathfrak{p}^\perp(\gamma)}}{\det\left(e^{\sqrt{\underline{F}}/2} - e^{-\sqrt{\underline{F}}/2}\mathrm{Ad}\left(k^{-1}\right)\right)\big|_{\mathfrak{p}^\perp(\gamma)}}. \quad (5.3.45)$$

If $\mathrm{Ad}(k)$ is equal to the identity, the operator $-J^{\mathfrak{p}^\perp(\gamma),2}$ is no longer invertible, and the constants span its kernel. Still $-J^{\mathfrak{p}^\perp(\gamma),2}$ is invertible on the vector space of the L_2 functions that are orthogonal to the constants. We denote by $\det^*$ the determinant of operators acting precisely on this vector space. By [BL08, equation (7.8.2)], we have the identity

$$\det\left(e^{\sqrt{F}/2} - e^{-\sqrt{F}/2}\right)\big|_{\mathfrak{p}^\perp(\gamma)} = \det{}^{1/2}\left[F\right]\big|_{\mathfrak{p}^\perp(\gamma)}\det{}^{*1/2}\left[1 - FJ^{\mathfrak{p}^\perp(\gamma),-2}\right]. \quad (5.3.46)$$

By (5.3.46), we still get (5.3.45) when $\mathrm{Ad}(k) = 1$. Therefore (5.3.45) holds in full generality.

By (5.3.43) and (5.3.45), we get

$$
\int_{\mathfrak{p}^\perp(\gamma) \times \mathfrak{p}^\perp(\gamma)} T^{\mathfrak{p}^\perp(\gamma) \times \mathfrak{p}^\perp(\gamma)} \left(Z_0, \mathrm{Ad}\left(k^{-1}\right) Z_0 \right) dZ_0
$$

$$
= \frac{1}{\left| \det\left(1 - d^\perp \gamma^{-1} \varphi_1\right) \right|} \frac{\det\left(e^{\sqrt{F}/2} - e^{-\sqrt{F}/2} \mathrm{Ad}\left(k^{-1}\right) \right) |_{\mathfrak{p}^\perp(\gamma)}}{\det\left(e^{\sqrt{E}/2} - e^{-\sqrt{E}/2} \mathrm{Ad}\left(k^{-1}\right) \right) |_{\mathfrak{p}^\perp(\gamma)}}. \tag{5.3.47}
$$

Now we will evaluate the contribution of $\mathfrak{k}^\perp(\gamma)$ to the trace in (5.3.9). The relevant operator $S^{\mathfrak{k}^\perp(\gamma)}$ acting on $\mathfrak{k}^\perp(\gamma)$ is given by

$$
S^{\mathfrak{k}^\perp(\gamma)} = \frac{1}{2}\left(-\Delta^{\mathfrak{k}^\perp(\gamma)} + \left| \left[Y^{\mathfrak{k}^\perp(\gamma)}, a \right] \right|^2 \right). \tag{5.3.48}
$$

Let $T^{\mathfrak{k}^\perp(\gamma)}\left(Y^{\mathfrak{k}^\perp(\gamma)}, Y^{\mathfrak{k}^\perp(\gamma)\prime} \right)$ be the smooth heat kernel for $\exp\left(-S^{\mathfrak{k}^\perp(\gamma)} \right)$. By proceeding as in (5.3.16), one finds easily that

$$
\int_{\mathfrak{k}^\perp(\gamma)} T^{\mathfrak{k}^\perp(\gamma)} \left(Y^{\mathfrak{k}^\perp(\gamma)}, \mathrm{Ad}\left(k^{-1}\right) Y^{\mathfrak{k}^\perp(\gamma)} \right) dY^{\mathfrak{k}^\perp(\gamma)}
$$

$$
= \frac{1}{\det\left(e^{\sqrt{F}/2} - e^{-\sqrt{F}/2} \mathrm{Ad}\left(k^{-1}\right) \right) |_{\mathfrak{k}^\perp(\gamma)}}. \tag{5.3.49}
$$

By (5.3.4), (5.3.11), (5.3.47), (5.3.49), we obtain

$$
\int_{\mathfrak{p}^\perp(\gamma) \times (\mathfrak{p} \oplus \mathfrak{k}^\perp(\gamma))} T_{Y_0^{\mathfrak{k}}} \left((y, Y^{\mathfrak{g}}), \mathrm{Ad}\left(k^{-1}\right)(y, Y^{\mathfrak{g}}) \right) dy dY^{\mathfrak{g}}
$$

$$
= (2\pi)^{-r/2} \widehat{A}^2 \left(i \mathrm{ad}\left(Y_0^{\mathfrak{k}} \right) |_{\mathfrak{p}(\gamma)} \right)
$$

$$
\frac{1}{\det\left(1 - \mathrm{Ad}\left(k^{-1}\right)\right)^2 |_{\mathfrak{p}_0^\perp(\gamma)} \left| \det\left(1 - \mathrm{Ad}(\gamma)\right) |_{\mathfrak{z}_0^\perp} \right|}
$$

$$
\frac{\det\left(e^{\sqrt{F}/2} - e^{-\sqrt{F}/2} \mathrm{Ad}\left(k^{-1}\right) \right) |_{\mathfrak{p}^\perp(\gamma)}}{\det\left(e^{\sqrt{E}/2} - e^{-\sqrt{E}/2} \mathrm{Ad}\left(k^{-1}\right) \right) |_{\mathfrak{p}^\perp(\gamma)}}
$$

$$
\frac{1}{\det\left(e^{\sqrt{F}/2} - e^{-\sqrt{F}/2} \mathrm{Ad}\left(k^{-1}\right) \right) |_{\mathfrak{k}^\perp(\gamma)}}. \tag{5.3.50}
$$

Since $\mathrm{ad}(a)$ is an invertible map from $\mathfrak{z}_0^\perp$ into itself that exchanges $\mathfrak{p}_0^\perp$ and $\mathfrak{k}_0^\perp$ and commutes with $\mathrm{Ad}\left(k^{-1}\right)$, we get

$$
\frac{\det\left(e^{\sqrt{F}/2} - e^{-\sqrt{F}/2} \mathrm{Ad}\left(k^{-1}\right) \right) |_{\mathfrak{p}^\perp(\gamma)}}{\det\left(e^{\sqrt{F}/2} - e^{-\sqrt{F}/2} \mathrm{Ad}\left(k^{-1}\right) \right) |_{\mathfrak{k}^\perp(\gamma)}} = \frac{\det\left(1 - \mathrm{Ad}\left(k^{-1}\right)\right) |_{\mathfrak{p}_0^\perp(\gamma)}}{\det\left(1 - \mathrm{Ad}\left(k^{-1}\right)\right) |_{\mathfrak{k}_0^\perp(\gamma)}}. \tag{5.3.51}
$$

Moreover, we have the trivial

$$\det\left(1 - \operatorname{Ad}\left(k^{-1}\right)\right)\big|_{\mathfrak{p}_0^\perp(\gamma)} \det\left(1 - \operatorname{Ad}\left(k^{-1}\right)\right)\big|_{\mathfrak{k}_0^\perp(\gamma)} \det\left(1 - \operatorname{Ad}(\gamma)\right)\big|_{\mathfrak{z}_0^\perp}$$

$$= \det\left(1 - \operatorname{Ad}(\gamma)\right)\big|_{\mathfrak{z}^\perp(\gamma)}. \quad (5.3.52)$$

By (5.3.50)–(5.3.52), we get (5.3.9). The proof of our theorem is completed.

□

5.4 Some formulas of linear algebra

Proposition 5.4.1. *The following identities hold:*

$$\det\left(e^{\sqrt{F}/2} - e^{-\sqrt{F}/2}\operatorname{Ad}\left(k^{-1}\right)\right)\big|_{\mathfrak{p}_0^\perp(\gamma)} > 0,$$

$$\det\left(e^{\sqrt{F}/2} - e^{-\sqrt{F}/2}\operatorname{Ad}\left(k^{-1}\right)\right)\big|_{\mathfrak{p}_0^\perp} > 0, \quad (5.4.1)$$

$$\det\left(e^{\sqrt{F}/2} - e^{-\sqrt{F}/2}\operatorname{Ad}\left(k^{-1}\right)\right)\big|_{\mathfrak{z}_0^\perp}$$

$$= \left[\det\left(e^{\sqrt{F}/2} - e^{-\sqrt{F}/2}\operatorname{Ad}\left(k^{-1}\right)\right)\big|_{\mathfrak{p}_0^\perp}\right]^2.$$

Moreover,

$$\det\left(1 - \exp\left(-\operatorname{ad}(a) - i\theta\operatorname{ad}\left(Y_0^\mathfrak{k}\right)\right)\operatorname{Ad}\left(k^{-1}\right)\right)\big|_{\mathfrak{z}_0^\perp}$$

$$= (-1)^{\dim \mathfrak{p}_0^\perp}\det\left(\operatorname{Ad}(k)\right)\big|_{\mathfrak{p}_0^\perp}\left[\det\left(e^{\sqrt{F}/2} - e^{-\sqrt{F}/2}\operatorname{Ad}\left(k^{-1}\right)\right)\big|_{\mathfrak{p}_0^\perp}\right]^2,$$

$$\left|\det\left(1 - \exp\left(-i\theta\operatorname{ad}\left(Y_0^\mathfrak{k}\right)\right)\operatorname{Ad}\left(k^{-1}\right)\right)\big|_{\mathfrak{z}_0^\perp(\gamma)}\right| \quad (5.4.2)$$

$$= \det\left(e^{\sqrt{F}/2} - e^{-\sqrt{F}/2}\operatorname{Ad}\left(k^{-1}\right)\right)\big|_{\mathfrak{z}_0^\perp(\gamma)}.$$

In particular,

$$\det\left(1 - \operatorname{Ad}(\gamma)\right)\big|_{\mathfrak{z}_0^\perp} = (-1)^{\dim \mathfrak{p}_0^\perp}\det\left(\operatorname{Ad}(k)\right)\big|_{\mathfrak{p}_0^\perp}$$

$$\left[\det\left(e^{\sqrt{F}/2} - e^{-\sqrt{F}/2}\operatorname{Ad}\left(k^{-1}\right)\right)\big|_{\mathfrak{p}_0^\perp}\right]^2. \quad (5.4.3)$$

Proof. Since $\underline{F}$ commutes with $\operatorname{Ad}(k)$ and is nonnegative on $\mathfrak{p}_0^\perp(\gamma)$, to establish the first identity in (5.4.1), it is enough to consider the case where $\underline{F}$ is a nonnegative constant, and $\operatorname{Ad}(k)$ is an isometry with no eigenvalue equal to 1. If $\operatorname{Ad}(k)$ is equal to -1, the first identity in (5.4.1) is obvious. If $\operatorname{Ad}(k)$ is a non trivial rotation in dimension 2, the first identity in (5.4.1) is also obvious, so that we get the first inequality in (5.4.1) in full generality. To establish the second inequality in (5.4.1), we may now assume that $\underline{F}$ is a positive constant. If $\operatorname{Ad}(k) = 1$, the inequality is obvious, and the other cases for $\operatorname{Ad}(k)$ have already been covered. Since $\operatorname{ad}(a)$ is one to one from $\mathfrak{p}_0^\perp$ into $\mathfrak{k}_0^\perp$, the third identity in (5.4.1) is obvious.

Now we establish (5.4.2). First, note that since $\operatorname{ad}(a)$ commutes with $\operatorname{ad}\left(Y_0^\mathfrak{k}\right)$ and anticommutes with θ, it anticommutes with $\theta\operatorname{ad}\left(Y_0^\mathfrak{k}\right)$, so that

$$\left(\operatorname{ad}(a) + i\theta\operatorname{ad}\left(Y_0^\mathfrak{k}\right)\right)^2 = \operatorname{ad}^2(a) - \operatorname{ad}^2\left(Y_0^\mathfrak{k}\right) = \underline{F}. \quad (5.4.4)$$

Clearly, $\mathrm{ad}\,(a)+i\theta\mathrm{ad}\,\left(Y_0^{\mathfrak{k}}\right)$ commutes with $\underline{F}$. Since $\mathrm{ad}\,(a)$ induces an isomorphism from $\mathfrak{p}_0^{\perp}$ into $\mathfrak{k}_0^{\perp}$, it follows that the restriction of $\mathrm{ad}\,(a)+i\theta\mathrm{ad}\,\left(Y_0^{\mathfrak{k}}\right)$ to $\mathfrak{z}_0^{\perp}$ is diagonalizable, and its eigenvalues coincide with the eigenvalues of $\pm\sqrt{\underline{F}}$. The first equality in (5.4.2) follows. On $\mathfrak{z}_0^{\perp}\,(\gamma)$, $\mathrm{ad}\,(a)$ vanishes, so that the above argument cannot be used any more. However, by splitting $\mathfrak{z}_0^{\perp}\,(\gamma)\otimes_{\mathbf{R}}\mathbf{C}$ according to the eigenvalues of $\mathrm{ad}\,\left(Y_0^{\mathfrak{k}}\right)$, we obtain easily the second identity in (5.4.2).

If $A\in\mathrm{End}\,\left(\mathfrak{z}_0^{\perp}\right)$, let A^* be its adjoint with respect to the scalar product of $\mathfrak{z}_0^{\perp}$. Since $\gamma=e^a k^{-1}$, we find that

$$\mathrm{Ad}\,(\gamma)\,|_{\mathfrak{g}}^{*-1}=\mathrm{Ad}\,\left(e^{-a}k^{-1}\right)\,|_{\mathfrak{g}}. \tag{5.4.5}$$

By the first identity in (5.4.2) and by (5.4.5), we get

$$\det\left(1-\mathrm{Ad}\,(\gamma)^{-1}\right)\,|_{\mathfrak{z}_0^{\perp}}$$
$$=(-1)^{\dim\,\mathfrak{p}_0^{\perp}}\det\,(\mathrm{Ad}\,(k))\,|_{\mathfrak{p}_0^{\perp}}\left[\det\left(e^{\sqrt{F}/2}-e^{-\sqrt{F}/2}\mathrm{Ad}\,\left(k^{-1}\right)\right)\,|_{\mathfrak{p}_0^{\perp}}\right]^2. \tag{5.4.6}$$

Moreover, the right-hand side of (5.4.6) is unchanged when replacing a by $-a$ and k by k^{-1}, which allows us to replace γ^{-1} by γ, so that we get (5.4.3). The proof of our proposition is completed. $\qquad\square$

Remark 5.4.2. Since when $Y_0^{\mathfrak{k}}=0$, $F=\underline{F}$, the results of Proposition 5.4.1 also hold when making $Y_0^{\mathfrak{k}}=0$ and when replacing $\underline{F}$ by F.

By (5.4.2), we get

$$\det\left(1-\exp\left(-\mathrm{ad}\,(a)-i\theta\mathrm{ad}\,\left(Y_0^{\mathfrak{k}}\right)\right)\mathrm{Ad}\,\left(k^{-1}\right)\right)\,|_{\mathfrak{z}_0^{\perp}}$$
$$\det\left(1-\exp\left(-\mathrm{ad}\,(a)\right)\mathrm{Ad}\,\left(k^{-1}\right)\right)\,|_{\mathfrak{z}_0^{\perp}}$$
$$=\left[\det\left(e^{\sqrt{\underline{F}}/2}-e^{-\sqrt{\underline{F}}/2}\mathrm{Ad}\,\left(k^{-1}\right)\right)\,|_{\mathfrak{p}_0^{\perp}}\right.$$
$$\left.\det\left(e^{\sqrt{F}/2}-e^{-\sqrt{F}/2}\mathrm{Ad}\,\left(k^{-1}\right)\right)\,|_{\mathfrak{p}_0^{\perp}}\right]^2. \tag{5.4.7}$$

Therefore the left-hand side of (5.4.7) has a natural positive square root, which is given by

$$\left[\det\left(1-\exp\left(-\mathrm{ad}\,(a)-i\theta\mathrm{ad}\,\left(Y_0^{\mathfrak{k}}\right)\right)\mathrm{Ad}\,\left(k^{-1}\right)\right)\,|_{\mathfrak{z}_0^{\perp}}\right.$$
$$\left.\det\left(1-\exp\left(-\mathrm{ad}\,(a)\right)\mathrm{Ad}\,\left(k^{-1}\right)\right)\,|_{\mathfrak{z}_0^{\perp}}\right]^{1/2}$$
$$=\det\left(e^{\sqrt{\underline{F}}/2}-e^{-\sqrt{\underline{F}}/2}\mathrm{Ad}\,\left(k^{-1}\right)\right)\,|_{\mathfrak{p}_0^{\perp}}$$
$$\det\left(e^{\sqrt{F}/2}-e^{-\sqrt{F}/2}\mathrm{Ad}\,\left(k^{-1}\right)\right)\,|_{\mathfrak{p}_0^{\perp}}. \tag{5.4.8}$$

Moreover,

$$\det \left(1 - \exp\left(-i\theta \mathrm{ad}\left(Y_0^{\mathfrak{k}}\right)\right) \mathrm{Ad}\left(k^{-1}\right)\right)_{\mathfrak{z}_0^\perp(\gamma)} \det\left(1 - \mathrm{Ad}\left(k^{-1}\right)\right)|_{\mathfrak{z}_0^\perp(\gamma)} \quad (5.4.9)$$

has a natural square root, which depends analytically on $Y_0^{\mathfrak{k}}$. Indeed no eigenvalue of $\mathrm{Ad}(k)$ acting on $\mathfrak{z}_0^\perp(\gamma)$ is equal to 1. Moreover, if $\mathfrak{z}_0^\perp(\gamma)$ is 1-dimensional, then $\mathrm{Ad}(k)|_{\mathfrak{z}_0^\perp(\gamma)} = -1$ and $\mathrm{ad}\left(Y_0^{\mathfrak{k}}\right)|_{\mathfrak{z}_0^\perp(\gamma)} = 0$, the square root is 2. If $\mathfrak{z}_0^\perp(\gamma)$ is of dimension 2, if $\mathrm{Ad}(k)$ is a rotation of angle ϕ and $\theta\mathrm{ad}\left(Y_0^{\mathfrak{k}}\right)|_{\mathfrak{z}_0^\perp(\gamma)}$ acts by an infinitesimal rotation of angle ϕ', such a square root is given by

$$4\sin\left(\frac{\phi}{2}\right)\sin\left(\frac{\phi + i\phi'}{2}\right). \quad (5.4.10)$$

We will denote the above square root by

$$\left[\det\left(1 - \exp\left(-i\theta\mathrm{ad}\left(Y_0^{\mathfrak{k}}\right)\right)\mathrm{Ad}\left(k^{-1}\right)\right)_{\mathfrak{z}_0^\perp(\gamma)}\det\left(1 - \mathrm{Ad}\left(k^{-1}\right)\right)|_{\mathfrak{z}_0^\perp(\gamma)}\right]^{1/2}.$$
$$(5.4.11)$$

Of course in (5.4.11), we may as well replace $\mathfrak{z}_0^\perp(\gamma)$ by $\mathfrak{p}_0^\perp(\gamma)$ or $\mathfrak{k}_0^\perp(\gamma)$.

It follows that

$$\det\left(1 - \exp\left(-\mathrm{ad}(a) - i\theta\mathrm{ad}\left(Y_0^{\mathfrak{k}}\right)\right)\mathrm{Ad}\left(k^{-1}\right)\right)|_{\mathfrak{z}^\perp(\gamma)}$$
$$\det\left(1 - \exp\left(-\mathrm{ad}(a)\right)\mathrm{Ad}\left(k^{-1}\right)\right)|_{\mathfrak{z}^\perp(\gamma)}$$

has a natural square root, which will be denoted by

$$\left[\det\left(1 - \exp\left(-\mathrm{ad}(a) - i\theta\mathrm{ad}\left(Y_0^{\mathfrak{k}}\right)\right)\mathrm{Ad}\left(k^{-1}\right)\right)|_{\mathfrak{z}^\perp(\gamma)}\right.$$

$$\left.\det\left(1 - \exp\left(-\mathrm{ad}(a)\right)\mathrm{Ad}\left(k^{-1}\right)\right)|_{\mathfrak{z}^\perp(\gamma)}\right]^{1/2}. \quad (5.4.12)$$

Incidentally observe that the all above quantities are unchanged when replacing a by $-a$. This is easily seen by conjugating the above matrices by θ.

Recall that $\widehat{\mathrm{Tr}_s}$ was defined in section 5.1.

Theorem 5.4.3. *The following identities hold:*

$$\mathrm{Tr}_s^{\Lambda^\cdot\left(\mathfrak{z}_0^{\perp,*}\right)}\left[\mathrm{Ad}\left(k^{-1}\right)\exp\left(\widehat{c}(\mathrm{ad}(a)) - c\left(\mathrm{ad}(a) + i\theta\mathrm{ad}\left(Y_0^{\mathfrak{k}}\right)\right)\right)\right]$$
$$= \det\left(e^{\sqrt{E}/2} - e^{-\sqrt{E}/2}\mathrm{Ad}\left(k^{-1}\right)\right)|_{\mathfrak{p}_0^\perp}\det\left(e^{\sqrt{F}/2} - e^{-\sqrt{F}/2}\mathrm{Ad}\left(k^{-1}\right)\right)|_{\mathfrak{p}_0^\perp},$$
$$(5.4.13)$$

$$\mathrm{Tr}_s^{\Lambda^\cdot\left(\mathfrak{z}_0^{\perp,*}(\gamma)\right)}\left[\mathrm{Ad}\left(k^{-1}\right)\exp\left(-c\left(i\theta\mathrm{ad}\left(Y_0^{\mathfrak{k}}\right)\right)\right)\right]$$
$$= \left[\det\left(1 - \exp\left(-i\theta\mathrm{ad}\left(Y_0^{\mathfrak{k}}\right)\right)\mathrm{Ad}\left(k^{-1}\right)\right)_{\mathfrak{z}_0^\perp(\gamma)}\det\left(1 - \mathrm{Ad}\left(k^{-1}\right)\right)|_{\mathfrak{z}_0^\perp(\gamma)}\right]^{1/2}.$$

Moreover,

$$\widehat{\mathrm{Tr}_s}\left[\exp\left(-\alpha - c\left(i\theta\mathrm{ad}\left(Y_0^{\mathfrak{k}}\right)|_{\mathfrak{z}(\gamma)}\right)\right)\right]$$
$$= \widehat{A}^{-1}\left(i\mathrm{ad}\left(Y_0^{\mathfrak{k}}\right)|_{\mathfrak{p}(\gamma)}\right)\widehat{A}^{-1}\left(i\mathrm{ad}\left(Y_0^{\mathfrak{k}}\right)|_{\mathfrak{k}(\gamma)}\right). \quad (5.4.14)$$

In particular,

$$\widehat{\mathrm{Tr}}_s \left[\mathrm{Ad}\left(k^{-1}\right) \exp\left(-\alpha + \widehat{c}\left(\mathrm{ad}\left(a\right)\right) - c\left(\mathrm{ad}\left(a\right) + i\theta \mathrm{ad}\left(Y_0^{\mathfrak{k}}\right)\right)\right) \right]$$
$$= \widehat{A}^{-1}\left(iad\left(Y_0^{\mathfrak{k}}\right)|_{\mathfrak{p}(\gamma)}\right) \widehat{A}^{-1}\left(iad\left(Y_0^{\mathfrak{k}}\right)|_{\mathfrak{k}(\gamma)}\right)$$

$$\left[\det\left(1 - \exp\left(-\mathrm{ad}\left(a\right) - i\theta\mathrm{ad}\left(Y_0^{\mathfrak{k}}\right)\right) \mathrm{Ad}\left(k^{-1}\right)\right)|_{\mathfrak{z}^{\perp}(\gamma)} \right.$$

$$\left. \det\left(1 - \exp\left(-\mathrm{ad}\left(a\right)\right) \mathrm{Ad}\left(k^{-1}\right)\right)|_{\mathfrak{z}^{\perp}(\gamma)} \right]^{1/2}. \qquad (5.4.15)$$

Proof. Observe that if $\mathbf{R}$ is the canonical Euclidean 1-dimensional vector space equipped with the canonical scalar product or with its negative, if $e = \pm 1$, among the monomials in the $c(e), \widehat{c}(e^*)$, up to permutation, $c(e)\widehat{c}(e^*)$ is the only monomial whose supertrace on $\Lambda^{\cdot}(\mathbf{R}^*)$ does not vanish, and moreover,

$$\mathrm{Tr}_s\left[c\left(e\right)\widehat{c}\left(e^*\right)\right] = -2. \qquad (5.4.16)$$

Similarly, the only monomial in $c(e), \widehat{c}(e^*)$ whose trace does not vanish is 1, and moreover,

$$\mathrm{Tr}\left[1\right] = 2. \qquad (5.4.17)$$

Also observe that $u = -1$ acts like 1 on $\Lambda^0\left(\mathbf{R}^*\right) \simeq \mathbf{R}$, and like -1 on $\Lambda^1\left(\mathbf{R}^*\right) \simeq \mathbf{R}$, so that if $C \in \mathrm{End}\left(\Lambda^{\cdot}\left(\mathbf{R}^*\right)\right)$,

$$\mathrm{Tr}_s^{\Lambda^{\cdot}(R^*)}\left[uC\right] = \mathrm{Tr}^{\Lambda^{\cdot}(\mathbf{R}^*)}\left[C\right]. \qquad (5.4.18)$$

To establish the first equation in (5.4.13), we split $\mathfrak{z}_0^{\perp} \otimes_{\mathbf{R}} \mathbf{C}$ according to the eigenvalues of $\mathrm{Ad}\left(k^{-1}\right)$. Assume that $\mathrm{Ad}\left(k^{-1}\right) = \pm 1$, and that $\mathfrak{p}_0^{\perp}, \mathfrak{k}_0^{\perp}$ are 1-dimensional, so that $\mathrm{ad}\left(Y_0^{\mathfrak{k}}\right)|_{\mathfrak{z}_0^{\perp}} = 0$. The matrix of $\mathrm{ad}\left(a\right)$ with respect to the splitting $\mathfrak{z}_0^{\perp} = \mathfrak{p}_0^{\perp} \oplus \mathfrak{k}_0^{\perp}$ and a corresponding basis e, f consisting of vectors of norm 1 is given by

$$\mathrm{ad}\left(a\right) = \begin{pmatrix} 0 & \lambda \\ \lambda & 0 \end{pmatrix}. \qquad (5.4.19)$$

Then

$$c\left(\mathrm{ad}\left(a\right)\right) = \frac{\lambda}{2}c\left(e\right)c\left(f\right), \qquad \widehat{c}\left(\mathrm{ad}\left(a\right)\right) = -\frac{\lambda}{2}\widehat{c}\left(e\right)\widehat{c}\left(f\right). \qquad (5.4.20)$$

Moreover,

$$\left(c\left(e\right)c\left(f\right)\right)^2 = 1, \qquad \left(\widehat{c}\left(e\right)\widehat{c}\left(f\right)\right)^2 = 1. \qquad (5.4.21)$$

By (5.4.20), (5.4.21), we get

$$\exp\left(-c\left(\mathrm{ad}\left(a\right)\right)\right) = \cosh\left(\lambda/2\right) - \sinh\left(\lambda/2\right)c\left(e\right)c\left(f\right), \qquad (5.4.22)$$
$$\exp\left(\widehat{c}\left(\mathrm{ad}\left(a\right)\right)\right) = \cosh\left(\lambda/2\right) - \sinh\left(\lambda/2\right)\widehat{c}\left(e\right)\widehat{c}\left(f\right).$$

By (5.4.16), (5.4.17), (5.4.22), we obtain

$$\mathrm{Tr}_s^{\Lambda^{\cdot}\left(\mathfrak{z}_0^{\perp,*}\right)}\left[\exp\left(\widehat{c}\left(\mathrm{ad}\left(a\right)\right) - c\left(\mathrm{ad}\left(a\right)\right)\right)\right] = 4\sinh^2\left(\lambda/2\right), \qquad (5.4.23)$$
$$\mathrm{Tr}^{\Lambda^{\cdot}\left(\mathfrak{z}_0^{\perp,*}\right)}\left[\exp\left(\widehat{c}\left(\mathrm{ad}\left(a\right)\right) - c\left(\mathrm{ad}\left(a\right)\right)\right)\right] = 4\cosh^2\left(\lambda/2\right).$$

By (5.4.18), both equations in (5.4.23) are compatible with the first identity in (5.4.13).

Let us now suppose that $\mathfrak{p}_0^\perp, \mathfrak{k}_0^\perp$ are both of dimension 2. Let e, e' and f, f' be orthonormal bases of $\mathfrak{p}_0^\perp, \mathfrak{k}_0^\perp$. Also we assume that $\mathrm{Ad}\,(k)$ is a rotation of angle ϕ on both vector spaces, that $\mathrm{ad}\,(a)$ maps e, e' into $\lambda f, \lambda f'$ and f, f' into $\lambda e, \lambda e'$, and that the matrix of $\mathrm{ad}\,(Y_0^{\mathfrak{k}})$ with respect to e, e' and f, f' is given by

$$\mathrm{ad}\,(Y_0^{\mathfrak{k}}) = \begin{pmatrix} 0 & -\phi' \\ \phi' & 0 \end{pmatrix}. \tag{5.4.24}$$

Incidentally observe that $\mathfrak{z}_0^\perp$ is canonically oriented by the basis e, e', f, f'.

By (5.4.20), we get

$$c\,(\mathrm{ad}\,(a)) = \frac{\lambda}{2}\,(c\,(e)\,c\,(f) + c\,(e')\,c\,(f')), \tag{5.4.25}$$

$$\widehat{c}\,(\mathrm{ad}\,(a)) = -\frac{\lambda}{2}\,(\widehat{c}\,(e)\,\widehat{c}\,(f) + \widehat{c}\,(e')\,\widehat{c}\,(f')).$$

Moreover,

$$c\,(\theta \mathrm{ad}\,(Y_0^{\mathfrak{k}})) = -\frac{\phi'}{2}\,(c\,(e)\,c\,(e') + c\,(f)\,c\,(f')). \tag{5.4.26}$$

By (5.4.22), (5.4.25), we get

$$\exp\,(\widehat{c}\,(\mathrm{ad}\,(a))) = (\cosh\,(\lambda/2) - \sinh\,(\lambda/2)\,\widehat{c}\,(e)\,\widehat{c}\,(f)) \\ (\cosh\,(\lambda/2) - \sinh\,(\lambda/2)\,\widehat{c}\,(e')\,\widehat{c}\,(f')). \tag{5.4.27}$$

Set

$$B = \frac{\phi}{2}\,(c\,(e)\,c\,(e') - \widehat{c}\,(e)\,\widehat{c}\,(e')) - \frac{\phi}{2}\,(c\,(f)\,c\,(f') - \widehat{c}\,(f)\,\widehat{c}\,(f')). \tag{5.4.28}$$

Then $\mathrm{Ad}\,(k^{-1})$ acts on $\Lambda^{\cdot}\,(\mathfrak{z}_0^{\perp *})$ like $\exp\,(-B)$.

By (5.4.16), the supertrace in the first identity in (5.4.13) is obtained as 16 times the coefficient of

$$c\,(e)\,c\,(e')\,c\,(f)\,c\,(f')\,\widehat{c}\,(e)\,\widehat{c}\,(e')\,\widehat{c}\,(f)\,\widehat{c}\,(f')$$

in the expansion of

$$\exp\,(-B + \widehat{c}\,(\mathrm{ad}\,(a)) - c\,(\mathrm{ad}\,(a) + i\theta \mathrm{ad}\,(Y_0^{\mathfrak{k}}))).$$

First, we make the c equal to 0. This way, we will get the coefficient of $\widehat{c}\,(e)\,\widehat{c}\,(e')\,\widehat{c}\,(f)\,\widehat{c}\,(f')$. By (5.4.27), (5.4.28), this coefficient is given by

$$-\cosh^2\,(\lambda/2)\sin^2\,(\phi/2) - \sinh^2\,(\lambda/2)\cos^2\,(\phi/2) = -\left|\sin\left(\frac{\phi + i\lambda}{2}\right)\right|^2. \tag{5.4.29}$$

Now we will obtain the coefficient of $c\,(e)\,c\,(e')\,c\,(f)\,c\,(f')$. Put

$$\underline{c}\,(f) = -ic\,(f), \underline{c}\,(f') = -ic\,(f'). \tag{5.4.30}$$

By (5.4.25), (5.4.26),

$$c\left(\mathrm{ad}\left(a\right)\right) = \frac{i\lambda}{2}\left(c\left(e\right)\underline{c}\left(f\right) + c\left(e'\right)\underline{c}\left(f'\right)\right), \tag{5.4.31}$$

$$c\left(i\theta\mathrm{ad}\left(Y_0^{\mathfrak{t}}\right)\right) = -\frac{i\phi'}{2}\left(c\left(e\right)c\left(e'\right) - \underline{c}\left(f\right)\underline{c}\left(f'\right)\right).$$

Moreover, if $c\left(B\right)$ is the component of B that only contains c variables, we get from (5.4.28),

$$c\left(B\right) = \frac{\phi}{2}\left(c\left(e\right)c\left(e'\right) + \underline{c}\left(f\right)\underline{c}\left(f'\right)\right). \tag{5.4.32}$$

Put

$$M = \begin{bmatrix} 0 & \phi - i\phi' & i\lambda & 0 \\ -\phi + i\phi' & 0 & 0 & i\lambda \\ -i\lambda & 0 & 0 & \phi + i\phi' \\ 0 & -i\lambda & -\phi - i\phi' & 0 \end{bmatrix}. \tag{5.4.33}$$

Let $\mathrm{End}^{\mathrm{as}}\left(\mathfrak{z}_0^{\perp}\right)$ be the set of antisymmetric endomorphisms of $\mathfrak{z}_0^{\perp}$ with respect to the scalar product of $\mathfrak{z}_0^{\perp}$. Then $M \in \mathrm{End}^{\mathrm{as}}\left(\mathfrak{z}_0^{\perp}\right) \otimes_{\mathbf{R}} \mathbf{C}$.

Let $c'\left(\mathfrak{z}_0^{\perp}\right)$ be the Clifford algebra associated with the Euclidean vector space $\mathfrak{z}_0^{\perp}$. The algebra $c'\left(\mathfrak{z}_0^{\perp}\right)$ is generated by $c\left(e\right), c\left(e'\right), \underline{c}\left(f\right), \underline{c}\left(f'\right)$. Let $c'\left(M\right) \in c\left(\mathfrak{z}_0^{\perp}\right) \otimes_{\mathbf{R}} \mathbf{C}$ be associated to the antisymmetric matrix M as in (1.1.9) with respect to the scalar product of $\mathfrak{z}_0^{\perp}$. By (5.4.31), (5.4.32), one verifies easily that

$$-c\left(B\right) - c\left(\mathrm{ad}\left(a\right) + i\theta\mathrm{ad}\left(Y_0^{\mathfrak{t}}\right)\right) = c'\left(M\right). \tag{5.4.34}$$

We have to evaluate the coefficient $\mathrm{Tr_s}^{c}\left[\exp\left(c'\left(M\right)\right)\right]$ of $c\left(e\right)c\left(e'\right)c\left(f\right)c\left(f'\right)$ in $\exp\left(c'\left(M\right)\right)$. To do this, we first consider the case of $N \in \mathrm{End}^{\mathrm{as}}\left(\mathfrak{z}_0^{\perp}\right)$. We will evaluate the coefficient $\mathrm{Tr_s}^{c}\left[\exp\left(c'\left(N\right)\right)\right]$ of $\exp\left(c'\left(N\right)\right)$, and obtain the corresponding coefficient for $\exp\left(c'\left(M\right)\right)$ by analytic continuation.

There is an orthonormal oriented basis i, j, k, l of $\mathfrak{z}_0^{\perp}$ such that the matrix ν of N in this basis is given by

$$\nu = \begin{bmatrix} 0 & -\alpha & 0 & 0 \\ \alpha & 0 & 0 & 0 \\ 0 & 0 & 0 & -\beta \\ 0 & 0 & \beta & 0 \end{bmatrix}. \tag{5.4.35}$$

Then

$$c'\left(\nu\right) = \frac{1}{2}\left(\alpha c'\left(i\right)c'\left(j\right) + \beta c'\left(k\right)c'\left(l\right)\right). \tag{5.4.36}$$

Also,

$$\exp\left(c'\left(\nu\right)\right) = \left(\cos\left(\alpha/2\right) + \sin\left(\alpha/2\right)c'\left(i\right)c'\left(j\right)\right)$$
$$\left(\cos\left(\beta/2\right) + \sin\left(\beta/2\right)c'\left(k\right)c'\left(l\right)\right). \tag{5.4.37}$$

Moreover, using (5.4.30) and the fact that i, j, k, l is an oriented basis, we get

$$c'\left(i\right)c'\left(j\right)c'\left(k\right)c'\left(l\right) = -c\left(e\right)c\left(e'\right)c\left(f\right)c\left(f'\right). \tag{5.4.38}$$

By (5.4.36)–(5.4.38) we get

$$\mathrm{Tr_s}^{c}\left[\exp\left(c'\left(N\right)\right)\right] = -\sin\left(\alpha/2\right)\sin\left(\beta/2\right). \tag{5.4.39}$$

We rewrite (5.4.39) in the form

$$4\mathrm{Tr_s}^{c}\left[\exp\left(c'\left(N\right)\right)\right] = -\alpha\beta\widehat{A}^{-1}\left(i\alpha\right)\widehat{A}^{-1}\left(i\beta\right). \tag{5.4.40}$$

Now the Pfaffian $\mathrm{Pf}\left[N\right]$ of N is given by

$$\mathrm{Pf}\left[N\right] = \alpha\beta. \tag{5.4.41}$$

Then (5.4.40) can be written in the form

$$4\mathrm{Tr_s}^{c}\left[\exp\left(c'\left(N\right)\right)\right] = -\mathrm{Pf}\left[N\right]\widehat{A}^{-1}\left(N\right). \tag{5.4.42}$$

By analytic continuation, we deduce from (5.4.42) that

$$4\mathrm{Tr_s}^{c}\left[\exp\left(c'\left(M\right)\right)\right] = -\mathrm{Pf}\left[M\right]\widehat{A}^{-1}\left(M\right). \tag{5.4.43}$$

As we already saw in the proof of Proposition 5.4.1, the eigenvalues of M are given by $\pm i\phi \pm \sqrt{\lambda^2 + \phi'^2}$, so that

$$\widehat{A}^{-1}\left(M\right) = \widehat{A}^{-1}\left(i\phi + \sqrt{\lambda^2 + \phi'^2}\right)\widehat{A}^{-1}\left(-i\phi + \sqrt{\lambda^2 + \phi'^2}\right). \tag{5.4.44}$$

We identify $\mathfrak{z}_0^{\perp}$ to $\mathfrak{z}_0^{\perp*}$ by its scalar product. In particular $\Lambda^{\cdot}\left(\mathfrak{z}_0^{\perp*}\right) \simeq \Lambda^{\cdot}\left(\mathfrak{z}_0^{\perp}\right)$. Let $\omega_M \in \Lambda^2\left(\mathfrak{z}_0^{\perp*}\right) \otimes_{\mathbf{R}} \mathbf{C}$ be such that if $U, V \in \mathfrak{z}_0^{\perp}$, then

$$\omega_M\left(U, V\right) = \langle U, MV \rangle. \tag{5.4.45}$$

A straightforward computation shows that

$$\frac{1}{2}\omega_M \wedge \omega_M = \left(\phi^2 + \lambda^2 + \phi'^2\right)e \wedge e' \wedge f \wedge f', \tag{5.4.46}$$

so that

$$\mathrm{Pf}\left[M\right] = \phi^2 + \lambda^2 + \phi'^2. \tag{5.4.47}$$

By (5.4.43), (5.4.44), and (5.4.47), we obtain

$$\mathrm{Tr_s}^{c}\left[\exp\left(c'\left(M\right)\right)\right] = -\left|\sin\left(\frac{\phi + i\sqrt{\lambda^2 + \phi'^2}}{2}\right)\right|^2. \tag{5.4.48}$$

By the considerations that follow (5.4.28), and by (5.4.29), (5.4.48), we get

$$\mathrm{Tr_s}^{\Lambda^{\cdot}\left(\mathfrak{z}_0^{\perp,*}\right)}\left[\mathrm{Ad}\left(k^{-1}\right)\exp\left(\widehat{c}\left(\mathrm{ad}\left(a\right)\right) - c\left(\mathrm{ad}\left(a\right) + i\theta\mathrm{ad}\left(Y_0^{\mathfrak{k}}\right)\right)\right)\right]$$
$$= 16\left|\sin\left(\frac{\phi + i\lambda}{2}\right)\right|^2\left|\sin\left(\frac{\phi + i\sqrt{\lambda^2 + \phi'^2}}{2}\right)\right|^2. \tag{5.4.49}$$

Again (5.4.49) fits with the first identity in (5.4.13).

To establish the second identity in (5.4.13), we only need to consider the contribution of $\mathfrak{p}_0^{\perp}\left(\gamma\right)$, the contribution of $\mathfrak{k}_0^{\perp}\left(\gamma\right)$ being of the same type. Clearly,

$$\mathrm{Tr_s}^{\Lambda^{\cdot}\left(\mathfrak{p}_0^{\perp*}\left(\gamma\right)\right)}\left[\mathrm{Ad}\left(k^{-1}\right)\right] = \det\left(1 - \mathrm{Ad}\left(k\right)\right)|_{\mathfrak{p}_0^{\perp}\left(\gamma\right)}$$
$$= \det\left(1 - \mathrm{Ad}\left(k^{-1}\right)\right)|_{\mathfrak{p}_0^{\perp}\left(\gamma\right)}. \tag{5.4.50}$$

Equation (5.4.50) is compatible with the second identity in (5.4.13) when $\mathfrak{p}_0^\perp(\gamma)$ is of dimension 1, and $\mathrm{Ad}(k)$ acts on it like -1.

Now we assume that $\mathfrak{p}_0^\perp(\gamma)$ is of dimension 2, and that $\mathrm{Ad}(k)$ acts on $\mathfrak{p}_0^\perp(\gamma)$ by a rotation of angle ϕ, and that $\theta\mathrm{ad}\left(Y_0^{\mathfrak{k}}\right)$ acts on $\mathfrak{p}_0^\perp(\gamma)$ by an infinitesimal rotation of angle ϕ'. Then one verifies easily that in this case,

$$\mathrm{Tr}_s^{\Lambda^{\cdot}\left(\mathfrak{p}_0^{\perp *}(\gamma)\right)}\left[\mathrm{Ad}\left(k^{-1}\right)\exp\left(-c\left(i\theta\mathrm{ad}\left(Y_0^{\mathfrak{k}}\right)\right)\right)\right] = 4\sin(\phi/2)\sin\left(\frac{\phi+i\phi'}{2}\right).$$
(5.4.51)

Comparing with (5.4.10), we also get the second equation in (5.4.13) in this case. We have completed the proof of this second equation.

To establish (5.4.14), we may as well assume that $q = 0$, so that $\mathfrak{z}(\gamma)$ is simply $\mathfrak{p}(\gamma)$. If $\mathfrak{p}(\gamma)$ is of dimension 1, then $\mathrm{ad}\left(Y_0^{\mathfrak{k}}\right)$ acts on $\mathfrak{p}(\gamma)$ like 0, and our identity is obvious. If $\mathfrak{p}(\gamma)$ is of dimension 2, our identity is an identity of Mathai-Quillen [MaQ86, eq. (2.13)]. In the general case, we obtain (5.4.14) by splitting $\mathfrak{p}(\gamma)\otimes_{\mathbf{R}}\mathbf{C}$ into the eigenspaces of $\mathrm{ad}\left(Y_0^{\mathfrak{k}}\right)$.

Using (5.4.8), (5.4.13) and (5.4.14), we get (5.4.15). The proof of our theorem is completed. □

5.5 A formula for $J_\gamma\left(Y_0^{\mathfrak{k}}\right)$

Set

$$A = \frac{1}{\det\left(1 - \mathrm{Ad}\left(k^{-1}\right)\right)|_{\mathfrak{z}_0^\perp(\gamma)}}\frac{\det\left(1 - \exp\left(-i\mathrm{ad}\left(Y_0^{\mathfrak{k}}\right)\right)\mathrm{Ad}\left(k^{-1}\right)\right)|_{\mathfrak{k}_0^\perp(\gamma)}}{\det\left(1 - \exp\left(-i\mathrm{ad}\left(Y_0^{\mathfrak{k}}\right)\right)\mathrm{Ad}\left(k^{-1}\right)\right)|_{\mathfrak{p}_0^\perp(\gamma)}}.$$
(5.5.1)

We claim that A has a natural square root. Indeed we have

$$\det\left(1 - \mathrm{Ad}\left(k^{-1}\right)\right)|_{\mathfrak{z}_0^\perp(\gamma)} = \det\left(1 - \mathrm{Ad}\left(k^{-1}\right)\right)|_{\mathfrak{p}_0^\perp(\gamma)}$$
$$\det\left(1 - \mathrm{Ad}\left(k^{-1}\right)\right)|_{\mathfrak{k}_0^\perp(\gamma)}. \quad (5.5.2)$$

As we already saw after (5.4.9) and (5.4.11),

$$\det\left(1 - \exp\left(-i\mathrm{ad}\left(Y_0^{\mathfrak{k}}\right)\right)\mathrm{Ad}\left(k^{-1}\right)\right)|_{\mathfrak{p}_0^\perp(\gamma)}\det\left(1 - \mathrm{Ad}\left(k^{-1}\right)\right)|_{\mathfrak{p}_0^\perp(\gamma)}$$

has a natural square root. The same is true when replacing $\mathfrak{p}_0^\perp(\gamma)$ by $\mathfrak{k}_0^\perp(\gamma)$. Also

$$\frac{\det\left(1 - \exp\left(-i\mathrm{ad}\left(Y_0^{\mathfrak{k}}\right)\right)\mathrm{Ad}\left(k^{-1}\right)\right)|_{\mathfrak{k}_0^\perp(\gamma)}}{\det\left(1 - \mathrm{Ad}\left(k^{-1}\right)\right)|_{\mathfrak{k}_0^\perp(\gamma)}}$$
$$= \frac{\det\left(1 - \exp\left(-i\mathrm{ad}\left(Y_0^{\mathfrak{k}}\right)\right)\mathrm{Ad}\left(k^{-1}\right)\right)|_{\mathfrak{k}_0^\perp(\gamma)}\det\left(1 - \mathrm{Ad}\left(k^{-1}\right)\right)|_{\mathfrak{k}_0^\perp(\gamma)}}{\left[\det\left(1 - \mathrm{Ad}\left(k^{-1}\right)\right)|_{\mathfrak{k}_0^\perp(\gamma)}\right]^2}.$$
(5.5.3)

Using again the results we mentioned, we see that (5.5.3) also has a natural square root.

From the above, we conclude that A has a natural square root. It will be denoted

$$A^{1/2} = \left[\frac{1}{\det\left(1 - \operatorname{Ad}\left(k^{-1}\right)\right)|_{\mathfrak{z}_0^\perp(\gamma)}} \right.$$

$$\left. \frac{\det\left(1 - \exp\left(-i\operatorname{ad}\left(Y_0^{\mathfrak{k}}\right)\right)\operatorname{Ad}\left(k^{-1}\right)\right)|_{\mathfrak{k}_0^\perp(\gamma)}}{\det\left(1 - \exp\left(-i\operatorname{ad}\left(Y_0^{\mathfrak{k}}\right)\right)\operatorname{Ad}\left(k^{-1}\right)\right)|_{\mathfrak{p}_0^\perp(\gamma)}} \right]^{1/2}. \tag{5.5.4}$$

Theorem 5.5.1. *The following identity holds:*

$$J_\gamma\left(Y_0^{\mathfrak{k}}\right) = \frac{1}{\left|\det\left(1 - \operatorname{Ad}\left(\gamma\right)\right)|_{\mathfrak{z}_0^\perp}\right|^{1/2}} \frac{\widehat{A}\left(i\operatorname{ad}\left(Y_0^{\mathfrak{k}}\right)|_{\mathfrak{p}(\gamma)}\right)}{\widehat{A}\left(i\operatorname{ad}\left(Y_0^{\mathfrak{k}}\right)_{\mathfrak{k}(\gamma)}\right)}$$

$$\left[\frac{1}{\det\left(1 - \operatorname{Ad}\left(k^{-1}\right)\right)|_{\mathfrak{z}_0^\perp(\gamma)}} \frac{\det\left(1 - \exp\left(-i\operatorname{ad}\left(Y_0^{\mathfrak{k}}\right)\right)\operatorname{Ad}\left(k^{-1}\right)\right)|_{\mathfrak{k}_0^\perp(\gamma)}}{\det\left(1 - \exp\left(-i\operatorname{ad}\left(Y_0^{\mathfrak{k}}\right)\right)\operatorname{Ad}\left(k^{-1}\right)\right)|_{\mathfrak{p}_0^\perp(\gamma)}} \right]^{1/2}. \tag{5.5.5}$$

Proof. By equations (5.1.11), (5.3.9), (5.4.3), (5.4.8), and (5.4.13)–(5.4.15), we get

$$J_\gamma\left(Y_0^{\mathfrak{k}}\right) = \frac{\widehat{A}\left(i\operatorname{ad}\left(Y_0^{\mathfrak{k}}\right)|_{\mathfrak{p}(\gamma)}\right)}{\widehat{A}\left(i\operatorname{ad}\left(Y_0^{\mathfrak{k}}\right)|_{\mathfrak{k}(\gamma)}\right)}$$

$$\frac{1}{\left|\det\left(1 - \operatorname{Ad}\left(\gamma\right)\right)|_{\mathfrak{z}_0^\perp}\right|^{1/2}\det\left(1 - \operatorname{Ad}\left(k^{-1}\right)\right)|_{\mathfrak{z}_0^\perp(\gamma)}}$$

$$\left[\det\left(1 - \exp\left(-i\theta\operatorname{ad}\left(Y_0^{\mathfrak{k}}\right)\right)\operatorname{Ad}\left(k^{-1}\right)\right)|_{\mathfrak{z}_0^\perp(\gamma)}\det\left(1 - \operatorname{Ad}\left(k^{-1}\right)\right)|_{\mathfrak{z}_0^\perp(\gamma)}\right]^{1/2}$$

$$\frac{1}{\det\left(e^{\sqrt{E/2}} - e^{-\sqrt{E/2}}\operatorname{Ad}\left(k^{-1}\right)\right)|_{\mathfrak{p}_0^\perp(\gamma)}}. \tag{5.5.6}$$

In the right-hand side of (5.5.6), the determinants of the square roots over $\mathfrak{z}_0^\perp(\gamma)$ can be factored as a product of determinants over $\mathfrak{k}_0^\perp(\gamma)$ and $\mathfrak{p}_0^\perp(\gamma)$. The contribution of $\mathfrak{k}_0^\perp(\gamma)$ is given by

$$\left[\frac{\det\left(1 - \exp\left(-i\theta\operatorname{ad}\left(Y_0^{\mathfrak{k}}\right)\right)\operatorname{Ad}\left(k^{-1}\right)\right)|_{\mathfrak{k}_0^\perp(\gamma)}}{\det\left(1 - \operatorname{Ad}\left(k^{-1}\right)\right)|_{\mathfrak{k}_0^\perp(\gamma)}}\right]^{1/2}. \tag{5.5.7}$$

Since $\theta = 1$ on $\mathfrak{k}$, we may as well remove θ in (5.5.7).

Using the second identity in (5.4.2) applied to $\mathfrak{p}_0^\perp(\gamma)$, the contribution of $\mathfrak{p}_0^\perp(\gamma)$ to the product of determinants in (5.5.6) is given by

$$\left[\frac{\det\left(1 - \exp\left(-i\theta\operatorname{ad}\left(Y_0^{\mathfrak{k}}\right)\right)\operatorname{Ad}\left(k^{-1}\right)\right)|_{\mathfrak{p}_0^\perp(\gamma)}}{\det\left(1 - \operatorname{Ad}\left(k^{-1}\right)\right)|_{\mathfrak{p}_0^\perp(\gamma)}}\right]^{1/2}$$

$$\left|\det\left(1 - \exp\left(-i\theta\operatorname{ad}\left(Y_0^{\mathfrak{k}}\right)\right)\operatorname{Ad}\left(k^{-1}\right)\right)|_{\mathfrak{p}_0^\perp(\gamma)}\right|^{-1}. \tag{5.5.8}$$

Clearly,

$$\left| \det \left(1 - \exp\left(-i\theta\mathrm{ad}\left(Y_0^{\mathfrak{k}}\right)\right) \mathrm{Ad}\left(k^{-1}\right)\right)|_{\mathfrak{p}_0^{\perp}(\gamma)} \right|$$

$$= \left[\det \left(1 - \exp\left(-i\theta\mathrm{ad}\left(Y_0^{\mathfrak{k}}\right)\right) \mathrm{Ad}\left(k^{-1}\right)\right)|_{\mathfrak{p}_0^{\perp}(\gamma)} \right.$$

$$\left. \det \left(1 - \exp\left(i\theta\mathrm{ad}\left(Y_0^{\mathfrak{k}}\right)\right) \mathrm{Ad}\left(k^{-1}\right)\right)|_{\mathfrak{p}_0^{\perp}(\gamma)} \right]^{1/2}. \quad (5.5.9)$$

Using (5.5.9), we conclude that (5.5.8) is given by

$$\left[\det \left(1 - \exp\left(i\theta\mathrm{ad}\left(Y_0^{\mathfrak{k}}\right)\right) \mathrm{Ad}\left(k^{-1}\right)\right)|_{\mathfrak{p}_0^{\perp}(\gamma)} \det \left(1 - \mathrm{Ad}\left(k^{-1}\right)\right)_{\mathfrak{p}_0^{\perp}(\gamma)} \right]^{-1/2}.$$
$$(5.5.10)$$

Since $\theta = -1$ on $\mathfrak{p}$, we can replace $i\theta$ by $-i$ in (5.5.10).

By (5.5.6)–(5.5.10), we get (5.5.5). The proof of our theorem is completed.
□

Remark 5.5.2. Observe that $J_\gamma\left(Y_0^{\mathfrak{k}}\right)$ is unchanged when replacing the bilinear form B by $B/t, t > 0$.

Note that $\mathfrak{z}(1) = \mathfrak{g}, \mathfrak{p}(1) = \mathfrak{p}, \mathfrak{k}(1) = \mathfrak{k}$. By (5.5.5), if $Y_0^{\mathfrak{k}} \in \mathfrak{k}$, we get

$$J_1\left(Y_0^{\mathfrak{k}}\right) = \widehat{A}\left(i\mathrm{ad}\left(Y_0\right)|_{\mathfrak{p}}\right) \widehat{A}^{-1}\left(i\mathrm{ad}\left(Y_0^{\mathfrak{k}}\right)|_{\mathfrak{k}}\right). \quad (5.5.11)$$

Chapter Six

A formula for semisimple orbital integrals

This chapter is the central part of the book. First, we give an explicit formula for the orbital integrals associated with the heat kernel of $\mathcal{L}_A^X$ in terms of a Gaussian integral on $\mathfrak{k}(\gamma)$. In chapter 9, this formula will be obtained by the explicit computation of the asymptotics as $b \to +\infty$ of the orbital integrals associated with $\mathcal{L}_{A,b}^X$. From the formula for the heat kernel, we derive a corresponding formula for the semisimple orbital integrals associated with the wave operator of $\mathcal{L}_A^X$.

This chapter is organized as follows. In section 6.1, we give the formula for the orbital integral associated with the heat kernel of $\mathcal{L}_A^X$.

In section 6.2, we obtain the formula for the orbital integral associated with $\mu\left(\sqrt{\mathcal{L}_A^X}\right)$, when $\mu \in \mathcal{S}(\mathbf{R})$ is even, and its Fourier transform $\widehat{\mu}$ decays like a Gaussian at infinity.

Finally, in section 6.3, we obtain a formula for the orbital integrals associated with the wave operator of $\mathcal{L}_A^X$.

In the whole chapter, $\gamma \in G$ is supposed to be semisimple, and is written as in (4.2.1). Also we use the notation of chapters 2–5.

6.1 Orbital integrals for the heat kernel

Recall that $Z(\gamma) \subset G$ is the centralizer of γ, and that $\mathfrak{z}(\gamma) = \mathfrak{p}(\gamma) \oplus \mathfrak{k}(\gamma)$ is its Lie algebra. As before, we use the notation $p = \dim \mathfrak{p}(\gamma), q = \dim \mathfrak{k}(\gamma), r = \dim \mathfrak{z}(\gamma)$. For $Y_0^{\mathfrak{k}} \in \mathfrak{k}(\gamma)$, $J_\gamma\left(Y_0^{\mathfrak{k}}\right)$ is given by equation (5.5.5) in Theorem 5.5.1.

By (5.5.5), there exist $c_\gamma > 0, C_\gamma > 0$ such that if $Y_0^{\mathfrak{k}} \in \mathfrak{k}(\gamma)$,

$$\left| J_\gamma\left(Y_0^{\mathfrak{k}}\right) \right| \le c_\gamma \exp\left(C_\gamma \left| Y_0^{\mathfrak{k}} \right|\right). \tag{6.1.1}$$

Theorem 6.1.1. *For any $t > 0$, the following identity holds:*

$$\mathrm{Tr}^{[\gamma]}\left[\exp\left(-t\mathcal{L}_A^X\right)\right] = \frac{\exp\left(-|a|^2/2t\right)}{(2\pi t)^{p/2}}$$

$$\int_{\mathfrak{k}(\gamma)} J_\gamma\left(Y_0^{\mathfrak{k}}\right) \mathrm{Tr}^E\left[\rho^E\left(k^{-1}\right) \exp\left(-i\rho^E\left(Y_0^{\mathfrak{k}}\right) - tA\right)\right]$$

$$\exp\left(-\left|Y_0^{\mathfrak{k}}\right|^2/2t\right) \frac{dY_0^{\mathfrak{k}}}{(2\pi t)^{q/2}}. \tag{6.1.2}$$

Proof. The proof of our theorem will be given in chapter 9. $\qquad\square$

6.2 A formula for general orbital integrals

Let $\Delta^{\mathfrak{z}(\gamma)}$ be the standard Laplacian on $\mathfrak{z}(\gamma) = \mathfrak{p}(\gamma) \oplus \mathfrak{k}(\gamma)$ with respect to the scalar product induced by the scalar product of $\mathfrak{g}$.

For $t > 0$, let $\exp\left(t\Delta^{\mathfrak{z}(\gamma)}/2\right)$ be the corresponding heat operator. We denote by $\exp\left(t\Delta^{\mathfrak{z}(\gamma)}/2\right)\left(\left(y, Y_0^{\mathfrak{k}}\right), \left(y', Y_0^{\mathfrak{k}\prime}\right)\right)$ the associated Gaussian heat kernel. Here the heat kernel is calculated with respect to the given volume element on $\mathfrak{z}(\gamma)$, which is fixed once and for all.

Then $J_\gamma\left(Y_0^{\mathfrak{k}}\right)\rho^E\left(k^{-1}\right)\exp\left(-i\rho^E\left(Y_0^{\mathfrak{k}}\right)\right)\delta_{y=a}$ is a distribution on $\mathfrak{z}(\gamma) = \mathfrak{p}(\gamma) \oplus \mathfrak{k}(\gamma)$ with values in $\mathrm{End}(E)$. The heat kernel $\exp\left(t\Delta^{\mathfrak{z}(\gamma)}/2 - tA\right)$ can be applied to this distribution, and the resulting smooth function over $\mathfrak{z}(\gamma)$ will be denoted by

$$\exp\left(t\Delta^{\mathfrak{z}(\gamma)}/2 - tA\right)\left[J_\gamma\left(Y_0^{\mathfrak{k}}\right)\rho^E\left(k^{-1}\right)\exp\left(-i\rho^E\left(Y_0^{\mathfrak{k}}\right)\right)\delta_{y=a}\right]. \quad (6.2.1)$$

When taking the trace of (6.2.1), we obtain a smooth function on $\mathfrak{z}(\gamma)$ with values in $\mathbf{C}$. This function can be evaluated at $0 \in \mathfrak{z}(\gamma)$.

Theorem 6.2.1. *For any $t > 0$, the following identity holds:*

$$\mathrm{Tr}^{[\gamma]}\left[\exp\left(-t\left(\mathcal{L}^X + A\right)\right)\right] = \mathrm{Tr}^E\left[\exp\left(t\Delta^{\mathfrak{z}(\gamma)}/2 - tA\right)\right.$$

$$\left.\left[J_\gamma\left(Y_0^{\mathfrak{k}}\right)\rho^E\left(k^{-1}\right)\exp\left(-i\rho^E\left(Y_0^{\mathfrak{k}}\right)\right)\delta_{y=a}\right]\right](0). \quad (6.2.2)$$

Proof. Observe that

$$\exp\left(t\Delta^{\mathfrak{z}(\gamma)/2}\right)\left((0,0), \left(y, Y_0^{\mathfrak{k}}\right)\right) = \frac{1}{(2\pi t)^{(p+q)/2}}\exp\left(-|y|^2/2t - \left|Y_0^{\mathfrak{k}}\right|^2/2t\right).$$

$$(6.2.3)$$

By (6.1.2), (6.2.3), we get (6.2.2). $\qquad\square$

Let $\mathcal{S}(\mathbf{R})$ be the Schwartz space of $\mathbf{R}$, let $\mathcal{S}^{\mathrm{even}}(\mathbf{R})$ be the space of even functions in $\mathcal{S}(\mathbf{R})$.

Take $\mu \in \mathcal{S}^{\mathrm{even}}(\mathbf{R})$. Let $\widehat{\mu} \in \mathcal{S}^{\mathrm{even}}(\mathbf{R})$ denote its Fourier transform, i.e.,

$$\widehat{\mu}(y) = \int_{\mathbf{R}} e^{-2i\pi yx}\mu(x)\,dx. \quad (6.2.4)$$

We will assume that there exists $C > 0$ such that for any $k \in \mathbf{N}$, there exists $c_k > 0$ such that

$$\left|\widehat{\mu}^{(k)}(y)\right| \le c_k\exp\left(-Cy^2\right). \quad (6.2.5)$$

Then $\mu\left(\sqrt{\mathcal{L}^X + A}\right)$ is self-adjoint. Let $\mu\left(\sqrt{\mathcal{L}^X + A}\right)(x, x'), x, x' \in X$ be the corresponding smooth kernel, which we will still denote $\mu\left(\sqrt{\mathcal{L}^X + A}\right)$.

The algebra of operators $\mathcal{Q}$ was defined in Definition 4.1.1. Using (6.2.5) and finite propagation speed for the wave operator $\cos\left(t\sqrt{\mathcal{L}^X + A}\right)$ [CP81, section 7.8], [T81, section 4.4], it can be easily shown that $\mu\left(\sqrt{\mathcal{L}^X + A}\right) \in \mathcal{Q}$. Therefore $\operatorname{Tr}^{[\gamma]}\left[\mu\left(\sqrt{\mathcal{L}^X + A}\right)\right]$ is well-defined.

Similarly from (6.2.5), one finds that the kernel of $\mu\left(\sqrt{-\Delta^{\mathfrak{z}(\gamma)} + A}\right)$ on $\mathfrak{z}(\gamma)$ has a Gaussian-like decay. Combining this with (6.1.1), we find that in (6.2.1), we can replace the function $\exp\left(-tx^2/2\right)$ by our function μ, and still obtain a continuous function on $\mathfrak{z}(\gamma)$.

We give the following extension of Theorem 6.2.1.

Theorem 6.2.2. *The following identity holds:*

$$\operatorname{Tr}^{[\gamma]}\left[\mu\left(\sqrt{\mathcal{L}^X + A}\right)\right] = \operatorname{Tr}^E\left[\mu\left(\sqrt{-\Delta^{\mathfrak{z}(\gamma)}/2 + A}\right) J_\gamma\left(Y_0^{\mathfrak{t}}\right)\right.$$
$$\left. \rho^E\left(k^{-1}\right) \exp\left(-i\rho^E\left(Y_0^{\mathfrak{t}}\right)\right) \delta_{y=a}\right](0). \quad (6.2.6)$$

Proof. For $t > 0$, if $\mu(x) = \exp\left(-tx^2/2\right)$, equation (6.2.6) is just (6.2.2). More generally, by differentiating (6.2.2) with respect to t, we find that (6.2.6) still holds for $\mu(x) = x^{2k}\exp\left(-tx^2/2\right), t > 0, k \in \mathbf{N}$. Using basic results on the harmonic oscillator, we know that given $t > 0$, linear combinations of such functions are dense in $\mathcal{S}^{\mathrm{even}}(\mathbf{R})$.

Set

$$\widehat{\nu}(y) = \exp\left(Cy^2/2\right)\widehat{\mu}(y). \quad (6.2.7)$$

Using (6.2.5), we find that $\widehat{\nu} \in \mathcal{S}^{\mathrm{even}}(\mathbf{R})$. From the above, given $M \in \mathbf{N}, \epsilon > 0$, there is a finite linear combination $\widehat{\nu}_{M,\epsilon}$ of functions of the type $x^{2k}\exp\left(-x^2/2\right), k \in \mathbf{N}$ such that for $p \in \mathbf{N}, p \leq M$,

$$\left|(\widehat{\nu} - \widehat{\nu}_{M,\epsilon})^{(p)}(y)\right| \leq \epsilon. \quad (6.2.8)$$

Set

$$\widehat{\mu}_{M,\epsilon}(y) = \exp\left(-Cy^2/2\right)\widehat{\nu}_{\epsilon,M}(y). \quad (6.2.9)$$

Let $\mu_{M,\epsilon} \in \mathcal{S}^{\mathrm{even}}(\mathbf{R})$ be the Fourier transform of $\widehat{\mu}_{M,\epsilon}$. By (6.2.7)–(6.2.9), for $p \in \mathbf{N}, p \leq M$,

$$\left|(\widehat{\mu} - \widehat{\mu}_{M,\epsilon})^{(p)}(y)\right| \leq c_M\epsilon\exp\left(-Cy^2/4\right). \quad (6.2.10)$$

As we saw above, equation (6.2.6) is valid for $\mu_{M,\epsilon}$. By taking $M \in \mathbf{N}$ large enough and $\epsilon > 0$ small enough, we find that the left- and right-hand sides of (6.2.6) for μ are approximated by the corresponding expressions for $\mu_{M,\epsilon}$, which turn out to be equal. Therefore (6.2.6) is valid for μ. The proof of our theorem is completed. $\square$

Remark 6.2.3. The density result used at the beginning of the proof of Theorem 6.2.2 follows from basic properties of the harmonic oscillator. The harmonic oscillator, which had just disappeared from the geometric picture, resurrects itself in the proof, with a completely different interpretation.

The deformation argument for the heat kernel given in Theorem 4.6.1 is not valid for the more general kernels considered in Theorem 6.2.2. Equivalently, the elliptic trace in the left-hand side of (6.2.6) cannot be replaced in general by a hypoelliptic supertrace, with $\mathcal{L}^X$ replaced by $\mathcal{L}_b^X$. This is especially true when $\widehat{\mu}$ has compact support.

6.3 The orbital integrals for the wave operator

Let $\mathrm{Tr}^{[\gamma]}\left[\cos\left(s\sqrt{\mathcal{L}^X + A}\right)\right]$ be the even distribution on $\mathbf{R}$ such that for any $\mu \in \mathcal{S}^{\mathrm{even}}(\mathbf{R})$ with $\widehat{\mu}$ having compact support,

$$\mathrm{Tr}^{[\gamma]}\left[\mu\left(\sqrt{\mathcal{L}^X + A}\right)\right] = \int_{\mathbf{R}} \widehat{\mu}(s)\,\mathrm{Tr}^{[\gamma]}\left[\cos\left(2\pi s\sqrt{\mathcal{L}^X + A}\right)\right] ds. \quad (6.3.1)$$

The wave operator $\cos\left(\sqrt{2}s\sqrt{\mathcal{L}^X + A}\right)$ defines a distribution on $\mathbf{R} \times X \times X$. By finite propagation speed for the wave equation [CP81, section 7.8], [T81, section 4.4], its support is included in (s, x, x'), $|s| \geq d(x, x')$. Recall that we have identified TX and T^*X by the metric, and that $s \in \mathbf{R} \to \varphi_s$ is the geodesic flow on $\mathcal{X} = \mathcal{X}^*$. Let τ be the variable dual to s. By [Hör85a, Theorem 23.1.4 and remark], the wave front set $\mathrm{WF}\left(\cos\left(\sqrt{2}s\sqrt{\mathcal{L}^X + A}\right)\right)$ of the distribution $\cos\left(\sqrt{2}s\sqrt{\mathcal{L}^X + A}\right)$ is the conic set in $\mathbf{R}^2 \times T^*X \times T^*X$ generated by $(x', -Y') = \varphi_{\pm s}(x, Y)$, $|Y| = 1$, $\tau = \pm 1$. Conic here means that the dilations by $\lambda > 0$ are applied to the variables Y, Y', τ.

As we saw in section 3.4, the map $f \in \mathfrak{p}^\perp(\gamma) \to \rho_\gamma(1, f) \in X$ identifies $\mathfrak{p}^\perp(\gamma)$ to a smooth submanifold $P^\perp(\gamma)$ of X. Let $N_{P^\perp(\gamma)/X}$ be the orthogonal bundle to $TP(\gamma)$ in TX.

Set

$$\Delta_X^\gamma = \left\{(x, \gamma x), x \in P^\perp(\gamma)\right\}. \quad (6.3.2)$$

Then Δ_X^γ is a smooth submanifold of $X \times X$. The conormal bundle to $\mathbf{R} \times \Delta_X^\gamma \subset \mathbf{R} \times X \times X$ can be identified with the set $((s, \tau), (x, Y), (x', Y')) \in \mathbf{R}^2 \times \mathcal{X} \times \mathcal{X}$ such that $\tau = 0$, $x \in P^\perp(\gamma)$, $x' = \gamma x$, $\gamma^* Y' + Y \in N_{P^\perp(\gamma)/X}$.

By [Hö83, Theorem 8.2.10], $\cos\left(\sqrt{2}s\sqrt{\mathcal{L}^X + A}\right)\Delta_X^\gamma$ is a well-defined distribution on $\mathbf{R} \times X \times X$, and its wave front set is the formal sum of the wave front sets of the two above distributions. Using (3.4.4) and the property of the support of $\cos\left(\sqrt{2}s\sqrt{\mathcal{L}^X + A}\right)$, which was given before, the push-forward of the distribution $\mathrm{Tr}^F\left[\gamma\cos\left(\sqrt{2}s\sqrt{\mathcal{L}^X + A}\right)\right]$ by the projection $\mathbf{R} \times X \times X \to \mathbf{R}$ is well-defined. It will be denoted

$$\int_{\Delta_X^\gamma} \mathrm{Tr}^F\left[\gamma\cos\left(\sqrt{2}s\sqrt{\mathcal{L}^X + A}\right)\right]. \quad (6.3.3)$$

This is an even distribution on $\mathbf{R}$.

Tautologically, we have the identity of even distributions on $\mathbf{R}$,

$$\mathrm{Tr}^{[\gamma]}\left[\cos\left(s\sqrt{\mathcal{L}^X + A}\right)\right] = \int_{\Delta_X^\gamma} \mathrm{Tr}^F\left[\gamma\cos\left(s\sqrt{\mathcal{L}^X + A}\right)\right]. \tag{6.3.4}$$

Proposition 6.3.1. *The singular support of* $\mathrm{Tr}^{[\gamma]}\left[\cos\left(\sqrt{2}s\sqrt{\mathcal{L}^X + A}\right)\right]$ *is included in* $s = \pm|a|$, *the ordinary support is included in* $\{s \in \mathbf{R}, |s| \geq |a|\}$. *For* $a = 0$, *if* $\mathfrak{p}(\gamma) = 0$, *the singular support of* $\mathrm{Tr}^{[\gamma]}\left[\cos\left(\sqrt{2}s\sqrt{\mathcal{L}^X + A}\right)\right]$ *is empty.*

Proof. The wave front set of the distribution (6.3.3) can be obtained from the wave front set of the product $\cos\left(\sqrt{2}s\sqrt{\mathcal{L}^X + A}\right)\Delta_X^\gamma$ by [Hö83, Theorem 8.2.13]. Using this result, we find that $s \in \mathbf{R}$ lies in the singular support of this distribution if and only if there is $x \in X, Y \in T_xX, |Y| = 1$ such that if

$$\varphi_s(x, Y) = (x', Y'), \tag{6.3.5}$$

then

$$x \in P(\gamma), \qquad x' = \gamma x, \qquad \gamma^*Y' - Y \in N_{P^\perp(\gamma)/X}. \tag{6.3.6}$$

By (6.3.5), we get

$$\varphi_1(x, sY) = (x', sY'). \tag{6.3.7}$$

By (6.3.6), (6.3.7), x is a critical point of the restriction of d_γ^2 to $P^\perp(\gamma)$. Since d_γ^2 is a convex function, we find that $x = p1$, and that $|s| = |a|$. This shows that the singular support of $\mathrm{Tr}^{[\gamma]}\left[\cos\left(\sqrt{2}s\sqrt{\mathcal{L}^X + A}\right)\right]$ is included in $\pm|a|$. If $a \neq 0$, note that in (6.3.5), $Y = \pm a/|a|$.

Since $d(x, \gamma x) = d_\gamma(x) \geq |a|$, using the considerations we made after equation (6.3.1), the support of $\mathrm{Tr}^{[\gamma]}\left[\cos\left(\sqrt{2}s\sqrt{\mathcal{L}^X + A}\right)\right]$ is included in $\{s \in \mathbf{R}, |s| \geq |a|\}$.

Assume now that $a = 0$, so that $\gamma = k^{-1}$. We already know that in (6.3.5), (6.3.6), $s = 0$, and so $Y' = Y$. By (6.3.6), $(\mathrm{Ad}(k) - 1)Y$ is fixed by $\mathrm{Ad}(k)$, so that

$$\mathrm{Ad}(k)Y = Y, \tag{6.3.8}$$

so that $Y \in \mathfrak{p}(\gamma)$. If $\mathfrak{p}(\gamma) = 0$, we cannot have $|Y| = 1$. This completes the proof of our proposition. $\qquad\square$

We define the even distribution on $\mathbf{R}$,

$$\mathrm{Tr}^E\left[\cos\left(s\sqrt{-\Delta^{\mathfrak{s}(\gamma)}/2 + A}\right)J_\gamma\left(Y_0^{\mathfrak{k}}\right)\rho^E\left(k^{-1}\right)\exp\left(-i\rho^E\left(Y_0^{\mathfrak{k}}\right)\right)\delta_{y=a}\right](0) \tag{6.3.9}$$

by the formula

$$\mathrm{Tr}^E \left[\mu \left(\sqrt{-\Delta^{\mathfrak{z}(\gamma)}/2 + A} \right) J_\gamma \left(Y_0^{\mathfrak{k}} \right) \rho^E \left(k^{-1} \right) \exp \left(-i\rho^E \left(Y_0^{\mathfrak{k}} \right) \right) \delta_{y=a} \right] (0)$$

$$= \int_{\mathbf{R}} \widehat{\mu} \left(s \right) \mathrm{Tr}^E \left[\cos \left(2\pi s \sqrt{-\Delta^{\mathfrak{z}(\gamma)}/2 + A} \right) J_\gamma \left(Y_0^{\mathfrak{k}} \right) \rho^E \left(k^{-1} \right) \right.$$

$$\left. \exp \left(-i\rho^E \left(Y_0^{\mathfrak{k}} \right) \right) \delta_{y=a} \right] (0). \quad (6.3.10)$$

We will denote by $z = \left(y, Y_0^{\mathfrak{k}} \right)$ the generic element of $\mathfrak{z}(\gamma) = \mathfrak{p}(\gamma) \oplus \mathfrak{k}(\gamma)$. Also we equip $\mathfrak{z}(\gamma)$ with its Euclidean norm.

By finite propagation speed for the wave equation [CP81, section 7.8], [T81, section 4.4], $\cos \left(\sqrt{2} s \sqrt{-\Delta^{\mathfrak{z}(\gamma)}/2 + A} \right)$ is a distribution on $\mathbf{R} \times \mathfrak{z}(\gamma) \times \mathfrak{z}(\gamma)$ whose support is included in (s, z, z'), $|s| \geq |z' - z|$. By [Hör85a, Theorem 23.1.4 and remark], its wave front set $\mathrm{WF} \left(\cos \left(\sqrt{2} s \sqrt{-\Delta^{\mathfrak{z}(\gamma)}/2 + A} \right) \right)$ is equal to the conic set associated with $(y', -Y') = (y \pm sY, Y)$, $|Y| = 1, \tau = \pm 1$. Conic set means again that the dilations by $\lambda > 0$ are applied to the variables Y, Y', τ.

Set

$$H^\gamma = \{0\} \times (a, \mathfrak{k}(\gamma)) \subset \mathfrak{z}(\gamma) \times \mathfrak{z}(\gamma). \quad (6.3.11)$$

The wave front set associated with $\mathbf{R} \times H^\gamma \subset \mathbf{R} \times \mathfrak{z}(\gamma) \times \mathfrak{z}(\gamma)$ is such that $Y'^{\mathfrak{k}(\gamma)} = 0, \tau = 0$, so that the product $\cos \left(s \sqrt{-\Delta^{\mathfrak{z}(\gamma)}/2 + A} \right) H^\gamma$ is well defined.

The function $J_\gamma \left(Y_0^{\mathfrak{k}} \right) \rho^E \left(k^{-1} \right) \exp \left(-i\rho^E \left(Y_0^{\mathfrak{k}} \right) \right)$ can be viewed as a smooth function on the second copy of $\mathfrak{z}(\gamma)$ in $\mathfrak{z}(\gamma) \times \mathfrak{z}(\gamma)$. It lifts to a smooth function on $\mathfrak{z}(\gamma) \times \mathfrak{z}(\gamma)$.

Therefore,

$$\mathrm{Tr_s}^E \left[\cos \left(s \sqrt{-\Delta^{\mathfrak{z}(\gamma)}/2 + A} \right) H^\gamma J_\gamma \left(Y_0^{\mathfrak{k}} \right) \rho^E \left(k^{-1} \right) \exp \left(-i\rho^E \left(Y_0^{\mathfrak{k}} \right) \right) \right]$$
$$(6.3.12)$$

is a well-defined distribution on $\mathbf{R} \times \mathfrak{z}(\gamma) \times \mathfrak{z}(\gamma)$. The pushforward of this distribution by the projection $\mathbf{R} \times \mathfrak{z}(\gamma) \times \mathfrak{z}(\gamma) \to \mathbf{R}$ will be denoted

$$\int_{H^\gamma} \mathrm{Tr_s}^E \left[\cos \left(s \sqrt{-\Delta^{\mathfrak{z}(\gamma)}/2 + A} \right) J_\gamma \left(Y_0^{\mathfrak{k}} \right) \rho^E \left(k^{-1} \right) \exp \left(-i\rho^E \left(Y_0^{\mathfrak{k}} \right) \right) \right].$$
$$(6.3.13)$$

This is an even distribution supported in $|s| \geq \sqrt{2} |a|$, with singular support included in $s = \pm\sqrt{2} |a|$. Note that if $a = 0$ and if $\mathfrak{p}(\gamma) = 0$, the singular support of this distribution is empty.

Tautologically,

$$
\mathrm{Tr}^{E}\left[\cos\left(s\sqrt{-\Delta^{\mathfrak{z}(\gamma)}/2+A}\right)J_{\gamma}\left(Y_{0}^{\mathfrak{k}}\right)\rho^{E}\left(k^{-1}\right)\exp\left(-i\rho^{E}\left(Y_{0}^{\mathfrak{k}}\right)\right)\delta_{y=a}\right](0)
$$
$$
=\int_{H^{\gamma}}\mathrm{Tr_{s}}^{E}\left[\cos\left(s\sqrt{-\Delta^{\mathfrak{z}(\gamma)}/2+A}\right)J_{\gamma}\left(Y_{0}^{\mathfrak{k}}\right)\right.
$$
$$
\left.\rho^{E}\left(k^{-1}\right)\exp\left(-i\rho^{E}\left(Y_{0}^{\mathfrak{k}}\right)\right)\right]. \quad (6.3.14)
$$

Theorem 6.3.2. *We have the identity of even distributions on* $\mathbf{R}$ *supported on* $|s|\geq\sqrt{2}\,|a|$ *with singular support included in* $\pm\sqrt{2}\,|a|$,

$$
\int_{\Delta_{X}^{\gamma}}\mathrm{Tr}^{F}\left[\gamma\cos\left(s\sqrt{\mathcal{L}^{X}+A}\right)\right]
$$
$$
=\int_{H^{\gamma}}\mathrm{Tr}^{E}\left[\cos\left(s\sqrt{-\Delta^{\mathfrak{z}(\gamma)}/2+A}\right)J_{\gamma}\left(Y_{0}^{\mathfrak{k}}\right)\right.
$$
$$
\left.\rho^{E}\left(k^{-1}\right)\exp\left(-i\rho^{E}\left(Y_{0}^{\mathfrak{k}}\right)\right)\right]. \quad (6.3.15)
$$

Proof. By (6.2.6), (6.3.1), (6.3.4), (6.3.10), and (6.3.14), we get (6.3.15). $\square$

Remark 6.3.3. Theorem 6.3.2 is a microlocal version of Theorem 6.2.2. Along the lines of Remark 6.2.3, note that although the hypoelliptic Laplacian does not have a wave operator, equation (6.3.15) has been obtained via the hypoelliptic Laplacian and its corresponding heat kernel.

Chapter Seven

An application to local index theory

The purpose of this chapter is to verify the compatibility of our formula in Theorem 6.1.1 for the orbital integrals of heat kernels to the index formula of Atiyah-Singer [AS68a, AS68b], to the fixed point formulas of Atiyah-Bott [ABo67, ABo68], and to the index formula for orbifolds of Kawasaki [Ka79]. Recall that the McKean-Singer formula [McKS67] expresses the index of a Dirac operator over a compact manifold Z as the supertrace of a heat kernel. If Z is the quotient of X by a cocompact torsion free group, this supertrace can be evaluated explicitly by the formulas we gave in chapter 6. Here we will directly check these formulas to be compatible with the index formulas.

This chapter is organized as follows. In section 7.1, we establish identities that relate the characteristic forms of TX to the characteristic forms of N.

In section 7.2, we construct the Dirac operator D^X acting on twisted spinors over X.

In section 7.3, we state the McKean-Singer formula for the index of a Dirac operator.

In section 7.4, we evaluate the orbital integrals associated with the index of a Dirac operator, when γ is the identity, and when γ is semisimple and nonelliptic.

In section 7.5, we establish the results of section 7.4.

In section 7.6, we establish a compatibility result when X is a complex manifold.

In section 7.7, we extend our results to the case when γ is elliptic.

In section 7.8, we consider the de Rham-Hodge operator $d^X + d^{X*}$.

Finally, in section 7.9, we evaluate the integrand that appears in the definition of the Ray-Singer analytic torsion [RS71] for the de Rham complex. In particular, we recover the vanishing results of Moscovici-Stanton [MoSt91] for this integrand.

7.1 Characteristic forms on X

We will work here on the manifold X, but the statements that follow can be easily obtained in more general situations.

Recall that as we saw in equation (2.2.2), the vector bundle $TX \oplus N$ is equipped with a flat connection $\nabla^{TX \oplus N, f}$, which preserves the bilinear form B. On the other hand, the connection $\nabla^{TX \oplus N}$ preserves the splitting, and also the bilinear form B. The connection $\nabla^{TX \oplus N}$ is a metric connection for the metric $g^{TX \oplus N} = g^{TX} \oplus g^N$ induced by B on $TX \oplus N$.

In this context, we can use the formalism of Bismut-Lott [BLo95, section 1(g)]. Namely, set

$$\omega \left(TX \oplus N, g^{TX \oplus N} \right) = \left(g^{TX \oplus N} \right)^{-1} \nabla^{TX \oplus N, f} g^{TX \oplus N}. \tag{7.1.1}$$

Then $\omega \left(TX \oplus N, g^{TX \oplus N} \right)$ is a 1-form with values in symmetric endomorphisms of $TX \oplus N$. One readily verifies that

$$\omega \left(TX \oplus N, g^{TX \oplus N} \right) = -2 \mathrm{ad}\,(\cdot). \tag{7.1.2}$$

By comparing with (2.2.2), we get

$$\nabla^{TX \oplus N} = \nabla^{TX \oplus N, f} + \frac{1}{2} \omega \left(TX \oplus N, g^{TX \oplus N} \right). \tag{7.1.3}$$

In the formalism of [BLo95], $\nabla^{TX \oplus N}$ is the metric preserving connection canonically associated with the flat connection $\nabla^{TX \oplus N, f}$ and the metric $g^{TX \oplus N}$.

If $\left(H, g^H, \nabla^H \right)$ is a complex Hermitian vector bundle equipped with a metric connection, we denote by $R^H = \nabla^{H, 2}$ the curvature of ∇^H. The Chern character form $\mathrm{ch}\left(H, \nabla^H \right)$ is given by

$$\mathrm{ch}\left(H, \nabla^H \right) = \mathrm{Tr}\left[\exp \left(-\frac{R^H}{2i\pi} \right) \right]. \tag{7.1.4}$$

If $\left(H', g^{H'}, \nabla^{H'} \right)$ is a real Euclidean vector bundle equipped with an Euclidean connection with curvature $R^{H'}$, the form $\widehat{A}\left(H', \nabla^{H'} \right)$ is given by

$$\widehat{A}\left(H', \nabla^{H'} \right) = \left[\det \left(\widehat{A}\left(-\frac{R^{H'}}{2i\pi} \right) \right) \right]^{1/2}. \tag{7.1.5}$$

Then the forms $\mathrm{ch}\left(H, \nabla^H \right), \widehat{A}\left(H', \nabla^{H'} \right)$ are real and closed, and the corresponding cohomology classes are denoted $\mathrm{ch}\,(H), \widehat{A}\,(H')$.

Proposition 7.1.1. *The following identities of closed forms hold on X:*

$$\mathrm{ch}\left(TX, \nabla^{TX} \right) + \mathrm{ch}\left(N, \nabla^N \right) = \dim \mathfrak{g}, \quad \widehat{A}\left(TX, \nabla^{TX} \right) \widehat{A}\left(N, \nabla^N \right) = 1.$$

$$\tag{7.1.6}$$

Proof. We can use (7.1.3) and obtain (7.1.6) as a consequence of [BLo95, Proposition 1.3]. We will give here a direct proof, by following [BLo95]. The curvature of ∇^{TX} and of ∇^N is obtained by restricting $\mathrm{ad}\,(\Omega) \in \Lambda^2 \left(T^* X \right) \otimes \mathrm{End}\,(\mathfrak{g})$ to $\mathfrak{p}$ and to $\mathfrak{k}$. By (2.1.10),

$$\mathrm{ad}\,(\Omega) = -\mathrm{ad}^2 \left(\omega^{\mathfrak{p}} \right). \tag{7.1.7}$$

Let φ be the endomorphism of $\Lambda^{\cdot} \left(T^* Z \right)$ given by $\alpha \to (2i\pi)^{-\deg \alpha / 2} \alpha$. We consider $\mathfrak{g} = \mathfrak{p} \oplus \mathfrak{k}$ as a trivially $\mathbf{Z}_2$-graded vector space, and $TX \oplus N$ as the corresponding trivially $\mathbf{Z}_2$-graded vector bundle. By (7.1.7),

$$\mathrm{ch}\left(TX, \nabla^{TX} \right) + \mathrm{ch}\left(N, \nabla^N \right) = \varphi \mathrm{Tr}\left[\exp \left(\mathrm{ad}^2 \left(\omega^{\mathfrak{p}} \right) \right) \right]. \tag{7.1.8}$$

Clearly,

$$\mathrm{ad}^2\left(\omega^{\mathfrak{p}}\right) = \frac{1}{2}\left[\mathrm{ad}\left(\omega^{\mathfrak{p}}\right), \mathrm{ad}\left(\omega^{\mathfrak{p}}\right)\right]. \tag{7.1.9}$$

Since traces vanish on supercommutators, we get

$$\frac{\partial}{\partial s}\mathrm{Tr}\left[\exp\left(s\mathrm{ad}^2\left(\omega^{\mathfrak{p}}\right)\right)\right] = \mathrm{Tr}\left[\mathrm{ad}^2\left(\omega^{\mathfrak{p}}\right)\exp\left(s\mathrm{ad}^2\left(\omega^{\mathfrak{p}}\right)\right)\right]$$

$$= \frac{1}{2}\mathrm{Tr}\left[\left[\mathrm{ad}\left(\omega^{\mathfrak{p}}\right), \mathrm{ad}\left(\omega^{\mathfrak{p}}\right)\exp\left(s\mathrm{ad}^2\left(\omega^{\mathfrak{p}}\right)\right)\right]\right] = 0. \tag{7.1.10}$$

Using (7.1.10), and making $s = 0$, we get the first identity in (7.1.6).

Similarly,

$$\widehat{A}\left(TX, \nabla^{TX}\right) = \varphi\left[\det\left(\widehat{A}\left(\mathrm{ad}^2\omega^{\mathfrak{p}}\right)|_{\mathfrak{p}}\right)\right]^{1/2}, \tag{7.1.11}$$

$$\widehat{A}\left(N, \nabla^{TN}\right) = \varphi\left[\det\left(\widehat{A}\left(\mathrm{ad}^2\omega^{\mathfrak{p}}\right)|_{\mathfrak{k}}\right)\right]^{1/2}.$$

By (7.1.11), we get

$$\log\left(\widehat{A}\left(TX, \nabla^{TX}\right)\widehat{A}\left(N, \nabla^{N}\right)\right) = \frac{1}{2}\varphi\mathrm{Tr}\left[\log\left(\widehat{A}\left(\mathrm{ad}^2\left(\omega^{\mathfrak{p}}\right)\right)\right)\right]. \tag{7.1.12}$$

By using the same deformation argument as before, from (7.1.12), we get the second identity in (7.1.6). $\qquad\square$

7.2 The vector bundle of spinors on X and the Dirac operator

Here we will assume K to be connected and simply connected, and that $\mathfrak{p}$ is even dimensional and oriented. Then the adjoint representation $K \to \mathrm{Aut}\left(\mathfrak{p}\right)$ preserves the orientation of $\mathfrak{p}$, and lifts to a unitary representation $\rho^{S^{\mathfrak{p}}} : K \to \mathrm{Aut}^{\mathrm{even}}\left(S^{\mathfrak{p}}\right)$, where $S^{\mathfrak{p}} = S_+^{\mathfrak{p}} \oplus S_-^{\mathfrak{p}}$ is the $\mathbf{Z}_2$-graded vector space of $\mathfrak{p}$-spinors. Then $S^{\mathfrak{p}}$ descends to the vector bundle $S^{TX} = S_+^{TX} \oplus S_-^{TX}$ of $\left(TX, g^{TX}\right)$ spinors.

Here $c\left(\mathfrak{p}\right)$ still denotes the Clifford algebra of $\left(\mathfrak{p}, B|_{\mathfrak{p}}\right)$. To avoid confusion with earlier parts of the book, if $e \in \mathfrak{p}$, $\overline{c}\left(e\right)$ will denote the action of $e \in c\left(\mathfrak{p}\right)$ on $S^{\mathfrak{p}}$. More generally, the objects attached to $c\left(\mathfrak{p}\right)$ that were considered in chapter 1 will be overlined when acting on $S^{\mathfrak{p}}$. By (1.1.9), if $f \in \mathfrak{k}$,

$$\rho^{S^{\mathfrak{p}}}\left(f\right) = \overline{c}\left(\mathrm{ad}\left(f\right)|_{\mathfrak{p}}\right). \tag{7.2.1}$$

Let D^X be the Dirac operator acting on $C^{\infty}\left(X, S^{TX} \otimes F\right)$. If $e_1, \ldots, e_m$ is an orthonormal basis of TX, then

$$D^X = \sum_{i=1}^{m}\overline{c}\left(e_i\right)\nabla_{e_i}^{S^{TX} \otimes F}. \tag{7.2.2}$$

The operator D^X can be written in matrix form with respect to the obvious splitting of $C^{\infty}\left(S^{TX} \otimes F\right)$, so that

$$D^X = \begin{bmatrix} 0 & D_-^X \\ D_+^X & 0 \end{bmatrix}. \tag{7.2.3}$$

Let $e_1, \ldots, e_{m+n}$ be a basis of $\mathfrak{g}$ taken as in (2.5.2). Let $C^{\mathfrak{g},X}$ be the Casimir operator acting on $C^\infty\left(X, S^{TX} \otimes F\right)$. Let $C^{\mathfrak{k},S^{\mathfrak{p}} \otimes E}$ be the Casimir operator for $\mathfrak{k}$ associated with the representation of K on $S^{\mathfrak{p}} \otimes E$. Of course this Casimir operator is still associated with the restriction of B to $\mathfrak{k}$, so that

$$C^{\mathfrak{k},S^{\mathfrak{p}} \otimes E} = \sum_{i=m+1}^{m+n} \left(\overline{c}\left(\mathrm{ad}\left(e_i\right)|_{\mathfrak{p}}\right) + \rho^E\left(e_i\right)\right)^2. \tag{7.2.4}$$

Recall that for $e \in \mathfrak{p}$, $\mathrm{ad}\,(e)$ acts as a symmetric endomorphism of $\mathfrak{g}$ with respect to its scalar product, which exchanges $\mathfrak{p}$ and $\mathfrak{k}$. Then we can rewrite (7.2.4) in the form

$$C^{\mathfrak{k},S^{\mathfrak{p}} \otimes E} = \sum_{i=m+1}^{m+n} \left(\frac{1}{4} \sum_{1 \le j,k \le m} \langle e_i, [e_j, e_k]\rangle \overline{c}\left(e_j\right) \overline{c}\left(e_k\right) - \rho^E\left(e_i\right)\right)^2. \tag{7.2.5}$$

As in the last identity in (2.12.17), we get

$$C^{\mathfrak{g},X} = C^{\mathfrak{g},H} + C^{\mathfrak{k},S^{\mathfrak{p}} \otimes E}. \tag{7.2.6}$$

Recall that $C^{\mathfrak{k},E} \in \mathrm{End}\,(E)$ was defined in (2.5.8).

Let $\Delta^{H,X}$ be the Bochner Laplacian acting on $C^\infty\left(X, S^{TX} \otimes F\right)$. We have the Lichnerowicz formula

$$D^{X,2} = -\Delta^{X,H} + \frac{R^X}{4} + \frac{1}{2} \sum_{1 \le i,j \le m} \overline{c}\left(e_i\right) \overline{c}\left(e_j\right) R^F\left(e_i, e_j\right). \tag{7.2.7}$$

Let $\mathcal{L}^X$ be the operator defined in (2.13.3), with E replaced by $S^{\mathfrak{p}} \otimes E$. Then

$$\mathcal{L}^X = \frac{1}{2} C^{\mathfrak{g},X} + \frac{1}{8} B^*\left(\kappa^{\mathfrak{g}}, \kappa^{\mathfrak{g}}\right). \tag{7.2.8}$$

Now we prove an identity in the spirit of an identity in [W88, section 9.3, Lemma].

Theorem 7.2.1. *The following identity holds:*

$$\frac{D^{X,2}}{2} = \mathcal{L}^X - \frac{1}{8} B^*\left(\kappa^{\mathfrak{k}}, \kappa^{\mathfrak{k}}\right) - \frac{1}{2} C^{\mathfrak{k},E}. \tag{7.2.9}$$

Proof. Let $D^{\mathfrak{g}}$ be the obvious analogue of the operator in (2.7.2), in which the Clifford variables $c\left(e_i\right)$ have been replaced by the $\overline{c}\left(e_i\right)$, which are now defined for $1 \le i \le m + n$. Put

$$D_H^{\mathfrak{g}} = \sum_{i=1}^{m} \overline{c}\left(e_i\right) e_i, \tag{7.2.10}$$

$$D_V^{\mathfrak{g}} = -\sum_{i=m+1}^{m+n} \overline{c}\left(e_i\right)\left(e_i + \overline{c}\left(\mathrm{ad}\left(e_i\right)|_{\mathfrak{p}}\right)\right) + \frac{1}{2}\overline{c}\left(\kappa^{\mathfrak{k}}\right),$$

so that as in (2.7.5), (2.7.6),

$$D^{\mathfrak{g}} = D_H^{\mathfrak{g}} + D_V^{\mathfrak{g}}. \tag{7.2.11}$$

By Theorem 2.7.2,

$$D^{\mathfrak{g},2} = C^{\mathfrak{g}} + \frac{1}{4}B^* \left(\kappa^{\mathfrak{g}}, \kappa^{\mathfrak{g}}\right), \qquad D^{\mathfrak{g},2} = D_H^{\mathfrak{g},2} + D_V^{\mathfrak{g},2}. \tag{7.2.12}$$

Clearly the operator $D_H^{\mathfrak{g}}$ descends to the operator D^X. On the other hand the operator $D_V^{\mathfrak{g}}$ descends to an operator $D_V^{\mathfrak{g},X}$ given by

$$D_V^{\mathfrak{g},X} = \sum_{i=m+1}^{m+n} \overline{c}\,(e_i)\, \rho^E\,(e_i) + \frac{1}{2}\overline{c}\left(\kappa^{\mathfrak{k}}\right). \tag{7.2.13}$$

By Theorem 2.7.2 applied to K, we get

$$D_V^{\mathfrak{g},X,2} = C^{\mathfrak{k},E} + \frac{1}{4}B^* \left(\kappa^{\mathfrak{k}}, \kappa^{\mathfrak{k}}\right). \tag{7.2.14}$$

By (7.2.8), (7.2.12) and (7.2.14), we get (7.2.9). The proof of our theorem is completed. $\square$

Put

$$\mathcal{A} = -\frac{1}{48}\mathrm{Tr}^{\mathfrak{k}}\left[C^{\mathfrak{k},\mathfrak{k}}\right] - \frac{1}{2}C^{\mathfrak{k},E}. \tag{7.2.15}$$

Then $\mathcal{A}$ has the properties of A given in section 4.4. By (2.6.7), (4.4.1), (7.2.9), and (7.2.15), we get

$$\frac{D^{X,2}}{2} = \mathcal{L}_{\mathcal{A}}^X. \tag{7.2.16}$$

7.3 The McKean-Singer formula on Z

This section serves only as a motivation for the next sections.

We make the assumptions of section 4.8 and we use the corresponding notation. In particular Z is a compact orbifold. The bundle of TX-spinors S^{TX} descends to the bundle of TZ-spinors S^{TZ}.

The Dirac operator D^X descends to the Dirac operator D^Z, which acts on $C^\infty\left(Z, S^{TZ} \otimes F\right)$. Similarly the operator $\mathcal{L}_{\mathcal{A}}^X$ descends to an operator $\mathcal{L}_{\mathcal{A}}^Z$. The operator D_+^Z is a Fredholm operator. When Γ does not contain elliptic elements, Z is smooth, and the index $\mathrm{Ind}\left(D_+^Z\right)$ of D_+^Z is given by the Atiyah-Singer index formula

$$\mathrm{Ind}\left(D_+^Z\right) = \int_Z \widehat{A}\,(TX)\,\mathrm{ch}\,(F). \tag{7.3.1}$$

In the case where Z is an orbifold, $\mathrm{Ind}\left(D_+^Z\right)$ is given by the Kawasaki formula of [Ka79]. This formula of Kawasaki is an elaboration of the fixed point formula of Atiyah-Bott [ABo67, ABo68].

The McKean-Singer formula [McKS67] for the index $\mathrm{ind}\left(D_+^Z\right)$ of D_+^Z asserts that for any $t > 0$,

$$\mathrm{Tr}_s\left[\exp\left(-tD^{Z,2}\right)\right] = \mathrm{Ind}\left(D_+^Z\right). \tag{7.3.2}$$

By (7.2.16), (7.3.2), for $t > 0$,

$$\mathrm{Tr}_s \left[\exp \left(-t \mathcal{L}_{\mathcal{A}}^Z \right) \right] = \mathrm{Ind} \left(D_+^Z \right), \tag{7.3.3}$$

and $\mathrm{Ind} \left(D_+^Z \right)$ is given by (7.3.1).

On the other hand, Theorem 6.1.1 gives us an explicit formula for the left-hand side of (7.3.3). In the next sections, we will verify that Theorem 6.1.1 is compatible with (7.3.3) and with the index formulas of Atiyah-Singer and Kawasaki.

7.4 Orbital integrals and the index theorem

We make again the same assumptions as in sections 7.1 and 7.2. We may disregard the considerations we made on Γ and the manifold Z.

Let η be the unit volume form on $\mathfrak{p}$ or on TX that defines the orientation of $\mathfrak{p}$ or of TX. If $\alpha \in \Lambda^{\cdot} (\mathfrak{p}^*)$ or if $\alpha \in \Lambda^{\cdot} (T^*X)$, for $0 \le p \le m$, let $\alpha^{(p)}$ be the component of α in $\Lambda^p (\mathfrak{p}^*)$ or in $\Lambda^p (T^*X)$. Let $\alpha^{\max} \in \mathbf{R}$ be defined by

$$\alpha^{(m)} = \alpha^{\max} \eta. \tag{7.4.1}$$

Let γ be a semisimple element of G as in (4.2.1).

Theorem 7.4.1. *If $\gamma \in G$ is nonelliptic, i.e., if $a \ne 0$, for $Y_0^{\mathfrak{k}} \in \mathfrak{k}(\gamma)$,*

$$\mathrm{Tr}_s^{S^{\mathfrak{p}}} \left[\varrho^{S^{\mathfrak{p}}} \left(k^{-1} \right) \exp \left(-i \rho^{S^{\mathfrak{p}}} \left(Y_0^{\mathfrak{k}} \right) \right) \right] = 0, \tag{7.4.2}$$

and for any $t > 0$,

$$\mathrm{Tr}_s^{[\gamma]} \left[\exp \left(-t D^{X,2}/2 \right) \right] = 0. \tag{7.4.3}$$

Moreover, for any $t > 0$,

$$\mathrm{Tr}_s^{[1]} \left[\exp \left(-t D^{X,2}/2 \right) \right] = \frac{1}{(2\pi t)^{m/2}} \int_{\mathfrak{k}} J_1 \left(Y_0^{\mathfrak{k}} \right)$$

$$\mathrm{Tr}_s^{S^{\mathfrak{p}} \otimes E} \left[\exp \left(-i \rho^{S^{\mathfrak{p}} \otimes E} \left(Y_0^{\mathfrak{k}} \right) - t \mathcal{A} \right) \right] \exp \left(-\left| Y_0^{\mathfrak{k}} \right|^2 / 2t \right) \frac{dY_0^{\mathfrak{k}}}{(2\pi t)^{n/2}}$$

$$= \left[\widehat{A} \left(TX, \nabla^{TX} \right) \mathrm{ch} \left(F, \nabla^F \right) \right]^{\max}. \tag{7.4.4}$$

Proof. Recall that $a \in \mathfrak{p}$ is fixed by $\mathrm{Ad}(k)$, and also that $\mathrm{ad}\left(Y_0^{\mathfrak{k}}\right)$ vanishes on a. Since $a \ne 0$, 1 is an eigenvalue of $\mathrm{Ad}\left(k^{-1}\right) \exp\left(-i\mathrm{ad}\left(Y_0^{\mathfrak{k}}\right)\right)$. Also it is well-known that

$$\mathrm{Tr}_s^{S^{\mathfrak{p}}} \left[\varrho^{S^{\mathfrak{p}}} \left(k^{-1} \right) \exp \left(-i \rho^{S^{\mathfrak{p}}} \left(Y_0^{\mathfrak{k}} \right) \right) \right]$$

is a square root of $\det \left(1 - \mathrm{Ad}\left(k^{-1}\right) \exp\left(-i\mathrm{ad}\left(Y_0^{\mathfrak{k}}\right)\right) \right)_{\mathfrak{p}}$, which vanishes. This proves (7.4.2). By equation (6.1.2) in Theorem 6.1.1 and by (7.4.2), we get (7.4.3).

The first part of (7.4.4) follows from Theorem 6.1.1. The proof of the second part is deferred to section 7.5. $\square$

7.5 A proof of (7.4.4)

First, we assume that K is connected, simply connected and simple. Let $T \subset K$ be a maximal torus in K and let $\mathfrak{t} \subset \mathfrak{k}$ be its Lie algebra. Let W be the Weyl group. Let $R \subset \mathfrak{t}^*$ be the root system, and let $\overline{CR} \subset \mathfrak{t}$ be the system of coroots, so that $T \simeq \mathfrak{t}/\overline{CR}$. Our conventions are such that if $\alpha \in R, h \in \overline{CR}, \langle \alpha, h \rangle \in \mathbf{Z}$. Let $\overline{CR}^* \subset \mathfrak{t}^*$ be the dual lattice to $\overline{CR}$, so that if $h \in \overline{CR}, \lambda \in \overline{CR}^*, \langle \lambda, h \rangle \in \mathbf{Z}$.

Let $R_+ \subset R$ be a system of positive roots, and let $C_+ \in \mathfrak{t}$ be the positive Weyl chamber. Let $P_{++} \subset \overline{CR}^*$ be the system of dominant weights.

Set

$$\rho = \frac{1}{2} \sum_{\alpha \in R_+} \alpha. \tag{7.5.1}$$

As is well-known, $\rho \in P_{++}$.

Let $\pi : \mathfrak{t} \to \mathbf{C}$ be the polynomial function

$$\pi(t) = \prod_{\alpha \in R_+} \langle 2i\pi\alpha, t \rangle. \tag{7.5.2}$$

Let $\sigma : T \to \mathbf{C}$ be the denominator in Weyl's character formula

$$\sigma(t) = \prod_{\alpha \in R_+} \left(\exp\left(i\pi \langle \alpha, t \rangle \right) - \exp\left(-i\pi \langle \alpha, t \rangle \right) \right). \tag{7.5.3}$$

Proposition 7.5.1. *The following identity holds:*

$$4\pi^2 |\rho|^2 = -\frac{1}{24} \mathrm{Tr}^{\mathfrak{k}} \left[C^{\mathfrak{k},\mathfrak{k}} \right] = -\frac{1}{4} B^* \left(\kappa^{\mathfrak{k}}, \kappa^{\mathfrak{k}} \right). \tag{7.5.4}$$

Proof. The first identity is Kostant's strange formula [Ko76]. The second one is just equation (2.6.7). $\square$

In the sequel, we will assume that ρ^E is an irreducible representation of K of highest weight $\lambda \in P_{++}$. Let $\chi_\lambda : K \to \mathbf{C}$ be the corresponding character, so that if $g \in K$,

$$\chi_\lambda(g) = \mathrm{Tr}^E \left[\rho^E(g) \right]. \tag{7.5.5}$$

Proposition 7.5.2. *The following identity holds:*

$$\mathcal{A} = 2\pi^2 |\rho + \lambda|^2. \tag{7.5.6}$$

Proof. Since ρ^E is the representation of K of highest weight λ, we get

$$C^{\mathfrak{k},E} = -4\pi^2 \left(|\rho + \lambda|^2 - |\rho|^2 \right). \tag{7.5.7}$$

By (7.2.15), (7.5.4), and (7.5.7), we get (7.5.6). $\square$

If $A \in \mathrm{End}(\mathfrak{p})$ is antisymmetric, let $\mathrm{Pf}[A]$ be the Pfaffian of A. Then $\mathrm{Pf}[A]$ is a polynomial function of A, which is a square root of $\det[A]$. The form $\omega_A \in \Lambda^2(\mathfrak{p}^*)$ associated with A is given by $U, V \in \mathfrak{p} \to \langle U, AV \rangle$. Then $\mathrm{Pf}[A]$ is given by

$$\mathrm{Pf}[A] = [\exp(\omega_A)]^{\mathrm{max}}. \tag{7.5.8}$$

Theorem 7.5.3. *For $t > 0$,*

$$\frac{\exp\left(-2\pi^2 t \, |\rho + \lambda|^2\right)}{(2\pi t)^{m/2}} \int_{\mathfrak{k}} J_1\left(Y_0^{\mathfrak{k}}\right) \mathrm{Tr_s}^{S^{\mathfrak{p}} \otimes E}\left[\exp\left(-i\rho^{S^{\mathfrak{p}} \otimes E}\left(Y_0^{\mathfrak{k}}\right)\right)\right]$$

$$\exp\left(-\left|Y_0^{\mathfrak{k}}\right|^2 / 2t\right) \frac{dY_0^{\mathfrak{k}}}{(2\pi t)^{n/2}} = \left[\widehat{A}\left(TX, \nabla^{TX}\right) \mathrm{ch}\left(F, \nabla^F\right)\right]^{\max}. \quad (7.5.9)$$

Proof. We may and we will assume that $t = 1$. Clearly,

$$\mathrm{Tr_s}^{S^{\mathfrak{p}} \otimes E}\left[\exp\left(-i\rho^{S^m \otimes E}\left(Y_0^{\mathfrak{k}}\right)\right)\right] = \mathrm{Tr_s}^{S^{\mathfrak{p}}}\left[\exp\left(-ic\left(\mathrm{ad}\left(Y_0^{\mathfrak{k}}\right)\right)\right)\right]$$

$$\mathrm{Tr}^E\left[\exp\left(-i\rho^E\left(Y_0^{\mathfrak{k}}\right)\right)\right]. \quad (7.5.10)$$

Also we have the fundamental identity

$$\mathrm{Tr_s}^{S^{\mathfrak{p}}}\left[\exp\left(-ic\left(\mathrm{ad}\left(Y_0^{\mathfrak{k}}\right)\right)\right)\right] = \mathrm{Pf}\left[\mathrm{ad}\left(Y_0^{\mathfrak{k}}\right)|_{\mathfrak{p}}\right] \widehat{A}^{-1}\left(i\mathrm{ad}\left(Y_0^{\mathfrak{k}}\right)|_{\mathfrak{p}}\right). \quad (7.5.11)$$

Set

$$L = \frac{1}{(2\pi)^{m/2}} \int_{\mathfrak{k}} J_1\left(Y_0^{\mathfrak{k}}\right) \mathrm{Tr_s}^{S^{\mathfrak{p}} \otimes E}\left[\exp\left(-i\rho^{S^{\mathfrak{p}} \otimes E}\left(Y_0^{\mathfrak{k}}\right)\right)\right]$$

$$\exp\left(-\left|Y_0^{\mathfrak{k}}\right|^2 / 2\right) \frac{dY_0^{\mathfrak{k}}}{(2\pi)^{n/2}}. \quad (7.5.12)$$

By (5.5.11) and (7.5.10)–(7.5.12), we get

$$L = \frac{1}{(2\pi)^{m/2}} \int_{\mathfrak{k}} \mathrm{Pf}\left[\mathrm{ad}\left(Y_0^{\mathfrak{k}}\right)|_{\mathfrak{p}}\right] \widehat{A}^{-1}\left(i\mathrm{ad}\left(Y_0^{\mathfrak{k}}\right)|_{\mathfrak{k}}\right) \mathrm{Tr}^E\left[\exp\left(-i\rho^E\left(Y_0^{\mathfrak{k}}\right)\right)\right]$$

$$\exp\left(-\left|Y_0^{\mathfrak{k}}\right|^2 / 2\right) \frac{dY_0^{\mathfrak{k}}}{(2\pi)^{n/2}}. \quad (7.5.13)$$

Let $e_1, \ldots, e_m$ be an oriented orthonormal basis of $\mathfrak{p}$. The 2-form $\omega_{\mathrm{ad}(Y_0^{\mathfrak{k}})}$ on $\mathfrak{p}$ associated with $\mathrm{ad}\left(Y_0^{\mathfrak{k}}\right)|_{\mathfrak{p}}$ is given by

$$\omega_{\mathrm{ad}(Y_0^{\mathfrak{k}})} = \frac{1}{2} \sum_{1 \leq i,j \leq m} \langle e_i, [Y_0^{\mathfrak{k}}, e_j] \rangle \, e^i \wedge e^j. \quad (7.5.14)$$

Recall that the curvature $\Omega \in \Lambda^2\left(\mathfrak{p}^*\right) \otimes \mathfrak{k}$ of the canonical connection on the K-principal bundle $p : G \to G/K$ is given by (2.1.10), so that

$$\Omega = -\frac{1}{2} \sum_{1 \leq i,j \leq m} e^i \wedge e^j \otimes [e_i, e_j]. \quad (7.5.15)$$

One can rewrite (7.5.14) in the form

$$\omega_{\mathrm{ad}(Y_0^{\mathfrak{k}})} = -\langle Y_0^{\mathfrak{k}}, \Omega \rangle. \quad (7.5.16)$$

In (7.5.16), the scalar product is taken in $\mathfrak{k}$ and only involves the $\mathfrak{k}$ component of Ω.

Using (7.5.8), (7.5.16), we get

$$\mathrm{Pf}\left[\mathrm{ad}\left(Y_0^{\mathfrak{k}}\right)|_{\mathfrak{p}}\right] = \left[\exp\left(-\left\langle Y_0^{\mathfrak{k}},\Omega\right\rangle\right)\right]^{\max}. \qquad (7.5.17)$$

In (7.5.17), the right-hand side is the coefficient of $e^1 \wedge \ldots \wedge e^m$ in the expansion of the given form.

By (7.5.13), (7.5.17), we obtain

$$L = \frac{\exp\left(|\Omega|^2/2\right)}{(2\pi)^{m/2}}\left[\int_{\mathfrak{k}} \widehat{A}^{-1}\left(i\,\mathrm{ad}\left(Y_0^{\mathfrak{k}}\right)|_{\mathfrak{k}}\right)\mathrm{Tr}^E\left[\exp\left(-i\rho^E\left(Y_0^{\mathfrak{k}}\right)\right)\right]\right.$$

$$\left.\exp\left(-\left|Y_0^{\mathfrak{k}} + \Omega\right|^2/2\right)\frac{dY_0^{\mathfrak{k}}}{(2\pi)^{n/2}}\right]^{\max}. \qquad (7.5.18)$$

In (7.5.18), $|\Omega|^2 \in \Lambda^{\mathrm{even}}(\mathfrak{p}^*)$ refers to the square of the norm of Ω where only the component of Ω in $\mathfrak{k}$ is taken into account, while the forms in $\Lambda^{\mathrm{even}}(\mathfrak{p}^*)$ are multiplied as usual.

We claim that

$$|\Omega|^2 = 0. \qquad (7.5.19)$$

Indeed

$$|\Omega|^2 = -\frac{1}{4}\sum_{1\leq i,j,k,l\leq m}\left\langle[[e_i,e_j],e_k],e_l\right\rangle e^i \wedge e^j \wedge e^k \wedge e^l. \qquad (7.5.20)$$

Using the Jacobi identity, we get (7.5.19).

Recall that $\Delta^{\mathfrak{k}}$ is the standard Laplacian in $\mathfrak{k}$. From (7.5.18), (7.5.19), we get

$$L = \frac{1}{(2\pi)^{m/2}}\left[\exp\left(\Delta^{\mathfrak{k}}/2\right)\left(\widehat{A}^{-1}\left(i\,\mathrm{ad}\left(Y_0^{\mathfrak{k}}\right)|_{\mathfrak{k}}\right)\chi_\lambda\left(\exp\left(-iY_0^{\mathfrak{k}}\right)\right)\right)(-\Omega)\right]^{\max}. \qquad (7.5.21)$$

Let $\Delta^{\mathfrak{t}}$ be the Laplacian on $\mathfrak{t}$. It is well-known that when acting on Ad-invariant functions on $\mathfrak{k}$, we have the identity of operators,

$$\Delta^{\mathfrak{k}} = \frac{1}{\pi}\Delta^{\mathfrak{t}}\pi. \qquad (7.5.22)$$

The function

$$\widehat{A}^{-1}\left(i\,\mathrm{ad}\left(Y_0^{\mathfrak{k}}\right)|_{\mathfrak{k}}\right)\chi_\lambda\left(\exp\left(-iY_0^{\mathfrak{k}}\right)\right)$$

is Ad-invariant. By (7.5.21), (7.5.22), we get

$$L = (2\pi)^{-m/2}\left[\frac{1}{\pi}\exp\left(\Delta^{\mathfrak{t}}/2\right)\right.$$

$$\left.\left(\pi(t)\,\widehat{A}^{-1}\left(i\,\mathrm{ad}(t)|_{\mathfrak{k}}\right)\chi_\lambda\left(\exp\left(-it\right)\right)\right)(-\Omega)\right]^{\max}. \qquad (7.5.23)$$

In the right-hand side of (7.5.23), the function appearing in the right-hand side of (7.5.23) is viewed as a function of $t \in \mathfrak{t}$, which is W-invariant, and lifts to a central function on $\mathfrak{k}$. Of course we do not claim that $-\Omega$ takes its values in $\mathfrak{t}$, but that since the right-hand side defines a central function on $\mathfrak{k}$, it can be evaluated at $-\Omega$. Equation (7.5.23) can also be obtained as a consequence of the Weyl integration formula, and of the localization formulas in equivariant cohomology of Duistermaat-Heckman [DH82, DH83], Berline-Vergne [BeVe83] over the generic coadjoint orbits of K.

Now if $t \in \mathfrak{t} \otimes_{\mathbf{R}} \mathbf{C}$, we have the obvious identity

$$\widehat{A}^{-1}\left(\mathrm{ad}\,(t)\,|_{\mathfrak{k}}\right) = \frac{\sigma}{\pi}\,(t)\,. \tag{7.5.24}$$

If $r_+ = |R_+|$, from (7.5.24), we get

$$\pi\,(t)\,\widehat{A}^{-1}\left(i\mathrm{ad}\,(t)\,|_{\mathfrak{k}}\right) = i^{r_+}\sigma\,(-it)\,. \tag{7.5.25}$$

If $w \in W$, let $\epsilon_w = \pm 1$ be the determinant of w acting on $\mathfrak{t}$. The Lefschetz formula asserts that

$$\sigma\,(-it)\,\chi_\lambda\,(\exp\,(-it)) = \sum_{w \in W} \epsilon_w \exp\left(\langle 2\pi\,\langle \rho + \lambda, wt\rangle\rangle\right)\,. \tag{7.5.26}$$

By (7.5.23)–(7.5.26), we finally obtain

$$L = (2\pi)^{-m/2}\exp\left(2\pi^2\,|\rho + \lambda|^2\right)\left[\widehat{A}^{-1}\left(-i\mathrm{ad}\,(\Omega)\,|_{\mathfrak{k}}\right)\chi_\lambda\,(\exp\,(i\Omega))\right]^{\max}\,. \tag{7.5.27}$$

Since $\widehat{A}\,(x)$ is an even function, we can rewrite (7.5.27) in the form

$$L = \exp\left(2\pi^2\,|\rho + \lambda|^2\right)\left[\widehat{A}^{-1}\left(-\frac{R^N}{2i\pi}\right)\mathrm{Tr}\left[\exp\left(-\frac{R^F}{2i\pi}\right)\right]\right]^{\max}, \tag{7.5.28}$$

which is just

$$L = \exp\left(2\pi^2\,|\rho + \lambda|^2\right)\left[\widehat{A}^{-1}\left(N, \nabla^N\right)\mathrm{ch}\,\left(F, \nabla^F\right)\right]^{\max}\,. \tag{7.5.29}$$

By the second identity in (7.1.6), by (7.5.12), and by (7.5.29), we get (7.5.9) for $t = 1$. The proof of our theorem is completed. $\qquad\square$

Remark 7.5.4. From (7.5.6) and (7.5.9), we claim that one can obtain the last identity in (7.4.4) when K is connected and simply connected. Indeed we can write K in the form

$$K = \prod_1^s K_i, \tag{7.5.30}$$

the K_i being simply connected and simple. We may and we will assume E to be an irreducible representation of K. Then E can be written in the form

$$E = \bigotimes_{i=1}^s E_i, \tag{7.5.31}$$

the E_i being irreducible representations of the K_i. By using (7.5.9) for each $i, 1 \leq i \leq s$, we obtain (7.4.4).

7.6 The case of complex symmetric spaces

For holomorphic locally symmetric spaces and bounded symmetric domains, we refer to Helgason [He78, Chapter VIII].

We will assume that $\mathfrak{p}$ is equipped with a complex structure $J^{\mathfrak{p}}$ such that if $a, b \in \mathfrak{p}$,

$$B\left(J^{\mathfrak{p}}a, J^{\mathfrak{p}}b\right) = B\left(a, b\right). \tag{7.6.1}$$

Let $\mathfrak{p}_+, \mathfrak{p}_- \in \mathfrak{p} \otimes_{\mathbf{R}} \mathbf{C}$ be the eigenspaces associated with the eigenvalues $i, -i$ of $J^{\mathfrak{p}}$, so that

$$\mathfrak{p} \otimes_{\mathbf{R}} \mathbf{C} = \mathfrak{p}_+ \oplus \mathfrak{p}_-. \tag{7.6.2}$$

In particular $\mathfrak{p}_+$ is now equipped with a Hermitian product. Also we assume that

$$[\mathfrak{p}_+, \mathfrak{p}_+] = 0, [\mathfrak{p}_-, \mathfrak{p}_-] = 0. \tag{7.6.3}$$

Finally, we assume that the action of K on $\mathfrak{p}$ preserves the complex structure $J^{\mathfrak{p}}$, so that K acts on $\mathfrak{p}_+$ by unitary automorphisms. Let $\rho^{\det(\mathfrak{p}_+)} : K \to S^1$ be the determinant of the representation.

The manifold $X = G/K$ is now a complex Kähler manifold. Indeed (7.6.2) guarantees that the almost complex structure on TX induced by $J^{\mathfrak{p}}$ is integrable. Since the connection ∇^{TX} is torsion free and preserves the complex structure, X is Kähler. Then $\mathfrak{p}_+$ descends to the holomorphic tangent space $T^{(1,0)}X$ of X.

As it should be, by (2.1.10) and (7.6.3), the curvature Ω is of complex type $(1,1)$. It follows that if $\rho^E = K \to \operatorname{Aut}(E)$ is a unitary representation of K, the corresponding vector bundle F on X is a holomorphic Hermitian vector bundle. Also G acts on the left by holomorphic isometries of X. If Γ is a cocompact torsion-free subgroup of G, $Z = \Gamma \backslash X$ is a compact Kähler manifold, if Γ is cocompact but not torsion free, Z is a Kähler orbifold.

We do not assume any more K to be simply connected. Still K acts unitarily on $\Lambda^{\cdot}\left(\mathfrak{p}_-^*\right)$ and preserves the grading. In the constructions of sections 7.2–7.3, we may as well replace $S^{\mathfrak{p}}$ by $\Lambda^{\cdot}\left(\mathfrak{p}_-^*\right)$. The manifolds X and Z are not necessarily spin, but they are equipped with a Spin^c structure.

We can still define the operators D^X as before. By [Hi74], we know that

$$D^X = \sqrt{2}\left(\overline{\partial}^X + \overline{\partial}^{X*}\right). \tag{7.6.4}$$

If Γ is a discrete group taken as before, a similar formula holds on Z.

Let $\chi(Z, F)$ be the Euler characteristic of F. Then $\chi(Z, F)$ is the index of the Dirac operator D^Z.

If Z is smooth, the Riemann-Roch-Hirzebruch formula asserts that

$$\chi(Z, F) = \int_Z \operatorname{Td}\left(T^{(1,0)}Z\right) \operatorname{ch}(F). \tag{7.6.5}$$

We claim that a full analogue of the results of sections 7.2–7.5 can be developed in this case. Let us briefly explain this point. Let $\rho^{1/2}_{\det(\mathfrak{p}_+)} \to S^1$

be the square root of $\rho^{\det(\mathfrak{p})}$, which is well-defined up to sign. Let $\lambda = \det\left(T^{(1,0)}Z\right)$ be the line bundle on Z corresponding to the representation $\rho^{\det(\mathfrak{p}_+)}$, and let $\lambda^{1/2}$ be a locally defined square root.

We still can define $\mathcal{A}$ as in (7.2.15), with E replaced by $E \otimes \det^{1/2}(\mathfrak{p}_+)$, since the Casimir operator is only obtained from the representation of the Lie algebra. The analogue of (7.4.2) is that if $a \neq 0, Y_0^{\mathfrak{k}} \in \mathfrak{k}(\gamma)$,

$$\mathrm{Tr_s}^{\Lambda^{\cdot}(\mathfrak{p}_-^*)}\left[\mathrm{Ad}\left(k^{-1}\right)\exp\left(-i\rho^{\Lambda^{\cdot}(\mathfrak{p}_-^*)}\left(Y_0^{\mathfrak{k}}\right)\right)\right] = 0, \qquad (7.6.6)$$

the proof being the same as in Theorem 7.4.1.

The analogue of (7.4.4) says that for any $t > 0$,

$$\mathrm{Tr_s}^{[1]}\left[\exp\left(-tD^{X,2}/2\right)\right] = \frac{1}{(2\pi t)^{m/2}}\int_{\mathfrak{k}} J_1\left(Y_0^{\mathfrak{k}}\right)$$

$$\mathrm{Tr_s}^{\Lambda^{\cdot}(\mathfrak{p}_-^*)\otimes E}\left[\exp\left(-i\rho^{\Lambda^{\cdot}(\mathfrak{p}_-^*)\otimes E}\left(Y_0^{\mathfrak{k}}\right) - t\mathcal{A}\right)\right]$$

$$\exp\left(-\left|Y_0^{\mathfrak{k}}\right|^2/2t\right)\frac{dY_0^{\mathfrak{k}}}{(2\pi t)^{n/2}} = \left[\mathrm{Td}\left(T^{(1,0)}X, \nabla^{T^{(1,0)}X}\right)\mathrm{ch}\left(F, \nabla^F\right)\right]^{\max}.$$

$$(7.6.7)$$

The proof essentially consists in extending the arguments of section 7.5. Note that K is no longer supposed to be simply connected. Let Z_K be the center of K and let $\mathfrak{z}_K$ be its Lie algebra. Let $\mathfrak{k}^s$ be the semisimple part of $\mathfrak{k}$, so that we have the orthogonal splitting of Lie algebras,

$$\mathfrak{k} = \mathfrak{z}_K \oplus \mathfrak{k}^s. \qquad (7.6.8)$$

The splitting (7.6.8) allows us to reduce the proof to the case where $K = T \times \prod_{i=1}^s K_i$, with T a torus and the K_i being simple groups. The proof proceeds as in Remark 7.5.4.

7.7 The case of an elliptic element

We still assume K to be simply connected.

Let $\gamma \in G$ be elliptic. We may and we will assume that $\gamma \in K$, so that $\gamma = k^{-1}, k \in K$. Then $X(\gamma) \subset X$ is just the fixed point set of k.

Note that γ acts naturally on $TX|_{X(\gamma)}$. Moreover, $TX(\gamma)$ is just the eigenspace of this action associated with the eigenvalue 1, and γ acts on $N_{X(\gamma)/X}$ as an isometry. The distinct angles $\pm\theta_1, \ldots, \pm\theta_s, 0 < \theta_i \leq \pi$ are exactly the nonzero angles of the action of $\mathrm{Ad}\left(k^{-1}\right)$ on $\mathfrak{p}$. Let $N_{X(\gamma)/X,\theta_i}, 1 \leq i \leq s$ be the part of $N_{X(\gamma)/X}$ on which $\mathrm{Ad}(\gamma)$ acts by a rotation of angle θ_i.

The action of γ on $TX|_{X(\gamma)}$ is parallel. Therefore ∇^{TX} induces metric connections on the above subbundles of $TX|_{X(\gamma)}$. Let $R^{TX(\gamma)}, R^{N_{X(\gamma)/X,\theta_i}}, 1 \leq i \leq s$ be the curvatures of these connections on $TX(\gamma), N_{X(\gamma)/X,\theta_i}, 1 \leq i \leq s$.

If $\theta \in \mathbf{R} \setminus 2\pi\mathbf{Z}$, set

$$\widehat{A}^{\theta}(x) = \frac{1}{2\sinh\left(\frac{x+i\theta}{2}\right)}. \qquad (7.7.1)$$

Given θ, we identify $\widehat{A}^\theta(x)$ with the corresponding multiplicative genus. Put

$$\widehat{A}^\gamma\left(TX|_{X(\gamma)}, \nabla^{TX|_{X_\gamma}}\right) = \widehat{A}\left(TX(\gamma), \nabla^{TX(\gamma)}\right)$$

$$\prod_{i=1}^r \widehat{A}^{\theta_i}\left(N_{X(\gamma)/X,\theta_i}, \nabla^{N_{X(\gamma)/X,\theta_i}}\right). \quad (7.7.2)$$

We can also define the closed form $\widehat{A}^\gamma\left(N|_{X(\gamma)}, \nabla^{N|_{X(\gamma)}}\right)$ by a similar formula. Let $\widehat{A}^{\gamma|_\mathfrak{p}}(0)$ be the component of degree 0 of $\widehat{A}^\gamma\left(TX|_{X(\gamma)}, \nabla^{TX|_{X_\gamma}}\right)$, and let $\widehat{A}^{\gamma|_\mathfrak{k}}(0)$ be the component of degree 0 of $\widehat{A}\left(N|_{X(\gamma)}, \nabla^{N|_{X(\gamma)}}\right)$. These are constants on $X(\gamma)$. Put

$$\widehat{A}^\gamma(0) = \widehat{A}^{\gamma|_\mathfrak{p}}(0)\,\widehat{A}^{\gamma|_\mathfrak{k}}(0). \quad (7.7.3)$$

Similarly, set

$$\mathrm{ch}^\gamma\left(F|_{X(\gamma)}, \nabla^{F|_{X(\gamma)}}\right) = \mathrm{Tr}\left[\rho^E\left(k^{-1}\right)\exp\left(-\frac{R^F|_{X(\gamma)}}{2i\pi}\right)\right]. \quad (7.7.4)$$

The closed forms in (7.7.2), (7.7.4) on $X(\gamma)$ are exactly the ones that appear in the Lefschetz fixed point formula of Atiyah-Bott [ABo67, ABo68]. Note that there are questions of signs to be taken care of, because of the need to distinguish between θ_i and $-\theta_i$. We refer to the above references for more detail.

By proceeding as in the proof of Proposition 7.1.1, one can easily prove the analogue of (7.1.6), i.e., we get the identity of differential forms on $X(\gamma)$,

$$\mathrm{ch}^\gamma\left(TX|_{X(\gamma)}, \nabla^{TX|_{X(\gamma)}}\right) + \mathrm{ch}^\gamma\left(N|_{X(\gamma)}, \nabla^{N|_{X(\gamma)}}\right) = \mathrm{Tr}^{\mathfrak{g}}\left[\mathrm{Ad}\left(k^{-1}\right)\right],$$

$$\quad (7.7.5)$$

$$\widehat{A}^\gamma\left(TX|_{X(\gamma)}, \nabla^{TX|_{X(\gamma)}}\right)\widehat{A}^\gamma\left(N|_{X(\gamma)}, \nabla^{N|_{X(\gamma)}}\right) = \widehat{A}^\gamma(0).$$

Now we prove an analogue of Theorem 7.4.1. Here the notation ^{max} refers to forms on $X(\gamma)$.

Theorem 7.7.1. *If $\gamma = k^{-1}, k \in K$, for any $t > 0$,*

$$\mathrm{Tr}_s^{[\gamma]}\left[\exp\left(-tD^{X,2}/2\right)\right]$$

$$= \frac{1}{(2\pi t)^{p/2}}\int_{\mathfrak{k}(\gamma)} J_\gamma\left(Y_0^\mathfrak{k}\right)\mathrm{Tr}_s^{S^\mathfrak{p}\otimes E}\left[\rho^{S^\mathfrak{p}\otimes E}\left(k^{-1}\right)\exp\left(-i\rho^{S^\mathfrak{p}\otimes E}\left(Y_0^\mathfrak{k}\right) - t\mathcal{A}\right)\right]$$

$$\exp\left(-\left|Y_0^\mathfrak{k}\right|^2/2t\right)\frac{dY_0^\mathfrak{k}}{(2\pi t)^{q/2}}$$

$$= \left[\widehat{A}^\gamma\left(TX|_{X(\gamma)}, \nabla^{TX|_{X(\gamma)}}\right)\mathrm{ch}^\gamma\left(F, \nabla^F\right)\right]^{\mathrm{max}}. \quad (7.7.6)$$

Proof. The first part of the identity follows from Theorem 6.1.1. So we concentrate on the proof of the second part. We may and we will assume that

$t = 1$. The proof proceeds very much as the proof of Theorems 7.4.1 and 7.5.3.

First, we work under the assumptions of Theorem 7.5.3. We may and we will assume that k lies in a maximal torus $T \subset K$, whose Lie algebra is still denoted by t. Let L be the obvious analogue of the expression in (7.5.12).

Instead of (7.5.11), we now have the identity

$$\mathrm{Tr}_s^{S^{\mathfrak{p}}} \left[\rho^{S^{\mathfrak{p}}} \left(k^{-1} \right) \exp \left(-ic \left(\mathrm{ad} \left(Y_0^{\mathfrak{k}} \right) \right) \right) \right]$$
$$= \mathrm{Pf} \left[\mathrm{ad} \left(Y_0^{\mathfrak{k}} \right) |_{\mathfrak{p}(\gamma)} \right] \widehat{A}^{-1} \left(iad \left(Y_0^{\mathfrak{k}} \right) |_{\mathfrak{p}(\gamma)} \right) \left(\widehat{A}^{ke^{iY_0^{\mathfrak{k}}}} |_{\mathfrak{p}^\perp(\gamma)} (0) \right)^{-1}. \quad (7.7.7)$$

By (5.4.10), (5.4.11), (5.5.5), (7.5.12), and (7.7.7), we get

$$L = \frac{(-1)^{\dim \mathfrak{p}^\perp(\gamma)/2}}{(2\pi)^{p/2}} \int_{\mathfrak{k}(\gamma)} \mathrm{Pf} \left[\mathrm{ad} \left(Y_0^{\mathfrak{k}} \right) |_{\mathfrak{p}(\gamma)} \right]$$
$$\widehat{A}^{-1} \left(iad \left(Y_0^{\mathfrak{k}} \right) |_{\mathfrak{k}(\gamma)} \right) \widehat{A}^{k} |_{\mathfrak{p}^\perp(\gamma)} (0)$$
$$\left[\frac{\det \left(1 - \exp \left(-iad \left(Y_0^{\mathfrak{k}} \right) \right) \mathrm{Ad} \left(k^{-1} \right) \right) |_{\mathfrak{k}^\perp(\gamma)}}{\det \left(1 - \mathrm{Ad} \left(k^{-1} \right) \right) |_{\mathfrak{k}^\perp(\gamma)}} \right]^{1/2}$$
$$\mathrm{Tr}^E \left[\rho^E \left(k^{-1} \right) \exp \left(-i\rho^E \left(Y_0^{\mathfrak{k}} \right) \right) \right] \exp \left(- \left| Y_0^{\mathfrak{k}} \right|^2 /2 \right) \frac{dY_0^{\mathfrak{k}}}{(2\pi)^{q/2}}. \quad (7.7.8)$$

Note that the factor $(-1)^{\dim \mathfrak{p}^\perp(\gamma)/2}$ in the right-hand side of (7.7.8) comes from an identity for the expression in (5.4.10), which gives

$$4\sin \left(\frac{\phi}{2} \right) \sin \left(\frac{\phi + i\phi'}{2} \right) = -4\sinh \left(\frac{i\phi}{2} \right) \sinh \left(\frac{i\phi - \phi'}{2} \right). \quad (7.7.9)$$

Let $\Omega^{\mathfrak{z}(\gamma)}$ be the analogue of Ω in (2.1.10), when replacing $\mathfrak{g}$ by $\mathfrak{z}(\gamma)$. By proceeding as in the proof of Theorem 7.5.3, from (5.4.10) and (7.7.8), we get the following analogue of (7.5.27),

$$L = (-1)^{\dim \mathfrak{p}^\perp(\gamma)/2} \frac{\exp \left(2\pi^2 \left| \rho + \lambda \right|^2 \right)}{(2\pi)^{p/2}}$$
$$\left[\widehat{A}^k (0) \left(\widehat{A}^k \right)^{-1} \left(-iad \left(\Omega^{\mathfrak{z}(\gamma)} \right) |_{\mathfrak{k}} \right) \chi_\lambda \left[k^{-1} \exp \left(i\Omega^{\mathfrak{z}(\gamma)} \right) \right] \right]^{\max}. \quad (7.7.10)$$

By the second identity in (7.7.5), we get

$$\widehat{A}^k (0) \left(\widehat{A}^k \right)^{-1} \left(-iad \left(\Omega^{\mathfrak{z}(\gamma)} \right) |_{\mathfrak{k}} \right) = \widehat{A}^k \left(-iad \left(\Omega^{\mathfrak{z}(\gamma)} \right) |_{\mathfrak{p}} \right). \quad (7.7.11)$$

Finally, we have the trivial

$$(-1)^{\dim \mathfrak{p}^\perp(\gamma)/2} \widehat{A}^k \left(-iad \left(\Omega^{\mathfrak{z}(\gamma)} \right) |_{\mathfrak{p}} \right) = \widehat{A}^{k^{-1}} \left(iad \left(\Omega^{\mathfrak{z}(\gamma)} \right) |_{\mathfrak{p}} \right). \quad (7.7.12)$$

By (7.7.10)–(7.7.12), we obtain

$$L = \frac{\exp \left(2\pi^2 \left| \rho + \lambda \right|^2 \right)}{(2\pi)^{p/2}} \left[\widehat{A}^\gamma \left(iad \left(\Omega^{\mathfrak{z}(\gamma)} \right) |_{\mathfrak{p}} \right) \chi_\lambda \left[k^{-1} \exp \left(i\Omega^{\mathfrak{z}(\gamma)} \right) \right] \right]^{\max}.$$
$$\quad (7.7.13)$$

By (7.7.13), we get (7.7.6).

To complete the proof of (7.7.6) in full generality, we proceed as in Remark 7.5.4. The proof of our theorem is completed. □

Remark 7.7.2. Using equation (4.8.26), we find that Theorem 7.7.1 is compatible with the Kawasaki formula [Ka79].

7.8 The de Rham-Hodge operator

Let $\left(\Omega^{\cdot}\left(X\right),d^{X}\right)$ be the de Rham complex of smooth forms on X with compact support. Let d^{X*} be the formal adjoint with respect to the obvious L_2 product. Set

$$D^X = d^X + d^{X*}. \tag{7.8.1}$$

Then $D^{X,2} = \left[d^X, d^{X*}\right]$ is the Hodge Laplacian of X.

The operator $\mathcal{L}^X$ associated with $E = \Lambda^{\cdot}\left(\mathfrak{p}^*\right)$ acts on $\Omega^{\cdot}\left(X\right)$. First, we establish a special case of Theorem 7.2.1.

Proposition 7.8.1. *The following identity holds:*

$$\frac{D^{X,2}}{2} = \mathcal{L}^X - \frac{1}{8}B^*\left(\kappa^{\mathfrak{k}},\kappa^{\mathfrak{k}}\right) - \frac{1}{16}\mathrm{Tr}^{\mathfrak{p}}\left[C^{\mathfrak{k},\mathfrak{p}}\right]. \tag{7.8.2}$$

Proof. We will assume temporarily that K is simply connected, and that $\mathfrak{p}$ is even dimensional. Then

$$\Lambda^{\cdot}\left(\mathfrak{p}^*\right) = S^{\mathfrak{p}}\widehat{\otimes}S^{\mathfrak{p}^*}. \tag{7.8.3}$$

It follows that D^X is exactly a Dirac operator of the kind that was considered in section 7.2, with $E = S^{\mathfrak{p}^*}$. Moreover,

$$C^{\mathfrak{k},S^{\mathfrak{p}^*}} = \sum_{i=m+1}^{m+n} \widehat{c}(\mathrm{ad}\,(e_i)\,|_{\mathfrak{p}})^2. \tag{7.8.4}$$

By (1.1.11) and (7.8.4), we get

$$C^{\mathfrak{k},S^{\mathfrak{p}*}} = \frac{1}{16}\sum_{i=m+1}^{m+n}\sum_{\substack{1\leq i,j\leq m \\ 1\leq k,l\leq m}} \langle [e_i,e_j],[e_k,e_l]\rangle\,\widehat{c}(e_i)\,\widehat{c}(e_j)\,\widehat{c}(e_k)\,\widehat{c}(e_l). \tag{7.8.5}$$

Using (2.6.8) and the Jacobi identity in (7.8.5), we finally obtain

$$C^{\mathfrak{k},S^{\mathfrak{p}*}} = \frac{1}{8}\mathrm{Tr}^{\mathfrak{p}}\left[C^{\mathfrak{k},\mathfrak{p}}\right]. \tag{7.8.6}$$

From (7.2.9) and (7.8.6), we get (7.8.2). When K is not simply connected, the above arguments remain valid.

When $\mathfrak{p}$ is odd dimensional, the proof of (7.8.2) is essentially the same. □

Put

$$\beta = -\frac{1}{48} \mathrm{Tr}^{\mathfrak{k}} \left[C^{\mathfrak{k},\mathfrak{k}} \right] - \frac{1}{16} \mathrm{Tr}^{\mathfrak{p}} \left[C^{\mathfrak{k},\mathfrak{p}} \right]. \tag{7.8.7}$$

By (7.8.6), β is just $\mathcal{A}$ in (7.2.15) when $E = S^{\mathfrak{p}*}$. By (7.2.16), (7.8.6), equation (7.8.2) can be written in the form

$$\frac{D^{X,2}}{2} = \mathcal{L}_\beta^X. \tag{7.8.8}$$

Let $e\left(TX, \nabla^{TX}\right)$ be the Euler form of TX that is associated with the connection ∇^{TX}. If $\mathfrak{p}$ is even-dimensional, then

$$e\left(TX, \nabla^{TX}\right) = \mathrm{Pf} \left[\frac{R^{TX}}{2\pi} \right]. \tag{7.8.9}$$

If $\mathfrak{p}$ is odd dimensional, then $e\left(TX, \nabla^{TX}\right)$ vanishes identically.

Now we establish a special case of Theorems 7.4.1 and 7.7.1, when $S^{\mathfrak{p}} \otimes E$ is replaced by $\Lambda^{\cdot}(\mathfrak{p}^*)$. In view of (7.8.3), this means that when K is simply connected and $\mathfrak{p}$ is even dimensional, $E = S^{\mathfrak{p}*}$.

Note that if $Y_0^{\mathfrak{k}} \in \mathfrak{k}$, then

$$\mathrm{Tr_s}^{\Lambda^{\cdot}(\mathfrak{p}^*)} \left[\exp\left(-i\rho^{\Lambda^{\cdot}(\mathfrak{p}*)}\left(Y_0^{\mathfrak{k}}\right) \right) \right] = \det\left(1 - \exp\left(i\mathrm{ad}\left(Y_0^{\mathfrak{k}}\right)\right)\right)|_{\mathfrak{p}}. \tag{7.8.10}$$

In particular if $\mathfrak{p}$ is odd dimensional, (7.8.10) vanishes identically.

Let γ be a semisimple element of G as in (4.2.1).

Theorem 7.8.2. *If $\gamma \in G$ is nonelliptic, for any $t > 0$,*

$$\mathrm{Tr_s}^{[\gamma]} \left[\exp\left(-tD^{X,2}/2 \right) \right] = 0. \tag{7.8.11}$$

For any $t > 0$, the following identities hold:

$$\begin{aligned}
\mathrm{Tr_s}^{[1]} &\left[\exp\left(-tD^{X,2}/2 \right) \right] \\
&= \frac{\exp(-t\beta)}{(2\pi t)^{m/2}} \int_{\mathfrak{k}} J_1\left(Y_0^{\mathfrak{k}}\right) \det\left(1 - \exp\left(i\mathrm{ad}\left(Y_0^{\mathfrak{k}}\right)\right)\right)|_{\mathfrak{p}} \\
&\qquad \exp\left(-\left|Y_0^{\mathfrak{k}}\right|/2t \right) \frac{dY_0^{\mathfrak{k}}}{(2\pi)^{n/2}} = \left[e\left(TX, \nabla^{TX}\right) \right]^{\max}. \tag{7.8.12}
\end{aligned}$$

If $\gamma \in G$ is elliptic, for any $t > 0$,

$$\mathrm{Tr_s}^{[\gamma]} \left[\exp\left(-tD^{X,2}/2 \right) \right] = \left[e\left(TX\left(\gamma\right), \nabla^{TX(\gamma)}\right) \right]^{\max}. \tag{7.8.13}$$

Proof. By proceeding as in the proof of Theorem 7.4.1, we get (7.8.11). If $\mathfrak{p}$ is odd dimensional, the last two expressions in (7.8.12) vanish identically. If $\mathfrak{p}$ is even dimensional, then (7.8.12) is a consequence of (7.4.4), (7.8.8), and (7.8.10). Equation (7.8.13) follows from Theorem 7.7.1. □

7.9 The integrand of de Rham torsion

Recall that $N^{\Lambda^{\cdot}(\mathfrak{p}^*)}$ is the number operator of $\Lambda^{\cdot}(\mathfrak{p}^*)$, which acts by multiplication by j on $\Lambda^j(\mathfrak{p}^*)$.

If g is an isometry of $\mathfrak{p}$,

$$\mathrm{Tr}_s^{\Lambda^{\cdot}(\mathfrak{p}^*)}[g] = \det\left(1 - g^{-1}\right), \tag{7.9.1}$$

$$\mathrm{Tr}_s^{\Lambda^{\cdot}(\mathfrak{p}^*)}\left[N^{\Lambda^{\cdot}(\mathfrak{p}^*)}g\right] = \frac{\partial}{\partial b}\det\left(1 - g^{-1}e^b\right)(0).$$

If the eigenspace associated with the eigenvalue 1 is of dimension ≥ 1, the first quantity in (7.9.1) vanishes. If it is of dimension ≥ 2, the second expression in (7.9.1) also vanishes. Also if m is even and g preserves the orientation, then

$$\mathrm{Tr}_s^{\Lambda^{\cdot}(\mathfrak{p}^*)}\left[\left(N^{\Lambda^{\cdot}(\mathfrak{p}^*)} - \frac{m}{2}\right)g\right] = 0. \tag{7.9.2}$$

Now we use the notation of section 7.8. In particular $D^X = d^X + d^{X*}$. Let $\gamma \in G$ be a semisimple element, which is written as in (4.2.1). By equation (6.1.2) in Theorem 6.1.1, we get

$$\mathrm{Tr}_s^{[\gamma]}\left[\left(N^{\Lambda^{\cdot}(T^*X)} - \frac{m}{2}\right)\exp\left(-tD^{X,2}/2\right)\right]$$
$$= \frac{1}{(2\pi t)^{p/2}}\exp\left(-t\beta - |a|^2/2t\right)\int_{\mathfrak{k}(\gamma)} J_\gamma\left(Y_0^{\mathfrak{k}}\right)$$
$$\mathrm{Tr}_s^{\Lambda^{\cdot}(\mathfrak{p}^*)}\left[\left(N^{\Lambda^{\cdot}(\mathfrak{p}^*)} - \frac{m}{2}\right)\exp\left(-i\mathrm{ad}\left(Y_0^{\mathfrak{k}}\right)\right)\mathrm{Ad}\left(k^{-1}\right)\right]\frac{dY_0^{\mathfrak{k}}}{(2\pi t)^{q/2}}. \tag{7.9.3}$$

Let $T \subset K$ be a maximal torus, and let $\mathfrak{t} \subset \mathfrak{k}$ be its Lie algebra. Set

$$\mathfrak{b} = \{e \in \mathfrak{p}, [e, \mathfrak{t}] = 0\}. \tag{7.9.4}$$

Put

$$\mathfrak{h} = \mathfrak{b} \oplus \mathfrak{t}. \tag{7.9.5}$$

By [Kn86, p. 129], $\mathfrak{h}$ is a Cartan subalgebra of $\mathfrak{g}$. Also $\dim \mathfrak{t}$ is the complex rank of K, and $\dim \mathfrak{h}$ is the complex rank of G.

If m is odd, then $\mathfrak{b}$ is of odd dimension ≥ 1.

Theorem 7.9.1. *If m is even, or if m is odd and $\dim \mathfrak{b} \geq 3$, for any $t > 0$,*

$$\mathrm{Tr}_s^{[\gamma]}\left[\left(N^{\Lambda^{\cdot}(T^*X)} - \frac{m}{2}\right)\exp\left(-tD^{X,2}/2\right)\right] = 0. \tag{7.9.6}$$

Proof. This follows from the considerations which follow (7.9.1), from (7.9.2), and (7.9.3). □

Remark 7.9.2. For $p, q \in \mathbf{N}$, let $\mathrm{SO}^0(p, q)$ be the connected component of the identity in the real group $\mathrm{SO}(p, q)$. By [He78, Table V p. 518] and [Kn86, Table C1 p. 713, and Table C2 p. 714], among the noncompact simple connected complex groups such that m is odd and $\dim \mathfrak{b} = 1$, there is only

$SL_2(\mathbf{C})$, and among the noncompact simple real connected groups, there are only $SL_3(\mathbf{R})$, $SL_4(\mathbf{R})$, $SL_2(\mathbb{H})$, and $SO^0(p,q)$ with pq odd > 1. Also by [He78, pp. 519, 520], $sl_2(\mathbf{C}) = so(3,1)$, $sl_4(\mathbf{R}) = so(3,3)$, and $sl_2(\mathbb{H}) = so(5,1)$. Therefore the above list can be reduced to $SL_3(\mathbf{R})$ and $SO^0(p,q)$ with pq odd > 1.[1]

Let Z be a compact locally symmetric space as in section 4.8. Let D^Z be the analogue of D^X. The Ray-Singer analytic torsion of Z is the derivative at 0 of the Mellin transform of $-\mathrm{Tr}_s\left[\left(N^{\Lambda^{\cdot}(T^*Z)} - \frac{m}{2}\right)\exp\left(-tD^{Z,2}/2\right)\right]$.

Now we recover a result of Moscovici-Stanton [MoSt91, Corollary 2.2].

Theorem 7.9.3. *If m is even, or if m is odd and $\dim \mathfrak{b} \geq 3$, for any $t > 0$,*

$$\mathrm{Tr}_s\left[\left(N^{\Lambda^{\cdot}(T^*Z)} - \frac{m}{2}\right)\exp\left(-tD^{Z,2}/2\right)\right] = 0. \qquad (7.9.7)$$

Proof. This follows from equations (4.8.12), (4.8.26), from Theorem 7.9.1, and Remark 7.9.2. $\qquad\qquad\qquad\qquad\qquad\qquad\qquad\qquad\qquad\qquad\square$

---

[1]I am indebted to Yves Benoist for providing the above information.

Chapter Eight

The case where $[\mathfrak{k}(\gamma), \mathfrak{p}_0] = 0$

The purpose of this chapter is to evaluate explicitly the Gaussian integral that appears in the right-hand side of our formula in (6.1.2) for the orbital integrals of the heat kernel, when γ is nonelliptic and $[\mathfrak{k}(\gamma), \mathfrak{p}_0] = 0$. Our computations can be easily extended to the more general kernels considered in chapter 6. It is remarkable that the index formulas of chapter 7 play here a key role.

This chapter is organized as follows. In section 8.1, we consider the case where $G = K$.

In section 8.2, we compute explicitly the Gaussian integral when γ is nonelliptic and $[\mathfrak{k}(\gamma), \mathfrak{p}_0] = 0$.

Finally, in section 8.3, the case where $G = \mathrm{SL}_2(\mathbf{R})$ is worked out. We recover the evaluation in [McK72] of the trace of the scalar heat kernel in Selberg's trace formula.

We make the same assumptions as in chapters 6 and 7, and we use the corresponding notation.

8.1 The case where $G = K$

In this section, we assume that $G = K$, so that $\mathfrak{g} = \mathfrak{k}, \mathfrak{p} = 0$. Otherwise we make the same assumptions as in chapter 6, and we use the corresponding notation.

Here $X = G/K$ is reduced to a point. If $\gamma = 1$,

$$\mathfrak{k}(\gamma) = \mathfrak{k}, \qquad\qquad \mathfrak{k}^{\perp}(\gamma) = 0. \qquad (8.1.1)$$

By (5.5.11), if $Y_0^{\mathfrak{k}} \in \mathfrak{k}$,

$$J_1\left(Y_0^{\mathfrak{k}}\right) = \widehat{A}^{-1}\left(i\mathrm{ad}\left(Y_0^{\mathfrak{k}}\right)\right). \qquad (8.1.2)$$

We still define $\mathcal{A}$ as in (7.2.15).

Proposition 8.1.1. *For any $t > 0$, the following identity holds:*

$$\int_{\mathfrak{k}} J_1\left(Y_0^{\mathfrak{k}}\right) \mathrm{Tr}^{E}\left[\exp\left(-i\rho^{E}\left(Y_0^{\mathfrak{k}}\right) - t\mathcal{A}\right)\right] \exp\left(-\left|Y_0^{\mathfrak{k}}\right|^2/2t\right) \frac{dY_0^{\mathfrak{k}}}{(2\pi t)^{n/2}}$$
$$= \dim E. \quad (8.1.3)$$

Proof. This is a special case of equation (7.4.4) in Theorem 7.4.1. $\qquad\square$

Now we make the same assumptions and we use the same notation as before, except that K is no longer supposed to be connected. Let K^0 be the connected component of the identity. Let $\gamma = k^{-1} \in K$.

For $Y_0^{\mathfrak{k}} \in \mathfrak{k}(\gamma)$, we define $J_\gamma\left(Y_0^{\mathfrak{k}}\right)$ as in (5.5.5), with $a = 0, \mathfrak{p} = 0$. Namely,

$$J_\gamma\left(Y_0^{\mathfrak{k}}\right) = \widehat{A}^{-1}\left(i\mathrm{ad}\left(Y_0^{\mathfrak{k}}\right)|_{\mathfrak{k}(\gamma)}\right)$$
$$\left[\frac{\det\left(1 - \exp\left(-i\mathrm{ad}\left(Y_0^{\mathfrak{k}}\right)\right)\mathrm{Ad}\left(k^{-1}\right)\right)|_{\mathfrak{k}^\perp(\gamma)}}{\det\left(1 - \mathrm{Ad}\left(k^{-1}\right)\right)|_{\mathfrak{k}^\perp(\gamma)}}\right]^{1/2}. \tag{8.1.4}$$

By (5.4.10), we can rewrite (8.1.4) in the form

$$J_\gamma\left(Y_0^{\mathfrak{k}}\right) = \frac{\left(\widehat{A}^k\right)^{-1}\left(i\mathrm{ad}\left(Y_0^{\mathfrak{k}}\right)\right)}{\left(\widehat{A}^k\right)^{-1}(0)}. \tag{8.1.5}$$

Let $\rho^E : K \to \mathrm{Aut}(E)$ be a finite dimensional unitary representation of K. We still define $\mathcal{A}$ as in (7.2.15).

Now we establish the following extension of Proposition 8.1.1.

Theorem 8.1.2. *For any $t > 0$,*

$$\int_{\mathfrak{k}(\gamma)} J_\gamma\left(Y_0^{\mathfrak{k}}\right) \mathrm{Tr}^E\left[\rho^E\left(k^{-1}\right)\exp\left(-i\rho^E\left(Y_0^{\mathfrak{k}}\right) - t\mathcal{A}\right)\right]$$
$$\exp\left(-\left|Y_0^{\mathfrak{k}}\right|^2/2t\right)\frac{dY_0^{\mathfrak{k}}}{(2\pi t)^{q/2}} = \mathrm{Tr}^E\left[\rho^E\left(k^{-1}\right)\right]. \tag{8.1.6}$$

Proof. If $k \in K^0$, then (8.1.6) is just a special case of (7.7.6) in Theorem 7.7.1.

Now we will consider the general case, where k does not necessarily lie in K^0. We will indeed use the formalism of chapter 2 in the case where $G = K$, so that $X = G/K$ is reduced to a point. Clearly,

$$\mathcal{L}^X = 0. \tag{8.1.7}$$

Then (8.1.6) is just a special case of (6.1.2) in Theorem 6.1.1. Indeed none of the arguments going into the proof ever uses the fact that K is connected. $\square$

8.2 The case $a \neq 0, [\mathfrak{k}(\gamma), \mathfrak{p}_0] = 0$

We take $\gamma \in G$ semisimple as in (4.2.1). We assume that γ is nonelliptic, i.e., $a \neq 0$. We also assume that

$$[\mathfrak{k}(\gamma), \mathfrak{p}_0] = 0. \tag{8.2.1}$$

Note that if G is of real rank 1, then $\mathfrak{p}_0$ is the vector subspace generated by a, so that (8.2.1) holds.

Set

$$K_0 = K \cap Z(a). \tag{8.2.2}$$

Then K_0 is a Lie subgroup of K with Lie algebra $\mathfrak{k}_0$. Moreover, $k \in K_0$.
By (5.5.5), (8.2.1), we get

$$
J_\gamma \left(Y_0^{\mathfrak{k}} \right) = \frac{1}{\left| \det \left(1 - \operatorname{Ad} \left(\gamma \right) \right) \big|_{\mathfrak{z}_0^\perp} \right|^{1/2}} \frac{1}{\det \left(1 - \operatorname{Ad} \left(k^{-1} \right) \right) \big|_{\mathfrak{p}_0^\perp (\gamma)}}
$$

$$
\widehat{A}^{-1} \left(i\operatorname{ad} \left(Y_0^{\mathfrak{k}} \right) \big|_{\mathfrak{k}(\gamma)} \right) \left[\frac{\det \left(1 - \exp \left(-i\operatorname{ad} \left(Y_0^{\mathfrak{k}} \right) \right) \operatorname{Ad} \left(k^{-1} \right) \right) \big|_{\mathfrak{k}_0^\perp (\gamma)}}{\det \left(1 - \operatorname{Ad} \left(k^{-1} \right) \right) \big|_{\mathfrak{k}_0^\perp (\gamma)}} \right]^{1/2} .
$$

$$(8.2.3)$$

We equip $\mathfrak{k}_0$ with the bilinear form $B_{\mathfrak{k}_0}$. This form is negative on $\mathfrak{k}_0$. Let $C^{\mathfrak{k}_0}$ be the corresponding Casimir. Since K_0 acts on $\mathfrak{k}_0$, there is a corresponding operator $C^{\mathfrak{k}_0, \mathfrak{k}_0} \in \operatorname{End}(\mathfrak{k}_0)$. In particular $\operatorname{Tr}^{\mathfrak{k}_0} \left[C^{\mathfrak{k}_0, \mathfrak{k}_0} \right]$ is well-defined. Similarly, we can also define the operator $C^{\mathfrak{k}_0, E}$. Let $\mathcal{A}_{[\gamma]}$ be the analogue of $\mathcal{A}$ for the group K_0. By (7.2.15), we get

$$
\mathcal{A}_{[\gamma]} = -\frac{1}{48} \operatorname{Tr}^{\mathfrak{k}_0} \left[C^{\mathfrak{k}_0, \mathfrak{k}_0} \right] - \frac{1}{2} C^{\mathfrak{k}_0, E} . \tag{8.2.4}
$$

Theorem 8.2.1. *For any $t > 0$,*

$$
\operatorname{Tr}^{[\gamma]} \left[\exp \left(-t \mathcal{L}_A^X \right) \right] = \frac{\exp \left(-|a|^2 / 2t \right)}{\left| \det \left(1 - \operatorname{Ad} \left(\gamma \right) \right) \big|_{\mathfrak{z}_0^\perp} \right|^{1/2}} \frac{1}{\det \left(1 - \operatorname{Ad} \left(k^{-1} \right) \right) \big|_{\mathfrak{p}_0^\perp (\gamma)}}
$$

$$
\frac{1}{(2\pi t)^{p/2}} \operatorname{Tr}^E \left[\rho^E \left(k^{-1} \right) \exp \left(-t \left(A - \mathcal{A}_{[\gamma]} \right) \right) \right] . \quad (8.2.5)
$$

Proof. Again, we use equation (6.1.2) in Theorem 6.1.1, and also equation (8.2.3). We combine these results with equation (8.1.6) in Theorem 8.1.2 applied to the Lie group K_0, and we get (8.2.5). $\qquad\square$

8.3 The case where $G = \operatorname{SL}_2 (\mathbf{R})$

Assume that $G = \operatorname{SL}_2 (\mathbf{R})$, $K = S^1$. We equip the Lie algebra $\mathfrak{g}$ with the form B that is half of the Killing form. If $\alpha \in \mathbf{R}$, set

$$
a = \begin{pmatrix} \alpha/2 & 0 \\ 0 & -\alpha/2 \end{pmatrix} . \tag{8.3.1}
$$

Then

$$
|a|^2 = \alpha^2 . \tag{8.3.2}
$$

If $\gamma = e^a$, then

$$
\left| \det \left(1 - \operatorname{Ad} \left(\gamma \right) \right) \big|_{\mathfrak{z}_0^\perp} \right|^{1/2} = 2 \sinh \left(|\alpha| / 2 \right) . \tag{8.3.3}
$$

Let $X = \operatorname{SL}_2 (\mathbf{R}) / S^1$ be the symmetric space associated with $\operatorname{SL}_2 (\mathbf{R})$. Then X is the upper-half space in $\mathbf{C}$. With the above conventions, its scalar curvature S^X is equal to -2. By (2.6.11),

$$
B^* \left(\kappa^{\mathfrak{g}}, \kappa^{\mathfrak{g}} \right) = -1 . \tag{8.3.4}
$$

We take E to be the trivial representation. Let Δ^X be the Laplace-Beltrami operator on X. By (2.13.2), (2.13.3), and (8.3.4),

$$\mathcal{L}^X = -\frac{1}{2}\Delta^X - \frac{1}{8}. \tag{8.3.5}$$

By (5.5.11), when $Y_0^{\mathfrak{k}} \in \mathfrak{k} = \mathbf{R}$, we get

$$J_1\left(Y_0^{\mathfrak{k}}\right) = \widehat{A}\left(Y_0^{\mathfrak{k}}\right). \tag{8.3.6}$$

By equation (6.1.2) in Theorem 6.1.1, by (8.3.5), and (8.3.6), we obtain

$$\mathrm{Tr}^{[1]}\left[\exp\left(t\Delta^X/2\right)\right] = \frac{\exp\left(-t/8\right)}{2\pi t} \int_{\mathbf{R}} \exp\left(-\left|Y_0^{\mathfrak{k}}\right|^2/2t\right) \widehat{A}\left(Y_0^{\mathfrak{k}}\right) \frac{dY_0^{\mathfrak{k}}}{\sqrt{2\pi t}}. \tag{8.3.7}$$

By (8.2.5) in Theorem 8.2.1, and by (8.3.2)–(8.3.5), if $\alpha \neq 0$, we get

$$\mathrm{Tr}^{[\gamma]}\left[\exp\left(t\Delta^X/2\right)\right] = \frac{1}{\sqrt{2\pi t}} \frac{\exp\left(-\alpha^2/2t - t/8\right)}{2\sinh\left(|\alpha|/2\right)}. \tag{8.3.8}$$

Equations (8.3.7) and (8.3.8) fit with the evaluation of Selberg's trace formula for the heat kernel in [McK72, page 233].

Chapter Nine

A proof of the main identity

The purpose of this chapter is to establish Theorem 6.1.1. The proof consists in making $b \to +\infty$ in equation (4.6.1), that is, we evaluate the limit as $b \to +\infty$ of $\mathrm{Tr_s}^{[\gamma]} \left[\exp \left(-\mathcal{L}_{A,b}^X \right) \right]$.

This chapter is organized as follows. In section 9.1, we state various estimates on the hypoelliptic heat kernels, which are valid for $b \geq 1$. The proofs of these estimates are deferred to chapter 15. They will be used for dominated convergence in the hypoelliptic orbital integrals as $b \to +\infty$.

In section 9.2, we make a natural rescaling on the coordinates parametrizing $\widehat{\mathcal{X}}$.

In section 9.3, we introduce a conjugation on the Clifford variables, which is an analogue of the Getzler rescaling [Ge86] in local index theory.

In section 9.4, we show that the norm of the term defining the conjugation can be adequately controlled.

In section 9.5, we introduce a conjugate $\mathfrak{L}_{A,b}^X$ of $\mathcal{L}_{A,b}^X$ and its associated heat kernel.

In section 9.6, we obtain the limit as $b \to +\infty$ of the rescaled heat kernel. The proof is deferred to sections 9.8–9.11.

In section 9.7, we establish Theorem 6.1.1.

In section 9.8, we make the change of variables $Y^{TX} \to a^{TX} + Y^{TX}$ in the operator $\mathfrak{L}_{A,b}^X$, and we obtain a new operator $\mathcal{O}_{a,A,b}^X$.

In section 9.9, we choose a coordinate system on $\widehat{\mathcal{X}}$ that is based at $x_0 = p1$ and we trivialize our vector bundles. We obtain this way an operator $\mathcal{P}_{a,A,b,Y_0^t}^X$.

In section 9.10, we show that as $b \to +\infty$, $\mathcal{P}_{a,A,b,Y_0^t}^X$ converges in the proper sense to an operator $\mathcal{P}_{a,A,\infty,Y_0^t}^X$ acting over $\mathfrak{p} \times \mathfrak{g}$. This operator is closely related to the operator $\mathcal{P}_{a,Y_0^t}$ that was considered in chapter 5.

Finally, in section 9.11, we state a result on convergence of heat kernels, which implies the convergence result of section 9.6. The proof is deferred to chapter 15.

9.1 Estimates on the heat kernel $q_{b,t}^X$ away from $\widehat{i}_a \mathcal{N} \left(k^{-1} \right)$

Recall that $\widehat{\pi} : \widehat{\mathcal{X}} \to X$ is the obvious projection, and $q_{b,t}^X \left((x,Y), (x',Y') \right)$ is the smooth kernel associated with $\exp \left(-t \mathcal{L}_{A,b}^X \right)$.

For $b > 0, s(x, Y) \in C^\infty \left(\widehat{\mathcal{X}}, \widehat{\pi}^* \left(\Lambda^\cdot (T^*X \oplus N^*) \otimes F \right) \right)$, set

$$F_b s(x, Y) = s(x, -bY). \tag{9.1.1}$$

Put

$$\mathcal{L}_{A,b}^X = F_b \mathcal{L}_{A,b}^X F_b^{-1}. \tag{9.1.2}$$

By (2.13.7) and (4.5.1), we get

$$\mathcal{L}_{A,b}^X = \frac{b^4}{2} \left| [Y^N, Y^{TX}] \right|^2$$
$$+ \frac{1}{2} \left(-\frac{\Delta^{TX \oplus N}}{b^4} + |Y|^2 - \frac{1}{b^2}(m+n) \right) + \frac{N^{\Lambda^\cdot (T^*X \oplus N^*)}}{b^2}$$
$$- \left(\nabla_{Y^{TX}}^{C^\infty \left(TX \oplus N, \widehat{\pi}^* (\Lambda^\cdot (T^*X \oplus N^*) \widehat{\otimes} F) \right), f*, \widehat{f}} - c \left(i\theta \mathrm{ad} \left(Y^N \right) \right) - i\rho^E \left(Y^N \right) \right)$$
$$+ A. \tag{9.1.3}$$

Let $\underline{q}_{b,t}^X \left((x, Y), (x', Y') \right)$ be the kernel associated with $\exp \left(-t\mathcal{L}_{A,b}^X \right)$. When $t = 1$, we will write $\underline{q}_b^X$ instead of $\underline{q}_{b,1}^X$.

Take $\eta > 0, C' > 0, C'' > 0$. Then there exists $c > 0$ such that for $Y^{TX} \in TX, b \geq 1$,

$$C' \left| Y^{TX} \right|^2 + C'' \exp \left(-2\eta \left| Y^{TX} \right| \right) b^4 \geq \frac{C'}{2} \left| Y^{TX} \right|^2 + c \log^2(b). \tag{9.1.4}$$

Indeed (9.1.4) is obviously true for $\left| Y^{TX} \right| > \log(b)/\eta$, and also for $\left| Y^{TX} \right| \leq \log(b)/\eta$. By (9.1.4), we find that if $C''' > 0$ is another constant, there exists $d > 0$ such that

$$C' \left| Y^{TX} \right|^2 + C'' \exp \left(-2\eta \left| Y^{TX} \right| \right) b^4 - C''' \log(b) \geq -d. \tag{9.1.5}$$

By (9.1.4), (9.1.5), we deduce that there exist $c > 0, d > 0, e > 0$ such that

$$C' \left| Y^{TX} \right|^2 + C'' \exp \left(-2\eta \left| Y^{TX} \right| \right) b^4 - C''' \log(b) \geq \frac{C'}{2} \left| Y^{TX} \right|^2$$
$$+ \frac{C''}{2} \exp \left(-2\eta \left| Y^{TX} \right| \right) b^4 - c \geq \frac{C'}{4} \left| Y^{TX} \right|^2 + e \log^2(b) - d. \tag{9.1.6}$$

In the sequel, the notation of chapter 3 will be in force. Recall the map ρ_γ was defined in Theorem 3.4.1. We use the same trivialization of N as the one before Theorem 3.9.5. Namely, recall that $\mathrm{Ad}\left(k^{-1} \right)$ acts on $N|_{X(\gamma)}$. Along geodesics normal to $X(\gamma)$, we trivialize the vector bundle N using the Euclidean connection ∇^N, so that $\mathrm{Ad}\left(k^{-1} \right)$ acts on N fibrewise.

Theorem 9.1.1. *Given $\epsilon > 0, M > 0, \epsilon \leq M$, there exist $C_{\epsilon,M} > 0, C'_{\epsilon,M} > 0$ such that for $b \geq 1, \epsilon \leq t \leq M, (x, Y), (x', Y') \in \widehat{\mathcal{X}}$,*

$$\left| \underline{q}_{b,t}^X \left((x, Y), (x', Y') \right) \right|$$
$$\leq C_{\epsilon,M} b^{4m+2n} \exp \left(-C'_{\epsilon,M} \left(d^2(x, x') + |Y|^2 + |Y'|^2 \right) \right). \tag{9.1.7}$$

Given $\beta > 0, \epsilon > 0, M > 0, \epsilon \leq M$, there exist $\eta_M > 0, C_{\epsilon,M} > 0, C'_{\epsilon,M} > 0, C''_{\gamma,\beta,M} > 0$ such that for $b \geq 1, \epsilon \leq t \leq M, (x,Y) \in \widehat{\mathcal{X}}$, if $d(x, X(\gamma)) \geq \beta$,

$$\left| q^X_{b,t}((x,Y), \gamma(x,Y)) \right| \leq C_{\epsilon,M} b^{4m+2n}$$

$$\exp\left(-C'_{\epsilon,M}\left(d^2_\gamma(x) + |Y|^2 \right) - C''_{\gamma,\beta,M} \exp\left(-2\eta_M \left| Y^{TX} \right| \right) b^4 \right). \quad (9.1.8)$$

There exists $\eta > 0$ such that given $\beta > 0, \mu > 0$, there exist $C > 0, C' > 0, C''_{\gamma,\beta,\mu} > 0$ such that for $b \geq 1, (x,Y) \in \widehat{\mathcal{X}}$, if $d(x, X(\gamma)) \leq \beta, \left| Y^{TX} - a^{TX} \right| \geq \mu$,

$$\left| q^X_b((x,Y), \gamma(x,Y)) \right| \leq C b^{4m+2n}$$

$$\exp\left(-C' |Y|^2 - C''_{\gamma,\beta,\mu} \exp\left(-2\eta \left| Y^{TX} \right| \right) b^4 \right). \quad (9.1.9)$$

There exist $c > 0, C > 0, C'_\gamma > 0$ such that for $b \geq 1, f \in \mathfrak{p}^\perp(\gamma), |f| \leq 1, x = \rho_\gamma(1, f), Y \in (TX \oplus N)_x, \left| Y^{TX} - a^{TX} \right| \leq 1$,

$$\left| q^X_b((x,Y), \gamma(x,Y)) \right| \leq c b^{4m+2n} \exp\left(-C \left| Y^N \right|^2 \right.$$

$$- C'_\gamma\left(|f|^2 + \left| Y^{TX} - a^{TX} \right|^2 \right) b^4 - C'_\gamma \left| \left(\mathrm{Ad}\left(k^{-1} \right) - 1 \right) Y^N \right| b^2$$

$$\left. - C'_\gamma \left| \left[a^{TX}, Y^N \right] \right| b^2 \right). \quad (9.1.10)$$

Proof. The proof of our theorem is deferred to section 15.7. □

Definition 9.1.2. A kernel $\mathcal{K}((x,Y), (x',Y'))$ acting on

$$C^b\left(\widehat{\mathcal{X}}, \Lambda^{\cdot}\left(T^*X \oplus N^* \right) \otimes F \right)$$

is said to be rapidly decreasing if for any $k \in \mathbf{N}$,

$$\left(1 + |Y|^k + |Y'|^k \right) \left| \mathcal{K}((x,Y), (x',Y')) \right| \quad (9.1.11)$$

is uniformly bounded. If $\mathcal{K}$ also depends on $b > 0, t > 0$, if $D \in \mathbf{R}^*_+ \times \mathbf{R}^*_+$, we say that $\mathcal{K}$ is uniformly rapidly decreasing for $(b,t) \in D$ if the previous bounds are uniform.

If k is a multi-index in $1, \ldots, 2m+n$, $\nabla^{\Lambda^{\cdot}(T^*X \oplus N^*) \otimes F, k}_{x',Y'}$ denotes the covariant derivative of order k with respect to the connection $\nabla^{\Lambda^{\cdot}(T^*X \oplus N^*) \otimes F}$. Also $| \ |$ denotes the obvious norm.

Theorem 9.1.3. *For $b \geq 1, \epsilon \leq t \leq M$, for any multi-index k,*

$$\left| \nabla^{\Lambda^{\cdot}(T^*X \oplus N^*) \otimes F, k}_{(x',Y')} q^X_{b,t}((x,Y), (x',Y')) / b^{4m+2n+|2k|} \right| \quad (9.1.12)$$

is uniformly rapidly decreasing on $\widehat{\mathcal{X}} \times \widehat{\mathcal{X}}$.

Proof. This result will be established in section 15.8. □

9.2 A rescaling on the coordinates (f, Y)

By equation (2.14.4) and by Remark 5.5.2, it is enough to prove (6.1.2) with $t = 1$.

We start from identity (4.6.1) in Theorem 4.6.1, which says that for $b > 0$, $\mathrm{Tr}_s^{[\gamma]}\left[\exp\left(-\mathcal{L}_{A,b}^X\right)\right]$ is independent of b and equal to $\mathrm{Tr}^{[\gamma]}\left[\exp\left(-\mathcal{L}_A^X\right)\right]$. The proof of (6.1.2) will consist in getting the asymptotics of $\mathrm{Tr}_s\left[\exp\left(-\mathcal{L}_{A,b}^X\right)\right]$ as $b \to +\infty$.

When $f \in \mathfrak{p}^\perp(\gamma)$, we identify e^f and $e^f p1$. Using the same notation as in equation (4.3.10), we get

$$
\mathrm{Tr}_s^{[\gamma]}\left[\exp\left(-\mathcal{L}_{A,b}^X\right)\right]
$$
$$
= \int_{\widehat{\pi}^{-1}\mathfrak{p}^\perp(\gamma)} \mathrm{Tr}_s^{\Lambda^\cdot(T^*X\oplus N^*)\otimes F}\left[\gamma \underline{q}_b^X\left(\left(e^f, Y\right), \gamma\left(e^f, Y\right)\right)\right] r\left(f\right) dY df.
$$
$$
(9.2.1)
$$

Take $\beta \in\,]0, 1]$. By (3.4.4), (3.4.36), (9.1.6), and by equation (9.1.8) in Theorem 9.1.1, as $b \to +\infty$,

$$
\int_{\substack{(f,Y)\in\widehat{\pi}^{-1}\mathfrak{p}^\perp(\gamma)\\|f|>\beta}} \mathrm{Tr}_s^{\Lambda^\cdot(T^*X\oplus N^*)\otimes F}\left[\gamma \underline{q}_b^X\left(\left(e^f, Y\right), \gamma\left(e^f, Y\right)\right)\right] r\left(f\right) dY df
$$
$$
\to 0. \quad (9.2.2)
$$

By (9.1.6), (9.1.9), given $\beta > 0, \mu > 0$, we may as well obtain a similar result for the integral of

$$
\mathrm{Tr}_s^{\Lambda^\cdot(T^*X\oplus N^*)\otimes F}\left[\gamma \underline{q}_b^X\left(\left(e^f, Y\right), \gamma\left(e^f, Y\right)\right)\right] r\left(f\right)
$$

over the region considered in (9.1.9).

As we saw in section 3.10, the vector bundle $N(\gamma)$ on $X(\gamma)$ is the analogue of N on X, and moreover, $N(\gamma) \subset N|_{X(\gamma)}$. Let $N^\perp(\gamma)$ be the orthogonal to $N(\gamma)$ in $N|_{X(\gamma)}$. Clearly,

$$
N^\perp(\gamma) = Z^0(\gamma) \times_{K^0(\gamma)} \mathfrak{k}^\perp(\gamma). \qquad (9.2.3)
$$

We trivialize the vector bundles TX, N by parallel transport along the geodesics orthogonal to $X(\gamma)$ with respect to the connections ∇^{TX}, ∇^N, so that TX, N can be identified with $p_\gamma^* TX|_{X(\gamma)}, p_\gamma^* N|_{X(\gamma)}$. At $x = p1$,

$$
N(\gamma) = \mathfrak{k}(\gamma), \qquad\qquad N^\perp(\gamma) = \mathfrak{k}^\perp(\gamma). \qquad (9.2.4)
$$

Therefore at $\rho_\gamma(1, f)$, we may write $Y^N \in N$ in the form

$$
Y^N = Y_0^{\mathfrak{k}} + Y^{N,\perp}, \qquad Y_0^{\mathfrak{k}} \in \mathfrak{k}(\gamma), Y^{N,\perp} \in \mathfrak{k}^\perp(\gamma). \qquad (9.2.5)
$$

Let $dY_0^{\mathfrak{k}}, dY^{N,\perp}$ be the volume elements on $\mathfrak{k}(\gamma), \mathfrak{k}^\perp(\gamma)$, so that

$$
dY^N = dY_0^{\mathfrak{k}} dY^{N,\perp}. \qquad (9.2.6)
$$

Ultimately, to evaluate the limit of (9.2.1) as $b \to +\infty$, given $\beta > 0$, we may as well consider the integral

$$
\int_{\substack{|f| \le \beta \\ |Y^{TX} - a^{TX}| \le \beta}} \mathrm{Tr_s}^{\Lambda^{\cdot}(T^*X \oplus N^*) \otimes F} \left[\gamma \underline{q}_b^X \left((e^f, Y), \gamma (e^f, Y) \right) \right]
$$

$$
r(f) \, dY^{TX} dY^N df = b^{-4m-2n+2r} \int_{\substack{|f| \le \beta b^2 \\ |Y^{TX}| \le \beta b^2}} \mathrm{Tr_s}^{\Lambda^{\cdot}(T^*X \oplus N^*) \otimes F}
$$

$$
\left[\gamma \underline{q}_b^X \left(\left(e^{f/b^2}, a^{TX} + Y^{TX}/b^2, Y_0^{\mathfrak{e}} + Y^{N,\perp}/b^2 \right), \right. \right.
$$

$$
\left. \left. \gamma \left(e^{f/b^2}, a^{TX} + Y^{TX}/b^2, Y_0^{\mathfrak{e}} + Y^{N,\perp}/b^2 \right) \right) \right] r \left(f/b^2 \right) dY^{TX} dY_0^{\mathfrak{e}} dY^{N,\perp} df.
$$

$$
(9.2.7)
$$

By (9.1.10), for $|f| \le \beta b^2, \left| Y^{TX} \right| \le \beta b^2$, we get

$$
b^{-4m-2n} \left| q_b^X \left(\left(e^{f/b^2}, a^{TX} + Y^{TX}/b^2, Y_0^{\mathfrak{e}} + Y^{N,\perp}/b^2 \right), \right. \right.
$$

$$
\left. \left. \gamma \left(e^{f/b^2}, a^{TX} + Y^{TX}/b^2, Y_0^{\mathfrak{e}} + Y^{N,\perp}/b^2 \right) \right) \right| \le C \exp \left(-C' \left| Y_0^{\mathfrak{e}} + \frac{Y^{N,\perp}}{b^2} \right|^2 \right.
$$

$$
- C_\gamma' \left(|f|^2 + \left| Y^{TX} \right|^2 \right) - C_\gamma' \left| (\mathrm{Ad}\,(k^{-1}) - 1) \, Y^{N,\perp} \right|
$$

$$
\left. - C_\gamma' \left| \left[a_{e^{f/b^2}}^{TX}, b^2 Y_0^{\mathfrak{e}} + Y^{N,\perp} \right] \right| \right). \quad (9.2.8)
$$

Clearly,

$$
\left| Y_0^{\mathfrak{e}} + \frac{Y^{N,\perp}}{b^2} \right|^2 = \left| Y_0^{\mathfrak{e}} \right|^2 + \frac{\left| Y^{N,\perp} \right|^2}{b^4}. \quad (9.2.9)
$$

By (3.2.8), in the given trivialization of TX, $|f| \le \beta b^2$,

$$
a_{e^{f/b^2}}^{TX} = a + \mathcal{O} \left(|f|^2 / b^4 \right). \quad (9.2.10)
$$

By (9.2.10), since $\left[a, Y_0^{\mathfrak{e}} \right] = 0$,

$$
\left[a_{e^{f/b^2}}^{TX}, b^2 Y_0^{\mathfrak{e}} + Y^{N,\perp} \right] = [a, Y^{N,\perp}] + \beta \mathcal{O} \left(|f| \left(|Y_0^{\mathfrak{e}}| + \frac{|Y^{N,\perp}|}{b^2} \right) \right).
$$

$$
(9.2.11)
$$

By (9.2.9), (9.2.11), for $\beta > 0$ small enough, the last term can be absorbed by the other terms appearing in the right-hand side of (9.2.8), so that for

$$|f| \le \beta b^2, |Y^{TX}| \le \beta b^2,$$

$$b^{-4m-2n} \left| q_b^X \left(\left(e^{f/b^2}, a^{TX} + Y^{TX}/b^2, Y_0^{\mathfrak{k}} + Y^{N,\perp}/b^2 \right), \right.\right.$$

$$\left.\left. \gamma \left(e^{f/b^2}, a^{TX} + Y^{TX}/b^2, Y_0^{\mathfrak{k}} + Y^{N,\perp}/b^2 \right) \right) \right|$$

$$\le C \exp \left(-C' \left| Y_0^{\mathfrak{k}} \right|^2 \right.$$

$$\left. - C'_\gamma \left(|f|^2 + |Y^{TX}|^2 \right) - C'_\gamma \left| \left(\mathrm{Ad} \left(k^{-1} \right) - 1 \right) Y^{N,\perp} \right| - C'_\gamma \left| [a, Y^{N,\perp}] \right| \right).$$

$$(9.2.12)$$

The marvelous fact in the right-hand side of (9.2.12) is that it is integrable.

Still we have the extra diverging term b^{2r} in the right-hand side of (9.2.7). This term will be dealt with in the next section.

9.3 A conjugation of the Clifford variables

We use the notation of section 1.1 for $(\mathfrak{g}, B)$, and also the notation of section 5.1. Let $e_1, \dots, e_{m+n}$ be a basis of $\mathfrak{g}$, and let $e_1^*, \dots, e_{m+n}^*$ be the dual basis of $\mathfrak{g}$ with respect to B. By (1.1.15),

$$N^{\Lambda^{\cdot}(\mathfrak{g}^*)} - \frac{m+n}{2} = \frac{1}{2} \sum_{i=1}^{m+n} c(e_i^*) \hat{c}(e_i). \qquad (9.3.1)$$

Recall that

$$r = \dim \mathfrak{z}(\gamma). \qquad (9.3.2)$$

By (5.1.3),

$$r = p + q. \qquad (9.3.3)$$

Let $i = \mathfrak{z}(\gamma) \to \mathfrak{g}$ be the obvious embedding, and let $i^* : \mathfrak{g}^* \to \mathfrak{z}(\gamma)^*$ be its adjoint.

Recall that $\underline{\mathfrak{z}}(\gamma)$ is another copy of $\mathfrak{z}(\gamma)$, and that $\underline{\mathfrak{z}}(\gamma)^*$ is the corresponding copy of the dual of $\mathfrak{z}(\gamma)$. Also if $u \in \mathfrak{z}(\gamma)^*$, we denote by $\underline{u}$ the corresponding element in $\underline{\mathfrak{z}}(\gamma)^*$. Let $e_1, \dots, e_r$ be a basis of $\mathfrak{z}(\gamma)$, let $e^1, \dots, e^r$ be the corresponding dual basis of $\mathfrak{z}(\gamma)^*$.

Set

$$\alpha = \sum_{i=1}^r c(e_i) \underline{e}^i, \qquad \hat{\alpha} = \sum_1^r \hat{c}(e_i) \underline{e}^i. \qquad (9.3.4)$$

Then

$$\alpha \in c(\mathfrak{z}(\gamma)) \widehat{\otimes} \Lambda^{\cdot} \left(\underline{\mathfrak{z}}(\gamma)^* \right), \qquad \hat{\alpha} \in \hat{c}(\mathfrak{z}(\gamma)) \widehat{\otimes} \Lambda^{\cdot} \left(\underline{\mathfrak{z}}(\gamma)^* \right). \qquad (9.3.5)$$

Of course $\alpha, \widehat{\alpha}$ do not depend on the choice of the basis $e_1, \ldots, e_r$.

Clearly,

$$c\left(\mathfrak{z}\left(\gamma\right)\right) \subset c\left(\mathfrak{g}\right), \qquad\qquad \widehat{c}\left(\mathfrak{z}\left(\gamma\right)\right) \subset \widehat{c}\left(\mathfrak{g}\right). \qquad (9.3.6)$$

By (9.3.5), (9.3.6), we get

$$\alpha \in c\left(\mathfrak{g}\right) \widehat{\otimes} \Lambda^{\cdot}\left(\underline{\mathfrak{z}}\left(\gamma\right)^*\right), \qquad\qquad \widehat{\alpha} \in \widehat{c}\left(\mathfrak{g}\right) \widehat{\otimes} \Lambda^{\cdot}\left(\underline{\mathfrak{z}}\left(\gamma\right)^*\right). \qquad (9.3.7)$$

Recall that $\varphi : \mathfrak{g} \to \mathfrak{g}^*$ is defined by (1.1.2) with respect to $(\mathfrak{g}, B)$. As explained in section 1.1, $c\left(\mathfrak{g}\right), \widehat{c}\left(\mathfrak{g}\right)$ act on $\Lambda^{\cdot}\left(\mathfrak{g}^*\right)$. If $e \in \mathfrak{z}\left(\gamma\right)$, $c\left(e\right), \widehat{c}\left(e\right)$ will be identified with their corresponding actions on $\Lambda^{\cdot}\left(\mathfrak{g}^*\right)$.

Also note that when using the identification $TX \oplus N = \mathfrak{g}$, and when identifying $TX \oplus N$ to $T^*X \oplus N^*$ by its scalar product, then

$$\varphi = -\theta. \qquad (9.3.8)$$

By (1.1.3) and (9.3.4), we get

$$\alpha = \sum_{i=1}^{r}\left(\varphi e_i - i_{e_i}\right)\underline{e}^i, \qquad\qquad \widehat{\alpha} = \sum_{i=1}^{r}\left(\varphi e_i + i_{e_i}\right)\underline{e}^i. \qquad (9.3.9)$$

Put

$$B^{\gamma} = \sum_{i=1}^{r}\varphi e_i \wedge \underline{e}^i, \qquad\qquad J^{\gamma} = \sum_{i=1}^{r}\underline{e}^i i_{e_i}. \qquad (9.3.10)$$

We can rewrite (9.3.9) in the form

$$\alpha = B^{\gamma} + J^{\gamma}, \qquad\qquad \widehat{\alpha} = B^{\gamma} - J^{\gamma}. \qquad (9.3.11)$$

The group $Z\left(\gamma\right)$ acts on $c\left(\mathfrak{z}\left(\gamma\right)\right) \widehat{\otimes} \Lambda^{\cdot}\left(\underline{\mathfrak{z}}\left(\gamma\right)^*\right)$ and on $\widehat{c}\left(\mathfrak{z}\left(\gamma\right)\right) \widehat{\otimes} \Lambda^{\cdot}\left(\underline{\mathfrak{z}}\left(\gamma\right)^*\right)$ via its adjoint action on $\mathfrak{z}\left(\gamma\right)$ and $\underline{\mathfrak{z}}\left(\gamma\right)$. Clearly,

$$\mathrm{Ad}\left(\gamma\right)\alpha = \alpha, \qquad\qquad \mathrm{Ad}\left(\gamma\right)\widehat{\alpha} = \widehat{\alpha}, \qquad (9.3.12)$$
$$\mathrm{Ad}\left(k^{-1}\right)\alpha = \alpha, \qquad\qquad \mathrm{Ad}\left(k^{-1}\right)\widehat{\alpha} = \widehat{\alpha}.$$

Since $a \in \mathfrak{z}\left(\gamma\right)$, the Lie derivative operator L_a also acts on the above vector spaces. Since $\mathrm{ad}\left(a\right)$ vanishes on $\mathfrak{z}\left(\gamma\right)$, we get

$$L_a\alpha = 0, \qquad\qquad L_a\widehat{\alpha} = 0. \qquad (9.3.13)$$

Definition 9.3.1. For $b \geq 0, e \in \mathfrak{z}\left(\gamma\right)$, set

$$\widehat{c}_b\left(e\right) = \exp\left(-b^2\widehat{\alpha}\right)\widehat{c}\left(e\right)\exp\left(b^2\widehat{\alpha}\right). \qquad (9.3.14)$$

Using the commutation relations in (1.1.4), we get

$$\widehat{c}_b\left(e\right) = \widehat{c}\left(e\right) + 2b^2 i^*\varphi e \wedge . \qquad (9.3.15)$$

Moreover, if $e, f \in \mathfrak{g}$,

$$\left[\widehat{c}_b\left(e\right), \widehat{c}_b\left(f\right)\right] = 2B\left(e, f\right). \qquad (9.3.16)$$

Proposition 9.3.2. *The following identities hold:*

$$\exp\left(-b^2\widehat{\alpha}\right)N^{\Lambda^{\cdot}\left(\mathfrak{g}^*\right)}\exp\left(b^2\widehat{\alpha}\right) = N^{\Lambda^{\cdot}\left(\mathfrak{g}^*\right)} + b^2\alpha,$$
$$\exp\left(-b^2\widehat{\alpha}\right)\mathrm{Ad}\left(\gamma\right)\exp\left(b^2\widehat{\alpha}\right) = \mathrm{Ad}\left(\gamma\right), \qquad (9.3.17)$$
$$\exp\left(-b^2\widehat{\alpha}\right)L_a\exp\left(b^2\widehat{\alpha}\right) = L_a.$$

Proof. In (9.3.1), we may as well assume that $e_1, \ldots, e_r$ are taken as in (9.3.4), and that $e_{r+1}, \ldots, e_{m+n}$ is a basis of $\mathfrak{z}(\gamma)^\perp$. By (9.3.1), (9.3.15), we get

$$\exp\left(-b^2\widehat{\alpha}\right) N^{\Lambda^\cdot(\mathfrak{g}^*)} \exp\left(b^2\widehat{\alpha}\right) = N^{\Lambda^\cdot(\mathfrak{g}^*)} + b^2 \sum_{i=1}^{r} c\left(e_i^*\right) \underline{i^*\varphi e_i}, \qquad (9.3.18)$$

which is equivalent to the first identity in (9.3.17). The last two identities in (9.3.17) follow from (9.3.12), (9.3.13). The proof of our proposition is completed. $\qquad\square$

Let $e_1, \ldots, e_{m+n}$ be a basis of $\mathfrak{g}$. Then one verifies easily that up to permutation $c(e_1^*)\widehat{c}(e_1)\ldots c\left(e_{m+n}^*\right)\widehat{c}(e_{m+n})$ is the only monomial in the $c(e_i^*), \widehat{c}(e_i), 1 \leq i \leq m+n$ whose supertrace on $\Lambda^\cdot(\mathfrak{g}^*)$ is nonzero, and moreover,

$$\mathrm{Tr_s}^{\Lambda^\cdot(\mathfrak{g}^*)}\left[c\left(e_1^*\right)\widehat{c}(e_1)\ldots c\left(e_{m+n}^*\right)\widehat{c}(e_{m+n})\right] = (-2)^{m+n}. \qquad (9.3.19)$$

A basis $e_1, \ldots, e_{m+n}$ will be said to be B-unimodular if the determinant of the matrix of B on this basis is $(-1)^n$. If the basis $e_1, \ldots, e_{m+n}$ is unimodular, up to permutation, $c(e_1)\widehat{c}(e_1)\ldots c(e_{m+n})\widehat{c}(e_{m+n})$ is the only monomial whose supertrace is nonzero, and moreover,

$$\mathrm{Tr_s}^{\Lambda^\cdot(\mathfrak{g}^*)}\left[c(e_1)\widehat{c}(e_1)\ldots c(e_{m+n})\widehat{c}(e_{m+n})\right] = (-2)^{m+n}(-1)^n. \qquad (9.3.20)$$

Put

$$\mathcal{G} = \mathrm{End}\left(\Lambda^\cdot(\mathfrak{g}^*)\right)\widehat{\otimes}\Lambda^\cdot\left(\mathfrak{z}(\gamma)^*\right). \qquad (9.3.21)$$

Now we slightly redefine the map $\widehat{\mathrm{Tr}}_s$ that was considered in section 5.1. Let $e_1, \ldots, e_{m+n}$ be a unimodular basis of $\mathfrak{g}$, which is such that $e_1, \ldots, e_r$ is a basis of $\mathfrak{z}(\gamma)$. Let $\underline{e}^1, \ldots, \underline{e}^r$ be the basis of $\mathfrak{z}(\gamma)$ that is dual to $e_1, \ldots, e_r$. Let $\widehat{\mathrm{Tr}}_s$ be the linear map from $\mathcal{G}$ into $\mathbf{R}$ that, up to permutation, vanishes on all the monomials in the $c(e_i), \widehat{c}(e_i), 1 \leq i \leq m+n, \underline{e}^j, 1 \leq j \leq r$ except on $c(e_1)\underline{e}^1 \ldots c(e_r)\underline{e}^r c(e_{r+1})\widehat{c}(e_{r+1})\ldots c(e_{m+n})\widehat{c}(e_{m+n})$, and moreover,

$$\widehat{\mathrm{Tr}}_s\left[c(e_1)\underline{e}^1 \ldots c(e_r)\underline{e}^r c(e_{r+1})\widehat{c}(e_{r+1})\ldots c(e_{m+n})\widehat{c}(e_{m+n})\right]$$
$$= (-1)^r(-2)^{m+n-r}(-1)^{n-q}. \qquad (9.3.22)$$

If we assume that $e_{r+1}, \ldots, e_{m+n}$ is a basis of $\mathfrak{z}^\perp(\gamma)$, and that $e_{r+1}^*, \ldots, e_{m+n}^*$ the dual basis to $e_{r+1}, \ldots, e_{m+n}$ with respect to $B|_{\mathfrak{z}^\perp(\gamma)}$, then (9.3.22) can be replaced by

$$\widehat{\mathrm{Tr}}_s\left[c(e_1)\underline{e}^1 \ldots c(e_r)\underline{e}^r c\left(e_{r+1}^*\right)\widehat{c}(e_{r+1})\ldots c\left(e_{m+n}^*\right)\widehat{c}(e_{m+n})\right]$$
$$= (-1)^r(-2)^{m+n-r}. \qquad (9.3.23)$$

Set

$$\mathcal{H} = \mathcal{G}\otimes\mathrm{End}\left(E\right). \qquad (9.3.24)$$

We extend $\widehat{\mathrm{Tr}}_s$ to a linear map from $\mathcal{H}$ to $\mathbf{C}$, so that if $u \in \mathcal{G}, v \in \mathrm{End}(E)$,

$$\widehat{\mathrm{Tr}}_s[uv] = \widehat{\mathrm{Tr}}_s[u]\,\mathrm{Tr}^E[v]. \qquad (9.3.25)$$

Definition 9.3.3. For $b \geq 0, U \in \text{End}\,(\Lambda^{\cdot}\,(\mathfrak{g}^*))$, set

$$U_b = \exp\left(-b^2\widehat{\alpha}\right) U \exp\left(b^2\widehat{\alpha}\right). \tag{9.3.26}$$

Then $U_b \in \mathcal{G}$.

Proposition 9.3.4. *For $b > 0$, the following identity holds:*

$$\text{Tr}_s^{\Lambda^{\cdot}\,(\mathfrak{g}^*)}\,[U] = b^{-2r}\widehat{\text{Tr}}_s\,[U_b]. \tag{9.3.27}$$

Proof. Using (9.3.15), (9.3.19), (9.3.22), we get (9.3.27). $\qquad\square$

9.4 The norm of α

As we saw in section 3.10, the isomorphism $TX \oplus N = \mathfrak{g}$ induces a corresponding isomorphism $(TX \oplus N)\,(\gamma) = \mathfrak{z}\,(\gamma)$. Let $(\underline{T}X \oplus \underline{N})\,(\gamma)$ be the vector bundle corresponding to $\mathfrak{z}\,(\gamma)$. This vector bundle is another copy of $(TX \oplus N)\,(\gamma)$. We denote its dual by $(\underline{T}X \oplus \underline{N})\,(\gamma)^*$.

Then α is a section of $c\,(TX \oplus N)\,\widehat{\otimes}\Lambda^{\cdot}\left((\underline{T}X \oplus \underline{N})\,(\gamma)^*\right)$, and $\widehat{\alpha}$ is a section of $\widehat{c}\,(TX \oplus N)\,\widehat{\otimes}\Lambda^{\cdot}\left((\underline{T}X \oplus \underline{N})\,(\gamma)^*\right)$. Moreover, α and $\widehat{\alpha}$ can also be considered as sections of $\text{End}\,(\Lambda^{\cdot}\,(T^*X \oplus N^*))\,\widehat{\otimes}\Lambda^{\cdot}\left((\underline{T}X \oplus \underline{N})\,(\gamma)^*\right)$. We will denote by $\|\,\|$ the norm on this last vector bundle, which is induced by the metric of $TX \oplus N$.

From the flat connection $\nabla^{\Lambda^{\cdot}\,(T^*X \oplus N^*),f*,\widehat{f}}$ and from the trivial connection on $\Lambda^{\cdot}\,(\mathfrak{z}\,(\gamma)^*)$, we obtain a flat connection on $\Lambda^{\cdot}\,(T^*X \oplus N^*)\,\widehat{\otimes}\Lambda^{\cdot}\,(\mathfrak{z}\,(\gamma)^*)$, which is still denoted $\nabla^{\Lambda^{\cdot}\,(T^*X \oplus N^*),f*,\widehat{f}}$.

Proposition 9.4.1. *There is $C > 0$ such that*

$$\|\alpha\| \leq C, \qquad\qquad\qquad \|\widehat{\alpha}\| \leq C. \tag{9.4.1}$$

Moreover,

$$\nabla_{\cdot}^{\Lambda^{\cdot}\,(T^*X \oplus N^*),f*,\widehat{f}}\widehat{\alpha} = 0. \tag{9.4.2}$$

Proof. Equation (9.4.1) follows from (9.3.8), (9.3.10), and (9.3.11).

Recall that the flat connection $\nabla^{\Lambda^{\cdot}\,(T^*X \oplus N^*),f}$ on $\Lambda^{\cdot}\,(T^*X \oplus N^*)$ was defined in section 2.4, and that it is given by (2.4.1). One has the trivial

$$\nabla^{\Lambda^{\cdot}\,(T^*X \oplus N^*),f}\widehat{\alpha} = 0. \tag{9.4.3}$$

Comparing (2.4.1) and (2.4.5), from (9.4.3), we get (9.4.2). $\qquad\square$

9.5 A conjugation of the hypoelliptic Laplacian

We denote by $\text{Op}\left(\widehat{\mathcal{X}}, \widehat{\pi}^*F\right)$ the algebra of differential operators that act on $C^\infty\left(\widehat{\mathcal{X}}, \widehat{\pi}^*F\right)$. By (9.1.3),

$$\underline{\mathcal{L}}_{A,b}^X \in \text{Op}\left(\widehat{\mathcal{X}}, \widehat{\pi}^*F\right) \otimes \left(c\,(\mathfrak{g})\,\widehat{\otimes}\widehat{c}\,(\mathfrak{g})\right). \tag{9.5.1}$$

Definition 9.5.1. Set
$$\mathfrak{L}_{A,b}^X = \exp\left(-b^2\widehat{\alpha}\right)\underline{\mathcal{L}}_{A,b}^X\exp\left(b^2\widehat{\alpha}\right). \tag{9.5.2}$$
The operator $\mathfrak{L}_{A,b}^X$ acts on $C^\infty\left(\widehat{\mathcal{X}}, \widehat{\pi}^*\left(\Lambda^\cdot\left(T^*X \oplus N^*\right)\widehat{\otimes}F\widehat{\otimes}\Lambda^\cdot\left(\mathfrak{z}\left(\gamma\right)^*\right)\right)\right)$.

Theorem 9.5.2. *The following identity holds:*
$$\mathfrak{L}_{A,b}^X = \underline{\mathcal{L}}_{A,b}^X + \alpha. \tag{9.5.3}$$

Proof. Equation (9.5.3) follows from (9.1.3), (9.3.17), and (9.4.2). □

Definition 9.5.3. For $t > 0$, let $\mathfrak{q}_{b,t}^X\left(\left(x,Y\right),\left(x',Y'\right)\right)$ denote the smooth kernel associated with $\exp\left(-t\mathfrak{L}_{A,b}^X\right)$. Also we use the notation $\mathfrak{q}_b^X$ instead of $\mathfrak{q}_{b,1}^X$.

By (9.5.2), we get
$$\mathfrak{q}_b^X\left(\left(x,Y\right),\left(x',Y'\right)\right) = \exp\left(-b^2\widehat{\alpha}\right)\underline{\mathfrak{q}}_b^X\left(\left(x,Y\right),\left(x',Y'\right)\right)\exp\left(b^2\widehat{\alpha}\right). \tag{9.5.4}$$

Proposition 9.5.4. *For $b > 0$, the following identity holds:*
$$\mathrm{Tr}_s\left[\gamma\underline{\mathfrak{q}}_b^X\left(\left(x,Y\right),\gamma\left(x,Y\right)\right)\right] = b^{-2r}\widehat{\mathrm{Tr}}_s\left[\gamma\mathfrak{q}_b^X\left(\left(x,Y\right),\gamma\left(x,Y\right)\right)\right]. \tag{9.5.5}$$

Proof. Equation (9.5.5) follows from the second equation in (9.3.17) and from equation (9.3.27) in Proposition 9.3.4. □

Remark 9.5.5. Note that contrary to the standard situation in local index theory [Ge86], we have used a global Getzler rescaling method, which does not necessitate the choice of local coordinates. This can be done because $TX \oplus N = \mathfrak{g}$ is a flat vector bundle.

By (9.5.5), we get
$$b^{-4m-2n+2r}\mathrm{Tr}_s\left[\gamma\underline{\mathfrak{q}}_b^X\left(\left(x,Y\right),\gamma\left(x,Y\right)\right)\right]$$
$$= b^{-4m-2n}\widehat{\mathrm{Tr}}_s\left[\gamma\mathfrak{q}_b^X\left(\left(x,Y\right),\gamma\left(x,Y\right)\right)\right]. \tag{9.5.6}$$
In the sequel, the norm of $\mathfrak{q}_b^X\left(\left(x,Y\right),\left(x',Y'\right)\right)$ will be evaluated with respect to the norms of $\Lambda^\cdot\left(T^*X \oplus N^*\right), \Lambda^\cdot\left(\left(\underline{TX}\oplus\underline{N}\right)\left(\gamma\right)^*\right)$, and F.

Theorem 9.5.6. *Given $\beta > 0$, there exist $C > 0, C'_\gamma > 0$ such that for $b \geq 1, f \in \mathfrak{p}^\perp\left(\gamma\right), |f| \leq \beta b^2$, and $\left|Y^{TX}\right| \leq \beta b^2$,*

$$b^{-4m-2n}\left|\mathfrak{q}_b^X\left(\left(e^{f/b^2}, a^{TX} + Y^{TX}/b^2, Y_0^{\mathfrak{k}} + Y^{N,\perp}/b^2\right),\right.\right.$$

$$\left.\left.\gamma\left(e^{f/b^2}, a^{TX} + Y^{TX}/b^2, Y_0^{\mathfrak{k}} + Y^{N,\perp}/b^2\right)\right)\right|$$

$$\leq C\exp\left(-C'\left|Y_0^{\mathfrak{k}}\right|^2\right.$$

$$\left.-C'_\gamma\left(|f|^2 + \left|Y^{TX}\right|^2\right) - C'_\gamma\left|\left(\mathrm{Ad}\left(k^{-1}\right) - 1\right)Y^{N,\perp}\right| - C'_\gamma\left|\left[a, Y^{N,\perp}\right]\right|\right). \tag{9.5.7}$$

Proof. The proof of our theorem will be given in section 15.9. □

9.6 The limit of the rescaled heat kernel

By (9.2.7), (9.5.5), we get

$$
\int_{\substack{|f|\leq\beta \\ |Y^{TX}-a^{TX}|\leq\beta}} \mathrm{Tr_s}^{\Lambda^{\cdot}(T^*X\oplus N^*)\otimes F} \left[\gamma \mathfrak{q}_b^X \left((e^f, Y), \gamma(e^f, Y) \right) \right]
$$

$$
r(f)\, dY^{TX} dY^N df
$$

$$
= \int_{\substack{|f|\leq\beta b^2 \\ |Y^{TX}|\leq\beta b^2}} b^{-4m-2n} \widehat{\mathrm{Tr}}_s \left[\gamma \mathfrak{q}_b^X \left(\left(e^{f/b^2}, a^{TX} + Y^{TX}/b^2, Y_0^{\mathfrak{k}} + Y^{N,\perp}/b^2 \right), \right. \right.
$$

$$
\left. \left. \gamma\left(e^{f/b^2}, a^{TX} + Y^{TX}/b^2, Y_0^{\mathfrak{k}} + Y^{N,\perp}/b^2 \right) \right) \right] r\left(f/b^2\right) dY^{TX} dY_0^{\mathfrak{k}} dY^{N,\perp} df.
$$

$$(9.6.1)$$

The fundamental fact in (9.6.1) is that by (9.5.7), the integrand in the right-hand side of (9.6.1) is such that

$$
b^{-4m-2n} \left| \widehat{\mathrm{Tr}}_s \left[\gamma \mathfrak{q}_b^X \left(\left(e^{f/b^2}, a^{TX} + Y^{TX}/b^2, Y_0^{\mathfrak{k}} + Y^{N,\perp}/b^2 \right), \right. \right. \right.
$$

$$
\left. \left. \left. \gamma\left(e^{f/b^2}, a^{TX} + Y^{TX}/b^2, Y_0^{\mathfrak{k}} + Y^{N,\perp}/b^2 \right) \right) \right] \right|
$$

$$
\leq C\exp\left(-C' \left|Y_0^{\mathfrak{k}}\right|^2 \right.
$$

$$
\left. - C'_\gamma \left(|f|^2 + \left|Y^{TX}\right|^2 \right) - C'_\gamma \left| \left(\mathrm{Ad}\left(k^{-1}\right) - 1 \right) Y^{N,\perp} \right| - C'_\gamma \left| \left[a, Y^{N,\perp} \right] \right| \right).
$$

$$(9.6.2)$$

Recall that γ acts on $\Lambda^{\cdot}(\mathfrak{g}^*)$ like $\mathrm{Ad}(\gamma)$. Moreover, γ maps F_x into $F_{\gamma x}$. On $X(\gamma)$, we trivialize $F|_{X(\gamma)}, TX|_{X(\gamma)}, N|_{X(\gamma)}$ by parallel transport with respect to the connections $\nabla^F, \nabla^{TX}, \nabla^N$ along geodesics starting at $p1$, so that $TX|_{X(\gamma)} = \mathfrak{p}, N|_{X(\gamma)} = \mathfrak{k}$. In particular,

$$
\Lambda^{\cdot}(T^*X \oplus N^*)|_{X(\gamma)} \simeq \Lambda^{\cdot}(\mathfrak{p}^* \oplus \mathfrak{k}^*). \tag{9.6.3}
$$

Also along geodesics normal to $X(\gamma)$, we trivialize the given vector bundles with respect to their canonical Euclidean and Hermitian connections. Ultimately,

$$
\Lambda^{\cdot}(T^*X \oplus N^*) \simeq \Lambda^{\cdot}(\mathfrak{p}^* \oplus \mathfrak{k}^*). \tag{9.6.4}
$$

Recall that the kernel $R_{Y_0^{\mathfrak{k}}}$ on $\mathfrak{p} \times \mathfrak{g}$ was defined in section 5.1. Also $\mathrm{Ad}\left(k^{-1}\right)$ acts on $\Lambda^{\cdot}(\mathfrak{p}^* \oplus \mathfrak{k}^*)$.

Theorem 9.6.1. *As $b \to +\infty$,*

$$b^{-4m-2n}\gamma\mathfrak{q}_b^X\left(\left(e^{f/b^2}, a^{TX} + Y^{TX}/b^2, Y_0^{\mathfrak{k}} + Y^{N,\perp}/b^2\right),\right.$$

$$\left.\gamma\left(e^{f/b^2}, a^{TX} + Y^{TX}/b^2, Y_0^{\mathfrak{k}} + Y^{N,\perp}/b^2\right)\right)$$

$$\to \exp\left(-|a|^2/2 - \left|Y_0^{\mathfrak{k}}\right|^2/2\right) \mathrm{Ad}\left(k^{-1}\right)$$
$$R_{Y_0^{\mathfrak{k}}}\left((f,Y), \mathrm{Ad}\left(k^{-1}\right)(f,Y)\right)$$
$$\rho^E\left(k^{-1}\right)\exp\left(-i\rho^E\left(Y_0^{\mathfrak{k}}\right) - A\right). \quad (9.6.5)$$

Proof. Our theorem will be established in sections 9.8–9.11. □

9.7 A proof of Theorem 6.1.1

Using (3.4.35), (9.6.1), (9.6.2), and (9.6.5), we find that as $b \to +\infty$,

$$\int_{\substack{|f|\leq\beta \\ |Y^{TX}-a^{TX}|\leq\beta}} \mathrm{Tr}_s^{\Lambda^{\cdot}(T^*X\oplus N^*)\otimes F}\left[\gamma\underline{q}_b^X\left(\left(e^f, Y\right), \gamma\left(e^f, Y\right)\right)\right]$$

$$r(f)\, dY^{TX} dY^N df \to \exp\left(-|a|^2/2\right)\int_{\substack{(y,Y^{\mathfrak{g}},Y_0^{\mathfrak{k}}) \\ \in \mathfrak{p}^\perp(\gamma)\times\left(\mathfrak{p}\oplus\mathfrak{k}^\perp(\gamma)\right)\times\mathfrak{k}(\gamma)}}$$

$$\widehat{\mathrm{Tr}}_s\left[\mathrm{Ad}\left(k^{-1}\right)R_{Y_0^{\mathfrak{k}}}\left((y,Y^{\mathfrak{g}}), \mathrm{Ad}\left(k^{-1}\right)(y,Y^{\mathfrak{g}})\right)\right]$$

$$\mathrm{Tr}^E\left[\rho^E\left(k^{-1}\right)\exp\left(-i\rho^E\left(Y_0^{\mathfrak{k}}\right) - A\right)\right]\exp\left(-\left|Y_0^{\mathfrak{k}}\right|^2/2\right)dy\, dY^{\mathfrak{g}} dY_0^{\mathfrak{k}}. \quad (9.7.1)$$

By (5.1.11), (9.2.1), by (9.2.2) and the considerations that follow, and by (9.7.1), we get (6.1.2). This completes the proof of Theorem 6.1.1.

9.8 A translation on the variable Y^{TX}

Definition 9.8.1. Set

$$T_a s\left(x, Y^{TX}, Y^N\right) = s\left(x, a^{TX} + Y^{TX}, Y^N\right). \quad (9.8.1)$$

Put

$$\mathcal{N}_{a,A,b}^X = T_a \mathcal{L}_{A,b}^X T_a^{-1}. \quad (9.8.2)$$

Recall that if $e \in TX \oplus N$, ∇_e^V denotes differentiation along e.

Proposition 9.8.2. *The following identity holds:*

$$\mathcal{N}_{a,A,b}^{X} = \frac{b^4}{2} \left| \left[Y^N, a^{TX} + Y^{TX} \right] \right|^2$$

$$+ \frac{1}{2} \left(-\frac{\Delta^{TX \oplus N}}{b^4} + \left| a^{TX} + Y \right|^2 - \frac{1}{b^2}(m+n) \right)$$

$$+ \frac{N^{\Lambda^{\cdot}(T^*X \oplus N^*)}}{b^2} + \alpha - \left(\nabla_{a^{TX}+Y^{TX}}^{C^\infty\left(TX \oplus N, \widehat{\pi}^*\left(\Lambda^{\cdot}(T^*X \oplus N^*)\widehat{\otimes}F\right)\right)),f*,\widehat{f}} \right.$$

$$\left. - c\left(i\theta\mathrm{ad}\left(Y^N\right)\right) - i\rho^E\left(Y^N\right) \right) + \nabla_{[a^N, a^{TX}+Y^{TX}]}^{V} + A. \quad (9.8.3)$$

Proof. By (2.17.10), (9.1.3), and (9.5.3), we get (9.8.3). □

Put

$$\mathcal{O}_{a,A,b}^{X} = \mathcal{N}_{a,A,b}^{X} + L_a. \quad (9.8.4)$$

Proposition 9.8.3. *The following identity holds:*

$$\mathcal{O}_{a,A,b}^{X} = \frac{b^4}{2} \left| \left[Y^N, a^{TX} + Y^{TX} \right] \right|^2$$

$$+ \frac{1}{2} \left(-\frac{\Delta^{TX \oplus N}}{b^4} + \left| a^{TX} + Y \right|^2 - \frac{1}{b^2}(m+n) \right)$$

$$+ \frac{N^{\Lambda^{\cdot}(T^*X \oplus N^*)}}{b^2} + \alpha - \left(\nabla_{Y^{TX}}^{C^\infty\left(TX \oplus N, \widehat{\pi}^*\left(\Lambda^{\cdot}(T^*X \oplus N^*)\widehat{\otimes}F\right)\right)),f*,\widehat{f}} \right.$$

$$\left. - c\left(\mathrm{ad}\left(a^{TX} - a^N\right)\right) + \widehat{c}\left(\mathrm{ad}(a)\right) - c\left(i\theta\mathrm{ad}\left(Y^N\right)\right) - \rho^E\left(iY^N - a^N\right) \right)$$

$$+ \nabla_{[a^N, a^{TX}+2Y^{TX}+Y^N]}^{V} + A. \quad (9.8.5)$$

Proof. By (2.4.5), (2.18.1), (2.18.2), and (9.8.3), we get (9.8.5). Equation (9.8.5) also follows from (2.18.4). □

Remark 9.8.4. Recall that the operator $\mathcal{L}_{a,b}^{X}$ was defined in (2.18.3). Set

$$\mathcal{L}_{a,A,b}^{X} = \mathcal{L}_{a,b}^{X} + A. \quad (9.8.6)$$

By (2.17.10), (2.18.1), and (2.18.2), we get

$$T_a L_a T_a^{-1} = L_a. \quad (9.8.7)$$

From (9.3.17), (9.8.6), and (9.8.7), we obtain

$$\mathcal{O}_{a,A,b}^{X} = T_a \exp\left(-b^2\widehat{\alpha}\right) F_b \mathcal{L}_{a,A,b}^{X} F_b^{-1} \exp\left(b^2\widehat{\alpha}\right) T_a^{-1}. \quad (9.8.8)$$

Equation (9.8.5) can also be derived from equation (2.18.4) for $\mathcal{L}_{a,b}^{X}$ and from (9.8.6)–(9.8.8).

Also observe that by Theorem 2.18.2, $\mathcal{L}_{a,A,b}^{X}$ commutes with $Z(a)$. Therefore $\mathcal{O}_{a,A,b}^{X}$ also commutes with $Z(a)$.

Let $\mathfrak{o}_{a,b}^X\left((x,Y),(x',Y')\right)$ be the smooth kernel for $\exp\left(-\mathcal{O}_{a,A,b}^X\right)$. Then

$$\mathfrak{o}_{a,b}^X\left((x,Y),(x',Y')\right) = e^a \mathfrak{q}_b^X\left(e^{-a}\left(x,a^{TX}+Y\right),\left(x',a^{TX}+Y'\right)\right). \quad (9.8.9)$$

Since $\gamma = e^a k^{-1}$, from (9.8.9), we obtain

$$\gamma \mathfrak{q}_b^X\left((x,a^{TX}+Y),\gamma\left(x,a^{TX}+Y\right)\right)$$
$$= k^{-1}\mathfrak{o}_{a,b}^X\left(e^a\left(x,Y\right),e^a k^{-1}\left(x,Y\right)\right). \quad (9.8.10)$$

Definition 9.8.5. Set

$$\underline{\mathcal{O}}_{a,A,b}^X = \exp\left(-b^4\left\langle\left[a^N,a^{TX}\right],Y^{TX}\right\rangle\right)\mathcal{O}_{a,A,b}^X\exp\left(b^4\left\langle\left[a^N,a^{TX}\right],Y^{TX}\right\rangle\right). \quad (9.8.11)$$

Note that $\left[a,Y^{TX}\right] = \left[a^{TX},Y^{TX}\right] + \left[a^N,Y^{TX}\right]$ is a section of $TX\oplus N$.

Proposition 9.8.6. *The following identity holds:*

$$\underline{\mathcal{O}}_{a,A,b}^X = \frac{b^4}{2}\left|\left[Y^N,a^{TX}+Y^{TX}\right]\right|^2$$

$$+\frac{1}{2}\left(-\frac{\Delta^{TX\oplus N}}{b^4}+\left|a^{TX}+Y\right|^2-\frac{1}{b^2}(m+n)\right)$$

$$+\frac{N^{\Lambda^{\cdot}(T^*X\oplus N^*)}}{b^2}+\alpha+\frac{b^4}{2}\left|\left[a^N,a^{TX}\right]\right|^2+b^4\left|\left[a,Y^{TX}\right]\right|^2$$

$$+b^4\left\langle\left[a^N,a^{TX}\right],\left[a^N,2Y^{TX}\right]\right\rangle-\left(\nabla_{Y^{TX}}^{C^\infty(TX\oplus N,\pi^*(\Lambda^{\cdot}(T^*X\oplus N^*)\otimes F)),f^*,\widehat{f}}\right.$$

$$\left.-c\left(\mathrm{ad}\left(a^{TX}-a^N\right)\right)+\widehat{c}\left(\mathrm{ad}\left(a\right)\right)-c\left(i\theta\mathrm{ad}\left(Y^N\right)\right)-\rho^E\left(iY^N-a^N\right)\right)$$

$$+\nabla_{\left[a^N,2Y^{TX}+Y^N\right]}^V+A. \quad (9.8.12)$$

Proof. We start from equation (9.8.5) for $\mathcal{O}_{a,A,b}^X$. Let Δ^{TX} be the Laplacian along the fibre TX. Let $e_1,\ldots,e_m$ be an orthonormal basis of TX. Clearly,

$$-\frac{1}{2b^4}\Delta^{TX}+\nabla_{\left[a^N,a^{TX}\right]}^V = -\frac{1}{2b^4}\sum_{1\le i\le m}\left(\nabla_{e_i}^V - b^4\left\langle\left[a^N,a^{TX}\right],e_i\right\rangle\right)^2$$

$$+\frac{b^4}{2}\left|\left[a^N,a^{TX}\right]\right|^2. \quad (9.8.13)$$

By (9.8.13), we get

$$\exp\left(-b^4\left\langle\left[a^N,a^{TX}\right],Y^{TX}\right\rangle\right)\left(-\frac{1}{2b^4}\Delta^{TX\oplus N}+\nabla_{\left[a^N,a^{TX}\right]}^V\right)$$

$$\exp\left(b^4\left\langle\left[a^N,a^{TX}\right],Y^{TX}\right\rangle\right) = -\frac{1}{2b^4}\Delta^{TX\oplus N}+\frac{b^4}{2}\left|\left[a^N,a^{TX}\right]\right|^2. \quad (9.8.14)$$

Also by equation (3.2.3) in Proposition 3.2.1,

$$\left\langle\left[a^N,a^{TX}\right],Y^{TX}\right\rangle = -\nabla_{Y^{TX}}\frac{\left|a^{TX}\right|^2}{2}. \quad (9.8.15)$$

From (9.8.15), we deduce that

$$\exp\left(-b^4\left\langle\left[a^N,a^{TX}\right],Y^{TX}\right\rangle\right)\nabla_{Y^{TX}}\exp\left(b^4\left\langle\left[a^N,a^{TX}\right],Y^{TX}\right\rangle\right)$$
$$=\nabla_{Y^{TX}}-b^4\nabla_{Y^{TX}}^{TX}\nabla_{Y^{TX}}\frac{\left|a^{TX}\right|^2}{2}. \quad (9.8.16)$$

By (3.2.4) and (9.8.16), we get

$$\exp\left(-b^4\left\langle\left[a^N,a^{TX}\right],Y^{TX}\right\rangle\right)\nabla_{Y^{TX}}\exp\left(b^4\left\langle\left[a^N,a^{TX}\right],Y^{TX}\right\rangle\right)$$
$$=\nabla_{Y^{TX}}-b^4\left|\left[a,Y^{TX}\right]\right|^2. \quad (9.8.17)$$

Also,

$$\exp\left(-b^4\left\langle\left[a^N,a^{TX}\right],Y^{TX}\right\rangle\right)\nabla_{\left[a^N,2Y^{TX}\right]}^V\exp\left(b^4\left\langle\left[a^N,a^{TX}\right],Y^{TX}\right\rangle\right)$$
$$=\nabla_{\left[a^N,2Y^{TX}\right]}^V+b^4\left\langle\left[a^N,a^{TX}\right],\left[a^N,2Y^{TX}\right]\right\rangle. \quad (9.8.18)$$

By (9.8.5), (9.8.11), (9.8.14), (9.8.17), and (9.8.18), we get (9.8.12). The proof of our proposition is completed. $\qquad\square$

9.9 A coordinate system and a trivialization of the vector bundles

Put

$$x_0=p1. \quad (9.9.1)$$

The map $y\in\mathfrak{p}\to pe^y\in X$ gives a coordinate system on X. In this coordinate system, $\mathfrak{z}(\gamma)$ is identified with $X(\gamma)$. Moreover,

$$k^{-1}pe^y=pe^{\text{Ad}\left(k^{-1}\right)y}. \quad (9.9.2)$$

The same procedure produces natural coordinates on $\widehat{\mathcal{X}}$, the total space of $TX\oplus N$. This coordinate system is given by $\widehat{\psi}:(y,Y^{\mathfrak{g}})\in\mathfrak{p}\times\mathfrak{g}\to(e^y,Y^{\mathfrak{g}})\in\widehat{\mathcal{X}}$. Equivalently, this is the coordinate system on $\widehat{\mathcal{X}}$ one obtains by parallel transport with respect to the connection $\nabla^{TX\oplus N}$ along the geodesics $t\in\mathbf{R}\to pe^{ty},y\in\mathfrak{p}$. Moreover,

$$k^{-1}\widehat{\psi}\left(y,Y^{\mathfrak{g}}\right)=\widehat{\psi}\left(\text{Ad}\left(k^{-1}\right)y,\text{Ad}\left(k^{-1}\right)Y^{\mathfrak{g}}\right). \quad (9.9.3)$$

We trivialize F along the geodesic $t\to pe^{ty}$ by parallel transport with respect to the connection ∇^F, and we trivialize $\Lambda^\cdot\left(T^*X\oplus N^*\right)$ along this geodesic by parallel transport with respect to $\nabla^{TX\oplus N}$. Also $\Lambda^\cdot\left(\mathfrak{z}(\gamma)^*\right)$ is already naturally trivialized. On X, $\Lambda^\cdot\left(T^*X\oplus N^*\right)\widehat{\otimes}\Lambda^\cdot\left((\underline{T}X\oplus\underline{N})(\gamma)^*\right)\otimes F$ is identified with $\Lambda^\cdot\left(\mathfrak{g}^*\right)\widehat{\otimes}\Lambda^\cdot\left(\mathfrak{z}(\gamma)^*\right)\otimes E$. The operator $\mathcal{O}_{a,A,b}^X$ now acts on $C^\infty\left(\mathfrak{p}\times\mathfrak{g},\Lambda^\cdot\left(\mathfrak{g}^*\right)\widehat{\otimes}\Lambda^\cdot\left(\mathfrak{z}(\gamma)^*\right)\otimes E\right)$.

Definition 9.9.1. If $Y_0^{\mathfrak{k}}\in\mathfrak{k}^\perp(\gamma)$, set

$$H_{b,Y_0^{\mathfrak{k}}}s(y,Y)=s\left(y/b^2,Y_0^{\mathfrak{k}}+Y/b^2\right). \quad (9.9.4)$$

Put

$$\mathcal{P}^X_{a,A,b,Y_0^{\mathfrak{t}}} = H_{b,Y_0^{\mathfrak{t}}} \mathcal{O}^X_{a,A,b} H^{-1}_{b,Y_0^{\mathfrak{t}}}, \qquad \underline{\mathcal{P}}^X_{a,A,b,Y_0^{\mathfrak{t}}} = H_{b,Y_0^{\mathfrak{t}}} \underline{\mathcal{O}}^X_{a,A,b} H^{-1}_{b,Y_0^{\mathfrak{t}}}. \qquad (9.9.5)$$

The operators in (9.9.5) also act on $C^\infty \left(\mathfrak{p} \times \mathfrak{g}, \Lambda^{\cdot}(\mathfrak{g}^*) \widehat{\otimes} \Lambda^{\cdot}(\mathfrak{z}(\gamma)^*) \otimes E \right)$.

Let $\mathfrak{p}^X_{a,b,Y_0^{\mathfrak{t}}} \left((y,Y), (y',Y') \right)$ be the smooth kernel for $\exp\left(-\mathcal{P}^X_{a,A,b,Y_0^{\mathfrak{t}}} \right)$ with respect to the volume $dy'dY'$. Recall that the function $\delta(Y^{\mathfrak{p}})$ was defined in (4.1.11). Then

$$b^{-4m-2n} \mathfrak{o}^X_{a,b} \left(\left(e^{y/b^2}, Y_0^{\mathfrak{t}} + Y/b^2 \right), \left(e^{y'/b^2}, Y_0^{\mathfrak{t}} + Y'/b^2 \right) \right)$$
$$\delta\left(y'/b^2 \right) = \mathfrak{p}^X_{a,b,Y_0^{\mathfrak{t}}} \left((y,Y), (y',Y') \right). \qquad (9.9.6)$$

By (9.8.9), (9.9.6), we get

$$\mathfrak{p}^X_{a,b,Y_0^{\mathfrak{t}}} \left(b^2(y,Y), b^2(y',Y') \right)$$
$$= b^{-4m-2n} e^a \mathfrak{q}^X_b \left(e^{-a} \left(e^y, a^{TX} + Y_0^{\mathfrak{t}} + Y \right), \left(e^{y'}, a^{TX} + Y_0^{\mathfrak{t}} + Y' \right) \right) \delta(y'). \qquad (9.9.7)$$

Since $\mathscr{L}^X_{A,b}$ commutes with e^a, we get

$$\mathfrak{q}_b \left((x,Y), (x',Y') \right) = e^a \mathfrak{q}_b \left(e^{-a}(x,Y), e^{-a}(x',Y') \right) e^{-a}. \qquad (9.9.8)$$

In (9.9.8), e^{-a} maps $\left(\Lambda^{\cdot}(T^*X \oplus N^*) \otimes F \right)_{x'}$ into $\left(\Lambda^{\cdot}(T^*X \oplus N^*) \otimes F \right)_{e^{-a}x'}$, and e^a maps $\left(\Lambda^{\cdot}(T^*X \oplus N^*) \otimes F \right)_{e^{-a}x}$ into $\left(\Lambda^{\cdot}(T^*X \oplus N^*) \otimes F \right)_x$.

By (9.8.10), (9.9.6), and (9.9.8), we obtain

$$b^{-4m-2n} \gamma \mathfrak{q}^X_b \left(\left(e^{f/b^2}, a^{TX} + Y^{TX}/b^2, Y_0^{\mathfrak{t}} + Y^{N,\perp}/b^2 \right), \right.$$

$$\left. \gamma \left(e^{f/b^2}, a^{TX} + Y^{TX}/b^2, Y_0^{\mathfrak{t}} + Y^{N,\perp}/b^2 \right) \right) \delta\left(f/b^2 \right)$$

$$= e^a k^{-1} \mathfrak{p}^X_{a,b,Y_0^{\mathfrak{t}}} \left((f,Y), k^{-1}(f,Y) \right) e^{-a}. \qquad (9.9.9)$$

In the given coordinate system,

$$k^{-1}(f,Y) = \mathrm{Ad}\left(k^{-1} \right)(f,Y). \qquad (9.9.10)$$

Moreover, in the trivialization of $\Lambda^{\cdot}(T^*X \oplus N^*) \otimes F$ that was considered before, the action of k^{-1} on $\Lambda^{\cdot}(T^*X \oplus N^*) \otimes F$ is just given by $\mathrm{Ad}\left(k^{-1} \right) \otimes \rho^E\left(k^{-1} \right)$. Also the action of e^a on $\left(\Lambda^{\cdot}(T^*X \oplus N^*) \otimes F \right)_{x_0}$ is the parallel transport along the geodesic $t \in [0,1] \to e^{ta}x_0 \in X(\gamma)$ with respect to the obvious connection. In the above trivialization of $\Lambda^{\cdot}(T^*X \oplus N^*) \otimes F$, we can rewrite (9.9.9) in the form

$$b^{-4m-2n} \gamma \mathfrak{q}^X_b \left(\left(e^{f/b^2}, a^{TX} + Y^{TX}/b^2, Y_0^{\mathfrak{t}} + Y^{N,\perp}/b^2 \right), \right.$$

$$\left. \gamma \left(e^{f/b^2}, a^{TX} + Y^{TX}/b^2, Y_0^{\mathfrak{t}} + Y^{N,\perp}/b^2 \right) \right) \delta\left(f/b^2 \right)$$

$$= \mathrm{Ad}\left(k^{-1} \right) \otimes \rho^E\left(k^{-1} \right) \mathfrak{p}^X_{a,b,Y_0^{\mathfrak{t}}} \left((f,Y), \mathrm{Ad}\left(k^{-1} \right)(f,Y) \right). \qquad (9.9.11)$$

9.10 The asymptotics of the operator $\mathcal{P}^X_{a,A,b,Y_0^{\mathfrak{e}}}$ as $b \to +\infty$

Definition 9.10.1. Put

$$
\mathcal{P}^X_{a,A,\infty,Y_0^{\mathfrak{e}}} = \frac{1}{2} \left| [Y^{\mathfrak{e}}, a] + [Y_0^{\mathfrak{e}}, Y^{\mathfrak{p}}] \right|^2 + \frac{1}{2} \left(-\Delta^{\mathfrak{p} \oplus \mathfrak{e}} + |a|^2 + \left| Y_0^{\mathfrak{e}} \right|^2 \right)
$$
$$
+ \alpha - \nabla^H_{Y^{\mathfrak{p}}} - \nabla^V_{[a + Y_0^{\mathfrak{e}}, [a,y]]} - \widehat{c}\,(\mathrm{ad}\,(a)) + c\left(\mathrm{ad}\,(a) + i\theta\mathrm{ad}\,(Y_0^{\mathfrak{e}})\right) + i\rho^E\left(Y_0^{\mathfrak{e}}\right) + A.
$$
$$
(9.10.1)
$$

Recall that the operator $\mathcal{P}_{a,Y_0^{\mathfrak{e}}}$ was defined in (5.1.5). By comparing (5.1.5) and (9.10.1), we obtain

$$
\mathcal{P}^X_{a,A,\infty,Y_0^{\mathfrak{e}}} = \mathcal{P}_{a,Y_0^{\mathfrak{e}}} + \frac{1}{2} \left(|a|^2 + \left| Y_0^{\mathfrak{e}} \right|^2 \right) + i\rho^E\left(Y_0^{\mathfrak{e}}\right) + A. \qquad (9.10.2)
$$

Take $\beta \in\,]0,1]$. In the asymptotic expansion that follows, we will assume that $|y|/b^2 \le \beta$.

Theorem 9.10.2. *As* $b \to +\infty$,

$$
\mathcal{P}^X_{a,A,b,Y_0^{\mathfrak{e}}} = \mathcal{P}^X_{a,A,\infty,Y_0^{\mathfrak{e}}} + \mathcal{O}\left((1 + |y|)/b^2\right) + \mathcal{O}\left(|Y|/b^2\right) + \mathcal{O}\left(\left|Y_0^{\mathfrak{e}}\right|^2 |y|^4/b^4\right)
$$
$$
+ \mathcal{O}\left(\left|Y^{\mathfrak{e}}\right|^2 |y|^4/b^8\right) + \mathcal{O}\left(\left|Y^{\mathfrak{e}}\right|^2 |Y^{\mathfrak{p}}|^2/b^4\right)
$$
$$
+ \mathcal{O}\left(\left|[Y^{\mathfrak{e}}, a] + [Y_0^{\mathfrak{e}}, Y^{\mathfrak{p}}]\right|\left(\left|Y_0^{\mathfrak{e}}\right| |y|^2/b^2 + \left|Y^{\mathfrak{e}}\right| |y|^2/b^4 + \left|Y^{\mathfrak{e}}\right| |Y^{\mathfrak{p}}|/b^2\right)\right)
$$
$$
+ \mathcal{O}\left(|y|^4/b^8 + |Y|^2/b^4\right)
$$
$$
+ \mathcal{O}\left((|a| + \left|Y_0^{\mathfrak{e}}\right|)\left(|y|^2/b^4 + |Y|/b^2\right)\right) + \mathcal{O}\left(|y| |Y|^2/b^4\right)\nabla^V
$$
$$
+ \mathcal{O}\left(|Y| \left|Y_0^{\mathfrak{e}}\right| |y|/b^2\right)\nabla^V + \mathcal{O}\left(|y| |Y|/b^2\right)\nabla^V
$$
$$
+ \mathcal{O}\left(|y|^3\left(1 + \left|Y_0^{\mathfrak{e}}\right|\right)/b^4\right)\nabla^V. \qquad (9.10.3)
$$

Proof. We use equation (9.8.5) for $\mathcal{O}^X_{a,A,b}$. Note that by (2.17.10), since a^N vanishes on $X(\gamma)$, in the given trivialization of TX,

$$
a^{TX}_{y/b^2} = a + \mathcal{O}\left(|y|^2/b^4\right), \qquad a^N_{y/b^2} = [a,y]/b^2 + \mathcal{O}\left(|y|^3/b^6\right). \qquad (9.10.4)
$$

Note that equation (9.10.4) also follows from (3.2.8).

Since $[a, Y_0^{\mathfrak{e}}] = 0$, from (9.10.4), we find that as $b \to +\infty$,

$$
\frac{b^4}{2} \left| \left[Y_0^{\mathfrak{e}} + Y^{\mathfrak{e}}/b^2, a^{TX}_{y/b^2} + Y^{\mathfrak{p}}/b^2 \right] \right|^2 = \frac{1}{2} \left| [Y^{\mathfrak{e}}, a] + [Y_0^{\mathfrak{e}}, Y^{\mathfrak{p}}] \right|^2
$$
$$
+ \mathcal{O}\left(\left|Y_0^{\mathfrak{e}}\right|^2 |y|^4/b^4\right) + \mathcal{O}\left(\left|Y^{\mathfrak{e}}\right|^2 |y|^4/b^8\right) + \mathcal{O}\left(\left|Y^{\mathfrak{e}}\right|^2 |Y^{\mathfrak{p}}|^2/b^4\right)
$$
$$
+ \mathcal{O}\left(\left|[Y^{\mathfrak{e}}, a] + [Y_0^{\mathfrak{e}}, Y^{\mathfrak{p}}]\right|\left(\left|Y_0^{\mathfrak{e}}\right| |y|^2/b^2 + \left|Y^{\mathfrak{e}}\right| |y|^2/b^4 + \left|Y^{\mathfrak{e}}\right| |Y^{\mathfrak{p}}|/b^2\right)\right).
$$
$$
(9.10.5)
$$

Also,

$$\left| a_{y/b^2}^{TX} + Y_0^{\mathfrak{e}} + Y/b^2 \right|^2 = |a|^2 + \left| Y_0^{\mathfrak{e}} \right|^2 + \mathcal{O}\left(|y|^4/b^8 + |Y|^2/b^4 \right)$$
$$+ \mathcal{O}\left(\left(|a| + \left| Y_0^{\mathfrak{e}} \right| \right) \left(|y|^2/b^4 + |Y|/b^2 \right) \right). \quad (9.10.6)$$

Moreover, as $b \to +\infty$,

$$H_{b,Y_0^{\mathfrak{e}}} \nabla_{Y^{TX}}^{C^\infty \left(TX \oplus N, \hat{\pi}^* \left(\Lambda \cdot (T^* X \oplus N^*) \hat{\otimes} F \right) \right), f*, \hat{f}} H_{b,Y_0^{\mathfrak{e}}}^{-1} = \nabla_{Y^{\mathfrak{p}}}^H$$
$$+ \mathcal{O}\left(|y| \, |Y|^2 /b^4 \right) \nabla^V + \mathcal{O}\left(|Y| \left| Y_0^{\mathfrak{e}} \right| |y|/b^2 \right) \nabla^V + \mathcal{O}\left(|Y|/b^2 \right). \quad (9.10.7)$$

In (9.10.7), we have used the fact that the connection form for $\nabla^{TX \oplus N}$ in the given trivialization vanishes at x_0, and so it gives a contribution of the order $\mathcal{O}\left(|y|/b^2 \right)$. For more details on this computation, we refer to equations (15.10.3), (15.10.4).

Moreover,

$$H_{b,Y_0^{\mathfrak{e}}} \nabla_{[a^N, a^{TX} + 2Y^{TX} + Y^N]}^V H_{b,Y_0^{\mathfrak{e}}}^{-1} = \nabla_{\left[a_{y/b^2}^N, 2Y^{\mathfrak{p}} + Y^{\mathfrak{e}} + b^2 Y_0^{\mathfrak{e}} + b^2 a_{y/b^2}^{TX} \right]}^V. \quad (9.10.8)$$

By (9.10.4), (9.10.8), we obtain

$$H_{b,Y_0^{\mathfrak{e}}} \nabla_{[a^N, a^{TX} + 2Y^{TX} + Y^N]}^V H_{b,Y_0^{\mathfrak{e}}}^{-1} = -\nabla_{[a+Y_0^{\mathfrak{e}}, [a,y]]}^V + \mathcal{O}\left(|y| \, |Y|/b^2 \right) \nabla^V$$
$$+ \mathcal{O}\left(|y|^3 \left(1 + \left| Y_0^{\mathfrak{e}} \right| \right) /b^4 \right) \nabla^V. \quad (9.10.9)$$

By (9.8.5), and (9.10.5)–(9.10.9), we get (9.10.3). The proof of our theorem is completed. □

Remark 9.10.3. A similar asymptotics can be obtained for $\underline{\mathcal{P}}_{a,A,b,Y_0^{\mathfrak{e}}}^X$. The limit operator $\underline{\mathcal{P}}_{a,A,\infty,Y_0^{\mathfrak{e}}}^X$ is closely related with the operator $\underline{\mathcal{Q}}_{a,Y_0^{\mathfrak{e}}}$ in equation (5.2.1). Also the conjugation in (9.8.11) that transforms $\mathcal{O}_{a,A,b}^X$ into $\underline{\mathcal{O}}_{a,A,b}^X$ is reflected in equation (5.2.1), where $\underline{\mathcal{Q}}_{a,Y_0^{\mathfrak{e}}}$ is obtained from $\mathcal{Q}_{a,Y_0^{\mathfrak{e}}}$ by a related conjugation.

Also the reason for the term $\mathcal{O}\left(|y| \, |Y|^2 /b^4 \right) \nabla^V$ to appear in (9.10.3) is that in any trivialization of TX, the scalar operator $\nabla_{Y^{TX}}$ is quadratic in $Y = Y^{TX} + Y^N$ in the vertical direction. This term is not as bad as it seems; since the connection ∇^{TX} preserves the norm $\left| Y^{TX} \right|$, the corresponding vector field is norm preserving. A similar question is discussed at length in [BL08, sections 15.4 and 17.8] in a different context. More precisely, we refer to [BL08, eqs. (15.4.38) and (17.8.6)] and to the references to these equations.

9.11 A proof of Theorem 9.6.1

By (9.10.2),

$$\exp\left(-\underline{\mathcal{P}}_{a,A,\infty,Y_0^{\mathfrak{e}}}^X \right) = \exp\left(-|a|^2/2 - \left| Y_0^{\mathfrak{e}} \right|^2/2 \right)$$
$$\exp\left(-\mathcal{P}_{a,Y_0^{\mathfrak{e}}} \right) \exp\left(-i\rho^E \left(Y_0^{\mathfrak{e}} \right) - A \right). \quad (9.11.1)$$

Let $\mathfrak{p}^X_{a,\infty,Y_0^\mathfrak{e}}$ be the smooth kernel associated with $\exp\left(-\mathcal{P}^X_{a,A,\infty,Y_0^\mathfrak{e}}\right)$. By (9.11.1), we get

$$\mathfrak{p}^X_{a,\infty,Y_0^\mathfrak{e}} = \exp\left(-\left|a\right|^2/2 - \left|Y_0^\mathfrak{e}\right|^2/2\right) R_{Y_0^\mathfrak{e}} \exp\left(-i\rho^E\left(Y_0^\mathfrak{e}\right) - A\right). \quad (9.11.2)$$

Then Theorem 9.6.1 is an obvious consequence of (9.9.11), (9.11.2), and of the following result.

Theorem 9.11.1. *As* $b \to +\infty$,

$$\mathfrak{p}^X_{a,b,Y_0^\mathfrak{e}}\left((y,Y),(y',Y')\right) \to \mathfrak{p}^X_{a,\infty,Y_0^\mathfrak{e}}\left((y,Y),(y',Y')\right). \quad (9.11.3)$$

Proof. The proof of our theorem will be given in section 15.10.   □

Chapter Ten

The action functional and the harmonic oscillator

The purpose of this chapter is to solve explicitly certain natural variational problems associated with a scalar hypoelliptic Laplacian, in the case where the underlying Riemannian manifold is an Euclidean vector space. The action is the one introduced in [B05, eq. (0.10)]. It depends on the parameter $b > 0$. The behavior of the minimum values as well as of the minimizing trajectories is studied when $b \to 0$ and when $b \to +\infty$. Finally, certain heat kernels are computed in terms of the minimum value of the action.

The above variational problem has already been considered by Lebeau in [L05, section 3.6] as a warm up to the more general problem on Riemannian manifolds, in order to study in detail the small time asymptotics of the hypoelliptic heat kernel.

The results of the present chapter will be used in chapter 12 in combination with the Malliavin calculus to obtain a uniform control of the regularity of the scalar hypoelliptic heat kernel for bounded $b > 0$, and also for $b \to +\infty$. The relevant tangent variational problems take place in the vector spaces $\mathfrak{p}$ and $\mathfrak{g} = \mathfrak{p} \oplus \mathfrak{k}$.

Also, when G is an Euclidean vector space, in which case the main results of the book are trivial, we give a direct computational verification of the intermediate steps.

This chapter is organized as follows. In section 10.1, we describe the variational problem over a Riemannian manifold.

In section 10.2, we obtain the relevant Hamiltonian version of the problem, by an application of Pontryagin's maximum principle.

In section 10.3, we consider the case of an Euclidean vector space E. We compute explicitly the solution to the variational problem as a function of b, and we study its behavior as $b \to 0$, as well as the behavior of the minimum value of the action.

In section 10.4, we state Mehler's formula for the heat kernel of the harmonic oscillator on E, which we relate to one of the variational problems considered before.

In section 10.5, we make a similar computation for the hypoelliptic heat kernel on $E \oplus E$.

In section 10.6, when $G = E, K = \{0\}, X = E$, we give a direct computational verification of the results contained in the book. In this case, our main result, which is Theorem 6.1.1, gives just the explicit formula for the standard heat kernel on E. The intermediate estimates used in the proof of Theorem 6.1.1 can be verified by hand. Even though the isometry group of E is not reductive, we also apply the approach used in the book to the orbital integrals associated with an isometry of E.

In section 10.7, we make simple computations involving the heat kernel of the harmonic oscillator. The obtained estimates will play an important role in the sequel.

Finally, in section 10.8, we construct the diffusion process associated with the harmonic oscillator, and we establish estimates on the L_p norm of certain random variables.

The results of sections 10.5 and 10.6 are not used in the rest of the book.

10.1 A variational problem

We will now describe a variational problem that comes naturally from the theory of the hypoelliptic Laplacian. Let X be a smooth complete Riemannian manifold, and let g^{TX} be a smooth Riemannian metric on TX. Let ∇^{TX} denote the Levi-Civita connection on TX, and let R^{TX} be its curvature.

Let $s \in [0,1] \to x_s \in X$ be a smooth path. Let $\frac{D}{Ds}$ denote covariant differentiation with respect to the Levi-Civita connection. In particular, we use the notation

$$\ddot{x} = \frac{D}{Ds}\dot{x}. \tag{10.1.1}$$

For $b > 0, t > 0$, set

$$H_{b,t}(x) = \frac{1}{2}\int_0^t \left(|\dot{x}|^2 + b^4|\ddot{x}|^2\right)ds, \qquad K_{b,t}(x) = \frac{1}{2}\int_0^t |\dot{x} + b^2\ddot{x}|^2 ds. \tag{10.1.2}$$

Note that

$$K_{b,t}(x) = H_{b,t}(x) + \frac{b^2}{2}\left(|\dot{x}_t|^2 - |\dot{x}_0|^2\right). \tag{10.1.3}$$

The action $H_{b,t}$ appears naturally in the functional analytic description of the hypoelliptic Laplacian in [B05].

Set

$$y_s = x_{b^2 s}. \tag{10.1.4}$$

Then

$$H_{b,t}(x) = \frac{1}{b^2}H_{1,t/b^2}(y), \qquad K_{b,t}(x) = \frac{1}{b^2}K_{1,t/b^2}(y). \tag{10.1.5}$$

In what follows, we will fix $(x_0, \dot{x}_0) = (x, Y)$ and $(x_t, \dot{x}_t) = (x', Y')$, and we will find the extrema of $H_{b,t}$ and of $K_{b,t}$, which by (10.1.3) are equivalent problems. This problem was considered in detail by Lebeau [L05, section 3].

When $b = +\infty$, the corresponding variational problem is associated with the functional

$$\mathbf{H}_\infty(x) = \frac{1}{2}\int_0^t |\ddot{x}|^2 ds. \tag{10.1.6}$$

The existence of minima for the above actions can be easily established. In the sequel, our paths will be supposed to verify the above constraints.

Definition 10.1.1. Given a smooth path x, let $\mathcal{L}$ be the differential operator acting along smooth sections J of TX along this path,

$$\mathcal{L}J = \frac{D^2 J}{Ds^2} + R^{TX}(J, \dot{x})\dot{x}. \tag{10.1.7}$$

Now we state a result in [L05, Theorem 3.2].

Theorem 10.1.2. *A necessary and sufficient condition for a path x to be an extremum of $H_{b,t}$ is that*

$$\left(1 - b^4 \mathcal{L}\right)\ddot{x} = 0. \tag{10.1.8}$$

Proof. Let $u \in \mathbf{R} \to x_{u,\cdot}$ be a smooth family of paths verifying the given boundary conditions with $x_{0,\cdot} = x.$ the given extremal path. Set

$$J = \frac{\partial}{\partial u} x_{u,\cdot}|_{u=0}, \qquad\qquad \dot{J} = \frac{D}{Ds}J. \tag{10.1.9}$$

Then

$$J_0 = \dot{J}_0 = 0, \qquad\qquad J_t = \dot{J}_t = 0. \tag{10.1.10}$$

An elementary computation shows that

$$\frac{d}{du} H_{b,t}(x_u)|_{u=0} = -\int_0^t \left\langle J, \left(1 - b^4 \mathcal{L}\right)\ddot{x}\right\rangle ds. \tag{10.1.11}$$

By (10.1.11), the vanishing of $H'_{b,t}(x)$ is equivalent to (10.1.8). The proof of our theorem is completed. $\square$

Remark 10.1.3. For $b = 0$, we get the equation of geodesics

$$\ddot{x} = 0, \tag{10.1.12}$$

and for $b = +\infty$, we get

$$\mathcal{L}\ddot{x} = 0. \tag{10.1.13}$$

We will now reformulate the above as a control problem. Let L_2 denote the vector space of square-integrable functions on $[0, t]$ with values in E.

If $Y.$ is a smooth section of TX over a path $x.$, we will use the notation $\dot{Y} = \frac{D}{Ds}Y$. For $u \in L_2$, consider the differential equation

$$\dot{x} = Y, \qquad\qquad \dot{Y} = u, \tag{10.1.14}$$
$$(x_0, Y_0) = (x, Y), \qquad\qquad (x_t, Y_t) = (x', Y').$$

Set

$$L((x, Y), u) = \frac{1}{2}|Y|^2 + \frac{b^4}{2}|u|^2. \tag{10.1.15}$$

Then

$$H_{b,t}(x) = \int_0^t L((x, Y), u)\, ds. \tag{10.1.16}$$

The problem of finding the extremals of $H_{b,t}$ becomes a control theoretic problem.

A similar treatment can be applied to the functional $K_{b,t}$. For $v \in L_2$, the controlled differential equation is now given by

$$\dot{x} = Y, \qquad\qquad \dot{Y} = \frac{1}{b^2}\left(-Y + v\right), \qquad (10.1.17)$$

$$(x_0, Y_0) = (x, Y), \qquad\qquad (x_t, Y_t) = (x', Y').$$

Set

$$M\left((x, Y), v\right) = \frac{|v|^2}{2}. \qquad (10.1.18)$$

Then

$$K_{b,t}(x) = \int_0^t M\left((x, Y), v\right) ds. \qquad (10.1.19)$$

Let $\mathcal{X}$ be the total space of TX. The connection ∇^{TX} induces the splitting $T\mathcal{X} \simeq TX \oplus TX$, the first copy of TX being the horizontal part of $T\mathcal{X}$, and the second copy of TX being the tangent bundle along the fibre.

Let $\mathcal{X}^*$ be the total space of the cotangent bundle of $\mathcal{X}$. Then $\mathcal{X}^*$ can be identified with the total space of $TX \oplus (T^*X \oplus T^*X)$. The first copy of T^*X consists of the pull back of 1-forms on the basis X, and the second copy of the 1-forms along the fibres TX of $\mathcal{X}$ that vanish horizontally. The generic element of $\mathcal{X}^*$ will be denoted $((x, Y), (p, q))$. Let $\pi : \mathcal{X}^* \to \mathcal{X}$ be the obvious projection.

Let $T\mathcal{X}^*$ be the total space of the tangent bundle of $\mathcal{X}^*$. Then the fibres of $T\mathcal{X}^*$ can be identified with $(TX \oplus TX) \oplus (T^*X \oplus T^*X)$. Here $TX \oplus TX$ is the horizontal part of $T\mathcal{X}^*$, and is identified with $T\mathcal{X}$, and $T^*X \oplus T^*X$ is the tangent bundle to the fibres.

10.2 The Pontryagin maximum principle

Now we will apply the Pontryagin maximum principle to the above control problems, which we will consider in succession. Put

$$\mathcal{H}_b\left((x, Y), (p, q)\right) = \sup_u \left\{\langle p, Y \rangle + \langle q, u \rangle - L\left((x, Y), u\right)\right\}. \qquad (10.2.1)$$

Then

$$\mathcal{H}_b\left((x, Y), (p, q)\right) = \langle p, Y \rangle - \frac{1}{2}|Y|^2 + \frac{1}{2b^4}|q|^2. \qquad (10.2.2)$$

In the sequel, we will identify TX and T^*X by the given metric g^{TX}. Let θ be the canonical 1-form on $\mathcal{X}^*$, and let $\omega = d\theta$ be the canonical symplectic form. With the above identifications, θ is given by

$$\theta = (p, q). \qquad (10.2.3)$$

Equivalently,

$$\theta = \langle p, dx \rangle + \langle q, DY \rangle. \qquad (10.2.4)$$

Then an easy computation shows that

$$\omega = \langle Dp, dx \rangle + \langle Dq, DY \rangle + \langle q, R^{TX} Y \rangle. \tag{10.2.5}$$

Let $Y^{\mathcal{H}_b}$ be the Hamiltonian vector field on $\mathcal{X}^*$ associated with the Hamiltonian $\mathcal{H}_b$, so that

$$d\mathcal{H}_b + i_{Y^{\mathcal{H}_b}} \omega = 0. \tag{10.2.6}$$

Using (10.2.2), (10.2.5), and (10.2.6), we find that the vector field $Y^{\mathcal{H}_b}$ is given by

$$Y^{\mathcal{H}_b} = \left(Y, q/b^4, R^{TX}(q, Y) Y, -p + Y \right). \tag{10.2.7}$$

Let $z = ((x, Y), (p, q))$ be the solution of the differential equation

$$\dot{z} = Y^{\mathcal{H}_b}. \tag{10.2.8}$$

The Pontryagin principle asserts that $(x, Y) = \pi z$ is an extremum of $H_{b,t}$ as long as it verifies the given boundary conditions, and any extremum can be obtained this way. Moreover, $\mathcal{H}_b$ is conserved along the above trajectory.

The Hamilton equation (10.2.8) on $z = (x, Y, p, q)$ can be written in the form

$$\dot{x} = Y, \qquad\qquad \dot{Y} = \frac{q}{b^4}, \tag{10.2.9}$$

$$\dot{p} = R^{TX}(q, Y) Y, \qquad\qquad \dot{q} = -p + Y.$$

By (10.2.9), we get

$$\ddot{x} - \frac{q}{b^4} = 0, \qquad \ddot{q} + R^{TX}(q, Y) Y - \frac{q}{b^4} = 0. \tag{10.2.10}$$

It is then obvious that (10.1.8) and (10.2.10) are equivalent. Moreover, $\mathcal{H}_b$ is preserved by the Hamiltonian flow. Using (10.2.2) and (10.2.9), we find that the conserved quantity is given by

$$-b^4 \langle \ddot{x}, \dot{x} \rangle + \frac{1}{2} |\dot{x}|^2 + \frac{b^4}{2} |\ddot{x}|^2. \tag{10.2.11}$$

Also by comparing (10.1.14) and (10.2.9), we get

$$u = \frac{q}{b^4}. \tag{10.2.12}$$

Now we consider the variational problem associated with the functional $K_{b,t}$. We define the associated Hamiltonian $\mathcal{K}_b$ as in (10.2.1), i.e.,

$$\mathcal{K}_b((x, Y), (p, q)) = \sup_v \left\{ \langle p, Y \rangle + \frac{1}{b^2} \langle q, -Y + v \rangle - M((x, Y), v) \right\}. \tag{10.2.13}$$

We get

$$\mathcal{K}_b((x, Y), (p, q)) = \left\langle p - \frac{q}{b^2}, Y \right\rangle + \frac{1}{2b^4} |q|^2. \tag{10.2.14}$$

The Hamiltonian vector field $Y^{\mathcal{K}_b}$ associated with $\mathcal{K}_b$ is given by

$$Y^{\mathcal{K}_b} = \left(Y, -\frac{Y}{b^2} + \frac{q}{b^4}, R^{TX}(q, Y) Y, -p + \frac{q}{b^2} \right). \tag{10.2.15}$$

Instead of (10.2.9), we now have

$$\dot{x} = Y, \qquad\qquad\qquad \dot{Y} = -\frac{Y}{b^2} + \frac{q}{b^4}, \qquad\qquad (10.2.16)$$

$$\dot{p} = R^{TX}(q, Y) Y, \qquad\qquad \dot{q} = -p + \frac{q}{b^2}.$$

By comparing (10.1.17) and (10.2.16), we obtain

$$v = \frac{q}{b^2}. \qquad\qquad (10.2.17)$$

Put

$$S(x, Y, p, q) = \left(x, Y, p, q - b^2 Y \right). \qquad\qquad (10.2.18)$$

By (10.2.5), we find that S is a symplectic transformation. The transformation S is associated with the generating function $\sigma(x, Y, p, q)$ given by

$$\sigma(x, Y, p, q) = -\frac{b^2}{2} |Y|^2, \qquad\qquad (10.2.19)$$

so that

$$S^* \theta - \theta = d\sigma. \qquad\qquad (10.2.20)$$

Moreover,

$$\mathcal{K}_b = \mathcal{H}_b \circ S. \qquad\qquad (10.2.21)$$

As should be the case, equation (10.2.16) is obtained from (10.2.9) by replacing q by $q - b^2 Y$.

The above results were obtained by Lebeau in [L05, Theorem 3.3].

10.3 The variational problem on an Euclidean vector space

With the notation in (10.1.14), (10.1.17),

$$H_{b,t}(x) = \frac{1}{2} \int_0^t \left(|Y|^2 + b^4 \left| \dot{Y} \right|^2 \right) ds, \qquad K_{b,t}(x) = \frac{1}{2} \int_0^t \left| Y + b^2 \dot{Y} \right|^2 ds.$$

$$(10.3.1)$$

Let E be an Euclidean vector space of dimension m. Here we will take $X = E$.

In a first step, we will assume that x_0 and x_t are free to vary, while $Y_0 = Y, Y_1 = Y'$. This means that we can entirely disregard x.

Let $\mathcal{H}_b^*(Y, q), \mathcal{K}_b^*(Y, q)$ be the Hamiltonians associated with the above variational problems. Then

$$\mathcal{H}_b^*(Y, q) = -\frac{1}{2} |Y|^2 + \frac{1}{2b^4} |q|^2, \qquad \mathcal{K}_b^*(Y, q) = -\left\langle \frac{q}{b^2}, Y \right\rangle + \frac{1}{2b^4} |q|^2.$$

$$(10.3.2)$$

The Hamiltonians in (10.3.2) are obtained from the ones in (10.2.2), (10.2.14) by making $p = 0$.

Making $p = 0$ in (10.2.9), (10.2.16), the solution to the variational problem associated with the actions in (10.3.1) is such that

$$b^4 \ddot{Y} - Y = 0. \tag{10.3.3}$$

By (10.3.3), we obtain

$$Y_s = A e^{s/b^2} + B e^{-s/b^2}. \tag{10.3.4}$$

Adjusting A, B for the conditions $Y_0 = Y, Y_t = Y'$ leads to the formulas,

$$A = \frac{Y - e^{t/b^2} Y'}{1 - e^{2t/b^2}}, \qquad B = \frac{Y - e^{-t/b^2} Y'}{1 - e^{-2t/b^2}}. \tag{10.3.5}$$

Let $H_{b,t}^* (Y, Y'), K_{b,t}^* (Y, Y')$ be the values of $H_{b,t}, K_{b,t}$ over the path Y. given by (10.3.4), (10.3.5). By (10.1.3),

$$K_{b,t}^* (Y, Y') = H_{b,t}^* (Y, Y') + \frac{b^2}{2} \left(|Y'|^2 - |Y|^2 \right). \tag{10.3.6}$$

Proposition 10.3.1. *The following identities hold:*

$$H_{b,t}^* (Y, Y') = \frac{b^2}{2} \left(\tanh \left(t/2b^2 \right) \left(|Y|^2 + |Y'|^2 \right) + \frac{|Y' - Y|^2}{\sinh \left(t/b^2 \right)} \right),$$

$$K_{b,t}^* (Y, Y') = \frac{b^2}{2} \left(\tanh \left(t/2b^2 \right) \left(|Y|^2 + |Y'|^2 \right) \right. \tag{10.3.7}$$

$$\left. + \frac{|Y' - Y|^2}{\sinh \left(t/b^2 \right)} + |Y'|^2 - |Y|^2 \right)$$

$$= \frac{b^2}{2 \sinh \left(t/b^2 \right)} \left| e^{-t/2b^2} Y - e^{t/2b^2} Y' \right|^2.$$

Also we have the Hamilton-Jacobi equations,

$$\frac{\partial}{\partial t} H_{b,t}^* (Y, Y') + \mathcal{H}_b^* \left(Y', \frac{\partial}{\partial Y'} H_{b,t}^* (Y, Y') \right) = 0, \tag{10.3.8}$$

$$\frac{\partial}{\partial t} K_{b,t}^* (Y, Y') + \mathcal{K}_b^* \left(Y', \frac{\partial}{\partial Y'} K_{b,t}^* (Y, Y') \right) = 0.$$

Proof. Equation (10.3.7) is a trivial consequence of (10.3.4)–(10.3.6). Equation (10.3.8) follows from general arguments in the calculus of variations. It can also be directly verified. □

By (10.3.7), we get

$$H_{b,t}^* (Y, Y') = \frac{1}{b^2} H_{1,t/b^2}^* \left(b^2 Y, b^2 Y' \right), \quad K_{b,t}^* (Y, Y') = \frac{1}{b^2} K_{1,t/b^2}^* \left(b^2 Y, b^2 Y' \right). \tag{10.3.9}$$

Equation (10.3.9) also follows from (10.1.4), (10.1.5).

By (10.1.17), (10.3.4),

$$v = b^2 \dot{Y} + Y = 2A e^{s/b^2}. \tag{10.3.10}$$

Moreover,

$$K_{b,t}^* (Y, Y') = \frac{1}{2} \int_0^t |v|^2 \, ds. \tag{10.3.11}$$

Observe that by (10.3.4), (10.3.5), when $b \to 0$, on compact subsets of $]0, t[$, $Y.$ converges uniformly to 0, while remaining equal to Y at $s = 0$ and to Y' at $s = t$. Moreover, as $b \to 0$,

$$H_{b,t}^* (Y, Y') \to 0, \qquad\qquad K_{b,t}^* (Y, Y') \to 0. \tag{10.3.12}$$

From (10.3.12), we find that as $b \to 0$,

$$Y. \to 0 \text{ in } L_2. \tag{10.3.13}$$

Also as $b \to 0$,

$$H_{b,t}^* (Y/b, Y'/b) \to \underline{H}_0^* (Y, Y') = \frac{1}{2} \left(|Y|^2 + |Y'|^2 \right), \tag{10.3.14}$$

$$K_{b,t}^* (Y/b, Y'/b) \to \underline{K}_0^* (Y, Y') = |Y'|^2.$$

By (10.3.4), (10.3.5), when replacing Y, Y' by $Y/b, Y'/b$, as $b \to 0$, $Y.$ still converges to 0 uniformly on compact subsets of $]0, t[$.

Note that

$$\frac{1}{b} \int_0^s e^{-u/b^2} \, du = b \left(1 - e^{-s/b^2} \right), \tag{10.3.15}$$

so that

$$\frac{1}{b} \int_0^s e^{-u/b^2} \, du \leq b. \tag{10.3.16}$$

We still replace Y, Y' by $Y/b, Y'/b$. For $1 \leq p < +\infty$, L_p denotes the obvious vector space of functions defined on $[0, t]$ with values in E. By (10.3.4), (10.3.5), and (10.3.16), as $b \to 0$,

$$Y. \to 0 \text{ in } L_p, \; 1 \leq p < 2, \qquad\quad Y. \to 0 \text{ weakly in } L_2. \tag{10.3.17}$$

By (10.3.4), (10.3.5), in general, $Y.$ does not converge strongly to 0 in L_2. More precisely, as $b \to 0$,

$$\frac{1}{2} \int_0^t |Y|^2 \, ds \to \frac{1}{4} \left(|Y|^2 + |Y'|^2 \right), \quad \frac{b^4}{2} \int_0^t |\dot{Y}|^2 \, ds \to \frac{1}{4} \left(|Y|^2 + |Y'|^2 \right), \tag{10.3.18}$$

which fits with (10.3.1), (10.3.14). Moreover, given $0 < \epsilon \leq M < +\infty$, as long as Y, Y' remain uniformly bounded in E, and $\epsilon \leq t \leq M$, the convergence of $Y.$ to 0 in $L_p, 1 \leq p < 2$ is uniform. Also, $Y.$ remains uniformly bounded in L_2, and $bY.$ remains uniformly bounded.

We fix again Y, Y'. When $b \to +\infty$, we have the uniform convergence on $[0, t]$,

$$Y_s \to \left(1 - \frac{s}{t} \right) Y + \frac{s}{t} Y'. \tag{10.3.19}$$

Moreover, as $b \to +\infty$,

$$\frac{H_{b,t}^*}{b^4}(Y,Y') \to \frac{1}{2t}|Y'-Y|^2, \qquad \frac{K_{b,t}^*}{b^4}(Y,Y') \to \frac{1}{2t}|Y'-Y|^2. \qquad (10.3.20)$$

Also as $b \to +\infty$,

$$H_{b,t}^*(Y,Y) \to \frac{t}{2}|Y|^2, \qquad K_{b,t}^*(Y,Y) \to \frac{t}{2}|Y|^2. \qquad (10.3.21)$$

Now we consider the original problem on $E \oplus E$. Set

$$\underline{\mathcal{H}}_b((x,Y),(p,q)) = \frac{1}{2}|p|^2 + \mathcal{H}_b^*(Y,q). \qquad (10.3.22)$$

By (10.3.2), (10.3.22), we get

$$\underline{\mathcal{H}}_b((x,Y),(p,q)) = \frac{1}{2}|p|^2 - \frac{1}{2}|Y|^2 + \frac{1}{2b^4}|q|^2. \qquad (10.3.23)$$

A remarkable feature of $\underline{\mathcal{H}}_b$ is that the variables (x,p) and (Y,q) are uncoupled, which is not the case for $\mathcal{H}_b$ in (10.2.2).

Recall that since $X = E$, then $\mathcal{X}^* = E \oplus E \oplus E^* \oplus E^*$. The vector space $\mathcal{X}^*$ is equipped with the obvious symplectic form ω in (10.2.5), where D should just be d, and $R^{TX} = 0$. Let T be the map

$$T(x,Y,p,q) = (x+q, Y+p, p, q). \qquad (10.3.24)$$

Set

$$\tau(x,Y,p,q) = \langle p,q \rangle. \qquad (10.3.25)$$

Then τ is a generating function for T, i.e.,

$$T^*\theta - \theta = d\tau. \qquad (10.3.26)$$

Moreover, one verifies easily that

$$\underline{\mathcal{H}}_b = \mathcal{H}_b \circ T. \qquad (10.3.27)$$

Let us now replace in the above (Y,q) by $(iY, -iq)$, so that the transformation T becomes the complex symplectic transformation T' given by

$$T'(x,Y,p,q) = (x-iq, Y-ip, p, q). \qquad (10.3.28)$$

If $\mathcal{H}_b', \underline{\mathcal{H}}_b'$ are obtained from $\mathcal{H}_b, \underline{\mathcal{H}}_b$ by replacing Y by iY, by (10.3.27), we get

$$\underline{\mathcal{H}}_b' = \mathcal{H}_b' \circ T'. \qquad (10.3.29)$$

Incidentally, note that (10.3.29) is correct in spite of the fact that we did not change q into $-iq$.

Equation (10.3.27) gives the Hamiltonian counterpart to the conjugation arguments developed in [B05, section 3.10] and in [B09a, section 1]. In these references, it is shown that when $X = S^1$, the scalar hypoelliptic Laplacian on S^1 is conjugate to the sum of the Laplacian on S^1 and of a scaled version of the harmonic oscillator along the fibre of $\mathcal{X}$, via an unbounded self-adjoint conjugation. This conjugation is a simple form of Egorov's theorem [Hör85b,

Theorem 25.3.5], [T81, chapter 8, p. 147] with a conjugating operator formally associated with a Fourier integral operator with real phase. Equations (10.5.7)–(10.5.8) will provide an explicit illustration to these somewhat mysterious considerations.

Let us now return to the full variational problem in (10.3.1), with $x_0 = x, Y_0 = Y, x_t = x', Y_t = Y'$. By (10.2.9) or (10.2.16), p is constant, but no longer 0. Instead of (10.3.3), we get

$$b^4 \ddot{Y} - Y + p = 0. \tag{10.3.30}$$

Of course, (10.3.30) is equivalent to

$$b^4 Y^{(3)} - Y' = 0, \tag{10.3.31}$$

which by (10.1.14) or (10.1.17) can be written in the form

$$b^4 x^{(4)} - \ddot{x} = 0. \tag{10.3.32}$$

Note that as it should be, (10.3.32) is just (10.1.8).

By (10.3.30),

$$Y_s = p + A e^{s/b^2} + B e^{-s/b^2}. \tag{10.3.33}$$

By (10.3.5), we get

$$A = \frac{Y - p - e^{t/b^2} (Y' - p)}{1 - e^{2t/b^2}}, \quad B = \frac{Y - p - e^{-t/b^2} (Y' - p)}{1 - e^{-2t/b^2}}. \tag{10.3.34}$$

Taking into account the fact that $\dot{x} = Y, x_0 = x, x_t = x'$, by (10.3.33) and (10.3.34), we get

$$\left(t/b^2 - 2\tanh\left(t/2b^2\right)\right) p = \left(x' - x\right)/b^2 - \tanh\left(t/2b^2\right)\left(Y + Y'\right). \tag{10.3.35}$$

Equation (10.3.35) determines p. Instead of (10.3.10), we get

$$v = p + 2A e^{s/b^2}. \tag{10.3.36}$$

Recall that $\mathcal{H}_b\left((x, Y), (p, q)\right), \mathcal{K}_b\left((x, Y), (p, q)\right)$ were defined in (10.2.2) and (10.2.14).

Let $H_{b,t}\left((x, Y), (x', Y')\right), K_{b,t}\left((x, Y), (x', Y')\right)$ be the value of $H_{b,t}, K_{b,t}$ on the trajectory $(x_., Y_.)$. By (10.1.3),

$$K_{b,t}\left((x, Y), (x', Y')\right) = H_{b,t}\left((x, Y), (x', Y')\right) + \frac{b^2}{2}\left(|Y'|^2 - |Y|^2\right). \tag{10.3.37}$$

The analogue of equation (10.3.11) still holds, i.e.,

$$K_{b,t}\left((x, Y), (x', Y')\right) = \frac{1}{2}\int_0^t |v|^2 \, ds. \tag{10.3.38}$$

We denote by $d_{x',Y'} H_{b,t}\left((x, Y), (x', Y')\right)$ the gradient of $H_{b,t}$ in the variables (x', Y'). A similar notation will be used for $K_{b,t}$.

Proposition 10.3.2. *The following identities hold:*

$$H_{b,t}\left((x,Y),(x',Y')\right) = \frac{b^2}{2}\left(\tanh\left(t/2b^2\right)\left(|Y|^2 + |Y'|^2\right) + \frac{|Y'-Y|^2}{\sinh\left(t/b^2\right)}\right)$$

$$+ \frac{1}{2\left(t - 2b^2\tanh\left(t/2b^2\right)\right)}\left|x' - x - b^2\tanh\left(t/2b^2\right)\left(Y+Y'\right)\right|^2, \quad (10.3.39)$$

$$K_{b,t}\left((x,Y),(x',Y')\right) = \frac{b^2}{2\sinh\left(t/b^2\right)}\left|e^{-t/2b^2}Y - e^{t/2b^2}Y'\right|^2$$

$$+ \frac{1}{2\left(t - 2b^2\tanh\left(t/2b^2\right)\right)}\left|x' - x - b^2\tanh\left(t/2b^2\right)\left(Y+Y'\right)\right|^2.$$

Moreover, we have the Hamilton-Jacobi equations,

$$\frac{\partial}{\partial t}H_{b,t}\left((x,Y),(x',Y')\right) + \mathcal{H}_b\left((x',Y'),d_{x',Y'}H_{b,t}\left((x,Y),(x',Y')\right)\right) = 0,$$
$$(10.3.40)$$

$$\frac{\partial}{\partial t}K_{b,t}\left((x,Y),(x',Y')\right) + \mathcal{K}_b\left((x',Y'),d_{x',Y'}K_{b,t}\left((x,Y),(x',Y')\right)\right) = 0.$$

Proof. By (10.3.7), (10.3.33), and (10.3.37), we get

$$H_{b,t}\left((x,Y),(x',Y')\right) = \frac{b^2}{2}\left(\tanh\left(t/2b^2\right)\left(|Y-p|^2 + |Y'-p|^2\right)\right. \quad (10.3.41)$$

$$\left. + \frac{|Y'-Y|^2}{\sinh\left(t/b^2\right)}\right) + \langle p, x' - x\rangle - \frac{1}{2}|p|^2 t,$$

$$K_{b,t}\left((x,Y),(x',Y')\right) = \frac{b^2}{2}\left(\tanh\left(t/2b^2\right)\left(|Y-p|^2 + |Y'-p|^2\right)\right. \quad (10.3.42)$$

$$\left. + \frac{|Y'-Y|^2}{\sinh\left(t/b^2\right)} + |Y'|^2 - |Y|^2\right) + \langle p, x' - x\rangle - \frac{1}{2}|p|^2 t.$$

By (10.3.35), (10.3.41), we get (10.3.39). As to (10.3.40), it follows from general arguments in the calculus of variations or from a direct computation. $\square$

By (10.3.39), we get

$$H_{b,t}\left((x,Y),(x',Y')\right) = \frac{1}{b^2}H_{1,t/b^2}\left((x,b^2Y),(x',b^2Y')\right), \quad (10.3.43)$$

$$K_{b,t}\left((x,Y),(x',Y')\right) = \frac{1}{b^2}K_{1,t/b^2}\left((x,b^2Y),(x',b^2Y')\right).$$

Equation (10.3.43) also follows from (10.1.4), (10.1.5).

By (10.3.35), as $b \to 0$,

$$p \to \frac{x'-x}{t}. \quad (10.3.44)$$

By (10.3.33), (10.3.34), (10.3.44), Y remains uniformly bounded on $[0,t]$, the uniform bound only depends on $x' - x, Y, Y'$. Moreover, over compact subsets of $]0, t[$, $Y_.$ converges uniformly to $\frac{x'-x}{t}$. It follows that as $b \to 0$, we have the uniform convergence on $[0, t]$,

$$x_s \to \left(1 - \frac{s}{t}\right) x + \frac{s}{t} x'. \qquad (10.3.45)$$

As $b \to 0$,

$$H_{b,t}\left((x, Y), (x', Y')\right) \to H_{0,t}\left((x, Y), (x', Y')\right) = \frac{1}{2t}\left|x' - x\right|^2, \qquad (10.3.46)$$

$$K_{b,t}\left((x, Y), (x', Y')\right) \to K_{0,t}\left((x, Y), (x', Y')\right) = \frac{1}{2t}\left|x' - x\right|^2. $$

By (10.3.36), one finds easily that as $b \to 0$,

$$v \to \frac{x' - x}{t} \text{ in } L_2. \qquad (10.3.47)$$

Now we replace Y, Y' by $Y/b, Y'/b$. Inspection of (10.3.35) shows that (10.3.44) still holds. Also by (10.3.33), (10.3.35), $Y_.$ converges uniformly to $\frac{x'-x}{t}$ on compact subsets of $]0, t[$. However, in general, $Y_.$ does not remain uniformly bounded. By (10.3.33), (10.3.34), and by the analogue of (10.3.44), we conclude easily that as $b \to 0$,

$$Y_. \to \frac{x' - x}{t} \text{ in } L_p, \ 1 \le p < 2. \qquad (10.3.48)$$

Moreover, given $0 < \epsilon \le M < +\infty$, as long as $x' - x, Y, Y'$ remain uniformly bounded, and $\epsilon \le t \le M$, as $b \to 0$, the convergence in (10.3.48) is uniform, and $bY_.$ remains uniformly bounded on $[0, t]$. By (10.3.48), we conclude that the uniform convergence in (10.3.45) still holds, and that it is uniform as long as $\epsilon \le t \le M$, and $x' - x, Y, Y'$ remain bounded. Under the same conditions, $Y_.$ remains uniformly bounded in L_2.

Moreover, as $b \to 0$,

$$H_{b,t}\left((x, Y/b), (x', Y'/b)\right) \to \underline{H}_{0,t}\left((x, Y), (x', Y')\right) = \frac{1}{2t}\left|x' - x\right|^2 \quad (10.3.49)$$
$$+ \frac{1}{2}\left(|Y|^2 + |Y'|^2\right),$$

$$K_{b,t}\left((x, Y/b), (x', Y'/b)\right) \to \underline{K}_{0,t}\left((x, Y), (x', Y')\right) = \frac{1}{2t}\left|x' - x\right|^2 + |Y'|^2. $$

Inspection of (10.3.34)–(10.3.36) shows that as $b \to 0$,

$$v \to \frac{x' - x}{t} \text{ in } L_p, \ 1 \le p < 2, \qquad v \to \frac{x' - x}{t} \text{ weakly in } L_2. \qquad (10.3.50)$$

By (10.3.38) and (10.3.49), in general, the convergence in (10.3.50) is not a strong convergence in L_2. For $\epsilon \le t \le M$, under uniform bounds on $x' - x, Y, Y'$, for $0 < b \le 1$, v remains bounded in L_2. Also as $b \to 0$,

$$\frac{1}{2}\int_0^t |Y|^2 \, ds \to \frac{1}{2t}\left|x' - x\right|^2 + \frac{1}{4}\left(|Y|^2 + |Y'|^2\right), \qquad (10.3.51)$$

$$\frac{b^4}{2}\int_0^t \left|\dot{Y}\right|^2 \, ds \to \frac{1}{4}\left(|Y|^2 + |Y'|^2\right),$$

which fits with (10.3.49).

Proposition 10.3.3. *Given $\epsilon > 0, M > 0, \epsilon \leq M$, there exists $C_{\epsilon,M} > 0$ such that for $\epsilon \leq t \leq M, 0 < b \leq M$, and $x, x', Y, Y' \in E$,*

$$H_{b,t}\left((x, Y/b), (x', Y'/b)\right) \geq C_{\epsilon,M}\left(|x' - x|^2 + |Y|^2 + |Y'|^2\right). \quad (10.3.52)$$

Proof. This is an easy consequence of (10.3.39). $\qquad \square$

When $b \to +\infty$, a necessary and sufficient condition for p in (10.3.35) to converge is that

$$x' - x = \frac{t}{2}\left(Y + Y'\right), \quad (10.3.53)$$

in which case,

$$p = \frac{x' - x}{t} = \frac{1}{2}\left(Y + Y'\right). \quad (10.3.54)$$

If (10.3.53) is verified, by (10.3.19), (10.3.33), and (10.3.54), as $b \to +\infty$, we have the uniform convergence over $[0, t]$,

$$Y_s \to \left(1 - \frac{s}{t}\right) Y + \frac{s}{t} Y'. \quad (10.3.55)$$

By (10.3.55), as $b \to +\infty$, we have the uniform convergence over $[0, t]$,

$$x_s \to x + sY + \frac{s^2}{2t}\left(Y' - Y\right). \quad (10.3.56)$$

Under (10.3.53), if $Y = Y'$, as $b \to +\infty$,

$$H_{b,t}\left((x, Y), (x', Y)\right) \to \frac{1}{2t}|x' - x|^2, \quad K_{b,t}\left((x, Y), (x', Y)\right) \to \frac{1}{2t}|x' - x|^2. \quad (10.3.57)$$

We no longer assume (10.3.53) to hold. By (10.3.39), as $b \to +\infty$,

$$\frac{H_{b,t}}{b^4}\left((x, Y), (x', Y')\right) \to \frac{1}{2t}|Y' - Y|^2 + \frac{6}{t^3}\left|x' - x - \frac{t}{2}\left(Y + Y'\right)\right|^2, \quad (10.3.58)$$

$$\frac{K_{b,t}}{b^4}\left((x, Y), (x', Y')\right) \to \frac{1}{2t}|Y' - Y|^2 + \frac{6}{t^3}\left|x' - x - \frac{t}{2}\left(Y + Y'\right)\right|^2.$$

10.4 Mehler's formula

Let E still be an Euclidean vector space of dimension m, and let Y be the generic element of E. Let Δ^E be the Laplacian on E. Let O^E be the harmonic oscillator

$$O^E = \frac{1}{2}\left(-\Delta^E + |Y|^2 - m\right). \quad (10.4.1)$$

Given $t > 0$, let $h_t^E(Y, Y')$ be the smooth kernel associated with $\exp\left(-tO^E\right)$. By Mehler's formula [GlJ87], we get

$$
h_t^E(Y, Y') = \left(\frac{e^t}{2\pi \sinh(t)}\right)^{m/2}
$$
$$
\exp\left(-\frac{\tanh(t/2)}{2}\left(|Y|^2 + |Y'|^2\right) - \frac{1}{2\sinh(t)}|Y' - Y|^2\right). \quad (10.4.2)
$$

Let P^E be the operator

$$
P^E = \exp\left(|Y|^2/2\right)O^E \exp\left(-|Y|^2/2\right). \quad (10.4.3)
$$

Then

$$
P^E = \frac{1}{2}\left(-\Delta^E + 2\nabla_Y\right). \quad (10.4.4)
$$

The corresponding heat kernel $k_t^E(Y, Y')$ is given by

$$
k_t^E(Y, Y') = \exp\left(\frac{1}{2}\left(|Y|^2 - |Y'|^2\right)\right)h_t^E(Y, Y'). \quad (10.4.5)
$$

Equivalently,

$$
k_t^E(Y, Y') = \left(\frac{e^t}{2\pi \sinh(t)}\right)^{m/2}\exp\left(-\frac{1}{2\sinh(t)}\left|e^{t/2}Y' - e^{-t/2}Y\right|^2\right).
$$
$$
(10.4.6)
$$

Equation (10.4.6) indicates that given Y, $k_t^E(Y, Y')\,dY'$ is the probability law of a Gaussian variable, centered at $e^{-t}Y$, with variance given by $\left(1 - e^{-2t}\right)/2$.

If $f : E \to \mathbf{R}$ is smooth, for $a > 0$, set

$$
K_a f(Y) = f(aY). \quad (10.4.7)
$$

Set

$$
O_b^E = K_b \frac{O^E}{b^2}K_b^{-1}, \qquad P_b^E = K_b\frac{P^E}{b^2}K_b^{-1}. \quad (10.4.8)
$$

Then

$$
O_b^E = \frac{1}{2}\left(-\frac{\Delta^E}{b^4} + |Y|^2 - \frac{m}{b^2}\right), \qquad P_b^E = -\frac{1}{2b^4}\Delta^E + \frac{1}{b^2}\nabla_Y. \quad (10.4.9)
$$

Also by (10.4.3), we get

$$
P_b^E = \exp\left(b^2|Y|^2/2\right)O_b^E \exp\left(-b^2|Y|^2/2\right). \quad (10.4.10)
$$

Let $h_{b,t}^E(Y, Y'), k_{b,t}^E(Y, Y')$ be the smooth kernels associated with the operators $\exp\left(-tO_b^E\right), \exp\left(-tP_b^E\right)$. By (10.4.8), we get

$$
h_{b,t}^E(Y, Y') = b^m h_{t/b^2}^E(bY, bY'), \quad k_{b,t}^E(Y, Y') = b^m k_{t/b^2}^E(bY, bY'). \quad (10.4.11)
$$

Proposition 10.4.1. *The following identities hold:*

$$h_{b,t}^E (Y, Y') = \left[\frac{b^2 e^{t/b^2}}{2\pi \sinh(t/b^2)} \right]^{m/2} \exp\left(-H_{b,t}^* (Y, Y') \right), \qquad (10.4.12)$$

$$k_{b,t}^E (Y, Y') = \left[\frac{b^2 e^{t/b^2}}{2\pi \sinh(t/b^2)} \right]^{m/2} \exp\left(-K_{b,t}^* (Y, Y') \right).$$

Proof. This follows from (10.3.7), (10.4.2), (10.4.6), and (10.4.11). $\qquad \square$

Remark 10.4.2. Equation (10.4.11) can also be viewed as a consequence of (10.3.9) and (10.4.12).

10.5 The hypoelliptic heat kernel on an Euclidean vector space

Now we replace E by $E \oplus E$, whose generic element is denoted (x, Y). The operators O_b^E, P_b^E still act on the variable Y. We denote by ∇^H differentiation in x. Also $\Delta^{E,H}, \Delta^{E,V}$ denote the Laplacians on the first and second copies of E.

Let Q_b^E, R_b^E be the operators on $E \oplus E$,

$$Q_b^E = O_b^E - \nabla_Y^H, \qquad\qquad R_b^E = P_b^E - \nabla_Y^H. \qquad (10.5.1)$$

By (10.4.10), we get

$$R_b^E = \exp\left(b^2 |Y|^2 / 2 \right) Q_b^E \exp\left(-b^2 |Y|^2 / 2 \right). \qquad (10.5.2)$$

Let $h_{b,t}^E ((x, Y), (x', Y')), k_{b,t}^E ((x, Y), (x', Y'))$ be the smooth kernels associated with $\exp\left(-t Q_b^E \right), \exp\left(-t R_b^E \right)$. By (10.5.2),

$$k_{b,t}^E ((x, Y), (x', Y')) = \exp\left(\frac{b^2}{2} \left(|Y|^2 - |Y'|^2 \right) \right) h_{b,t}^E ((x, Y), (x', Y')). \qquad (10.5.3)$$

Also one verifies easily that

$$h_{b,t}^E ((x, Y), (x', Y')) = h_{1,t/b^2}^E ((x/b, bY), (x'/b, bY')), \qquad (10.5.4)$$

$$k_{b,t}^E ((x, Y), (x', Y')) = k_{1,t/b^2}^E ((x/b, bY), (x'/b, bY')).$$

Clearly,

$$Q_b^E = \frac{1}{2} \left(-\frac{\Delta^{E,V}}{b^4} + |Y - \nabla_{\cdot}^H|^2 - \frac{m}{b^2} \right) - \frac{1}{2} \Delta^{E,H}. \qquad (10.5.5)$$

Let $e_1, \ldots, e_m$ be an orthonormal basis of E, and let $x^i, Y^i, 1 \le i \le m$ be the coordinates of x, Y with respect to this basis. Set

$$N^E = \sum_{i=1}^m \frac{\partial^2}{\partial x^i \partial Y^i}. \qquad (10.5.6)$$

The operator $\exp\left(N^E\right)$ acts on functions $f\left(x,Y\right)$ that are polynomial in the variables x,Y.

Using (10.5.5), a formal calculation shows that

$$Q_b^E = \exp\left(-N^E\right)\left(O_b^E - \frac{1}{2}\Delta^{E,H}\right)\exp\left(N^E\right). \qquad (10.5.7)$$

Equation (10.5.7) can be viewed as the quantization of equation (10.3.29). Here we use the correspondence $p \to i\nabla^H$. Equivalently, (10.5.7) is an exact form of Egorov's theorem, in which the conjugating operator is unbounded. Sense can be made of the conjugation, once we use the fact that the eigenfunctions of the harmonic oscillator are analytic in the variable Y. This issue is discussed in detail in [B05, section 3.10], [B08c, section 2.9], and [B09a, section 1.2].

By (10.5.7), we get

$$\exp\left(-tQ_b^E\right) = \exp\left(-N^E\right)\exp\left(-t\left(O_b^E - \frac{1}{2}\Delta^{E,H}\right)\right)\exp\left(N^E\right). \quad (10.5.8)$$

Proposition 10.5.1. *The following identities hold:*

$$h_{b,t}^E\left((x,Y),(x',Y')\right) = \left[\frac{b^2 e^{t/b^2}}{4\pi^2 \sinh\left(t/b^2\right)\left(t - 2b^2\tanh\left(t/2b^2\right)\right)}\right]^{m/2}$$
$$\exp\left(-H_{b,t}\left((x,Y),(x',Y')\right)\right),$$

$$k_{b,t}^E\left((x,Y),(x',Y')\right) = \left[\frac{b^2 e^{t/b^2}}{4\pi^2 \sinh\left(t/b^2\right)\left(t - 2b^2\tanh\left(t/2b^2\right)\right)}\right]^{m/2}$$
$$\exp\left(-K_{b,t}\left((x,Y),(x',Y')\right)\right). \quad (10.5.9)$$

Proof. By (10.5.8), we deduce that

$$h_{b,t}^E\left((x,Y),(x',Y')\right) = h_{b,t}^E\left(Y - \nabla_\cdot^H, Y' - \nabla_\cdot^H\right)$$
$$\frac{1}{(2\pi t)^{m/2}}\exp\left(-\left|x' - x\right|^2/2t\right). \quad (10.5.10)$$

By (10.4.12) and (10.5.10), we obtain

$$h_{b,t}^E\left((x,Y),(x',Y')\right) = \left[\frac{b^2 e^{t/b^2}}{2\pi \sinh\left(t/b^2\right)}\right]^{m/2}$$
$$\exp\left(-H_{b,t}^*\left(Y - \nabla_\cdot^H, Y' - \nabla_\cdot^H\right) + t\Delta^{E,H}/2\right)(x,x'). \quad (10.5.11)$$

Moreover, using (10.3.7), we get

$$H_{b,t}^*\left(Y - \nabla_\cdot^H, Y' - \nabla_\cdot^H\right) - t\Delta^{E,H}/2 = H_{b,t}^*\left(Y,Y'\right)$$
$$- \frac{b^2}{2}\left(t/b^2 - 2\tanh\left(t/2b^2\right)\right)\Delta^{E,H} - b^2\tanh\left(t/2b^2\right)\nabla_{Y+Y'}^H. \quad (10.5.12)$$

By (10.5.11), (10.5.12), we obtain

$$h_{b,t}^E \left((x,Y),(x',Y')\right) = \left[\frac{b^2 e^{t/b^2}}{4\pi^2 \sinh\left(t/b^2\right)\left(t - 2b^2 \tanh\left(t/2b^2\right)\right)} \right]^{m/2}$$

$$\exp\Biggl(-H_{b,t}^*\left(Y,Y'\right) - \frac{1}{2\left(t - 2b^2 \tanh\left(t/2b^2\right)\right)}$$

$$\left| x' - x - b^2 \tanh\left(t/2b^2\right)\left(Y + Y'\right)\right|^2 \Biggr). \quad (10.5.13)$$

Using (10.3.7), (10.3.39), and (10.5.13), we get the first identity in (10.5.9). The second identity then follows from (10.3.37), (10.5.3), and from the first identity. $\qquad\Box$

Remark 10.5.2. The fact that $H_{b,t}^*\left(Y,Y'\right)$ and $H_{b,t}\left((x,Y),(x',Y')\right)$ appear in equations (10.4.12) for $h_{b,t}^E\left(Y,Y'\right)$ and (10.5.9) for $h_{b,t}^E\left((x,Y),(x',Y')\right)$ can also be viewed as a consequence of the Hamilton-Jacobi equations in (10.3.8), (10.3.40). Since

$$\int_{E\oplus E} k_{b,t}^E\left((x,Y),(x',Y')\right)dx'dY' = 1, \quad (10.5.14)$$

the normalizing factor in the right-hand side of (10.5.9) reflects the explicit form of $K_{b,t}\left((x,Y),(x',Y')\right)$ in (10.3.39). Also (10.5.4) can be viewed as a consequence of (10.3.43) and (10.5.9).

By combining (10.3.52) and (10.5.9), we find that given $\epsilon > 0, M > 0, \epsilon \leq M$, there exist $c_{\epsilon,M} > 0, C_{\epsilon,M} > 0$ such that for $\epsilon \leq t \leq M, 0 < b \leq M$,

$$b^{-m} h_{b,t}^E\left((x,Y/b),(x',Y'/b)\right)$$

$$\leq c_{\epsilon,M} \exp\left(-C_{\epsilon,M}\left(\left|x' - x\right|^2 + \left|Y\right|^2 + \left|Y'\right|^2\right)\right). \quad (10.5.15)$$

Equation (10.5.15) is just a version of the estimate in (4.5.3). Moreover, by (10.3.58), (10.5.9), as $b \to +\infty$, the kernels $h_{b,t}^E, k_{b,t}^E$ concentrate along the trajectories of the geodesic flow of E in $E \oplus E$.

10.6 Orbital integrals on an Euclidean vector space

The results contained in this section will not be needed later. We use again the formalism of chapters 2, 4, and 6, in the case where $G = E$. Multiplication in G will be denoted additively. Then $K = \{0\}$. The Lie algebra $\mathfrak{g}$ of G is given by

$$\mathfrak{g} = E, \quad (10.6.1)$$

so that

$$\mathfrak{p} = E, \qquad\qquad \mathfrak{k} = 0. \quad (10.6.2)$$

Also the scalar product of E is defined to be the nondegenerate bilinear symmetric form B on $\mathfrak{g}$. The symmetric space X is just E, and G acts on X by translations. Moreover, $TX = E, N = \{0\}$, and the Euclidean and flat connections on TX coincide. Clearly, $\mathcal{X} = \widehat{\mathcal{X}}$, and moreover,

$$\mathcal{X} = E \oplus E. \tag{10.6.3}$$

In (10.6.3), the first copy of E is identified with X, and the second copy with TX. Since $K = \{0\}$, the quotient procedures of section 2.12 become irrelevant.

We use the notation of section 10.5. In particular, (x, Y) still denotes the generic section of $E \oplus E$. In the sequel, the operators that will be constructed act on $C^\infty (E \oplus E, \Lambda^\cdot (E^*))$.

Let $e_1, \ldots, e_m$ be an orthonormal basis of E. By (2.7.2), we get

$$D^\mathfrak{g} = \sum_{i=1}^m c(e_i) \nabla_{e_i}^H, \qquad \widehat{D}^\mathfrak{g} = \sum_{i=1}^m \widehat{c}(e_i) \nabla_{e_i}^H. \tag{10.6.4}$$

By (2.7.7),

$$D^{\mathfrak{g},2} = \Delta^{E,H}. \tag{10.6.5}$$

Since $K = \{0\}$, we may as well take $F = \mathbf{R}$. Then $\widehat{D}^\mathfrak{g}$ and $\widehat{D}^{\mathfrak{g},X}$ coincide. By (2.9.4), (2.9.5), or by (2.12.19), we get

$$\mathfrak{D}_b^X = \widehat{D}^{\mathfrak{g},X} + \frac{1}{b}\left(\mathcal{D}^E + \mathcal{E}^E\right) = \widehat{D}^{\mathfrak{g},X} + \frac{\sqrt{2}}{b}\left(\overline{d}^E + \overline{d}^{E*}\right). \tag{10.6.6}$$

By (2.11.4) or by (2.12.19), we get

$$\frac{1}{2}\mathfrak{D}_b^{X,2} = \frac{1}{2}\Delta^{E,H} + \frac{1}{2b^2}\left(-\Delta^{E,V} + |Y|^2 - m\right) + \frac{N^{\Lambda^\cdot(E^*)}}{b^2} + \frac{1}{b}\nabla_Y^H. \tag{10.6.7}$$

By (2.13.3),

$$\mathcal{L}^X = -\frac{1}{2}\Delta^{E,H}. \tag{10.6.8}$$

By (2.13.5),

$$\mathcal{L}_b^X = \frac{1}{2b^2}\left(-\Delta^{E,V} + |Y|^2 - m\right) + \frac{N^{\Lambda^\cdot(E^*)}}{b^2} + \frac{1}{b}\nabla_Y^H. \tag{10.6.9}$$

For $t > 0$, let $p_t(x, x')$ be the smooth heat kernel on E associated with $\exp(t\Delta^E/2)$. Then

$$p_t(x, x') = \frac{1}{(2\pi t)^{m/2}}\exp\left(-\frac{1}{2t}|x' - x|^2\right). \tag{10.6.10}$$

For $b > 0, t > 0$, let $q_{b,t}^X((x, Y), (x', Y'))$ be the smooth kernel associated with $\exp(-t\mathcal{L}_b^X)$.

Proposition 10.6.1. *For $b > 0, t > 0$, the following identity holds:*

$$q_{b,t}^X((x, Y), (x', Y'))$$
$$= b^{-m}h_{b,t}^E((x, -Y/b), (x', -Y'/b))\exp\left(-tN^{\Lambda^\cdot(E^*)}/b^2\right). \tag{10.6.11}$$

Proof. This follows from (10.4.9), (10.5.1), and (10.6.9).   □

Let $a \in \mathfrak{p} = E$. Then $\gamma = e^a$ acts by translation by a on the first copy of $E \oplus E$. Since $G = E$ is commutative, we will use the notation γ instead of $[\gamma]$.

Proposition 10.6.2. *For $b > 0, t > 0$, the following identities hold:*

$$\mathrm{Tr}^\gamma \left[\exp\left(t\Delta^{E,H}/2\right)\right] = p_t(0, a), \tag{10.6.12}$$

$$\mathrm{Tr_s}^\gamma \left[\exp\left(-t\mathcal{L}_b^X\right)\right] = \left(1 - e^{-t/b^2}\right)^m \int_E h_{b,t}^E\left((0, Y), (a, Y)\right) dY.$$

Proof. The first identity in (10.6.12) is trivial, and the second is a consequence of equation (10.6.11) in Proposition 10.6.1.   □

The crucial identity (4.6.1) in Theorem 4.6.1 asserts that

$$\mathrm{Tr_s}^\gamma \left[\exp\left(-t\mathcal{L}_b^X\right)\right] = \mathrm{Tr}^\gamma \left[\exp\left(t\Delta^{E,H}/2\right)\right]. \tag{10.6.13}$$

In the special case we are considering, equation (10.6.13) can also be obtained as a consequence of (10.5.8).

Let us give a computational proof of (10.6.13). By equation (10.3.39) in Proposition 10.3.2 for $H_{b,t}$, we get easily

$$H_{b,t}\left((0, Y), (a, Y)\right) = \frac{b^2 t \tanh\left(t/2b^2\right)}{t - 2b^2 \tanh\left(t/2b^2\right)} \left|Y - \frac{a}{t}\right|^2 + \frac{|a|^2}{2t}. \tag{10.6.14}$$

By (10.6.14), we obtain

$$\int_E \exp\left(-H_{b,t}\left((0, Y), (a, Y)\right)\right) dY$$

$$= \left(\pi \frac{t - 2b^2 \tanh\left(t/2b^2\right)}{b^2 t \tanh\left(t/2b^2\right)}\right)^{m/2} \exp\left(-\frac{|a|^2}{2t}\right). \tag{10.6.15}$$

By equation (10.5.9) in Proposition 10.5.1 and by (10.6.15), we get

$$\int_E h_{b,t}^E\left((0, Y), (a, Y)\right) dY$$

$$= \left(\frac{e^{t/b^2}}{4\pi t \tanh\left(t/2b^2\right) \sinh\left(t/b^2\right)}\right)^{m/2} \exp\left(-\frac{|a|^2}{2t}\right). \tag{10.6.16}$$

Equivalently,

$$\int_E h_{b,t}^E\left((0, Y), (a, Y)\right) dY = \left(1 - e^{-t/b^2}\right)^{-m} (2\pi t)^{-m/2} \exp\left(-\frac{|a|^2}{2t}\right).$$

$$\tag{10.6.17}$$

Equation (10.6.13) follows from (10.6.12), (10.6.17).

Finally, let us reinterpret equation (10.6.13), by getting rid of γ. The enveloping algebra $U(\mathfrak{g})$ can be identified with the algebra of differential operators with constant coefficients on E. The operator $\mathcal{L}_b^X$ can be viewed as acting on $C^\infty(E, \Lambda^\cdot(E^*) \otimes U(\mathfrak{g}))$. For $t > 0$, $\mathrm{Tr}_s\left[\exp\left(-t\mathcal{L}_b^X\right)\right]$ is the supertrace of a trace class operator, with values in a completion of $U(\mathfrak{g})$. Equation (10.6.13) then says that for any $b > 0, t > 0$,

$$\mathrm{Tr}_s\left[\exp\left(-t\mathcal{L}_b^X\right)\right] = \exp\left(t\Delta^{E,H}/2\right). \qquad (10.6.18)$$

Equation (10.6.18) can also be obtained as a consequence of (10.5.8).

Equation (10.6.18) should be thought of as an analogue of the McKean-Singer formula [McKS67] for the index of a Dirac operator. In this case, for a given $t > 0$, the index is the operator $\exp\left(t\Delta^{E,H}/2\right)$. The fact that for $t > 0$, the right-hand side of (10.6.18) does not depend on $b > 0$ is one key feature of a McKean-Singer formula. From the point of view of the present book, as $b \to 0$, $\mathrm{Tr}_s\left[\exp\left(-t\mathcal{L}_b^X\right)\right]$ converges to the right-hand side, which is an operator. As $b \to +\infty$, $\mathrm{Tr}_s\left[\exp\left(-t\mathcal{L}_b^X\right)\right]$ converges to the explicit formula for the heat kernel of E. In index theory, this is called a local index theorem [ABoP73, Gi73, Gi84].

Let $O(E)$ be the orthogonal group of E, let $I(E)$ be the group of isometries of E. Then $I(E)$ is the semidirect product

$$I(E) = E \rtimes O(E). \qquad (10.6.19)$$

Clearly, we have the identification of symmetric spaces,

$$E = I(E)/O(E). \qquad (10.6.20)$$

The group $I(E)$ is not reductive, so that, strictly speaking, the methods and results contained in this book do not apply. However, if $\gamma \in I(E)$, for $t > 0$, the orbital integral $\mathrm{Tr}^{[\gamma]}\left[\exp\left(t\Delta^E/2\right)\right]$ is well-defined, and its computation is elementary.

The objects introduced in equations (10.6.4)–(10.6.9) are still adequate to study the above orbital integral by the methods used in this book. In particular we still take $\mathfrak{k} = 0$, although this is not the Lie algebra of $O(E)$, and equation (4.6.1) in Theorem 4.6.1 still holds, i.e., for $t > 0$,

$$\mathrm{Tr}_s^{[\gamma]}\left[\exp\left(-t\mathcal{L}_b^X\right)\right] = \mathrm{Tr}^{[\gamma]}\left[\exp\left(t\Delta^{E,H}/2\right)\right]. \qquad (10.6.21)$$

We will now verify (10.6.21) directly. With respect to what we did before, it is enough to take $k \in O(E)$ with no eigenvalue equal to 1, and to assume that $\gamma = k^{-1}$.

Proposition 10.6.3. *For $b > 0, t > 0$, the following identities hold:*

$$\mathrm{Tr}^{[\gamma]}\left[\exp\left(t\Delta^{E,H}/2\right)\right] = \frac{1}{\det\left(1 - k^{-1}\right)},$$

$$\mathrm{Tr}_s^{[\gamma]}\left[\exp\left(-t\mathcal{L}_b^X\right)\right] = \det\left(1 - e^{-t/b^2}k^{-1}\right) \qquad (10.6.22)$$

$$\int_{E \times E} k_{b,t}^E\left((x, Y), k^{-1}(x, Y)\right) dx dY.$$

Proof. Note that with $X = E$, then $X(\gamma)$ is reduced to 0. By (4.2.12), we get

$$\mathrm{Tr}^{[\gamma]} \exp\left(t\Delta^{E,H}/2\right) = \int_E p_t\left(x, k^{-1}x\right) dx. \tag{10.6.23}$$

By (10.6.10), (10.6.23), we obtain the first identity in (10.6.22). Moreover,

$$\mathrm{Tr}_s^{[\gamma]}\left[\exp\left(-t\mathcal{L}_b^X\right)\right] = \int_{E\times E} \mathrm{Tr}_s\left[k^{-1}q_{b,t}^X\left((x,Y), k^{-1}(x,Y)\right)\right] dxdY. \tag{10.6.24}$$

Also by (10.5.3), and by equation (10.6.11) in Proposition 10.6.1, we get

$$q_{b,t}^X\left((x,Y), k^{-1}(x,Y)\right)$$
$$= b^{-m}k_{b,t}^E\left((x, -Y/b), k^{-1}(x, -Y/b)\right)\exp\left(-tN^{\Lambda^{\cdot}(E^*)}/b^2\right). \tag{10.6.25}$$

The second identity in (10.6.22) follows from (10.6.24), (10.6.25). □

Remark 10.6.4. With the proviso indicated before, the methods used in the proof of Theorem 6.1.1 still apply. In particular, the first identity in (10.6.22) can be viewed as a consequence of equation (6.1.2) in Theorem 6.1.1.

Now we will verify (10.6.21) directly using Proposition 10.6.3. By equation (10.3.39) in Proposition 10.3.2, we get

$$K_{b,t}\left((x,Y), k^{-1}(x,Y)\right) = \frac{b^2 e^{t/b^2}}{2\sinh\left(t/b^2\right)}\left|\left(1 - e^{-t/b^2}k\right)Y\right|^2$$
$$+ \frac{1}{2\left(t - 2b^2\tanh\left(t/2b^2\right)\right)}\left|\left(1 - k^{-1}\right)x + b^2\tanh\left(t/2b^2\right)\left(1 + k^{-1}\right)Y\right|^2. \tag{10.6.26}$$

By (10.6.26), we get

$$\int_{E\times E} \exp\left(-K_{b,t}\left((x,Y), k^{-1}(x,Y)\right)\right) dxdY$$
$$= \frac{1}{\det\left[\left(1 - k^{-1}\right)\left(1 - e^{-t/b^2}k\right)\right]}$$
$$\left[4\pi^2 b^{-2}e^{-t/b^2}\sinh\left(t/b^2\right)\left(t - 2b^2\tanh\left(t/2b^2\right)\right)\right]^{m/2}. \tag{10.6.27}$$

By (10.5.9), (10.6.27), we obtain

$$\int_{E\times E} k_{b,t}^E\left((x,Y), k^{-1}(x,Y)\right) dxdY = \frac{1}{\det\left[\left(1 - k^{-1}\right)\left(1 - e^{-t/b^2}k\right)\right]}. \tag{10.6.28}$$

Since k is an isometry, we can replace $1 - e^{-t/b^2}k$ by $1 - e^{-t/b^2}k^{-1}$ in the right-hand side of (10.6.28).

By the second identity in (10.6.22) and by (10.6.28), we get

$$\mathrm{Tr}_s^{[\gamma]}\left[\exp\left(-t\mathcal{L}_b^X\right)\right] = \frac{1}{\det\left(1 - k^{-1}\right)}. \tag{10.6.29}$$

Equation (10.6.29) fits with (10.6.21) and (10.6.22).

The behavior of the integral in the right-hand side of the second identity in (10.6.22) says much about the hypoelliptic orbital integrals on symmetric spaces. One can directly observe the concentration as $b \to +\infty$ of this integral near $X(\gamma) = \{0\}$.

10.7 Some computations involving Mehler's formula

By (10.4.6), if $q < e^t / \sinh(t)$, we get

$$\int_E k_t^E(Y, Y') \exp\left(\frac{q}{2} |Y'|^2\right) dY' = \frac{1}{(1 - qe^{-t} \sinh(t))^{m/2}}$$
$$\exp\left(\frac{q}{2} e^{-2t} |Y|^2 / \left(1 - qe^{-t} \sinh(t)\right)\right). \quad (10.7.1)$$

In particular, from (10.7.1), we deduce that if $q < 2$,

$$\int_E k_t^E(Y, Y') \exp\left(\frac{q}{2} |Y'|^2\right) dY' \leq \frac{1}{(1 - q/2)^{m/2}} \exp\left(\frac{q}{2 - q} e^{-2t} |Y|^2\right).$$
$$(10.7.2)$$

By (10.7.1), we find that if $q \leq e^t / \cosh(t)$,

$$\int_E k_t^E(Y, Y') \exp\left(\frac{q}{2} |Y'|^2\right) dY' \leq \frac{1}{(1 - q/2)^{m/2}} \exp\left(|Y|^2 / 2\right). \quad (10.7.3)$$

Take $\epsilon > 0$. By (10.7.3), for $t \geq \epsilon, q = e^\epsilon / \cosh(\epsilon)$,

$$\int_E k_t^E(Y, Y') \exp\left(\frac{q}{2} |Y'|^2\right) dY' \leq \frac{1}{(1 - q/2)^{m/2}} \exp\left(|Y|^2 / 2\right). \quad (10.7.4)$$

From (10.7.4), we deduce that for $p \in \mathbf{N}, t \geq \epsilon$,

$$\exp\left(-|Y|^2 / 2\right) \int_E k_t^E(Y, Y') \exp\left(|Y'|^2 / 2\right) |Y'|^p dY' \leq C_{\epsilon, p}. \quad (10.7.5)$$

By taking $q = 1 + u$ in (10.7.1), with $u < \coth(t)$, we get

$$\exp\left(-|Y|^2 / 2\right) \int_E k_t^E(Y, Y') \exp\left(\frac{1 + u}{2} |Y'|^2\right) dY'$$
$$= \left(\frac{e^t}{\cosh(t) - u \sinh(t)}\right)^{m/2} \exp\left(\frac{1}{2} \frac{u \coth(t) - 1}{\coth(t) - u} |Y|^2\right). \quad (10.7.6)$$

By (10.7.6), if $0 \leq u < 1, t \geq \epsilon$,

$$\exp\left(-|Y|^2 / 2\right) \int_E k_t^E(Y, Y') \exp\left(\frac{1}{2} (1 + u) |Y'|^2\right) dY'$$
$$\leq \left(\frac{2}{1 - u}\right)^{m/2} \exp\left(-\frac{1 - u \coth(\epsilon)}{1 - u \tanh(\epsilon)} \tanh(\epsilon) |Y|^2 / 2\right). \quad (10.7.7)$$

Note that (10.7.7) does not follow from (10.7.2) as can be seen by making $q = 1$, or $u = 0$.

From (10.7.7), we deduce that for $t \geq \epsilon, p \in \mathbf{N}$,

$$\exp\left(-|Y|^2/2\right) \int_E k_t^E(Y, Y') \exp\left(|Y'|^2/2\right) |Y'|^p \, dY'$$

$$\leq C_{\epsilon,p} \exp\left(-\frac{1}{4}\tanh(\epsilon)|Y|^2\right). \quad (10.7.8)$$

By (10.7.1),

$$\exp\left(-|Y|^2/2\right) \int_E k_t^E(Y, Y') \exp\left(|Y'|^2/2\right) dY'$$

$$= \left(\frac{e^t}{\cosh(t)}\right)^{m/2} \exp\left(-\frac{\tanh(t)}{2}|Y|^2\right) \leq 2^{m/2}. \quad (10.7.9)$$

By (10.4.5), (10.7.6), we find that if $u < \coth(t)$,

$$\int_E h_t^E(Y, Y') \exp\left(u|Y'|^2/2\right) dY' = \left(\frac{e^t}{\cosh(t) - u\sinh(t)}\right)^{m/2}$$

$$\exp\left(\frac{1}{2}\frac{u\coth(t) - 1}{\coth(t) - u}|Y|^2\right). \quad (10.7.10)$$

By (10.7.10), we deduce that for $u < \coth(t)$,

$$\int_E h_t^E(Y, Y') \exp\left(u|Y'|^2/2\right) dY'$$

$$\leq \left(\frac{2}{1 - u\tanh(t)}\right)^{m/2} \exp\left(\frac{1}{2}\frac{u\coth(t) - 1}{\coth(t) - u}|Y|^2\right). \quad (10.7.11)$$

By (10.7.10), we get

$$\int_E h_t^E(Y, Y') \, dY' = \left(\frac{e^t}{\cosh(t)}\right)^{m/2} \exp\left(-\frac{\tanh(t)}{2}|Y|^2\right), \quad (10.7.12)$$

and the right-hand side of (10.7.12) is uniformly bounded for $t > 0$. By (10.4.5), (10.7.8), given $\epsilon > 0, p \in \mathbf{N}$, there is $C_{\epsilon,p} > 0$ such that for $t \geq \epsilon$,

$$\int_E h_t^E(Y, Y') |Y'|^p \, dY' \leq C_{\epsilon,p} \exp\left(-\frac{\tanh(\epsilon)}{4}|Y|^2\right). \quad (10.7.13)$$

10.8 The probabilistic interpretation of the harmonic oscillator

Let P be the probability measure on $\mathcal{C}(\mathbf{R}_+, E)$ of the Brownian motion $s \in \mathbf{R}_+ \to w_s \in E$, with $w_0 = 0$.

For $1 \leq p < +\infty$, $\|\,\|_p$ denotes the standard norm on the vector space L_p associated with the measure P on $\mathcal{C}(\mathbf{R}_+, E)$.

For $Y \in E$, consider the stochastic differential equation

$$\dot{Y} = -Y + \dot{w}, \qquad\qquad Y_0 = Y. \qquad (10.8.1)$$

This equation can be written in the form

$$Y_t = Y - \int_0^t Y_s ds + w_t. \tag{10.8.2}$$

Note that (10.8.2) can be integrated for any $w. \in \mathcal{C}(\mathbf{R}_+, E)$. The explicit solution of (10.8.2) is given by

$$Y_t = e^{-t}Y + e^{-t} \int_0^t e^s \delta w_s. \tag{10.8.3}$$

In the right-hand side of (10.8.3), the last term is an Itô integral. In particular, from (10.8.3), we deduce that the probability law of Y_t is a Gaussian centered at $e^{-t}Y$ with variance $\left(1 - e^{-2t}\right)/2$, i.e., the probability law of Y_t is given by $k_t^E(Y, Y') dY'$. Note that since e^s is a smooth function,

$$\int_0^t e^s \delta w_s = e^t w_t - \int_0^t e^s w_s ds, \tag{10.8.4}$$

so that the machinery of stochastic integration is not needed to make sense of the integral in (10.8.4).

Let E be the expectation operator with respect to P. It is a standard consequence of the Itô calculus that if $f \in C^{\infty,c}(E, \mathbf{R})$, for $t \geq 0$,

$$\exp\left(-tP^E\right) f(Y) = E\left[f(Y_t)\right]. \tag{10.8.5}$$

By (10.8.5), we recover the above result on the probability law of Y_t.

Let Y^* be the solution of

$$\dot{Y}^* = \dot{w}, \qquad\qquad Y_0^* = Y, \tag{10.8.6}$$

so that

$$Y_t^* = Y + w_t. \tag{10.8.7}$$

Let $f \in C^{\infty,c}(E, \mathbf{R})$. Then using the Itô calculus and the Feynman-Kac formula, we get

$$\exp\left(-tO^E\right) f(Y) = E\left[\exp\left(\frac{mt}{2} - \frac{1}{2}\int_0^t |Y_s^*|^2 ds\right) f(Y_t^*)\right]. \tag{10.8.8}$$

By (10.4.3), (10.8.5), we also obtain

$$\exp\left(-tO^E\right) f(Y) = \exp\left(-|Y|^2/2\right) E\left[\exp\left(|Y_t|^2/2\right) f(Y_t)\right]. \tag{10.8.9}$$

One can verify directly that the right-hand side of (10.8.9) is finite. Indeed by (10.7.1), given Y, for any $t \geq 0$, $\exp\left(|Y_t|^2/2\right) \in L_2$, and its L_2 norm remains uniformly bounded for bounded t.

Proposition 10.8.1. *Given $p \in [1, +\infty[$ there is $C_p > 0$ such that for $t \geq 0$,*

$$\|Y_t\|_p \leq e^{-t}|Y| + C_p. \tag{10.8.10}$$

Given $M \geq 0, p \in [1, +\infty[$, there is $C_{M,p} > 0$ such that for $0 \leq t \leq M, b > 0$,

$$\left\|\left[\int_0^t |Y_{s/b^2}|^4 ds\right]^{1/2}\right\|_p \leq C_{M,p}\left(1 + b|Y|^2\right). \tag{10.8.11}$$

For $t \geq 0$, *and* $1 \leq q \leq \frac{e^t}{\cosh(t)}$,

$$\left\| \exp\left(|Y_t|^2 / 2 \right) \right\|_q \leq \frac{1}{(1 - q/2)^{m/2q}} \exp\left(|Y|^2 / 2q \right). \tag{10.8.12}$$

Proof. Equation (10.8.10) is an easy consequence of the above results on the probability law of Y_t.

Let Y^0 be the solution of (10.8.1) with $Y_0 = 0$, so that in (10.8.3),

$$Y_t = e^{-t} Y + Y_t^0. \tag{10.8.13}$$

Then

$$\left[\int_0^t \left| Y_{s/b^2} \right|^4 ds \right]^{1/2} \leq C \left(b \, |Y|^2 + \left[\int_0^t \left| Y_{s/b^2}^0 \right|^4 ds \right]^{1/2} \right). \tag{10.8.14}$$

Moreover,

$$\left[\int_0^t \left| Y_{s/b^2}^0 \right|^4 ds \right]^{1/2} \leq \frac{1}{2} \left(1 + \int_0^t \left| Y_{s/b^2}^0 \right|^4 ds \right). \tag{10.8.15}$$

By Hölder's inequality, for $0 \leq t \leq M$,

$$\int_0^t \left| Y_{s/b^2}^0 \right|^4 ds \leq C_{M,p} \left[\int_0^t \left| Y_{s/b^2}^0 \right|^{4p} ds \right]^{1/p}. \tag{10.8.16}$$

By (10.8.16), we get

$$\left\| \int_0^t \left| Y_{s/b^2}^0 \right|^4 ds \right\|_p \leq C_{M,p} \left[\int_0^t E \left[\left| Y_{s/b^2}^0 \right|^{4p} \right] ds \right]^{1/p}. \tag{10.8.17}$$

By (10.8.10) with $Y = 0$, for $s \geq 0$, $E \left[\left| Y_s^0 \right|^{4p} \right]$ is uniformly bounded, so that from (10.8.17), for $0 \leq t \leq M$,

$$\left\| \int_0^t \left| Y_{s/b^2}^0 \right|^4 ds \right\|_p \leq C_{M,p}. \tag{10.8.18}$$

By (10.8.14)–(10.8.18), we get (10.8.11).

Since the probability law of Y_t is equal to $k_t^E (Y, Y') dY'$, we get

$$E \left[\exp\left(q \, |Y_t|^2 / 2 \right) \right] = \int_E k_t^E (Y, Y') \exp\left(q \, |Y'|^2 / 2 \right) dY'. \tag{10.8.19}$$

By (10.7.3) and (10.8.19), we get (10.8.12). The proof of our proposition is completed. $\square$

For $b > 0$, instead of equation (10.8.1), we consider the stochastic differential equation

$$\dot{Y}^b = -\frac{Y^b}{b^2} + \frac{\dot{w}}{b}, \qquad Y_0^b = Y. \tag{10.8.20}$$

The solution of (10.8.20) is given by

$$Y_t^b = e^{-t/b^2} Y + e^{-t/b^2} \int_0^t e^{s/b^2} \frac{\dot{w}_s}{b} ds. \tag{10.8.21}$$

Instead of (10.8.5), we have the more general,

$$\exp\left(-\frac{t}{b^2}P^E\right)f(Y) = E\left[f\left(Y_t^b\right)\right].\tag{10.8.22}$$

There is still another method to construct the probability law of the process $Y_\cdot^b$. Set

$$w_\cdot^b = bw_{\cdot/b^2}.\tag{10.8.23}$$

Then it is well-known that $w_\cdot^b$ is still a Brownian motion. Put

$$Y_{b,t} = Y_{t/b^2}.\tag{10.8.24}$$

Then

$$\dot{Y}_b = -\frac{Y_b}{b^2} + \frac{\dot{w}^b}{b}, \qquad\qquad Y_{b,0} = Y.\tag{10.8.25}$$

By (10.8.1) and (10.8.25), the processes $Y_\cdot^b$ and $Y_{b,\cdot}$ have the same probability law.

Proposition 10.8.2. *Given $p \in [1, +\infty[$ there is $C_p > 0$ such that for $t \geq 0$,*

$$\left\|Y_t^b\right\|_p \leq e^{-t/b^2}|Y| + C_p.\tag{10.8.26}$$

Given $M \geq 0, p \in [1, +\infty[$, there is $C_{M,p} > 0$ such that for $0 \leq t \leq M, b > 0$,

$$\left\|\left[\int_0^t |Y_s^b|^4\, ds\right]^{1/2}\right\|_p \leq C_{M,p}\left(1 + b|Y|^2\right).\tag{10.8.27}$$

Proof. This is an obvious consequence of Proposition 10.8.1 and of the fact that we may as well replace $Y_\cdot^b$ by $Y_{\cdot/b^2}$ in the above inequalities. A direct proof can also be given along the lines of the proof of Proposition 10.8.1. $\square$

Chapter Eleven

The analysis of the hypoelliptic Laplacian

The purpose of this chapter is to construct a functional analytic machinery that is adapted to the analysis of the hypoelliptic Laplacian $\mathcal{L}^X_{A,b}$.

Our constructions are inspired by our previous work with Lebeau [BL08, chapter 15]. There are two differences with [BL08]. The first difference is that the symmetric space X is noncompact, while the base manifold in [BL08] was assumed to be compact. A second difference is that in $\mathcal{L}^X_{A,b}$, the quartic term $\frac{1}{2}\left|\left[Y^N, Y^{TX}\right]\right|^2$ appears. Strictly speaking, it is not directly accessible to the methods of [BL08].

The analysis of the hypoelliptic Laplacian essentially consists in the construction of Sobolev spaces on which the operators $\mathcal{L}^X_{A,b}$ act as unbounded operators, and in the proof of regularizing properties of their resolvents and of their heat operators. The heat operators are shown to be given by smooth kernels.

In [BL08, chapter 15], when the base X is compact, one idea is to construct a Littlewood-Paley decomposition of functions along the fibres of the vector bundle TX, and to adapt the method developed by Kohn [Koh73] in his treatment of hypoelliptic second order operators. While still using the results of [BL08, chapter 15], a new ingredient in the present chapter is the construction of a Littlewood-Paley decomposition of functions also on the base X.

To make the analysis easier, we first work with a scalar hypoelliptic operator $\mathcal{A}^X_b$ over $\mathcal{X}$, to which part of the analysis of [BL08] can be adapted. This operator does not contain a quartic term. The results on $\mathcal{A}^X_b$ then easily extend to a scalar operator $\mathfrak{A}^X_b$ acting over $\widehat{\mathcal{X}}$, which also does not contain the quartic term. A scalar operator A^X_b on $\widehat{\mathcal{X}}$ containing the quartic term is introduced. Finally, we extend the analysis to the operator $\mathcal{L}^X_{A,b}$.

This chapter is organized as follows. In section 11.1, we define the scalar operator $\mathcal{A}^X_b$ over $\mathcal{X}$, and a conjugate operator $\mathcal{B}^X_b$.

In section 11.2, in the case where the base manifold X is compact, we describe a Littlewood-Paley decomposition of functions along the fibres TX, and the chain of Sobolev spaces constructed in [BL08, chapter 15].

In section 11.3, we construct another Littlewood-Paley decomposition of functions on X.

In section 11.4, we combine the Littlewood-Paley decompositions of sections 11.2 and 11.3, and we obtain a Littlewood-Paley decomposition of functions on $\mathcal{X}$, and a corresponding chain of Sobolev spaces. Also, along the lines of [BL08], we establish key estimates on the resolvent of $\mathcal{A}_b^X$ with respect to the Sobolev spaces.

In section 11.5, using the results of sections 11.2–11.4, we construct the heat operators associated with $\mathcal{A}_b^X, \mathcal{B}_b^X$ and the corresponding smooth heat kernels $r_{b,t}^X, s_{b,t}^X$.

In section 11.6, we obtain corresponding results for the scalar operators $\mathfrak{A}_b^X, \mathfrak{B}_b^X$ over $\widehat{\mathcal{X}}$.

In section 11.7, we extend these results to a scalar operator A_b^X over $\widehat{\mathcal{X}}$ containing the extra quartic term $\frac{1}{2}\left|\left[Y^N, Y^{TX}\right]\right|^2$. The main difficulty is that the methods of [BL08] have to be modified to accommodate this term.

Finally, in section 11.8, we prove corresponding results for the hypoelliptic Laplacian $\mathcal{L}_{A,b}^X$ and its heat kernel.

As explained in the introduction, except in the present chapter, most of the analysis in the book is based on the probabilistic construction of the heat kernels. In chapter 12, we establish the nontrivial fact that the heat kernels that are obtained via probability theory coincide with the heat kernels that are obtained here.

In the whole chapter, we fix the parameter $b > 0$.

11.1 The scalar operators $\mathcal{A}_b^X, \mathcal{B}_b^X$ on $\mathcal{X}$

In this section, we will consider scalar differential operators acting on $\mathcal{X}$. Recall that Δ^{TX} denotes the Laplacian acting along the fibres TX of $\mathcal{X}$. Also $\nabla_{Y^{TX}}$ denotes the first order operator associated with the vector field V, which is the generator of the geodesic flow on $\mathcal{X}$. The notation is compatible with the notation of chapter 2.

Let $\mathcal{A}_b^X$ be the scalar differential operator on $\mathcal{X}$,

$$\mathcal{A}_b^X = \frac{1}{2b^2}\left(-\Delta^{TX} + \left|Y^{TX}\right|^2 - m\right) - \frac{1}{b}\nabla_{Y^{TX}}. \qquad (11.1.1)$$

Set

$$\mathcal{B}_b^X = \exp\left(\left|Y^{TX}\right|^2/2\right)\mathcal{A}_b^X \exp\left(-\left|Y^{TX}\right|^2/2\right). \qquad (11.1.2)$$

Then

$$\mathcal{B}_b^X = \frac{1}{2b^2}\left(-\Delta^{TX} + 2\nabla_{Y^{TX}}^V\right) - \frac{1}{b}\nabla_{Y^{TX}}. \qquad (11.1.3)$$

There is an associated operator $\mathcal{B}_b^G$ on $G \times \mathfrak{p}$. Indeed if $Y^{\mathfrak{p}} \in \mathfrak{p}$, we still denote by $\nabla_{Y^{\mathfrak{p}}}$ the associated left-invariant vector field on G. Also $\nabla_{Y^{\mathfrak{p}}}^V$ denotes the obvious radial vector field along $\mathfrak{p}$. Set

$$\mathcal{B}_b^G = \frac{1}{2b^2}\left(-\Delta^{\mathfrak{p}} + 2\nabla_{Y^{\mathfrak{p}}}^V\right) - \frac{1}{b}\nabla_{Y^{\mathfrak{p}}}. \qquad (11.1.4)$$

Let p still denote the obvious projection $G \times \mathfrak{p} \to \mathcal{X}$. If $F \in C^\infty(X, \mathbf{R})$, then

$$\mathcal{B}_b^G p^* F = p^* \mathcal{B}_b^X F. \qquad (11.1.5)$$

11.2 The Littlewood-Paley decomposition along the fibres TX

Recall that $b > 0$ is fixed once and for all. In the sequel, positive constants may well depend on b, but the dependence will not be written explicitly.

For $x \in X, r > 0$, let $B(x, r)$ be the open ball of center x and radius r, and let $\overline{B}(x, r)$ be the corresponding closed ball.

We use the notation $x_0 = p1$. Let Z be the m-dimensional sphere obtained from $\overline{B}(x_0, 3)$ by collapsing the boundary to a point. The compact group K acts on Z. Let g^{TZ} be a K-invariant Riemannian metric on TZ, which restricts to the metric g^{TX} on $B(0, 2)$. Let $\mathcal{Z}$ be the total space of the vector bundle TZ over Z, and let π still denote the projection $\mathcal{Z} \to Z$. Let ∇^{TZ} be the Levi-Civita connection on TZ. Let $T^H \mathcal{Z} \subset T\mathcal{Z}$ be the horizontal subbundle associated with ∇^{TZ}. Clearly $T^H \mathcal{Z} \simeq TZ$, so that $T\mathcal{Z} = TZ \oplus TZ$. Let $g^{T\mathcal{Z}}$ denote the obvious direct sum metric on $T\mathcal{Z}$.

Let H be the Hilbert space of real square-integrable functions on $\mathcal{Z}$, and let $|\ |_H$ be the corresponding L_2 norm.

Now we follow [BL08, section 15.2]. Take $r_0 \in]1, 2[$. Let $\phi(r)$ be a smooth function defined on $\mathbf{R}_+$ with values in $[0, 1]$, which is decreasing and is such that $\phi(r) = 1$ if $r \le \frac{1}{r_0}$, and $\phi(r) = 0$ if $r \ge 1$. Set

$$\chi(r) = \phi(r/2) - \phi(r). \tag{11.2.1}$$

The support of χ is included in $[\frac{1}{r_0}, 2]$. For $j \in \mathbf{N}$, put

$$\chi_j(r) = \chi(2^{-j} r). \tag{11.2.2}$$

Clearly,

$$\phi(r) + \sum_{j=0}^{\infty} \chi_j(r) = 1. \tag{11.2.3}$$

By (11.2.3),

$$\sum_{j=0}^{+\infty} \chi_j^2 \le 1. \tag{11.2.4}$$

If $Y^{TZ} \in TZ$, put

$$< Y^{TZ} > = \left(1 + \left|Y^{TZ}\right|^2\right)^{1/2}. \tag{11.2.5}$$

For $u \in C^\infty(\mathcal{Z}, \mathbf{R}), j \in \mathbf{N}$, set

$$\delta_j(u) = \chi_j\left(< Y^{TZ} >\right) u. \tag{11.2.6}$$

By (11.2.3), (11.2.6), we obtain the Littlewood-Paley decomposition,

$$u = \sum_{j=0}^{\infty} \delta_j(u). \tag{11.2.7}$$

Set

$$\mathcal{B} = \left\{ Y^{TZ} \in TZ, \left|Y^{TZ}\right|^2 \le 3 \right\}. \tag{11.2.8}$$

The support of $\delta_0(u)$ is included in the ball $\mathcal{B}$. For $j \geq 1$, the support of the $\delta_j(u)$ is included in the annulus C_j given by

$$C_j = \left\{ Y^{TZ}, < Y^{TZ} >\in [2^j/r_0 , \ 2^{j+1}] \right\}. \tag{11.2.9}$$

Note that

$$C_j \cap C_{j+2} = \emptyset. \tag{11.2.10}$$

Using (11.2.4), (11.2.7) and (11.2.10), we get

$$\sum_{j=0}^{+\infty} |\delta_j(u)|_H^2 \leq |u|_H^2 \leq 3 \sum_{j=0}^{+\infty} |\delta_j(u)|_H^2. \tag{11.2.11}$$

Definition 11.2.1. For $u \in C^\infty(\mathcal{Z}, \mathbf{R})$, set

$$U_j(x, Y^{TZ}) = \delta_j(u)(x, 2^j Y^{TZ}). \tag{11.2.12}$$

Let $\mathcal{R}$ be the annulus,

$$\mathcal{R} = \left\{ Y^{TZ}, |Y^{TZ}|^2 \in [\frac{1}{r_0^2} - \frac{1}{4}, 4] \right\}. \tag{11.2.13}$$

For any $j \in \mathbf{N}$, $U_j \in C^{\infty, c}(\mathcal{Z}, \mathbf{R})$. Moreover, the support of U_0 is included in the ball $\mathcal{B}$, and for $j \geq 1$, the support of the U_j is included in $\mathcal{R}$. We recover u from U by the formula

$$u(x, Y^{TZ}) = \sum_{j=0}^{\infty} U_j(x, 2^{-j} Y^{TZ}). \tag{11.2.14}$$

Put

$$\mathcal{B}_0 = \left\{ Y^{TZ} \in TZ, |Y^{TZ}|^2 \leq 5 \right\}. \tag{11.2.15}$$

Let Y be the total space of the projectivization $\mathbf{P}(TZ \oplus \mathbf{R})$ of the vector bundle TZ over Z. Then K acts on Y. Let g^{TY} be a K-invariant metric on Y, which coincides with g^{TZ} on $\mathcal{B}_0$. Let Δ^Y be the Laplace-Beltrami operator on Y. As in [BL08, eq. (15.3.1)], set

$$\mathbb{S} = -\Delta^Y + 1. \tag{11.2.16}$$

As in [BL08, eq. (15.4.11)], for $\tau > 0$, set

$$\Lambda_\tau = \left(\mathbb{S} + \tau^{-4} \right)^{1/2}. \tag{11.2.17}$$

Now we recall a definition given in [BL08, eqs. (15.3.3) and (15.4.12)].

Definition 11.2.2. Let H be the Hilbert space of real square-integrable functions on Y, and let $||_{\mathsf{H}}$ be the corresponding L_2 norm. For $s \in \mathbf{R}, U \in C^\infty(Y, \mathbf{R})$, put

$$|U|_{\tau, s} = \tau^{-m/2} |\Lambda_\tau^s U|_{\mathsf{H}}. \tag{11.2.18}$$

The completion of $C^\infty(Y, \mathbf{R})$ for the norm $||_{\tau, s}$ is the Sobolev space H^s on Y.

Note that in [BL08, eq. (14.4.12)], the factor $\tau^{-m/2}$ has been omitted, while it is present in [BL08, eq. (15.3.3)] when $\tau = 2^{-j}$. All the computations in [BL08] are correct even without the factor $\tau^{-m/2}$ anyway.

Take $u \in C^{\infty,c}(\mathcal{Z}, \mathbf{R})$. For $j \in \mathbf{N}$, the support of U_j is included in $\mathcal{B}_0$, so that U_j can be viewed as an element of $C^\infty(Y, \mathbf{R})$.

Now we follow [BL08, Definition 15.3.1 and eqs. (15.5.16), (15.5.20)].

Definition 11.2.3. If $u \in C^{\infty,c}(\mathcal{Z}, \mathbf{R})$, if the U_j are defined as in (11.2.12), put

$$\|u\|_s^2 = \sum_{j=0}^\infty |U_j|_{2^{-j},s}^2 \,, \tag{11.2.19}$$

$$|||u|||_s^2 = \sum_{j=0}^{+\infty} \left(2^{3j/2} |U_j|_{2^{-j},s}^2 + 2^{-5j/4} |\nabla^V U_j|_{2^{-j},s-1/8}^2 \right).$$

We denote by $\mathcal{H}^s, \mathcal{W}^s$ the completions of $C^{\infty,c}(\mathcal{Z}, \mathbf{R})$ with respect to the norms $\| \, \|_s, |||u|||_s$. Note that $\mathcal{W}^s$ is denoted $\mathcal{W}^{s-1/4}$ in [BL08].

Observe that $\tau^{-1} \le \Lambda_\tau^{1/2}$, so that

$$\tau^{-3/4} \Lambda_\tau^s \le \Lambda_\tau^{s+3/8}. \tag{11.2.20}$$

By (11.2.19), (11.2.20), we get

$$\|u\|_s \le |||u|||_s \le C \|u\|_{s+7/8} \,. \tag{11.2.21}$$

From (11.2.21), we find that

$$\mathcal{H}^{s+7/8} \subset \mathcal{W}^s \subset \mathcal{H}^s. \tag{11.2.22}$$

Also, by (11.2.11), $\mathcal{H}^0 = H$ and there exist $C > 0, C' > 0$ such that if $u \in C^{\infty,c}(\mathcal{Z}, \mathbf{R})$,

$$C |u|_H \le \|u\|_0 \le C' |u|_H \,. \tag{11.2.23}$$

As explained in [BL08, Remark 15.3.2], the $\mathcal{H}^s$ form a chain of Sobolev spaces, and if $s' > s$, the embedding of $\mathcal{H}^{s'}$ in $\mathcal{H}^s$ is compact. Put

$$\mathcal{H}^\infty = \cap_{s \in \mathbf{R}} \mathcal{H}^s. \tag{11.2.24}$$

Let $\mathcal{S}(\mathcal{Z})$ be the vector space of smooth functions on $\mathcal{Z}$ that are rapidly decreasing in the fibre direction TZ together with their derivatives of arbitrary order. Equivalently $\mathcal{S}(\mathcal{Z})$ consists of the $u \in C^\infty(\mathcal{Z}, \mathbf{R})$, such that for any multi-index k, and $k' \in \mathbf{N}$,

$$\left(1 + \left| Y^{TZ} \right|^{k'} \right) \left| \nabla_{\cdot}^{TZ,k} u \right| \tag{11.2.25}$$

is uniformly bounded. By [BL08, eq. (15.3.7)],

$$\mathcal{H}^\infty = \mathcal{S}(\mathcal{Z}). \tag{11.2.26}$$

In [BL08, section 15.2], global Sobolev spaces $H^s, s \in \mathbf{R}$ are also defined on $\mathcal{Z}$. By [BL08, Remark 15.3.2] for any $s \in \mathbf{R}$, $\mathcal{H}^s \subset H^s$. Let H_{loc}^s be the local Sobolev space of index s on $\mathcal{Z}$. By the above, we get

$$\mathcal{H}^s \subset H_{\text{loc}}^s, \tag{11.2.27}$$

and the embedding in (11.2.27) is continuous.

Over $\mathcal{Z}$, we can define a scalar operator $\mathcal{A}_b^Z$ by a formula strictly similar to (11.1.1). Over $\pi^{-1}B(x_0, 2)$, it restricts to the operator $\mathcal{A}_b^X$. For $\lambda_0 > 0$, set

$$P_b = \mathcal{A}_b^Z + \lambda_0. \tag{11.2.28}$$

For $s \in \mathbf{R}$, let D_s be the domain of $\mathcal{A}_b^Z$ in $\mathcal{H}^s$, i.e.,

$$D_s = \{u \in \mathcal{H}^s, \mathcal{A}_b^Z u \in \mathcal{H}^s\}. \tag{11.2.29}$$

We now have a result stated in [BL08, Theorem 15.5.1 and eq. (15.5.22)].

Theorem 11.2.4. *Take* $\lambda_0 > 0$ *large enough. Then if* $s \in \mathbf{R}$, *if* u *is a tempered distribution on* $\mathcal{Z}$ *such that* $P_b u \in \mathcal{H}^s$, *then* $u \in \mathcal{W}^{s+1/4}$, *and there exists* $C_s > 0$ *such that for* u *taken as before,*

$$|||u|||_{s+1/4} \leq C_s \|P_b u\|_s. \tag{11.2.30}$$

Moreover, $C^{\infty, c}(\mathcal{Z}, \mathbf{R})$ *is dense in* D_s, *i.e., for any* $u \in D_s$, *there exists a sequence* $u_k \in C^{\infty, c}(\mathcal{Z}, \mathbf{R})$, $k \in \mathbf{N}$ *such that as* $k \to +\infty$,

$$\|u - u_k\|_s + \|P_b(u - u_k)\|_s \to 0. \tag{11.2.31}$$

11.3 The Littlewood-Paley decomposition on X

For $x \in X, r > 0$, the open balls $B(x, r) \subset X$ are isometric to the ball $B(x_0, r)$.

Let $x_n \in X, n \in \mathbf{N}$ be a sequence in X such that the open balls $B(x_n, 1/2)$ form an open covering of X. If x_n, x_m are such that $d(x_m, x_n) \leq 1/2$, then $B(x_m, 1/2) \subset B(x_n, 1)$. Therefore we may and we will assume that for $m \neq n, d(x_m, x_n) \geq 1/2$, and the balls $B(x_n, 1)$ form an open covering of X. In particular, $\{x_n, n \in \mathbf{N}\}$ is a discrete subset of X. As the notation suggests, we assume that x_0 is precisely the element of X defined in section 11.2, i.e., $x_0 = p1$. After renumbering, we may and we will assume that $d(x_0, x_n)$ increases with n.

We claim that given $n \in \mathbf{N}$, there is a uniformly bounded number of $m \in \mathbf{N}$ such that $B(x_n, 2) \cap B(x_m, 2)$ is nonempty. Indeed, each such $B(x_m, 2)$ is included in $B(x_n, 4)$. Moreover, the corresponding $B(x_m, 1/4)$ are mutually disjoint and have the same volume. Since the volume of $B(x_n, 4)$ does not depend on $n \in \mathbf{N}$, the number of such m is uniformly bounded.

Let $\varphi(r) : \mathbf{R}_+ \to [0, 1]$ be a smooth function such that $\varphi(r) = 1$ for $r \leq 1$, and $\phi(r) = 0$ for $r \geq 2$. For $n \in \mathbf{N}$, put

$$\varphi_n(x) = \varphi(d(x_n, x)). \tag{11.3.1}$$

If $\varphi_n(x) > 0$, then $x_n \in B(x, 2)$. Since the $B(x_n, 1/4)$ do not intersect and have the same volume, given $x \in X$, the number of $n \in \mathbf{N}$ such that $\varphi_n(x) > 0$ is uniformly bounded. Therefore there is $N > 1$ such that

$$1 \leq \sum_{n=0}^{+\infty} \varphi_n \leq N. \tag{11.3.2}$$

In the sequel, covariant derivatives will be taken with respect to the connection ∇^{TX}. By (11.3.1), the φ_n are uniformly bounded together with their covariant derivatives of arbitrary order. It follows from the above that $\sum_{n=0}^{+\infty} \varphi_n$ and its covariant derivatives of arbitrary order are also uniformly bounded.

For $n \in \mathbf{N}$, set

$$\psi_n = \frac{\varphi_n}{\sum_{n=0}^{+\infty} \varphi_n}. \tag{11.3.3}$$

Then

$$\sum_{n=0}^{+\infty} \psi_n = 1. \tag{11.3.4}$$

More precisely the $\psi_n, n \in \mathbf{N}$ form a partition of unity of X that is subordinated to the open covering $B(x_n, 2), n \in \mathbf{N}$. By the considerations we made after (11.3.2), the ψ_n and their covariant derivatives of arbitrary order are uniformly bounded. By (11.3.4), we get

$$\sum_{n=0}^{+\infty} \psi_n^2 \leq 1. \tag{11.3.5}$$

If $f \in C^\infty(X, \mathbf{R})$, for $n \in \mathbf{N}$, set

$$\epsilon_n(f) = \psi_n f. \tag{11.3.6}$$

By (11.3.4),

$$f = \sum_{n=0}^{+\infty} \epsilon_n(f). \tag{11.3.7}$$

Let $\mathbf{H}$ be the Hilbert space of real square-integrable functions on X, and let $|\ |_{\mathbf{H}}$ denote the corresponding norm. By the above results and in particular by (11.3.5), (11.3.7), if $f \in C^{\infty,c}(X, \mathbf{R})$, then

$$\sum_1^{+\infty} |\epsilon_n(f)|_{\mathbf{H}}^2 \leq |f|_{\mathbf{H}}^2 \leq C \sum_{n=0}^{+\infty} |\epsilon_n(f)|_{\mathbf{H}}^2. \tag{11.3.8}$$

11.4 The Littlewood Paley decomposition on $\mathcal{X}$

If $u \in C^\infty(\mathcal{X}, \mathbf{R})$, we define $\epsilon_n(u) \in C^\infty(\mathcal{X}, \mathbf{R})$ by the formula

$$\epsilon_n(u) = (\pi^* \psi_n) u. \tag{11.4.1}$$

The obvious analogue of (11.3.7) holds, i.e.,

$$u = \sum_{n=0}^{+\infty} \epsilon_n(u). \tag{11.4.2}$$

Let $\mathcal{H}'$ be the Hilbert space of real square integrable functions on $\mathcal{X}$, and let $||_{\mathcal{H}'}$ denote the corresponding norm. If $u \in C^{\infty,c}(\mathcal{X}, \mathbf{R})$, as in (11.3.8), we get

$$\sum_{n=0}^{+\infty} |\epsilon_n(u)|^2_{\mathcal{H}'} \leq |u|^2_{\mathcal{H}'} \leq C \sum_{n=0}^{+\infty} |\epsilon_n(u)|^2_{\mathcal{H}'}. \tag{11.4.3}$$

Moreover, by (11.2.23), (11.4.3), if $u \in C^{\infty,c}(\mathcal{X}, \mathbf{R})$, we have the analogue of (11.3.8),

$$\sum_{j,n \in \mathbf{N}} |\psi_n U_j|^2_{2-j,0} \leq |u|^2_{\mathcal{H}'} \leq C' \sum_{j,n \in \mathbf{N}} |\psi_n U_j|^2_{2-j,0}. \tag{11.4.4}$$

If $u \in C^{\infty,c}(\mathcal{X}, \mathbf{R})$, $\epsilon_n(u)$ can be viewed as an element of $C^{\infty,c}(\mathcal{Z}, \mathbf{R})$.

Definition 11.4.1. For $u \in C^{\infty,c}(\mathcal{X}, \mathbf{R})$, set

$$\|u\|'^2_s = \sum_{n=0}^{+\infty} \|\epsilon_n(u)\|^2_s, \qquad \||u|\|'^2_s = \sum_{n=0}^{+\infty} \||\epsilon_n(u)|\|^2_s. \tag{11.4.5}$$

By (11.2.21), (11.4.5), we get

$$\|u\|'_s \leq \||u|\|'_s \leq C \|u\|'_{s+7/8}. \tag{11.4.6}$$

Let $\mathcal{H}'^s, \mathcal{W}'^s$ be the completions of $C^{\infty,c}(\mathcal{X}, \mathbf{R})$ for the norms $\| \|'_s, \||u|\|'_s$. By (11.4.6), we get

$$\mathcal{H}'^{s+7/8} \subset \mathcal{W}'^s \subset \mathcal{H}'^s. \tag{11.4.7}$$

The vector spaces $\mathcal{H}'^s, \mathcal{W}'^s$ still form families of Sobolev spaces. However, for $s' > s$, the embeddings of $\mathcal{H}'^{s'}, \mathcal{W}'^{s'}$ into $\mathcal{H}'^s, \mathcal{W}'^s$ are no longer compact. For $u \in C^{\infty,c}(\mathcal{X}, \mathbf{R})$, equation (11.4.4) can be rewritten in the form

$$\|u\|'^2_0 \leq |u|^2_{\mathcal{H}'} \leq C' \|u\|'^2_0. \tag{11.4.8}$$

By (11.4.8), we deduce that

$$\mathcal{H}'^0 = \mathcal{H}'. \tag{11.4.9}$$

For $s \in \mathbf{R}$, let H'^s_{loc} be the local Sobolev space of index s on $\mathcal{X}$. Using (11.2.27), we get

$$\mathcal{H}'^s \subset H'^s_{\text{loc}}. \tag{11.4.10}$$

Also the embedding in (11.4.10) is continuous.

Set

$$\mathcal{H}'^\infty = \cap_{s \in \mathbf{R}} \mathcal{H}'^s. \tag{11.4.11}$$

By (11.4.10),

$$\mathcal{H}'^\infty \subset C^\infty(\mathcal{X}, \mathbf{R}). \tag{11.4.12}$$

Let $\mathcal{S}(\mathcal{X})$ be the vector space of the $u \in C^\infty(\mathcal{X}, \mathbf{R})$ that are rapidly decreasing along the fibres TX together with their derivatives of arbitrary order. The definition just consists in replacing ∇^{TZ} by ∇^{TX} in (11.2.25). Note that here, the rapid decay is uniform over X. By (11.2.26), we get the following refinement of (11.4.12),

$$\mathcal{H}'^{\infty} \subset \mathcal{S}(\mathcal{X}). \tag{11.4.13}$$

For $\lambda_0 > 0$, set

$$P_b = \mathcal{A}_b^X + \lambda_0. \tag{11.4.14}$$

For $s \in \mathbf{R}$, put

$$D_s' = \left\{ u \in \mathcal{H}'^s, \mathcal{A}_b^X u \in \mathcal{H}'^s \right\}. \tag{11.4.15}$$

We establish an extension of [BL08, Theorem 15.5.1 and eq. (15.5.22)], which was stated here as Theorem 11.2.4.

Theorem 11.4.2. *For $\lambda_0 > 0$ large enough, for any $s \in \mathbf{R}$, there exists $C_s > 0$ such that if $u \in D_s'$, then $u \in \mathcal{W}'^{s+1/4}$, and moreover,*

$$|||u|||'_{s+1/4} \leq C_s \|P_b u\|'_s. \tag{11.4.16}$$

Also, $C^{\infty,c}(\mathcal{X}, \mathbf{R})$ is dense in D_s', i.e., for $s \in \mathbf{R}, u \in D_s'$, there is a sequence $u_p \in C^{\infty,c}(\mathcal{X}, \mathbf{R}), p \in \mathbf{N}$ such that as $p \to +\infty$,

$$\|u - u_p\|'_s + \|P_b (u - u_p)\|'_s \to 0. \tag{11.4.17}$$

Proof. For simplicity, we first assume that $u \in C^{\infty,c}(\mathcal{X}, \mathbf{R})$. This will guarantee that all the expressions we deal with are finite. By equation (11.2.30) in Theorem 11.2.4, for λ_0 large enough, for any $n \in \mathbf{N}$,

$$|||\epsilon_n(u)|||_{s+1/4} \leq C_s \|P_b \epsilon_n(u)\|_s. \tag{11.4.18}$$

Moreover,

$$P_b \epsilon_n(u) = \epsilon_n(P_b u) + [P_b, \psi_n] u. \tag{11.4.19}$$

By (11.1.1),

$$[P_b, \psi_n] = -\frac{1}{b} \nabla_{Y^{TX}} \psi_n. \tag{11.4.20}$$

By (11.4.19), (11.4.20), we get

$$\|P_b \epsilon_n(u)\|_s \leq \|\epsilon_n(P_b u)\|_s + \frac{1}{b} \|(\nabla_{Y^{TX}} \psi_n) u\|_s. \tag{11.4.21}$$

By (11.2.19), we have the identity

$$\|(\nabla_{Y^{TX}} \psi_n) u\|_s^2 = \sum_{j=0}^{+\infty} 2^{2j} |\nabla_{Y^{TX}} \psi_n U_j|_{2^{-j},s}^2. \tag{11.4.22}$$

Also given $m \in \mathbf{N}, n \in \mathbf{N}$, if $\psi_m d\psi_n \neq 0$, then $d(x_m, x_n) \leq 4$. Given $m \in \mathbf{N}$, there is a uniformly bounded family of $n \in \mathbf{N}$ such that $\psi_m d\psi_n \neq 0$. Combining this fact with (11.3.4) and (11.4.22), we get

$$\sum_{n \in \mathbf{N}} \|(\nabla_{Y^{TX}} \psi_n) u\|_s^2 \leq C_s \sum_{j,n \in \mathbf{N}} 2^{2j} |\psi_n U_j|_{2^{-j},s}^2. \tag{11.4.23}$$

By (11.4.5), (11.4.18)–(11.4.23), we obtain

$$|||u|||'^2_{s+1/4} \leq C_s \left(\|P_b u\|'^2_s + \sum_{j,n \in \mathbf{N}} 2^{2j} |\psi_n U_j|^2_{2^{-j},s} \right). \tag{11.4.24}$$

By (11.2.19) and (11.4.5), we get

$$|||u|||'^2_{s+1/4} = \sum_{j,n \in \mathbf{N}} \left(2^{3j/2} |\psi_n U_j|^2_{2^{-j},s+1/4} + 2^{-5j/4} |\psi_n \nabla^V U_j|^2_{2^{-j},s+1/8} \right). \tag{11.4.25}$$

From (11.2.17), (11.2.18), we obtain

$$|\psi_n U_j|^2_{2^{-j},s+1/4} \geq 2^j |\psi_n U_j|^2_{2^{-j},s}. \tag{11.4.26}$$

By (11.4.24)–(11.4.26), we obtain

$$\sum_{j,n \in \mathbf{N}} \left(\frac{1}{2} 2^{5j/2} |\psi_n U_j|^2_{2^{-j},s} + \frac{1}{2} \sum_{j,n \in \mathbf{N}} 2^{3j/2} |\psi_n U_j|^2_{2^{-j},s+1/4} \right.$$
$$\left. + 2^{-5j/4} |\psi_n \nabla^V U_j|^2_{2^{-j},s+1/8} \right) \leq C_s \left(\|P_b u\|'^2_s + \sum_{j,n \in \mathbf{N}} 2^{2j} |\psi_n U_j|^2_{2^{-j},s} \right). \tag{11.4.27}$$

From (11.4.25), (11.4.27), we conclude that there is $j_s \in \mathbf{N}$ such that

$$|||u|||'^2_{s+1/4} \leq C_s \left(\|P_b u\|'^2_s + \sum_{\substack{0 \leq j \leq j_s \\ n \in \mathbf{N}}} |\psi_n U_j|^2_{2^{-j},s} \right). \tag{11.4.28}$$

By (11.2.19), (11.4.5), we get

$$\sum_{\substack{0 \leq j \leq j_s \\ n \in \mathbf{N}}} |\psi_n U_j|^2_{2^{-j},s} \leq \sum_{j,n \in \mathbf{N}} |\psi_n U_j|^2_{2^{-j},s} = \|u\|'^2_s. \tag{11.4.29}$$

By (11.4.28), (11.4.29), we find that

$$|||u|||'^2_{s+1/4} \leq C_s \left(\|P_b u\|'^2_s + \|u\|'^2_s \right). \tag{11.4.30}$$

For $\alpha > 0, A > 0, s \geq 0$,

$$\alpha^{2s} \leq A^{-1/2} \alpha^{2s+1/2} + A^{2s}. \tag{11.4.31}$$

By (11.2.18), (11.2.19), (11.4.5), and (11.4.6), we get

$$\|u\|'^2_s \leq A^{-1/2} \|u\|'^2_{s+1/4} + A^{2s} \|u\|'^2_0 \leq A^{-1/2} |||u|||'^2_{s+1/4} + A^{2s} \|u\|'^2_0. \tag{11.4.32}$$

By (11.4.30), and by taking A large enough in (11.4.32), we get

$$|||u|||'^2_{s+1/4} \leq C_s \left(\|P_b u\|'^2_s + \|u\|'^2_0 \right). \tag{11.4.33}$$

By [BL08, eqs. (15.4.7)–(15.4.9)], for $\lambda_0 > 0$ large enough,

$$|u|^2_{\mathcal{H}'} \le C |P_b u|^2_{\mathcal{H}'}.$$
(11.4.34)

By (11.4.8), we can rewrite (11.4.34) in the form

$$\|u\|'^2_0 \le C \|P_b u\|'^2_0.$$
(11.4.35)

By (11.4.33), (11.4.35), we obtain

$$\||u|\|'^2_{s+1/4} \le C_s \left(\|P_b u\|'^2_s + \|P_b u\|'^2_0 \right).$$
(11.4.36)

For $s \ge 0$,

$$\|P_b u\|'^2_0 \le \|P_b u\|'^2_s.$$
(11.4.37)

Using (11.4.36), (11.4.37), for $s \ge 0$, we get (11.4.16) when $u \in C^{\infty,c}(\mathcal{X}, \mathbf{R})$.

We will now establish (11.4.16) for $s \ge 0$, when only assuming that $u \in D'_s$. For the moment, we take $s \in \mathbf{R}$. We still have equation (11.4.24). However, it is no longer clear that any of the two sides in this equation is finite. Of course (11.4.24) remains valid for $t \le s$, i.e.,

$$\||u|\|'^2_{t+1/4} \le C_t \left(\|P_b u\|'^2_t + \sum_{j,n \in \mathbf{N}} 2^{2j} |\psi_n U_j|^2_{2^{-j},t} \right).$$
(11.4.38)

Again both sides of (11.4.38) may well be infinite.

Since $\Lambda_\tau \ge \tau^{-2}$,

$$\tau^{-1} \le \tau^{-3/4} \Lambda_\tau^{1/8}.$$
(11.4.39)

From (11.2.19), (11.4.5), and (11.4.39), we deduce that

$$\sum_{j,n \in \mathbf{N}} 2^{2j} |\psi_n U_j|^2_{2^{-j},t} \le \||u|\|'^2_{t+1/8}.$$
(11.4.40)

Still there is no guarantee that the right-hand side of (11.4.40) is finite.

By (11.4.7), since $u \in \mathcal{H}'^s$, then $u \in \mathcal{W}'^{s-7/8}$. Therefore when making $t = s - 1$ in (11.4.40), both sides of the equation are finite. Therefore for $t = s - 1$, both sides of (11.4.38) are finite, so that $u \in \mathcal{W}'^{s-3/4}$. By iterating (11.4.38), we find that $u \in \mathcal{W}'^{s+1/4}$, and that (11.4.38) holds with $s = t$, which is equation (11.4.24), where both sides of the equation are known to be finite. We can then proceed as in (11.4.25)–(11.4.29), and obtain (11.4.30), while knowing that both sides of the equation are finite.

For $s \ge 0$, we can proceed as in (11.4.31)–(11.4.37), and we get (11.4.16) for $s \ge 0$.

Let $\mathcal{A}_b^{X*}$ be the formal L_2 adjoint of $\mathcal{A}_b^X$ with respect to the scalar product on $\mathcal{H}'$. This operator has the same structure as $\mathcal{A}_b^X$. Let P_b^* be the L_2 adjoint of P_b. For $\lambda_0 > 0$ large enough, for $s \ge 0$, we get the analogue of (11.4.16),

$$\||u|\|'_{s+1/4} \le C_s \|P_b^* u\|'_s.$$
(11.4.41)

By duality, from (11.4.41), for $s \ge 0$, we obtain

$$\|u\|'_{-s} \le C_s \|P_b u\|_{-s-1/4}.$$
(11.4.42)

By (11.4.42), (11.4.16) also holds for $s \leq -1/4$. By interpolation, we get (11.4.16) for any $s \in \mathbf{R}$.

Take $s \in \mathbf{R}$, and assume that $u \in D'_s$, so that $u \in \mathcal{W}'^{s+1/4}$. For $p \in \mathbf{N}$, set

$$u_p = \sum_{0 \leq j, n \leq p} \psi_n \delta_j (u). \qquad (11.4.43)$$

We claim that as $p \to +\infty$, u_p converges to u in $\mathcal{W}'^{s+1/4}$, and $P_b u_p$ converges to $P_b u$ in $\mathcal{H}'^s$. Note that

$$u - u_p = \sum_{\substack{j \geq p+1 \\ \text{or } n \geq p+1}} \psi_n \delta_j (u). \qquad (11.4.44)$$

Recall that $d(x_0, x_n)$ is an increasing sequence that tends to $+\infty$ as $n \to +\infty$. From the properties of the sequence x_n given in section 11.3, one deduces easily that as $p \to +\infty$, $u - u_p$ converges to 0 in $\mathcal{W}'^{s+1/4}$.

Clearly,

$$P_b u_p = \sum_{0 \leq j, n \leq p} \psi_n \delta_j (P_b u) + \left[P_b, \sum_{0 \leq j, n \leq p} \psi_n \chi_j \left(< Y^{TX} > \right) \right] u. \qquad (11.4.45)$$

By the same arguments as before, the first term in the right-hand side of (11.4.45) lies in $\mathcal{H}'^s$ and converges to $P_b u$ in $\mathcal{H}'^s$ as $p \to +\infty$. Also,

$$\left[P_b, \sum_{0 \leq j, n \leq p} \psi_n \chi_j \left(< Y^{TX} > \right) \right] = - \left[P_b, \sum_{\substack{j \geq p+1 \\ \text{or } n \geq p+1}} \psi_n \chi_j \left(< Y^{TX} > \right) \right]. \qquad (11.4.46)$$

Moreover, we have the obvious

$$\left[P_b, \psi_n \chi_j \left(< Y^{TX} > \right) \right] = \left[P_b, \psi_n \right] \chi_j \left(< Y^{TX} > \right) + \psi_n \left[P_b, \chi_j \left(< Y^{TX} > \right) \right]. \qquad (11.4.47)$$

The term $[P_b, \psi_n]$ has been computed in (11.4.20). Using the fact that $u \in \mathcal{W}'^{s+1/4}$, equation (11.4.20), and also the easy computation in [BL08, eq. (15.5.8)] of the commutator $\left[P_b, \chi_j \left(< Y^{TX} > \right) \right]$, we find that as $p \to +\infty$, $P_b u_p \to P_b u$ in $\mathcal{H}'^s$.

By the above, we find that to establish the last part of our theorem, we may as well assume that $u \in D'_s$ has compact support. Let $H^{s,c}$ be the standard Sobolev space with compact support on $\mathcal{X}$. The condition on $u \in D'_s$ just says that $u \in H^{s,c}$, $P_b u \in H^{s,c}$. Let $\Psi_p, p \in \mathbf{N}$ be a family of smooth regularizing pseudodifferential operators of order 0 on $\mathcal{X}$ that converges to the identity as $p \to +\infty$. We may and we will assume that the support of the $\Psi_p, p \in \mathbf{N}$ is arbitrarily close to the diagonal. Such a family of operators can easily be constructed using a regularization of the wave kernel on $\mathcal{X}$. Set

$$u_p = \Psi_p u. \qquad (11.4.48)$$

Then the u_p are smooth functions with uniformly bounded compact support, and as $p \to +\infty$, u_p converges to u in $H_{\mathcal{X}}^{s,c}$ so that u_p converges to u in $\mathcal{H}'^s$. Moreover,

$$P_b u_p = \Psi_p P_b u + [P_b, \Psi_p] u. \qquad (11.4.49)$$

Since $P_b u \in \mathcal{H}'^s$ has compact support, the same argument as before shows that as $p \to +\infty$, $\Psi_p P_b u$ converges to $P_b u$ in $\mathcal{H}'^s$. Also by (11.1.1), we can write $[P_b, \Psi_p]$ in the form

$$[P_b, \Psi_p] = A_p \nabla^V + B_p, \tag{11.4.50}$$

where A_p, B_p are pseudodifferential operators of order 0 with support arbitrarily close to the diagonal, which converge to 0 in the space of pseudodifferential operators of order 0. Clearly $B_p u \in H^{s,c}$, and $B_p u$ converges to 0 in $H^{s,c}$, so that $B_p u \to 0$ in $\mathcal{H}'^s$. Since $u \in \mathcal{W}'^{s+1/4}$, and since u has compact support, by (11.2.19) and (11.4.5), $\nabla^V u \in \mathcal{H}'^{s+1/8}$, so that $\nabla^V u \in H^{s,c}$. Therefore $A_p \nabla^V u$ converges to 0 in $H^{s,c}$, so that $A_p \nabla^V u$ converges to 0 in $\mathcal{H}'^s$. From the above, we conclude that $P_b u_p$ converges to $P_b u$ in $\mathcal{H}'^s$.

The proof of our theorem is completed. $\qquad\square$

Remark 11.4.3. It is important to observe that while the proof of [BL08, Theorems 15.4.2 and 15.5.1] uses the fact that the Sobolev spaces $\mathcal{H}^s$ are compactly embedded into each other, because here X is noncompact, such an argument cannot be used any more. As a substitute, we used instead (11.4.24)–(11.4.35).

We will use the notation $D' = D'_0$. We establish an analogue of [BL08, Theorem 15.6.1].

Theorem 11.4.4. *There exist* $\lambda_0 > 0, C_0 > 0$ *such that for any* $u \in D', \lambda \in \mathbf{C}, \mathrm{Re}(\lambda) \le -\lambda_0$,

$$|\lambda|^{1/6} \|u\|'_0 + \|u\|'_{1/4} \le C_0 \left\| \left(\mathcal{A}^X_b - \lambda \right) u \right\|'_0. \tag{11.4.51}$$

There exists $C_1 > 0$ *such that for* $\sigma, \tau \in \mathbf{R}, \sigma \le C_1 |\tau|^{1/6}$, *if* $\lambda = -\lambda_0 + \sigma + i\tau$, *then*

$$(1 + |\lambda|)^{1/6} \|u\|'_0 \le C_1 \left\| \left(\mathcal{A}^X_b - \lambda \right) u \right\|'_0. \tag{11.4.52}$$

Proof. We proceed exactly as in the proof of [BL08, Theorem 15.6.1], while using the concepts and techniques introduced in the proof of Theorem 11.4.2. As explained in [BL08], equation (11.4.52) is a consequence of (11.4.51). So we concentrate on the proof of (11.4.51).

As in [BL08, eqs. (15.6.7) and (15.6.15)], set

$$\Lambda_{\lambda,\tau} = \left(\mathbb{S} + \tau^{-4} + \tau^2 |\lambda|^2 \right)^{1/2}. \tag{11.4.53}$$

If $U \in C^\infty(Y, \mathbf{R})$, we define $|U|_{\lambda,\tau,s}$ as in (11.2.18), by replacing Λ_τ by $\Lambda_{\lambda,\tau}$. Similarly we define norms $\| \ \|_{\lambda,s}, \||\ \||_{\lambda,s}$ as in (11.2.19) on $C^{\infty,c}(\mathcal{Z}, \mathbf{R})$. The completions $\mathcal{H}^s, \mathcal{W}^s$ are the same as in Definition 11.2.3. Finally, we define the norms $\| \ \|'_{\lambda,s}, \||\ \||'_{\lambda,s}$ on $C^{\infty,c}(\mathcal{X}, \mathbf{R})$ as in Definition 11.4.1. The completions $\mathcal{H}'^s, \mathcal{W}'^s$ do not depend on λ.

Set

$$P_{b,\lambda} = \mathcal{A}^X_b - \lambda. \tag{11.4.54}$$

To prove (11.4.51), we take $\lambda_0 > 0$ large enough as in Theorems 11.2.4 and 11.4.2. We define P_b as in (11.4.14). If $\alpha \in \mathbf{R}_+, \beta \in \mathbf{R}$, let $\lambda \in \mathbf{C}$ be given by

$$\lambda = -\lambda_0 - \alpha + i\beta, \qquad (11.4.55)$$

so that

$$P_{b,\lambda} = P_b + \alpha - i\beta. \qquad (11.4.56)$$

The same argument as in [BL08, eqs. (15.6.9)–(15.6.11), proof of Theorem 15.6.1] shows that (11.4.51) is a consequence of the estimate,

$$|||u|||'_{\lambda,s+1/4} \le C_s \left\|P_{b,\lambda}u\right\|'_{\lambda,s}, \qquad (11.4.57)$$

for $s = 0$. As the notation indicates, the constant $C_s > 0$ should not depend on λ.

As explained in [BL08, p. 243], an analogue of Theorem 11.2.4 holds with the new norms indexed by λ, s, the crucial point being that the positive constants that appear only depend on s and not on λ. The proof of (11.4.57) then proceeds as the proof of equation (11.4.16) in Theorem 11.4.2, which completes the proof of our theorem. $\qquad \square$

In the sequel, the Hilbert space $\mathcal{H}'$ is equipped with its canonical scalar product $\langle \rangle_{\mathcal{H}'}$. The formal adjoint of the operator $\mathcal{A}_b^X$ is taken with respect to this scalar product. If U is a bounded operator acting on $\mathcal{H}'$, we denote by $\|U\|$ its norm with respect to the norm of $\mathcal{H}'$. We establish now an analogue of [BL08, Theorem 15.7.1].

Theorem 11.4.5. *The adjoint of the operator $\mathcal{A}_b^X$ acting on $\mathcal{H}'$ with domain D' is the formal adjoint $\mathcal{A}_b^{X*}$ of $\mathcal{A}_b^X$ acting on $C^{\infty,c}(\mathcal{X}, \mathbf{R})$, with domain $D'^* = \{u \in \mathcal{H}', \mathcal{A}_b^{X*}u \in \mathcal{H}'\}$.*

There exist $c_0 > 0, \lambda_0 > 0, C > 0$ such that if $\mathcal{U} \subset \mathbf{C}$ is given by

$$\mathcal{U} = \{\lambda = -\lambda_0 + \sigma + i\tau, \ \sigma, \tau \in \mathbf{R}, \sigma \le c_0 |\tau|^{1/6}\}, \qquad (11.4.58)$$

if $\lambda \in \mathcal{U}$, the operator $\left(\mathcal{A}_b^X - \lambda\right)^{-1}$ exists, and moreover

$$\left\|(\mathcal{A}_b^X - \lambda)^{-1}\right\| \le \frac{C}{(1 + |\lambda|)^{1/6}}. \qquad (11.4.59)$$

There exists $C > 0$ such that if $\lambda \in \mathbf{R}, \lambda \le -\lambda_0$, then

$$\left\|(\mathcal{A}_b^X - \lambda)^{-1}\right\| \le C (1 + |\lambda|)^{-1}. \qquad (11.4.60)$$

If $\lambda \in \mathcal{U}$, $\left(\mathcal{A}_b^X - \lambda\right)^{-1}$ maps $\mathcal{H}'^s$ into $\mathcal{H}'^{s+1/4}$. In particular $\left(\mathcal{A}_b^X - \lambda\right)^{-1}$ maps $\mathcal{H}'^\infty$ into itself. If $s \in \mathbf{R}$, there exists $C_s > 0$ such that for $\lambda \in \mathcal{U}, u \in \mathcal{H}'^s$,

$$\left\|(\mathcal{A}_b^X - \lambda)^{-1} u\right\|'_{s+1/4} \le C_s (1 + |\lambda|)^{4|s|+1} \|u\|'_s. \qquad (11.4.61)$$

Proof. By Theorem 11.4.2, $\left(\mathcal{A}_b^X, D'\right)$ is the closure of $\left(\mathcal{A}_b^X, C^{\infty,c}\left(\mathcal{X}, \mathbf{R}\right)\right)$, which implies the first part of our theorem. Using Theorems 11.4.2 and 11.4.4, the proof of (11.4.59) proceeds exactly as the proof of [BL08, Theorem 15.7.1].

To establish (11.4.60), we proceed as in [BL08]. We use the notation in the proof of Theorem 11.4.4, and we take

$$\lambda = -\lambda_0 - \alpha, \ \alpha \geq 0. \tag{11.4.62}$$

By (11.1.1), we get

$$P_{b,\lambda} = \frac{1}{2b^2}\left(-\Delta^{TX} + \left|Y^{TX}\right|^2 - m\right) + \lambda_0 + \alpha - \frac{1}{b}\nabla_{TX}. \tag{11.4.63}$$

By (11.4.63), for $\lambda_0 > 0$ large enough, there is $C > 0$ such that if $u \in C^{\infty,c}\left(\mathcal{X}, \mathbf{R}\right)$,

$$(C + \alpha)\left|u\right|_{\mathcal{H}'}^2 \leq \langle P_{b,\lambda} u, u \rangle_{\mathcal{H}'}. \tag{11.4.64}$$

From (11.4.64), we get

$$(1 + \alpha)\left|u\right|_{\mathcal{H}'} \leq C \left|P_{b,\lambda} u\right|_{\mathcal{H}'}. \tag{11.4.65}$$

Also if λ is as in (11.4.62), $\lambda \in \mathcal{U}$, so that $P_{b,\lambda}^{-1}$ exists. By (11.4.65), we get (11.4.60).

By (11.4.6), (11.4.16), if $u \in C^{\infty,c}\left(\mathcal{X}, \mathbf{R}\right)$, if $\lambda \in \mathbf{C}$,

$$\|u\|_{s+1/4}' \leq C_s\left(\left\|\left(\mathcal{A}_b^X - \lambda\right)u\right\|_s' + \left|\lambda + \lambda_0\right| \|u\|_s'\right). \tag{11.4.66}$$

By (11.4.32), for $A > 0, s \geq 0$, we get

$$\|u\|_s' \leq A^{-1/4}\|u\|_{s+1/4}' + A^s \|u\|_0'. \tag{11.4.67}$$

By taking $A = 2^4 C_s^4 \left|\lambda + \lambda_0\right|^4$ in (11.4.67), we deduce from (11.4.66) that for $s \geq 0$,

$$\|u\|_{s+1/4}' \leq C_s\left(\left\|\left(\mathcal{A}_b^X - \lambda\right)u\right\|_s' + \left|\lambda + \lambda_0\right|^{4s+1} \|u\|_0'\right). \tag{11.4.68}$$

By (11.4.59), (11.4.68), we find that if $\lambda \in \mathcal{U}$, for $s \geq 0$, $\left(\mathcal{A}_b^X - \lambda\right)^{-1}$ maps $\mathcal{H}'^s$ into $\mathcal{H}'^{s+1/4}$, and also that (11.4.61) holds when $u \in C^{\infty,c}\left(\mathcal{X}, \mathbf{R}\right)$. By density this equation extends to $u \in \mathcal{H}'^s$. The case of a general $s \in \mathbf{R}$ can be obtained using duality and an interpolation argument. The proof of our theorem is completed. $\qquad\square$

11.5 The heat kernels for $\mathcal{A}_b^X, \mathcal{B}_b^X$

Let γ be the contour in $\mathbf{C}$ in Figure 11.1. This contour separates the domains $\delta_\pm$, which contain $\pm\infty$. Let $c_0 > 0, \lambda_0 > 0$ be the constants that were defined in Theorem 11.4.5. These constants depend on $b > 0$. Then

$$\gamma = \left\{\lambda = -\lambda_0 + \sigma + i\tau, \sigma, \tau \in \mathbf{R}, \sigma = c_0 \left|\tau\right|^{1/6}\right\}. \tag{11.5.1}$$

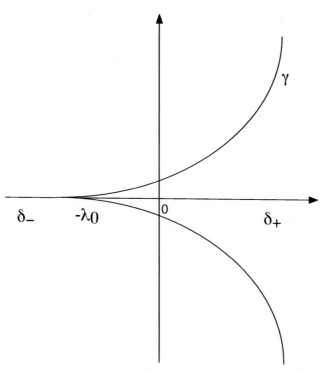

Figure 11.1

The domain of $\mathcal{A}_b^X$ is dense in $\mathcal{H}'$. Moreover, by equation (11.4.60) in Theorem 11.4.5 and by the theorem of Hille-Yosida [Y68, section IX-7, p. 246], there is a unique well-defined continuous semigroup $\exp\left(-t\mathcal{A}_b^X\right), t \geq 0$ acting on $\mathcal{H}'$.

We have the obvious analogue of [BL08, Proposition 3.3.1].

Proposition 11.5.1. *For $t > 0$, the heat operator $\exp\left(-t\mathcal{A}_b^X\right)$ is given by the contour integral,*

$$\exp\left(-t\mathcal{A}_b^X\right) = \frac{1}{2i\pi} \int_\gamma e^{-t\lambda} \left(\lambda - \mathcal{A}_b^X\right)^{-1} d\lambda. \tag{11.5.2}$$

Proof. Using Theorem 11.4.5, the proof of our proposition is the same as the proof of [BL08, Proposition 3.3.1]. □

Using (11.4.59), (11.5.2), and integration by parts, for $N \in \mathbf{N}$,

$$\exp\left(-t\mathcal{A}_b^X\right) = \frac{(-1)^N N!}{2i\pi t^N} \int_\gamma e^{-t\lambda} \left(\lambda - \mathcal{A}_b^X\right)^{-(N+1)} d\lambda. \tag{11.5.3}$$

By Theorem 11.4.5, if $\lambda \in \delta_-$, $\left(\mathcal{A}_b^X - \lambda\right)^{-1}$ maps $\mathcal{H}'^s$ into $\mathcal{H}'^{s+1/4}$ with a norm dominated by $C_s \left(1 + |\lambda|\right)^{4|s|+1}$. Therefore $\left(\mathcal{A}_b^X - \lambda\right)^{-N}$ maps $\mathcal{H}'^s$ into $\mathcal{H}'^{s+N/4}$ with a norm dominated by $C_{s,N} \left(1 + |\lambda|\right)^{(4|s|+1)N}$.

By (11.5.3), for $t > 0$, the operator $\exp\left(-t\mathcal{A}_b^X\right)$ maps $\mathcal{H}'^s$ into $\mathcal{H}'^\infty$. More precisely given $s, s' \in \mathbf{R}$, this operator is continuous from $\mathcal{H}'^s$ into $\mathcal{H}'^{s'}$ with a bounded norm. Combining these results with (11.4.13), standard arguments show that $\exp\left(-t\mathcal{A}_b^X\right)$ has a smooth kernel on $\mathcal{X}$ that is rapidly decreasing along the fibres $TX \times TX$ of $\mathcal{X} \times \mathcal{X}$ together with its derivatives of any order.

Let $F \in C^{\infty,c}\left(\mathcal{X}, \mathbf{R}\right)$. For $t \geq 0$, $\left(x, Y^{TX}\right) \in \mathcal{X}$, set

$$\Phi\left(t, \left(x, Y^{TX}\right)\right) = \exp\left(-t\mathcal{A}_b^X\right) F\left(x, Y^{TX}\right). \tag{11.5.4}$$

Proposition 11.5.2. *For any $t \geq 0$, the function $\Phi\left(t, \cdot\right)$ lies in $\mathcal{H}'^\infty$. As $t \to 0$, $\Phi\left(t, \cdot\right)$ converges to F in $\mathcal{H}'^\infty$. The function Φ is smooth on $\mathbf{R}_+ \times \mathcal{X}$, and is such that*

$$\frac{\partial}{\partial t}\Phi + \mathcal{A}_b^X \Phi = 0. \tag{11.5.5}$$

Finally, for bounded $t \geq 0$, $\Phi\left(t, \cdot\right)$ is uniformly rapidly decreasing along the fibres TX of $\mathcal{X}$ together with its covariant derivatives of any order in all variables.

Proof. For $t \geq 0$, $\exp\left(-t\mathcal{A}_b^X\right)$ maps $\mathcal{H}'^\infty$ into $\mathcal{H}'^\infty$, and so for $t \geq 0$, $\Phi\left(t, \cdot\right) \in \mathcal{H}'^\infty$. By (11.5.3), Φ is smooth on $]0, +\infty[\times\mathcal{X}$. Since $\exp\left(-t\mathcal{A}_b^X\right)$ is a continuous semigroup acting on $\mathcal{H}'$, as $t \to 0$, $\Phi\left(t, \cdot\right)$ converges to F in $\mathcal{H}'$, and $t \in \mathbf{R}_+ \to \Phi_t \in \mathcal{H}'$ is a smooth map.

Take $\lambda \in \mathcal{U}$. For any $k \in \mathbf{N}$,

$$\exp\left(-t\mathcal{A}_b^X\right) = \left(\mathcal{A}_b^X - \lambda\right)^{-k} \exp\left(-t\mathcal{A}_b^X\right) \left(\mathcal{A}_b^X - \lambda\right)^k F. \tag{11.5.6}$$

Since $\left(\mathcal{A}_b^X - \lambda\right)^k F \in C^{\infty,c}\left(\mathcal{X}, \mathbf{R}\right)$, $t \in \mathbf{R}_+ \to \exp\left(-t\mathcal{A}_b^X\right)\left(\mathcal{A}_b^X - \lambda\right)^k F$ is a smooth map with values in $\mathcal{H}'$. By equation (11.4.61) in Theorem 11.4.5 and by (11.5.6), for any $k \in \mathbf{N}$, $t \in \mathbf{R}_+ \to \exp\left(-t\mathcal{A}_b^X\right) F$ is a smooth map with values in $\mathcal{H}'^{k/4}$. By (11.4.10), we conclude that Φ is smooth on $\mathbf{R}_+ \times \mathcal{X}$, and also that $\Phi\left(t, \cdot\right)$ converges to F in $\mathcal{H}'^{\infty}$. Using (11.4.13), we conclude that for bounded $t \geq 0$, Φ is uniformly rapidly decreasing together with its derivatives of arbitrary order along the fibres TX. Since Φ is smooth, equation (11.5.5) follows from (11.5.4). The proof of our proposition is completed. $\square$

By (11.1.2), $\mathcal{B}_b^X$ is conjugated to $\mathcal{A}_b^X$. Therefore all the results we obtained for $\mathcal{A}_b^X$ can be tautologically transferred to $\mathcal{B}_b^X$. Of course the functional analytic machinery has to be adequately modified.

If $F \in C^{\infty,c}\left(\mathcal{X}, \mathbf{R}\right)$, set

$$\Psi\left(t, \left(x, Y^{TX}\right)\right) = \exp\left(-t\mathcal{B}_b^X\right) F\left(x, Y^{TX}\right). \qquad (11.5.7)$$

Let Φ be the function in (11.5.4), with F replaced by $\exp\left(-\left|Y^{TX}\right|^2/2\right) F$. By (11.1.2), we get

$$\Psi\left(t, \left(x, Y^{TX}\right)\right) = \exp\left(\left|Y^{TX}\right|^2/2\right) \Phi\left(t, \left(x, Y^{TX}\right)\right). \qquad (11.5.8)$$

By Proposition 11.5.2, the function Ψ is smooth on $\mathbf{R}_+ \times \mathcal{X}$, and is such that

$$\frac{\partial}{\partial t}\Psi + \mathcal{B}_b^X \Psi = 0. \qquad (11.5.9)$$

Definition 11.5.3. For $b > 0, t > 0$, we denote by

$$r_{b,t}^X\left(\left(x, Y^{TX}\right), \left(x', Y^{TX\prime}\right)\right), s_{b,t}^X\left(\left(x, Y^{TX}\right), \left(x', Y^{TX\prime}\right)\right)$$

the smooth kernels associated with $\exp\left(-t\mathcal{A}_b^X\right), \exp\left(-t\mathcal{B}_b^X\right)$. By (11.1.2),

$$s_{b,t}^X\left(\left(x, Y^{TX}\right), \left(x', Y^{TX\prime}\right)\right) = \exp\left(\left|Y^{TX}\right|^2/2\right)$$
$$r_{b,t}^X\left(\left(x, Y^{TX}\right), \left(x', Y^{TX\prime}\right)\right) \exp\left(-\left|Y^{TX\prime}\right|^2/2\right). \quad (11.5.10)$$

We know that given $b > 0, t > 0$, $r_{b,t}^X\left(\left(x, Y^{TX}\right), \left(x', Y^{TX\prime}\right)\right)$ is rapidly decreasing in the variables $Y^{TX}, Y^{TX\prime}$.

Remark 11.5.4. Using the estimate on $\Delta^X d^2\left(x_0, x\right)/2$ in Proposition 13.1.1, and combining this estimate with corresponding commutator estimates as in [BL91, section 11], given $b > 0, t > 0$, $r_{b,t}^X\left(\left(x, Y^{TX}\right), \left(x', Y^{TX\prime}\right)\right)$ can be shown to be rapidly decreasing together with its derivatives in the variables $\left(x', Y^{TX\prime}\right)$, the decay in the variable x' being measured via $d\left(x, x'\right)$. However, this result is superseded by the Gaussian estimate in Theorem 13.2.4, which is uniform for bounded $b > 0$.

11.6 The scalar hypoelliptic operators on $\widehat{\mathcal{X}}$

Here $\Delta^{TX\oplus N}$ still denotes the standard Laplacian acting along the fibres of $TX \oplus N$ of $\widehat{\mathcal{X}}$, and $Y = Y^{TX} + Y^N$ denotes the tautological section of $TX \oplus N$ over $\widehat{\mathcal{X}}$.

Let $\mathfrak{A}_b^X$ be the differential operator acting on $\widehat{\mathcal{X}}$,

$$\mathfrak{A}_b^X = \frac{1}{2b^2}\left(-\Delta^{TX\oplus N} + |Y|^2 - m - n\right) - \frac{1}{b}\nabla_{Y^{TX}}. \qquad (11.6.1)$$

Set

$$\mathfrak{B}_b^X = \exp\left(|Y|^2/2\right)\mathfrak{A}_b^X \exp\left(-|Y|^2/2\right). \qquad (11.6.2)$$

Then

$$\mathfrak{B}_b^X = \frac{1}{2b^2}\left(-\Delta^{TX\oplus N} + 2\nabla_Y^V\right) - \frac{1}{b}\nabla_{Y^{TX}}. \qquad (11.6.3)$$

Note that when acting on smooth functions on $\mathcal{X}$, the operators $\mathfrak{A}_b^X, \mathfrak{B}_b^X$ restrict to the operators $\mathcal{A}_b^X, \mathcal{B}_b^X$.

We claim that the results that were obtained before for the operators $\mathcal{A}_b^X, \mathcal{B}_b^X$ can be easily transferred to the operators $\mathfrak{A}_b^X, \mathfrak{B}_b^X$. First, the Euclidean vector bundle N with connection on $B(x_0, 2)$ extends to an Euclidean vector bundle with connection over Z, which we still denote by N. The action of K on Z extends to the vector bundles TZ, N. Let $\widehat{Z}$ be the total space of the vector bundle $TZ \oplus N$ over Z. The Euclidean connection $\nabla^{TZ\oplus N}$ induces a horizontal subbundle $T^H\widehat{Z}$ of $T\widehat{Z}$. Let $g^{T\widehat{Z}}$ be the obvious direct sum metric on $T\widehat{Z}$. Then $g^{T\widehat{Z}}$ is K-invariant.

We fix $r_0 \in]1, 2[$. Set

$$\mathcal{B}' = \left\{Y \in TX \oplus N, |Y|^2 \le 3\right\},$$

$$\mathcal{R}' = \{Y \in TZ \oplus N, |Y|^2 \in [\frac{1}{r_0^2} - \frac{1}{4}, 4]\}, \qquad (11.6.4)$$

$$\mathcal{B}'_0 = \left\{Y \in TZ \oplus N, |Y|^2 \le 5\right\}.$$

Let $\widehat{Y}$ be the total space of $\mathbf{P}(TZ \oplus N \oplus \mathbf{R})$ over Z. Let $g^{T\widehat{Y}}$ be a K-invariant metric on $T\widehat{Y}$ that coincides with $g^{T\widehat{Z}}$ on $\mathcal{B}'_0$. Let $\Delta^{\widehat{Y}}$ be the Laplace-Beltrami operator on $\widehat{Y}$. Let $\widehat{H}$ be the Hilbert space of square-integrable real functions on $\widehat{Y}$, and let $\|\ \|_{\widehat{H}}$ be the corresponding norm. We still define $\mathbb{S}$ as in (11.2.16), and Λ_τ as in (11.2.17), and the norms $\|\ \|_s$, $\|\|\ \|\|_s$ as in Definition 11.2.3.

The same techniques as in sections 11.2–11.5 can be used to establish properties of the resolvent of $\mathfrak{A}_b^X$ and of its heat kernel. In particular for $t > 0$, the heat operators for $\mathfrak{A}_b^X$ are well-defined, and are given by smooth kernels, which are rapidly decreasing on $\widehat{\mathcal{X}} \times \widehat{\mathcal{X}}$ together with their covariant derivatives of arbitrary order. By (12.2.10), the heat operators for $\mathfrak{B}_b^X$ are also given by smooth kernels.

Definition 11.6.1. For $t > 0$, let $\mathfrak{r}_{b,t}^X\left((x,Y),(x',Y')\right), \mathfrak{s}_{b,t}^X\left((x,Y),(x',Y')\right)$ be the smooth kernels associated with $\exp\left(-t\mathfrak{A}_b^X\right), \exp\left(-t\mathfrak{B}_b^X\right)$.

By (11.6.2), we get

$$\mathfrak{s}_{b,t}^X\left((x,Y),(x',Y')\right) = \exp\left(|Y|^2/2\right)$$
$$\mathfrak{r}_{b,t}^X\left((x,Y),(x',Y')\right)\exp\left(-|Y'|^2/2\right). \quad (11.6.5)$$

Remark 11.5.4 is still valid for the kernel $\mathfrak{r}_{b,t}^X$.

11.7 The scalar hypoelliptic operator on $\widehat{\mathcal{X}}$ with a quartic term

Again $b > 0$ is fixed. Set

$$A_b^X = \frac{1}{2}\left|\left[Y^N, Y^{TX}\right]\right|^2 + \frac{1}{2b^2}\left(-\Delta^{TX \oplus N} + |Y|^2 - m - n\right) - \frac{\nabla_{Y^{TX}}}{b}. \quad (11.7.1)$$

Then A_b^X is a scalar differential operator acting on $\widehat{\mathcal{X}}$. By (11.6.1) and (11.7.1), we get

$$A_b^X = \frac{1}{2}\left|\left[Y^N, Y^{TX}\right]\right|^2 + \mathfrak{A}_b^X. \quad (11.7.2)$$

Set

$$B_b^X = \exp\left(|Y|^2/2\right)A_b^X \exp\left(-|Y|^2/2\right). \quad (11.7.3)$$

Then

$$B_b^X = \frac{1}{2}\left|\left[Y^N, Y^{TX}\right]\right|^2 + \frac{1}{2b^2}\left(-\Delta^{TX \oplus N} + 2\nabla_Y^V\right) - \frac{\nabla_{Y^{TX}}}{b}. \quad (11.7.4)$$

In the analysis of the operator A_b^X, the arguments of sections 11.2–11.5 have to be adequately modified, because of the presence of the quartic term in (11.7.2).

There is an associated operator A_b^Z acting over $\widehat{\mathcal{Z}}$. Difficulties already appear in the proof of [BL08, Theorem 15.5.1 and eq. (15.5.22)], which is here Theorem 11.2.4.

For $\lambda_0 > 0$, instead of (11.2.28), set

$$Q_b = A_b^Z + \lambda_0. \quad (11.7.5)$$

Let us review some of the arguments given in [BL08, chapter 15]. Recall that for $a > 0$, K_a was defined in (2.14.2). We use the notation in [BL08, eq. (15.4.1)]. For $\tau > 0$, set

$$A_{b,\tau}^Z = K_\tau^{-1} A_b^Z K_\tau. \quad (11.7.6)$$

By (11.7.1), (11.7.6), we get

$$A_{b,\tau}^Z = \frac{\tau^{-4}}{2}\left|\left[Y^N, Y^{TZ}\right]\right|^2 + \frac{1}{2b^2}\left(-\tau^2\Delta^{TZ \oplus N} + \tau^{-2}|Y|^2 - m - n\right)$$
$$- \tau^{-1}\frac{\nabla_{Y^{TZ}}}{b}. \quad (11.7.7)$$

Put

$$Q_{b,\tau} = \mathsf{A}^Z_{b,\tau} + \lambda_0. \tag{11.7.8}$$

We now state an analogue of [BL08, Theorem 15.4.2].

Theorem 11.7.1. *If $\lambda_0 > 0$ is large enough, for any $s \in \mathbf{R}$, there exists $C_s > 0$ such that for any $j \in \mathbf{N}$, for any $U \in C^{\infty,c}\left(\widehat{\mathcal{Z}}, \mathbf{R}\right)$ whose support is included in the ball $\mathcal{B}'$ for $j = 0$ and in the annulus $\mathcal{R}'$ for $j \geq 1$, then*

$$2^{4j} |U|^2_{2-j,s} + \left|\nabla^V U\right|^2_{2-j,s} + 2^{3j/2} |U|^2_{2-j,s+1/4}$$
$$+ 2^{-5j/4} \left|\nabla^V U\right|^2_{2-j,s+1/8} \leq C_s \left|Q_{b,2-j}U\right|^2_{2-j,s}. \tag{11.7.9}$$

Proof. In the sequel, we will make $\tau = 2^{-j}, j \in \mathbf{N}$. Let $\widehat{H}$ be the Hilbert space of real square integrable functions on $\widehat{\mathcal{Z}}$, let $||\,|_{\widehat{H}}$ be the corresponding L_2 norm, and let $\langle\,\rangle_{\widehat{H}}$ be the scalar product. Let $\mathsf{A}^Z_{b,\tau,+}, \mathsf{A}^Z_{b,\tau,-}$ be the symmetric and antisymmetric parts of $\mathsf{A}^Z_{b,\tau}$ with respect to $\langle\,\rangle_{\widehat{H}}$. Then

$$\mathsf{A}^Z_{b,\tau,+} = \frac{\tau^{-4}}{2} \left|[Y^N, Y^{TX}]\right|^2 + \frac{1}{2b^2}\left(-\tau^2 \Delta^{TZ \oplus N} + \tau^{-2} |Y|^2 - m - n\right), \tag{11.7.10}$$

$$\mathsf{A}^Z_{b,\tau,-} = -\tau^{-1}\frac{\nabla_{Y^{TZ}}}{b}.$$

Let $Q_{b,\tau,+}, Q_{b,\tau,-}$ be the symmetric and antisymmetric parts of $Q_{b,\tau}$. By proceeding as in [BL08, eq. (15.4.7)], for $\lambda_0 > 0$ large enough, for any $U \in C^{\infty,c}\left(\widehat{\mathcal{Z}}, \mathbf{R}\right)$ with support in the annulus $\mathcal{R}'$, we get

$$\left|\nabla^V U\right|^2_{\widehat{H}} + \tau^{-4} |U|^2_{\widehat{H}} + \tau^{-6} \left|[Y^N, Y^{TX}]U\right|^2_{\widehat{H}} \leq C \left\langle Q_{b,\tau,+}U, \tau^{-2}U\right\rangle_{\widehat{H}}. \tag{11.7.11}$$

By proceeding as in [BL08, eqs. (15.4.8), (15.4.9)], from (11.7.11), we get

$$\left|\nabla^V U\right|^2_{\widehat{H}} + \tau^{-4} |U|^2_{\widehat{H}} + \tau^{-6} \left|[Y^N, Y^{TX}]U\right|^2_{\widehat{H}} \leq C |Q_{b,\tau}U|^2_{\widehat{H}}. \tag{11.7.12}$$

If $\tau = 1$, and if the support of U is included in $\mathcal{B}'$, (11.7.12) still holds for $\lambda_0 > 0$ large enough. The only difference with [BL08] is the presence of the third term in the left-hand side of (11.7.12).

Let $\rho(r) : \mathbf{R}_+ \to [0,1]$ be a smooth function, which is equal to 1 when $r^2 \leq 5$, and to 0 for $r^2 \geq 6$. Put

$$\theta_0(Y) = \rho(|Y|). \tag{11.7.13}$$

Set

$$R = \theta_0^2 Q_{b,\tau} \theta_0^2. \tag{11.7.14}$$

With respect to [BL08, eq. (15.4.18)], replacing θ_0 by θ_0^2 is done for reasons that will appear in (11.7.17).

If the support of U is included in $\mathcal{B}'_0$, then

$$Q_{b,\tau}U = RU. \tag{11.7.15}$$

Let R_+, R_- be the symmetric and antisymmetric components of R. Clearly,

$$R_+ = \theta_0^2 Q_{b,\tau,+}\theta_0^2, \qquad\qquad R_- = \theta_0^2 Q_{b,\tau,-}\theta_0^2. \qquad (11.7.16)$$

We define the algebra of pseudodifferential operators $\mathcal{E}^{\cdot}$ on $\widehat{Y}$ exactly as in [BL08, p. 231]. Let us briefly indicate that the weight function associated with this algebra is just $\left(\tau^{-4} + |\zeta|^2\right)^{1/2}, \zeta \in T^*\widehat{Y}$.

Take $E \in \mathcal{E}^d, d \in \mathbf{R}$. Instead of [BL08, eq. (15.4.23)], and using the fact that $\tau^{-2} \in \mathcal{E}^1$, we now have the commutation relation,

$$[R, E] \in \tau^2 \mathcal{E}^d \nabla^V + \tau^{-1}\mathcal{E}^d + \tau^{-4}\theta_0 \left[Y^N, Y^{TX}\right] \mathcal{E}^{d-1}. \qquad (11.7.17)$$

The first two terms in the right-hand side of (11.7.17) were already obtained in [BL08, eq. (15.4.23)]. Since $\tau^{-2} \in \mathcal{E}^1$, from (11.7.17), we get

$$[R, E] \in \tau^2 \mathcal{E}^d \nabla^V + \tau^{-1}\mathcal{E}^d + \tau^{-2}\theta_0 \left[Y^N, Y^{TX}\right] \mathcal{E}^d. \qquad (11.7.18)$$

Note that as it should be, we could as well have the factor $\theta_0 \left[Y^N, Y^{TX}\right]$ to the right of $\mathcal{E}^d$.

From (11.7.18), we deduce the weaker

$$[R, E] \in \tau^2 \mathcal{E}^d \nabla^V + \tau^{-2}\mathcal{E}^d. \qquad (11.7.19)$$

Equation (11.7.19) is exactly similar to [BL08, eq. (15.4.22)], which itself is weaker than [BL08, eq. (15.4.23)].

We claim that [BL08, Lemma 15.4.3] still holds. Indeed equation (11.7.12) is stronger than [BL08, eq. (15.4.25)], and the commutation relation (11.7.19) is the other main ingredient of the proof of this lemma.

By using the same argument as before, [BL08, Lemma 15.4.4 and eq. (15.4.51)] still hold.

Let $\theta_1(Y)$ be a smooth function of $|Y|$ with values in $[0, 1]$, which has compact support and is constructed in the same way as the function θ_0. We assume that θ_1 is equal to 1 near the annulus $\mathcal{R}'$ and to 0 near $Y = 0$. For $j \geq 1$, set $\theta_j = \theta_1$. Also θ_0 has been defined before. For $j \in \mathbf{N}$, we use the notation $\theta = \theta_j$.

Instead of [BL08, eq. (15.4.56)], because of (11.7.18), for $s \in \mathbf{R}$, we get

$$[R, \Lambda^s] \in \tau^2 \mathcal{E}^s \nabla^V + \tau^{-1}\mathcal{E}^s + \tau^{-2}\theta_0 \left[Y^N, Y^{TX}\right] \mathcal{E}^s. \qquad (11.7.20)$$

Using [BL08, (eq. (15.4.52)–(15.4.55)], and equation (11.7.20) instead of [BL08, eq. (15.4.56)], instead of [BL08, eq. (15.4.57)] we obtain

$$|U|_{\tau,s+1/4}^2 \leq \tau^{3/2}\left(|RU|_{\tau,s}^2 + C_s \tau^4 \left|\nabla^V U\right|_{\tau,s}^2 + C_s \tau^{-2} |U|_{\tau,s}^2\right.$$

$$\left. + C_s \tau^{-4} \left|\left[Y^N, Y^{TX}\right] U\right|_{\tau,s}^2\right). \qquad (11.7.21)$$

If we apply [BL08, eq. (15.4.51)] to $\theta\Lambda^{s+1/8}U$, by using the same commutation arguments as before, [BL08, eq. (15.4.58)] should be replaced by

$$\left|\nabla^V U\right|^2_{\tau,s+1/8} \le \delta\left(C\left|RU\right|^2_{\tau,s} + C_s\tau^4\left|\nabla^V U\right|^2_{\tau,s} + C_s\tau^{-2}\left|U\right|^2_{\tau,s}\right.$$

$$\left.+ C_s\tau^{-4}\left|\left[Y^N, Y^{TX}\right]U\right|^2_{\tau,s}\right) + \frac{\tau^{-4}}{\delta}\left(C\left|U\right|^2_{\tau,s+1/4} + C_s\tau^4\left|U\right|^2_{\tau,s}\right)$$

$$+ C_s\left|U\right|^2_{\tau,s+1/8}. \quad (11.7.22)$$

Using (11.7.21), (11.7.22) instead of [BL08, eqs. (15.4.57), (15.4.58)] and proceeding as in [BL08, eq. (15.4.59)], for $A > 0$, we get

$$\tau^{-3/2}\left|U\right|^2_{\tau,s+1/4} + A\tau^{5/4}\left|\nabla^V U\right|^2_{\tau,s+1/8} \le C\left|RU\right|^2_{\tau,s} + C_s\tau^4\left|\nabla^V U\right|^2_{\tau,s}$$

$$+ C_s\tau^{-2}\left|U\right|^2_{\tau,s} + C_s\tau^{-4}\left|\left[Y^N, Y^{TX}\right]U\right|^2_{\tau,s} + CA^2\tau^{-3/2}\left|U\right|^2_{\tau,s+1/4}$$

$$+ C_s\left(A^2\tau^{5/2}\left|U\right|^2_{\tau,s} + A\tau^{5/4}\left|U\right|^2_{\tau,s+1/8}\right). \quad (11.7.23)$$

By taking $A > 0$ small enough so that $CA^2 \le 1/2$, from (11.7.23), we deduce an analogue of [BL08, eq. (15.4.60)],

$$\tau^{-3/2}\left|U\right|^2_{\tau,s+1/4} + \tau^{5/4}\left|\nabla^V U\right|^2_{\tau,s+1/8} \le C\left|RU\right|^2_{\tau,s} + C_s\tau^4\left|\nabla^V U\right|^2_{\tau,s}$$

$$+ C_s\tau^{-2}\left|U\right|^2_{\tau,s} + C_s\tau^{-4}\left|\left[Y^N, Y^{TX}\right]U\right|^2_{\tau,s} + C_s\tau^{5/4}\left|U\right|^2_{\tau,s+1/8}. \quad (11.7.24)$$

Now we will use equation (11.7.12) applied to $\theta\Lambda^s U$, and we use again (11.7.20). Instead of [BL08, eq. (15.4.61)], we get

$$\left|\nabla^V U\right|^2_{\tau,s} + \tau^{-4}\left|U\right|^2_{\tau,s} + \tau^{-6}\left|\left[Y^N, Y^{TX}\right]U\right|^2_{\tau,s}$$

$$\le C\left|RU\right|^2_{\tau,s} + C_s\tau^4\left|\nabla^V U\right|^2_{\tau,s} + C_s\tau^{-2}\left|U\right|^2_{\tau,s} + C_s\tau^{-4}\left|\left[Y^N, Y^{TX}\right]U\right|^2_{\tau,s}. \quad (11.7.25)$$

By adding (11.7.24) and (11.7.25), instead of [BL08, eq. (15.4.62)], we obtain

$$\left|\nabla^V U\right|^2_{\tau,s} + \tau^{-4}\left|U\right|^2_{\tau,s} + \tau^{-3/2}\left|U\right|^2_{\tau,s+1/4} + \tau^{5/4}\left|\nabla^V U\right|^2_{\tau,s+1/8}$$

$$+ \tau^{-6}\left|\left[Y^N, Y^{TX}\right]U\right|^2_{\tau,s} \le C\left|RU\right|^2_{\tau,s} + C_s\tau^4\left|\nabla^V U\right|^2_{\tau,s} + C_s\tau^{-2}\left|U\right|^2_{\tau,s}$$

$$+ C_s\tau^{5/4}\left|U\right|^2_{\tau,s+1/8} + C_s\tau^{-4}\left|\left[Y^N, Y^{TX}\right]U\right|^2_{\tau,s}. \quad (11.7.26)$$

Moreover, given $\epsilon > 0$,

$$\tau^{-4} \le \epsilon\tau^{-6} + \frac{1}{\epsilon^2}. \quad (11.7.27)$$

By (11.7.26), (11.7.27), we get

$$\left|\nabla^V U\right|^2_{\tau,s} + \tau^{-4}\left|U\right|^2_{\tau,s} + \tau^{-3/2}\left|U\right|^2_{\tau,s+1/4} + \tau^{5/4}\left|\nabla^V U\right|^2_{\tau,s+1/8}$$

$$+ \tau^{-6}\left|\left[Y^N, Y^{TX}\right]U\right|^2_{\tau,s} \le C\left|RU\right|^2_{\tau,s} + C_s\tau^4\left|\nabla^V U\right|^2_{\tau,s}$$

$$+ C_s\tau^{-2}\left|U\right|^2_{\tau,s} + C_s\tau^{5/4}\left|U\right|^2_{\tau,s+1/8}. \quad (11.7.28)$$

The term containing τ^{-6} can be deleted in the left-hand side of (11.7.28). Then for $\tau = 2^{-j}$, for j large enough, (11.7.9) follows from (11.7.28). For the missing finite family of j, we proceed as in [BL08, proof of Theorem 15.4.2], or instead we use the same argument as in (11.4.32). The proof of our theorem is completed. $\square$

We can now deduce the obvious analogue for the operator Q_b of [BL08, Theorem 15.5.1 and eq. (15.5.22)], which was stated here as Theorem 11.2.4. Indeed the proof in [BL08] only uses Theorem 11.7.1, and also commutations in which the extra quartic term $\frac{1}{2}\left|\left[Y^N, Y^{TX}\right]\right|^2$ disappears. Similarly the results of [BL08, section 15.6] remain valid for the operator A_b^X.

One can then easily extend the arguments of sections 11.4–11.6 to the operator A_b^X.

Definition 11.7.2. For $t > 0$, let $\mathsf{r}_{b,t}^X\left((x,Y),(x',Y')\right), \mathsf{s}_{b,t}^X\left((x,Y),(x',Y')\right)$ be the smooth kernels associated with $\exp\left(-t\mathsf{A}_b^X\right), \exp\left(-t\mathsf{B}_b^X\right)$. By (11.7.3),

$$\mathsf{s}_{b,t}^X\left((x,Y),(x',Y')\right) = \exp\left(|Y|^2/2\right)$$
$$\mathsf{r}_{b,t}^X\left((x,Y),(x',Y')\right)\exp\left(-|Y'|^2/2\right). \quad (11.7.29)$$

Again $\mathsf{r}_{b,t}^X\left((x,Y),(x',Y')\right)$ is rapidly decreasing in Y, Y'. Also Remark 11.5.4 can still be applied to this heat kernel.

11.8 The heat kernel associated with the operator $\mathcal{L}_{A,b}^X$

We fix $b > 0$. We consider the hypoelliptic operator $\mathcal{L}_{A,b}^X$ given by (2.13.5), (4.5.1), which acts on $C^\infty\left(\widehat{\mathcal{X}}, \widehat{\pi}^*\left(\Lambda^{\cdot}\left(T^*X \oplus N^*\right) \otimes F\right)\right)$. Recall that by Theorem 2.13.2, except for the term $\frac{1}{b}\nabla_{Y^{TX}}^{C^\infty(TX \oplus N, \widehat{\pi}^*(\Lambda^{\cdot}(T^*X \oplus N^*)\otimes F))}$ that is skew-adjoint, the other components of $\mathcal{L}_{A,b}^X$ are formally self-adjoint with respect to the Hermitian product $\langle\rangle$ on $C^{\infty,c}\left(\widehat{\mathcal{X}}, \widehat{\pi}^*\left(\Lambda^{\cdot}\left(T^*X \oplus N^*\right) \otimes F\right)\right)$.

For simplicity, even though the operator $\mathcal{L}_{A,b}^X$ is no longer a scalar operator, we will use the same notation as in sections 11.2–11.7 for the function spaces defined there and for the corresponding norms. As in (11.7.5), for $\lambda_0 > 0$, set

$$Q_b = \mathcal{L}_{A,b}^X + \lambda_0. \quad (11.8.1)$$

We still denote by $\widehat{\pi}$ the projection $\widehat{Y} \to Z$. Let $\Delta^{H,\widehat{Y}}$ be the Bochner Laplacian acting on smooth sections of $C^\infty\left(\widehat{Y}, \widehat{\pi}\left(\Lambda^{\cdot}\left(T^*Z \oplus N^*\right) \otimes F\right)\right)$. Instead of (11.2.16), we define the operator $\mathbb{S}$ by

$$\mathbb{S} = -\Delta^{\widehat{Y}} + 1. \quad (11.8.2)$$

We claim that the obvious analogue of [BL08, Theorem 15.4.2], which was stated as Theorem 11.7.1, still holds. Indeed set

$$\epsilon(Y) = \widehat{c}\left(\mathrm{ad}\left(Y^{TX}\right)\right) - c\left(\mathrm{ad}\left(Y^{TX}\right) + i\theta\mathrm{ad}\left(Y^{N}\right)\right) - i\rho^{E}\left(Y^{N}\right). \quad (11.8.3)$$

Let $|\epsilon(Y)|$ be the norm of $\epsilon(Y)$ associated with the Hermitian product on $\Lambda^{\cdot}(T^{*}X \oplus N^{*}) \otimes F$. Then

$$\left|\frac{\epsilon(Y)}{b}\right| \le \frac{C}{b}|Y| \le \frac{|Y|^{2}}{4b^{2}} + C'. \quad (11.8.4)$$

Using (11.8.4), for $\lambda_0 > 0$ large enough, the obvious analogue of (11.7.11) still holds. The remainder of the proof of the analogue of Theorem 11.7.1 for $\mathcal{L}^{X}_{A,b}$ continues as in [BL08] and in the proof of Theorem 11.7.1.

The analysis of the operator $\mathcal{L}^{X}_{A,b}$ on $\widehat{\mathcal{X}}$ continues exactly as for the operator A^{X}_{b}. In particular for $t > 0$, the heat operator $\exp\left(-t\mathcal{L}^{X}_{A,b}\right)$ is well-defined.

Definition 11.8.1. For $t > 0$, let $q^{X}_{b,t}\left((x,Y),(x',Y')\right)$ denote the smooth kernel associated with the operator $\exp\left(-t\mathcal{L}^{X}_{A,b}\right)$.

Again this kernel is rapidly decreasing in the variables Y, Y'. Also the conclusions of Remark 11.5.4 still hold.

Chapter Twelve

Rough estimates on the scalar heat kernel

The purpose of this chapter is to establish rough estimates on the heat kernel $r_{b,t}^X$ for the scalar hypoelliptic operator $\mathcal{A}_b^X$ on $\mathcal{X}$ defined in chapter 11. By rough estimates, we mean just uniform bounds on the heat kernel, and not a Gaussian-like decay like in (4.5.3). Such refined estimates will be obtained in chapter 13 for bounded b, and in chapter 15 for b large. We will also obtain corresponding bounds for the heat kernels associated with operators $\mathfrak{A}_b^X, \mathsf{A}_b^X$ over $\widehat{\mathcal{X}}$. The case of $\mathcal{X}$ should be thought of as a warm-up for the case of $\widehat{\mathcal{X}}$, but the arguments used in both cases are essentially the same. Note that the operators $\mathcal{B}_b^X, \mathfrak{B}_b^X$, which are conjugate to $\mathcal{A}_b^X, \mathfrak{A}_b^X$, also play an important role in the present chapter.

Moreover, we give a probabilistic construction of the heat kernels. As explained in the introduction, most of the analytic results in the book will be obtained by probabilistic methods.

We also explain the relation of the heat equation for the hypoelliptic Laplacian on $\mathcal{X}$ to the wave equation on X. In this context, recall that the hypoelliptic Laplacian does not have a wave equation. However, we show that after integration in the fibre variable $Y^{TX} \in TX$, the projected heat kernel verifies a wave-like equation. This behavior becomes stronger and stronger as $b \to +\infty$. This idea is used to obtain the proper estimates on the hypoelliptic heat kernel in the present chapter as well as in chapter 13.

The rough bounds we mentioned before cannot be directly derived from our previous work with Lebeau [BL08], where only the case of a compact Riemannian manifold X is considered. Still the methods of [BL08] could lead to such rough estimates when $b > 0$ remains bounded and stays away from 0. In [BL08, chapter 17], the control of the behavior of the heat operator as $b \to 0$ is obtained by functional analytic arguments, from which uniform pointwise estimates cannot be easily extracted.

Recall that Malliavin [M78, M97] invented the Malliavin calculus to study the heat kernel associated with the second order hypoelliptic operators considered by Hörmander [Hö67]. Malliavin used the fact that the heat kernel can be obtained as the image of a classical flat Brownian measure P via the solution of a stochastic differential equation. Integration by parts on Wiener space ultimately explains the regularity of the heat kernel. Other methods of integration by parts on Wiener space have been given by Stroock [St81a, St81b] and ourselves [B81a, B84]. One key feature of Malliavin's approach is that it connects the calculus of variations with the analysis of the heat kernel for hypoelliptic operators. In particular, the estimates on the heat kernel rely on estimates of a certain Malliavin covariance matrix.

To obtain our rough estimates, we use the methods of the Malliavin calculus, part of which were already applied in [BL08, chapter 14] to the hypoelliptic Laplacian. It turns out that in the context of our hypoelliptic operators, estimating the Malliavin covariance matrix can be very easily done via the computations we made in chapter 10 of the explicit solutions to certain elementary variational problems. The parameter $b > 0$ entering explicitly in the computations, it is easy to obtain this way whatever uniformity is needed for the rough estimates.

We also prove that as $b \to 0$, the heat kernel $r_{b,t}^X$ converges to the standard heat kernel of X, by a method that is different from what is done in [BL08]. In [BL08], the convergence of the heat operators was obtained by functional analytic arguments, which did not produce pointwise estimates. Here the convergence will be obtained by dynamical arguments, involving a detailed analysis of the convergence as $b \to 0$ of the relevant hypoelliptic diffusion on X with parameter $b > 0$ to the standard Brownian motion of X.

The fact that the symmetric space X has constant curvature does not play a fundamental role in the arguments, except through the uniform control of the geometry of X.

This chapter is organized as follows. In section 12.1, we briefly recall the application of the Malliavin calculus given in [B84] to the Brownian motion on X and to the standard elliptic heat kernel, the key point being a simple integration by parts formula on Wiener space.

In section 12.2, we give a probabilistic description of the heat kernel for $\mathcal{B}_b^X$, by constructing an associated Markov diffusion $(x_., Y_.^{TX}) \in \mathcal{X}$.

In section 12.3, we relate the heat equation for $\mathcal{B}_b^X$ to the wave equation.

In section 12.4, along the lines of [BL08, chapter 14], we obtain an integration by parts formula for the diffusion $(x_., Y_.^{TX})$.

In section 12.5, we use the results of chapter 10 to obtain a more explicit integration by parts formula.

In section 12.6, we obtain a uniform control of the integration by parts formula as $b \to 0$.

In section 12.7, when $b > 0$ is bounded, we obtain the rough uniform estimates for $r_{b,t}^X$.

In section 12.8, we study the limit of $r_{b,t}^X$ as $b \to 0$.

In section 12.9, we obtain rough estimates on a rescaled version of $r_{b,t}^X$ as $b \to +\infty$.

In section 12.10, we extend the above results to the heat kernels associated with the scalar operator $\mathfrak{A}_b^X$ on $\widehat{\mathcal{X}}$.

Finally, in section 12.11, we obtain corresponding results for the heat kernels associated with the operator A_b^X over $\widehat{\mathcal{X}}$.

12.1 The Malliavin calculus for the Brownian motion on X

Here we assume again $X = G/K$ to be the symmetric space that was considered in the previous chapters of the book.

Let Δ^X be the Laplace-Beltrami operator on X. Then $-\Delta^X$ is just the action of the Casimir operator on $C^\infty(X, \mathbf{R})$.

We denote by $C^{\infty,c}(X, \mathbf{R})$ the vector space of smooth real functions on X that have compact support. Since X is complete, the restriction of Δ^X to $C^{\infty,c}(X, \mathbf{R})$ is essentially self-adjoint. For $t > 0$, let $p_t(x, x')$ be the smooth heat kernel associated with $\exp(t\Delta^X/2)$.

If $e \in \mathfrak{p}$, we identify e with the corresponding left invariant vector field on G. Let $e_1, \ldots, e_m$ be an orthonormal basis of $\mathfrak{p}$. Let $\Delta^{G,H}$ be the differential operator on G,

$$\Delta^{G,H} = \sum_{i=1}^{m} \nabla_{e_i}^2. \tag{12.1.1}$$

Then $\Delta^{G,H}$ is a self-adjoint nonelliptic operator on G. If $f \in C^\infty(X, \mathbf{R})$, it follows from the previous considerations that

$$\Delta^{G,H} p^* f = p^* \Delta^X f. \tag{12.1.2}$$

Let $\mathcal{C}(\mathbf{R}_+, \mathfrak{p})$ be the vector space of continuous functions from $\mathbf{R}_+$ into $\mathfrak{p}$, and let $w.$ be its generic element. For $t \in [0, +\infty[$, let $\mathcal{F}_t$ be the σ-algebra generated by $w_s, s \le t$. Let $\mathcal{F}_t|_{t \ge 0}$ denote the corresponding filtration. Let $\mathcal{F}_\infty$ be the σ-algebra generated by the $\mathcal{F}_t, t \ge 0$.

Let P be the Brownian measure on $\mathcal{C}(\mathbf{R}_+, \mathfrak{p})$, with $w_0 = 0$, and let E be the corresponding expectation operator. In the sequel, we use the probability space $(\mathcal{C}(\mathbf{R}_+, \mathfrak{p}), \mathcal{F}_\infty, P)$.

Consider the stochastic differential equation on X,

$$\dot{x} = \dot{w}, \qquad\qquad x_0 = p1. \tag{12.1.3}$$

In (12.1.3), $\dot{w}$ is identified with its parallel transport $\tau_t^0 \dot{w}$ along the trajectory $x..$ Recall that $w.$ is P a.s. nowhere differentiable. The theory of stochastic differential equations takes care of the difficulties in making sense of the above.

If we identify TG to $G \times \mathfrak{g}$ via the left-invariant vector fields, we may as well solve the stochastic differential equation on G,

$$\dot{g} = \dot{w}, \qquad\qquad g_0 = 1, \tag{12.1.4}$$

and set

$$x. = pg.. \tag{12.1.5}$$

Let $f \in C^{\infty,c}(X, \mathbf{R})$. The crucial link between the heat equation semigroup $\exp(t\Delta^X/2)$ and Brownian motion is that

$$\exp(t\Delta^X/2) f(x_0) = E[f(x_t)]. \tag{12.1.6}$$

Similarly, if $u \in C^{\infty,c}(G, \mathbf{R})$, then

$$\exp(t\Delta^{G,H}/2) u(1) = E[u(g_t)]. \tag{12.1.7}$$

Let h_t be a bounded process with values in $\mathfrak{p}$ that is predictable with respect to the filtration $\mathcal{F}_t|_{t\geq 0}$, let B_t be a bounded predictable process taking values in $\mathfrak{k}$. Now we describe the integration by parts formula of [B84, Theorem 2.2]. Let δw be the Itô differential of $w.$, let $dw = \dot{w}ds$ be its Stratonovitch differential.

For $\ell \in \mathbf{R}$, set

$$w_t^\ell = \int_0^t e^{\ell \operatorname{ad}(B_s)} \left(\delta w_s + \ell h_s ds\right). \qquad (12.1.8)$$

Then using the Girsanov formula [KSh91, Theorem 3.5.1 and Corollary 3.5.13] as in [B84, equation (2.18)], we know that for any $t \in \mathbf{R}_+$, the probability law $w_\cdot^\ell$ on $\mathcal{F}_t$ is equivalent to the probability law of $w.$. More precisely, put

$$Z_t^\ell = \exp\left(-\ell \int_0^t \langle h_s, \delta w_s\rangle - \frac{1}{2}\ell^2 \int_0^t |h_s|^2\, ds\right). \qquad (12.1.9)$$

Then Z_t^ℓ is an integrable martingale with respect to the filtration $\mathcal{F}_t|_{t\geq 0}$. Let P^ℓ be the probability measure on $\mathcal{C}(\mathbf{R}_+, \mathfrak{p})$, such that for any $t \in \mathbf{R}_+$,

$$\frac{dP^\ell}{dP}|_{\mathcal{F}_t} = Z_t^\ell. \qquad (12.1.10)$$

Using Girsanov's theorem and also the rotational invariance of Brownian motion, we find that under P^ℓ, the probability law of w^ℓ is equal to P.

In (12.1.3), we replace $w.$ by $w_\cdot^\ell$, and we denote by $x_\cdot^\ell$ the solution of the corresponding stochastic differential equation. Let $f : X \to \mathbf{R}$ be a smooth bounded function with bounded first derivative. By the above, we get

$$E\left[f\left(x_t^\ell\right) Z_t^\ell\right] = E\left[f\left(x_t\right)\right]. \qquad (12.1.11)$$

The integration by parts formula of [B84] is obtained by differentiating (12.1.11) at $\ell = 0$.

First, we calculate the differential of $x_\cdot^\ell$ with respect to ℓ at $\ell = 0$. It is a nontrivial fact that differentiation is legitimate. It is made possible via the theory of stochastic flows. Ultimately, and forgetting about technicalities, the rules of classical differential calculus do apply, as long as one uses Stratonovitch differentials.

Recall that the curvature Ω of the canonical connection on the principal bundle $p : G \to G/K$ is given by (2.1.10). Also the Ricci tensor R^X of X is given by (2.6.8). In particular R^X is parallel.

Consider the differential equation on the processes $(\vartheta_s, \varpi_s) \in \mathfrak{g} = \mathfrak{p} \oplus \mathfrak{k}$ along the path x_s,

$$d\vartheta = \left(-\frac{1}{2}R^X\vartheta + h\right) ds + [\varpi + B, \delta w], \qquad d\varpi = [\theta, dw], \qquad (12.1.12)$$

$$\vartheta_0 = 0, \qquad\qquad\qquad\qquad\qquad\qquad \varpi_0 = 0.$$

In (12.1.12), $\vartheta. \in \mathfrak{p}$ is identified to the corresponding section of TX over the path $x.$ that is obtained by parallel transport with respect to ∇^{TX}. Also in

(12.1.12), dw still denotes the Stratonovitch differential of w, and δw its Itô differential.

The integration by parts formula of [B84, Theorem 2.2] asserts in its simplest form that if $f : X \to \mathbf{R}$ is a smooth function, then

$$E\left[\langle f'(x_t), \vartheta_t \rangle\right] = E\left[f(x_t) \int_0^t \langle h_s, \delta w \rangle\right]. \qquad (12.1.13)$$

Consider the differential equation,

$$\dot{\vartheta} = -\frac{1}{2} R^X \vartheta + h, \qquad\qquad d\varpi = [\theta, dw], \qquad (12.1.14)$$

$$\vartheta_0 = 0, \qquad\qquad\qquad \varpi_0 = 0.$$

Observe that (12.1.14) is a special case of (12.1.12), by simply taking $B = -\varpi$. Then the key observation in [B84] is that (12.1.13) still holds.

From the above, we find that if ϑ. is a C^1 predictable path with values in $\mathfrak{p}$ with uniformly bounded first derivative such that $\vartheta_0 = 0$, then

$$E\left[\langle f'(x_t), \vartheta_t \rangle\right] = E\left[f(x_t) \int_0^t \left\langle \dot{\vartheta} + \frac{1}{2} R^X \vartheta, \delta w \right\rangle\right]. \qquad (12.1.15)$$

Observe that

$$E\left[\left|\int_0^t h \delta w\right|^2\right] = E\left[\int_0^t |h_s|^2\, ds\right]. \qquad (12.1.16)$$

More generally, by Doob's inequality [Do53, Chapter VII, Theorem 3.4], [ReY99, Theorem II (1.7)] and by the Burkholder-Davis-Gundy inequality [BuGu70], [ReY99, Theorem IV (4.1)], for $1 < p < +\infty$,

$$\left\| \int_0^t \langle h, \delta w \rangle \right\|_p \simeq \left\| \sup_{0 \le s \le t} \int_0^s \langle h, \delta w \rangle \right\|_p \simeq \left\| \left[\int_0^t |h|^2\, ds\right]^{1/2} \right\|_p. \qquad (12.1.17)$$

By (12.1.16), if h. is a nonrandom function,

$$E\left[\left|\int_0^t h \delta w\right|^2\right] = \int_0^t |h_s|^2\, ds. \qquad (12.1.18)$$

Now we fix $e \in \mathfrak{p}$, and we impose the condition $\vartheta_t = e$ in (12.1.14). It is natural to choose h so as to minimize the L_2 norm of h

$$|h|_{L_2} = \left[\int_0^t |h_s|^2\, ds\right]^{1/2}. \qquad (12.1.19)$$

Since R^X is a constant matrix, the first equation in (12.1.14) is deterministic and the question of finding the optimal h is a simple control theoretic problem. We get

$$\vartheta_t = \frac{\sinh\left(sR^X/2\right)}{\sinh\left(tR^X/2\right)} e, \qquad |h|_{L_2}^2 = \left\langle \frac{R^X}{1 - e^{-tR^X}} e, e \right\rangle. \qquad (12.1.20)$$

Of course (12.1.20) includes the case where R^X vanishes, and (12.1.20) becomes

$$\vartheta_t = \frac{s}{t} e, \qquad\qquad |h|_{L_2}^2 = \frac{|e|^2}{t}. \qquad (12.1.21)$$

Also note that if $\vartheta_s = se/t$, then $\vartheta.$ is a solution of the first equation in (12.1.14), with

$$h = \frac{1}{t}\left(1 + \frac{1}{2} s R^X\right) e. \qquad (12.1.22)$$

If h is given by (12.1.22), then

$$|h|_{L_2}^2 = \frac{1}{t}\left\langle \left(1 + \frac{t}{2} R^X + \frac{t^2}{12} R^{X,2}\right) e, e \right\rangle. \qquad (12.1.23)$$

The advantage of this choice of h is that it is valid even if R^X is non constant.

12.2 The probabilistic construction of $\exp\left(-t\mathcal{B}_b^X\right)$ over $\mathcal{X}$

Recall that the scalar operators $\mathcal{A}_b^X, \mathcal{B}_b^X$ over $\mathcal{X}$ were defined in (11.1.1)–(11.1.3), and the corresponding smooth heat kernels $r_{b,t}^X, s_{b,t}^X$ were introduced in Definition 11.5.3.

Again, we consider the probability space $(\mathcal{C}(\mathbf{R}_+, \mathfrak{p}), \mathcal{F}_\infty, P)$. Fix $Y^{TX} = Y^{\mathfrak{p}} \in \mathfrak{p}$. As explained in [B05, section 3.9] and in [BL08, section 14.2], the dynamics of the path $(x., Y^{TX}) \in \mathcal{X}$ associated with the operator $\mathcal{B}_b^X$ is given by

$$\dot{x} = \frac{Y^{TX}}{b}, \qquad\qquad \dot{Y}^{TX} = -\frac{Y^{TX}}{b^2} + \frac{\dot{w}}{b}, \qquad (12.2.1)$$
$$x_0 = p1, \qquad\qquad Y_0^{TX} = Y^{\mathfrak{p}}.$$

In (12.2.1), $\dot{Y}^{TX}$ denotes the covariant derivative of Y^{TX} with respect to ∇^{TX}. The first line of equation (12.2.1) can be rewritten in the form

$$b^2 \ddot{x} + \dot{x} = \dot{w}, \qquad\qquad Y^{TX} = b\dot{x}. \qquad (12.2.2)$$

Note that (12.2.1) is still a stochastic differential equation on $\mathcal{X}$. However, Y^{TX} is a continuous process, so that $x.$ is C^1. On a general smooth Riemannian manifold, there is no difficulty in defining parallel transport along such paths. Finally, observe that as $b \to 0$, the first equation in (12.2.2) degenerates formally to equation (12.1.3).

We said before that equations (12.2.1), (12.2.2) are stochastic differential equations. However, the probabilistic machinery is not needed to solve this equation, even on an arbitrary Riemannian manifold. More precisely, we claim that these equations can be solved for an arbitrary continuous path $w..$

Indeed let $x.$ be a C^1 path in X such that $x_0 = p1$. Let us fix a local coordinate system near x_0 on X, disregarding the fact that here, such a coordinate system can be chosen globally. Let Γ^{TX} be the connection form for ∇^{TX} in this coordinate system. The parallel transport τ_t^0 from $T_{x_0}X$ to $T_{x_t}X$ is obtained by solving the differential equation,

$$\dot{\tau}_t^0 + \Gamma_{x_t}^{TX}(\dot{x})\,\tau_t^0 = 0, \qquad\qquad \tau_0^0 = 1. \qquad (12.2.3)$$

Put

$$\tau_0^t = \left(\tau_t^0\right)^{-1}. \qquad (12.2.4)$$

Set

$$Z_t^{TX} = \tau_0^t Y_t^{TX}, \qquad (12.2.5)$$

so that Z_t^{TX} takes its values in $T_{x_0}X$. In local coordinates, equation (12.2.1) can be rewritten in the form

$$\dot{x} = \tau_t^0 \frac{Z^{TX}}{b}, \quad \dot{\tau}_t^0 + \Gamma_{x_t}^{TX}\left(\tau_t^0 \frac{Z^{TX}}{b}\right)\tau_t^0 = 0, \quad \dot{Z}^{TX} = -\frac{Z^{TX}}{b^2} + \frac{\dot{w}}{b},$$
$$(12.2.6)$$

$$x_0 = x, \qquad\qquad Z_0^{TX} = Y^{\mathfrak{p}}, \qquad\qquad \tau_0^0 = 1.$$

By rewriting the last equation in (12.2.6) in the form

$$Z_t^{TX} = Y^{\mathfrak{p}} - \int_0^t \frac{Z_s^{TX}}{b^2}\,ds + \frac{w_t}{b}, \qquad (12.2.7)$$

it is easy to see that at least locally, equation (12.2.6) has a unique solution for any continuous path $w_.$. This solution depends continuously on $w_.$ for the topology of uniform convergence of continuous functions over compact sets in $\mathbf{R}_+$. Such an argument can be globalized easily.

Let us give the probabilistic counterpart for the operator $\mathcal{B}_b^G$ in (11.1.4). Let $w_.$ be a Brownian motion in $\mathfrak{p}$. Then we consider the differential equation on $(g, Y^{\mathfrak{p}}) \in G \times \mathfrak{p}$,

$$\dot{g} = \frac{Y^{\mathfrak{p}}}{b}, \qquad\qquad \dot{Y}^{\mathfrak{p}} = -\frac{Y^{\mathfrak{p}}}{b^2} + \frac{\dot{w}}{b}, \qquad (12.2.8)$$

$$g_0 = 1, \qquad\qquad Y_0^{\mathfrak{p}} = Y^{\mathfrak{p}},$$

so that

$$x_t = pg_t, \qquad\qquad \left(x_t, Y_t^{TX}\right) = p\left(g_t, Y_t^{\mathfrak{p}}\right). \qquad (12.2.9)$$

By proceeding as in (12.2.7), we find that (12.2.8) makes sense for an arbitrary continuous function $w_.$, and also that the solution of (12.2.8) depends continuously on $w_.$ for the topology of uniform convergence over compact sets in $\mathbf{R}_+$.

Recall that in section 11.5, for $t \geq 0$, we have constructed the heat operators $\exp\left(-t\mathcal{A}_b^X\right), \exp\left(-t\mathcal{B}_b^X\right)$. Now we establish the obvious extension of (12.1.6), (12.1.7).

Theorem 12.2.1. *Let $F : \mathcal{X} \to \mathbf{R}$ be a smooth function with compact support. Then for $t \geq 0$,*

$$\exp\left(-t\mathcal{B}_b^X\right)F\left(x_0, Y^{\mathfrak{p}}\right) = E\left[F\left(x_t, Y_t^{TX}\right)\right]. \qquad (12.2.10)$$

Similarly, if $u : G \times \mathfrak{p} \to \mathbf{R}$ is a smooth function with compact support, then

$$\exp\left(-t\mathcal{B}_b^G\right)u\left(1, Y^{\mathfrak{p}}\right) = E\left[u\left(g_t, Y_t^{\mathfrak{p}}\right)\right]. \qquad (12.2.11)$$

Proof. Let $\Psi : \mathbf{R}_+ \times \mathcal{X} \to \mathbf{R}$ be given by (11.5.7). By Proposition 11.5.2 and by (11.5.8), this is a smooth function. Take $t > 0$. By (11.5.9), we know that $s \in [0, t] \to \Psi\left(t - s, \left(x_s, Y_s^{TX}\right)\right)$ is a local martingale. We will prove that it is a martingale.

By (11.5.8), we get

$$\Psi\left(t - s, \left(x_s, Y_s^{TX}\right)\right) = \exp\left(\left|Y_s^{TX}\right|^2 / 2\right) \Phi\left(t - s, \left(x_s, Y_s^{TX}\right)\right). \quad (12.2.12)$$

By Proposition 11.5.2, for $0 \le s \le t$, the function $\Phi\left(s, \left(x, Y^{TX}\right)\right)$ is smooth and uniformly bounded. Recall that the kernel $k_t^{\mathfrak{p}}$ on $\mathfrak{p}$ is given by (10.4.6). By the results of section 10.8 and by (12.2.1), for $s > 0$, $k_{s/b^2}^{\mathfrak{p}}\left(Y^{\mathfrak{p}}, Y^{\mathfrak{p}'}\right) dY^{\mathfrak{p}'}$ is the probability law of $Z_s^{TX} = \tau_0^s Y_s^{TX}$. By (10.7.1), given $Y^{\mathfrak{p}} \in \mathfrak{p}, t \ge 0$, $E\left[\exp\left(\left|Y_s^{TX}\right|^2\right)\right]$ is uniformly bounded for $0 \le s \le t$. Using (12.2.12) and the above, we find that the local martingale in (12.2.12) is a martingale. Since $\Psi\left(0, \cdot\right) = F$, we get (12.2.10). Equation (12.2.11) is just a trivial lift of (12.2.10). $\qquad\square$

Remark 12.2.2. Since G acts transitively on X, equation (12.2.10) is still valid for any $\left(x, Y^{TX}\right) \in \mathcal{X}$. An identity similar to (12.2.10) for the heat operator associated with a conjugate of $\mathcal{A}_b^X$ will also be given in equation (12.9.9).

The proof of Theorem 12.2.1 is the model of the proofs given in the book, which identify the analytically defined heat kernels of chapter 11 with their probabilistic counterpart.

12.3 The operator $\mathcal{B}_b^X$ and the wave equation

As before, Δ^{TX} denotes the Laplacian acting along the fibres TX of $\mathcal{X}$.

Let $e_1, \ldots, e_m$ be an orthonormal basis of TX. Let $\mathcal{C}$ be the differential operator on $\mathcal{X}$,

$$\mathcal{C} = \sum_{i=1}^{m} \nabla_{e_i} \nabla_{e_i}^V. \quad (12.3.1)$$

If $f \in C^\infty\left(X, \mathbf{R}\right)$, $\nabla_\cdot^{TX} \nabla.f$ denotes the Hessian of f.

Proposition 12.3.1. *The following identity holds:*

$$b^2 \mathcal{B}_b^{X,2} - \mathcal{B}_b^X = \frac{1}{4b^2}\left(-\Delta^{TX} + 2\nabla_{Y^{TX}}^V\right)^2 - \frac{1}{2b^2}\left(-\Delta^{TX} + 2\nabla_{Y^{TX}}^V\right)$$
$$- \frac{1}{b}\nabla_{Y^{TX}}\left(-\Delta^{TX} + 2\nabla_{Y^{TX}}^V\right) + \frac{\mathcal{C}}{b} + \nabla_{Y^{TX}}^{TX}\nabla_{Y^{TX}}. \quad (12.3.2)$$

In particular, if $f(x)$ is a smooth function on X, then

$$\left(b^2 \mathcal{B}_b^{X,2} - \mathcal{B}_b^X\right) f = \nabla_{Y^{TX}}^{TX} \nabla_{Y^{TX}} f. \quad (12.3.3)$$

Equation (12.3.3) can also be written in the form

$$-\mathcal{B}_b^X\left(f + b\nabla_{Y^{TX}} f\right) = \nabla_{Y^{TX}}^{TX} \nabla_{Y^{TX}} f. \quad (12.3.4)$$

Proof. Equation (12.3.2) follows from a trivial computation, which is left to the reader. By (12.3.2), we get (12.3.3). Also,

$$-b^2 \mathcal{B}_b^X f = b \nabla_{Y^{TX}} f, \tag{12.3.5}$$

so that (12.3.4) follows from (12.3.3). ∎

Remark 12.3.2. From the probabilistic point of view, equation (12.3.3) connects the hypoelliptic operator $\mathcal{B}_b^X$ with $-\Delta^X/2$. In fact it is because of equation (12.3.3) that as we shall see in section 12.8, as $b \to 0$, the operators $\mathcal{A}_b^X$ and $\mathcal{B}_b^X$ deform to the elliptic operator $-\Delta^X/2$. An identity very similar to (12.3.2) can be established for the operator $\mathcal{B}_b^G$. Finally, an identity similar to (12.3.4) is established in Proposition 14.2.1 for an operator $\mathcal{M}_b^X$ conjugate to the operator $\mathcal{L}_b^X$.

For $t > 0$, set

$$S_{b,t} = \exp\left(-t\mathcal{B}_b^X\right). \tag{12.3.6}$$

If $f(x)$ is a smooth bounded function on X, by (12.3.3), we get

$$\left(b^2 \frac{\partial^2}{\partial t^2} + \frac{\partial}{\partial t}\right) S_{b,t} f = S_{b,t} \nabla_{Y^{TX}}^{TX} \nabla_{Y^{TX}} f. \tag{12.3.7}$$

Recall that $s_{b,t}^X\left((x, Y^{TX}), (x', Y^{TX\prime})\right)$ is the smooth kernel associated with $S_{b,t}$. By Theorem 12.2.1, $s_{b,t}^X$ is nonnegative. We claim that $s_{b,t}^X$ is everywhere positive. Indeed by a theorem of Stroock and Varadhan [StV72], for $\left(x, Y^{TX}\right) \in \mathcal{X}$, the support of $s_{b,t}\left((x, Y^{TX}), (x', Y^{TX\prime})\right) dx' dY^{TX\prime}$ is equal to $\mathcal{X}$. By taking adjoints, the role of the couples $\left(x, Y^{TX}\right)$ and $\left(x', Y^{TX\prime}\right)$ can be reversed. Also,

$$s_{b,t}^X\left((x, Y^{TX}), (x', Y^{TX\prime})\right) = \int_{\mathcal{X}} s_{b,t/2}^X\left((x, Y), (z, Z^{TX})\right)$$
$$s_{b,t/2}^X\left((z, Z^{TX}), (x', Y^{TX\prime})\right) dz dZ^{TX}. \tag{12.3.8}$$

If the couples $\left(x, Y^{TX}\right), \left(x', Y^{TX\prime}\right)$ are such that (12.3.8) vanishes, by the above, as a function of $\left(z, Z^{TX}\right)$, $s_{b,t/2}^X\left((x, Y^{TX}), (z, Z^{TX})\right)$ vanishes identically. However, this is impossible in view of the fact that

$$\int_X s_{b,t/2}^X\left((x, Y^{TX}), (z, Z^{TX})\right) dz dZ^{TX} = 1. \tag{12.3.9}$$

Put

$$\sigma_{b,t}\left((x, Y^{TX}), x'\right) = \int_{T_{x'}X} s_{b,t}^X\left((x, Y^{TX}), (x', Y^{TX\prime})\right) dY^{TX\prime},$$

$$M_{b,t}\left((x, Y^{TX}), x'\right) = \frac{1}{\sigma_{b,t}\left((x, Y^{TX}), x'\right)} \int_{T_{x'}X} s_{b,t}^X\left((x, Y^{TX}), (x', Y^{TX\prime})\right)$$
$$\tag{12.3.10}$$

$$\left(Y^{TX\prime} \otimes Y^{TX\prime}\right) dY^{TX\prime}.$$

Note that $M_{b,t}\left(\left(x, Y^{TX}\right), x'\right)$ takes its values in symmetric positive endo-morphisms of $T_{x'}X$.

In the sequel, we fix $\left(x, Y^{TX}\right) \in \mathcal{X}$. We can associate to $M_{b,t}$ the second order elliptic operator acting on $C^{\infty}\left(X, \mathbf{R}\right)$,

$$\mathbf{M}_{b,t}\left(x, Y^{TX}\right) g\left(x'\right) = \left\langle \nabla^{TX}_{\cdot}\nabla_{\cdot}, M_{b,t}\left(\left(x, Y^{TX}\right), x'\right) g\left(x'\right)\right\rangle. \quad (12.3.11)$$

In (12.3.11), the operator $\nabla^{TX}_{\cdot}\nabla_{\cdot}$ acts on the variable x'.

By (12.3.7), (12.3.11), we get

$$\left(b^2 \frac{\partial^2}{\partial t^2} + \frac{\partial}{\partial t} - \mathbf{M}_{b,t}\left(x, Y^{TX}\right)\right) \sigma_{b,t}\left(\left(x, Y^{TX}\right), \cdot\right) = 0. \quad (12.3.12)$$

Equation (12.3.12) is a hyperbolic equation. Of course, it does not determine $\sigma_{b,t}$, since $\mathbf{M}_{b,t}$ depends on $s_{b,t}$.

As we shall see in (12.8.46), as $b \to 0$,

$$s^X_{b,t}\left(\left(x, Y^{TX}\right), \left(x', Y^{TX'}\right)\right) \to \pi^{-m/2} p_t\left(x, x'\right) \exp\left(-\left|Y^{TX'}\right|^2\right). \quad (12.3.13)$$

In (12.3.13), the convergence of the functions and their derivatives of arbitrary order is uniform over compact sets. Let $\mathbf{1}$ be the identity of TX. We identify $\mathbf{1}$ to the scalar product of TX. By (12.3.13), as $b \to 0$,

$$\sigma_{b,t}\left(\left(x, Y^{TX}\right), x'\right) \to p_t\left(x, x'\right), \quad M_{b,t}\left(\left(x, Y^{TX}\right), x'\right) \to \frac{1}{2}. \quad (12.3.14)$$

It is interesting to compare (12.3.12) with the heat equation,

$$\left(\frac{\partial}{\partial t} - \frac{\Delta^X}{2}\right) p_t\left(x, \cdot\right) = 0. \quad (12.3.15)$$

By (12.3.11), (12.3.13), and (12.3.14), as $b \to 0$, equation (12.3.12) converges in the proper sense to equation (12.3.15).

While equation (12.3.15) is parabolic, equation (12.3.12) is hyperbolic. This should give a sense that when only the variable x is considered, there is a wave equation quality to the heat operator for the hypoelliptic Laplacian. In that respect, it is tempting to also consider the genuinely hyperbolic operator

$$\mathcal{H}^p_b = b^2 \frac{\partial^2}{\partial t^2} + \frac{\partial}{\partial t} - \frac{\Delta^X}{2}. \quad (12.3.16)$$

This wave operator propagates at the speed $\frac{1}{\sqrt{2b}}$. When $b \to 0$, it converges in the proper sense to the parabolic heat operator $\frac{\partial}{\partial t} - \Delta^X/2$. However, it certainly does not interpolate between this parabolic operator and the geodesic flow, and so it cannot be used as a substitute to the hypoelliptic Laplacian.

The computations that follow are a path integral version of equation (12.3.4) in Proposition 12.3.1. Let $f : X \to \mathbf{R}$ be a smooth function. By (12.2.1), we get

$$f\left(x_t\right) = f\left(x_0\right) + \int_0^t \frac{\nabla_{Y^{TX}} f\left(x_s\right)}{b} ds. \quad (12.3.17)$$

Still using (12.2.1) and integrating by parts in (12.3.17), we obtain

$$\int_0^t \frac{\nabla_{Y^{TX}} f(x_s)}{b} ds = -b \left(\nabla_{Y_t^{TX}} f(x_t) - \nabla_{Y^{\mathfrak{p}}} f(x_0) \right)$$
$$+ \int_0^t \nabla_{Y_s^{TX}}^{TX} \nabla_{Y_s^{TX}} f(x_s) ds + \int_0^t \nabla_{\delta w_s} f(x_s). \quad (12.3.18)$$

In (12.3.18), the last term is an Itô integral with respect to the Brownian motion $w.$. By (12.3.17), (12.3.18), we get

$$f(x_t) = f(x_0) - b \left(\nabla_{Y_t^{TX}} f(x_t) - \nabla_{Y^{\mathfrak{p}}} f(x_0) \right)$$
$$+ \int_0^t \nabla_{Y_s^{TX}}^{TX} \nabla_{Y_s^{TX}} f(x_s) ds + \int_0^t \nabla_{\delta w_s} f(x_s). \quad (12.3.19)$$

Note that if f has uniformly bounded first and second derivatives, when taking the expectation in (12.3.19), and using (12.2.10), we recover (12.3.4).

Recall that if $e \in \mathfrak{g}$, e is identified with the corresponding left-invariant vector field on G. If $u : G \to \mathbf{R}$ is a smooth function, if $g., Y^{\mathfrak{p}}$ are taken as in (12.2.8), instead of (12.3.19), we have the identity

$$u(g_t) = u(1) - b \left(\nabla_{Y_t^{\mathfrak{p}}} u(g_t) - \nabla_{Y^{\mathfrak{p}}} u(1) \right)$$
$$+ \int_0^t \nabla_{Y_s^{\mathfrak{p}}} \nabla_{Y_s^{\mathfrak{p}}} u(g_s) ds + \int_0^t \nabla_{\delta w_s} u(g_s). \quad (12.3.20)$$

Of course, if $u = p^* f$, (12.3.20) is equivalent to (12.3.19).

12.4 The Malliavin calculus for the operator $\mathcal{B}_b^X$

In this section, we follow Bismut-Lebeau [BL08, chapter 14]. In that reference, the Malliavin calculus is developed when X is instead a compact Riemannian manifold.

We will obtain a formula of integration by parts for the process $(x., Y^{TX})$ in (12.2.1).

As in section 12.1, in equation (12.2.1), we replace $w.$ by w^ℓ given by (12.1.8). Let $\left(x^\ell., Y^{TX,\ell} \right)$ be the corresponding solution of (12.2.1). We will compute the differential of this solution with respect to ℓ at $\ell = 0$. In the sequel, $\frac{D}{Ds}$ denotes covariant differentiation with respect to the Levi-Civita connection along $x..$

Set

$$J_s^{TX} = \frac{\partial}{\partial \ell} x_s^\ell. \quad (12.4.1)$$

We use the notation $\dot{J}^{TX}, \ddot{J}^{TX}$ instead of $\frac{D}{Ds} J^{TX}, \frac{D^2}{Ds^2} J^{TX}$. By the first equation in (12.2.1),

$$\dot{J}_s^{TX} = \frac{1}{b} \frac{D}{D\ell} Y_s^{TX,\ell}. \quad (12.4.2)$$

By (12.2.1), (12.2.2), we obtain

$$b^2 \ddot{J}^{TX} + \dot{J}^{TX} + b^2 R^{TX} \left(J^{TX}, \dot{x}\right) \dot{x} = h + \left[B + \varpi, \frac{\delta w}{ds}\right], \qquad (12.4.3)$$

$$\dot{\varpi} = \left[J^{TX}, \dot{x}\right], \quad J_0^{TX} = 0, \quad \dot{J}_0^{TX} = 0, \quad \varpi_0 = 0.$$

Here $J^{TX} \in \mathfrak{p}$ is identified with its parallel transport along x. with respect to ∇^{TX}. Note that by the arguments given in section 12.2, differentiation with respect to ℓ does not necessitate sophisticated probabilistic arguments.

Now we state a result established in [BL08, section 14.2]. This result gives an analogue of equation (12.1.13), i.e., it is a formula of integration by parts.

Theorem 12.4.1. *Let $F : \mathcal{X} \to \mathbf{R}$ be a smooth function with compact support. Then*

$$E\left[\left\langle dF\left(x_t, Y_t^{TX}\right), \left(J_t^{TX}, b\dot{J}_t^{TX}\right)\right\rangle\right] = E\left[F\left(x_t, Y_t^{TX}\right) \int_0^t \langle h, \delta w\rangle\right].$$
$$(12.4.4)$$

Proof. The proof of (12.4.4) still relies on Girsanov's transformation, and can be given by following the same arguments as in section 12.1. The same argument as in (12.1.11) shows that

$$E\left[F\left(x_t^\ell, Y_t^{TX,\ell}\right) Z_t^\ell\right] = E\left[F\left(x_t, Y_t^{TX}\right)\right]. \qquad (12.4.5)$$

Using (12.4.1), (12.4.2) and (12.4.5), we get (12.4.4).   □

Now we proceed as in [BL08, section 14.2]. Consider the differential equation

$$b^2 \ddot{J}^{TX} + \dot{J}^{TX} + b^2 R^{TX}\left(J^{TX}, \dot{x}\right)\dot{x} = h, \qquad \dot{\varpi} = \left[J^{TX}, \dot{x}\right], \qquad (12.4.6)$$

$$J_0^{TX} = 0, \quad \dot{J}_0^{TX} = 0, \qquad\qquad\qquad\qquad \varpi_0 = 0.$$

Then (12.4.6) is a special case of (12.4.3), by taking

$$B = -\varpi. \qquad (12.4.7)$$

In particular Theorem 12.4.1 applies to the solution of (12.4.6).

Remark 12.4.2. The presence of the expression $b^2 R^{TX}\left(J^{TX}, \dot{x}\right)\dot{x}$, which is quadratic in $\dot{x}$, is still another reflection of the considerations we made in Remark 9.10.3. The facts outlined in this remark are also obvious in (12.2.6).

12.5 The tangent variational problem and integration by parts

Consider equation (12.4.6). It has been obtained by differentiating equation (12.2.2) in the parameter ℓ. Except for the term containing the curvature R^{TX}, when replacing formally h by $\dot{w}$, we recover equation (12.2.2).

Take $e, f \in \mathfrak{p}$. Let J^{TX} be any smooth deterministic function defined on $[0, t]$ with values in $\mathfrak{p}$, such that

$$J_0^{TX} = 0, \qquad \dot{J}_0^{TX} = 0, \qquad J_t^{TX} = e, \qquad \dot{J}_t^{TX} = f/b. \qquad (12.5.1)$$

Then equation (12.4.6) determines a predictable h to which (12.4.4) can be applied. This principle was used in [BL08, section 14.3], with a very simple choice of $J^{TX}_{\cdot}$. However, the choice made in [BL08] is not adequate for our purpose, because with this choice, the norm $|h|_{L_2}$ diverges as $b \to 0$.

So instead, when $E = \mathfrak{p}$, we consider the variational problem of section 10.1 for the functional $K_{b,t}$ in (10.1.2). Also we use the notation in (10.1.17)–(10.1.19). We fix $e, f \in \mathfrak{p}$. Given a control $v_{\cdot} \in L_2$, consider the differential equation

$$b^2 \ddot{J}^{TX} + \dot{J}^{TX} = v, \qquad J^{TX}_0 = 0, \qquad \dot{J}^{TX}_0 = 0, \qquad (12.5.2)$$
$$J^{TX}_t = e, \qquad\qquad\qquad \dot{J}^{TX}_t = f/b.$$

The control problem consists in finding v with minimal norm $|v|_{L_2}$.

This question was solved in section 10.3, in equations (10.3.30)–(10.3.39), in which we make $x = 0, Y = 0, x' = e, Y' = f/b$. Moreover, by (10.3.38),

$$\frac{1}{2} |v|^2_{L_2} = K_{b,t} \left((0,0), (e, f/b) \right). \qquad (12.5.3)$$

By the considerations that follow (10.3.47), as $b \to 0$, we have the uniform convergence on $[0, t]$,

$$J^{TX}_s \to \frac{s}{t} e. \qquad (12.5.4)$$

Moreover, by (10.3.49), (10.3.50), and (12.5.3), as $b \to 0$,

$$\frac{1}{2} |v|^2_{L_2} \to \frac{1}{2t} |e|^2 + |f|^2, \qquad v \to \frac{e}{t} \text{ weakly in } L_2. \qquad (12.5.5)$$

Take $\epsilon > 0, M > 0$ such that $0 < \epsilon \le M$. Using again the considerations after (10.3.47), we know that as long as $\epsilon \le t \le M$, and e, f remain uniformly bounded, the above convergence results are uniform with respect to t, e, f. Also $b \dot{J}^{TX}$ remains uniformly bounded.

Let τ^0_t be the parallel transport operator from $T_{x_0} X$ into $T_{x_t} X$ with respect to ∇^{TX}. We fix unit vectors $e, f \in \mathfrak{p}$, and for a given $b > 0$, we denote by v the optimal control that has been determined above. By Theorem 12.4.1, and by (12.4.6), (12.5.2), we get

$$E \left[\langle dF \left(x_t, Y^{TX}_t \right), \tau^0_t (e, f) \rangle \right]$$
$$= E \left[F \left(x_t, Y^{TX}_t \right) \int_0^t \langle v + b^2 R^{TX} \left(J^{TX}, \dot{x} \right) \dot{x}, \delta w \rangle \right]. \qquad (12.5.6)$$

Equation (12.5.6) is still not enough for our purpose.

If $a \in \mathfrak{g}$, recall that a^{TX} denotes the corresponding Killing vector field on X.

We fix $c, d \in \mathfrak{g}$. Let $e_1, \ldots, e_m$ be an orthonormal basis of $\mathfrak{p}$. Given $b > 0$, we denote by $v_i, \underline{v}_i$ the controls determined above, with $e = e_i, f = 0$, and $e = 0, f = e_i$, and by $J^{TX}_i, \underline{J}^{TX}_i$ the associated solutions of (12.5.2). Finally, we define $\varpi_i, \underline{\varpi}_i$ as in (12.4.6), i.e.,

$$\dot{\varpi}_i = \left[J^{TX}_i, \dot{x} \right], \qquad\qquad \varpi_{i,0} = 0, \qquad (12.5.7)$$
$$\dot{\underline{\varpi}}_i = \left[\underline{J}^{TX}_i, \dot{x} \right], \qquad\qquad \underline{\varpi}_{i,0} = 0.$$

Theorem 12.5.1. *Let* $F : \mathcal{X} \to \mathbf{R}$ *be a smooth function with compact support. The following identity holds:*

$$E\left[\left\langle dF\left(x_t, Y_t^{TX}\right), \left(c_{x_t}^{TX}, d_{x_t}^{TX}\right)\right\rangle\right]$$

$$= \sum_{1 \le i \le m} E\left[F\left(x_t, Y_t^{TX}\right)\left(\left\langle c_{x_t}^{TX}, \tau_t^0 e_i\right\rangle \int_0^t \left\langle v_i + b^2 R^{TX}\left(J_i^{TX}, \dot{x}\right)\dot{x}, \delta w\right\rangle\right.\right.$$

$$\left.\left. + \left\langle d_{x_t}^{TX}, \tau_t^0 e_i\right\rangle \int_0^t \left\langle \underline{v}_i + b^2 R^{TX}\left(\underline{J}_i^{TX}, \dot{x}\right)\dot{x}, \delta w\right\rangle\right)\right]$$

$$- \sum_{1 \le i \le m} E\left[F\left(x_t, Y_t^{TX}\right)\left(\left\langle c_{x_t}^{TX}, \tau_t^0 \varpi_{i,t} e_i\right\rangle + \left\langle d_{x_t}^{TX}, \tau_t^0 \underline{\varpi}_{i,t} e_i\right\rangle\right)\right]. \quad (12.5.8)$$

Proof. Clearly, we have the identities

$$c_{x_t}^{TX} = \sum_{1 \le i \le m} \left\langle c_{x_t}^{TX}, \tau_t^0 e_i\right\rangle \tau_t^0 e_i, \quad d_{x_t}^{TX} = \sum_{1 \le j \le m} \left\langle d_{x_t}^{TX}, \tau_t^0 e_j\right\rangle \tau_t^0 e_j. \quad (12.5.9)$$

Set

$$U_{i,t} = F\left(x_t, Y_t^{TX}\right)\left\langle c_{x_t}^{TX}, \tau_t^0 e_i\right\rangle, \quad (12.5.10)$$

$$\underline{U}_{i,t} = F\left(x_t, Y_t^{TX}\right)\left\langle d_{x_t}^{TX}, \tau_t^0 e_i\right\rangle.$$

Let h be a given bounded predictable process. Let $U_{i,t}^\ell, \underline{U}_{i,t}^\ell$ be the random variables $U_{i,t}, \underline{U}_{i,t}$ in (12.5.10) in which $w.$ has been replaced by w^ℓ given by (12.1.8). Then the obvious analogue of (12.1.11), (12.4.5) is that

$$E\left[U_{i,t}^\ell Z_t^\ell\right] = E\left[U_{i,t}\right], \qquad E\left[\underline{U}_{i,t}^\ell Z_t^\ell\right] = E\left[\underline{U}_{i,t}\right]. \quad (12.5.11)$$

In the first equation, we replace h by $v_i + b^2 R^{TX}\left(J_i^{TX}, \dot{x}\right)\dot{x}$, and in the second equation, we replace h by $\underline{v}_i + b^2 R^{TX}\left(\underline{J}_i^{TX}, \dot{x}\right)\dot{x}$. We take the differential of the identities in (12.5.11) at $\ell = 0$, and we sum them in $i, 1 \le i \le m$. Also we use (12.5.9) and the fact that since c^{TX} is a Killing vector field,

$$\left\langle \nabla_{\tau_t^0 e_i}^{TX} c^{TX}, \tau_t^0 e_i\right\rangle = 0, \quad (12.5.12)$$

and we get (12.1.15). The proof of our theorem is completed. $\qquad \square$

The principle of the Malliavin calculus is to iterate a formula like (12.5.8), in order to obtain on the left-hand side of (12.5.8) the expectation of a differential operator of arbitrary order in the vector fields c^{TX}, d^{TX} applied to F at $\left(x_t, Y_t^{TX}\right)$ in terms of expectations where only $F\left(x_t, Y_t^{TX}\right)$ appears. Equation (12.5.8) is the model of such an equation in the case where the order of the operator is 1.

Let us explain how to do this when the order is 2. We use (12.5.8) with $F\left(x, Y^{TX}\right)$ replaced by $u_{c,d}\left(x, Y^{TX}\right) = \left\langle dF\left(x, Y^{TX}\right), \left(c_x^{TX}, d_x^{TX}\right)\right\rangle$. On the right-hand side of (12.5.8), the function $u_{c,d}\left(x, Y^{TX}\right)$ appears. Then the same method as before can be used on this right-hand side so as to obtain another expression where only the function $F\left(x, Y^{TX}\right)$ appears. Of course extra terms have now to be differentiated. However, they are algebraically very simple. Moreover, since F has compact support, except for the terms containing $\dot{x}$ and δw, they are uniformly bounded on the domain where $F\left(x_t, Y_t^{TX}\right)$ is nonzero.

12.6 A uniform control of the integration by parts formula as $b \to 0$

Let K be the compact support of F. Over K, the norms of c^{TX}, d^{TX} are uniformly bounded. We will write (12.5.8) in the form

$$E\left[\langle dF\left(x_t, Y_t^{TX}\right), \left(c_{x_t}^{TX}, d_{x_t}^{TX}\right)\rangle\right] = E\left[F\left(x_t, Y_t^{TX}\right) M_t^{c,d}\right]. \qquad (12.6.1)$$

The random variable $M_t^{c,d}$ is explicitly determined in (12.5.8). Note that $M_t^{c,d}$ is linear in $c_{x_t}^{TX}, d_{x_t}^{TX}$. We denote by M_t the corresponding linear form on $T_{x_t}X \times T_{x_t}X$.

Theorem 12.6.1. *Given $\epsilon > 0, M > 0, \epsilon \leq M, p \in [1, +\infty[$, there is $C_{\epsilon,M,p} > 0$ such that for $0 < b \leq M, \epsilon \leq t \leq M$,*

$$\|M_t\|_p \leq C_{\epsilon,M,p}\left(1 + |Y^{\mathrm{p}}|^2\right). \qquad (12.6.2)$$

Proof. As we saw in section 12.5, for $0 < b \leq M, \epsilon \leq t \leq M$, the $v_i, \underline{v}_i$ are deterministic, and their L_2 norm is uniformly bounded. By (12.1.17), for $1 < p < +\infty$, we get a uniform control of the L_p norm of $\int_0^t \langle v_i, \delta w\rangle, \int_0^t \langle \underline{v}_i, \delta w\rangle$.

In the sequel, K_i^{TX} denotes one of the $J_i^{TX}, \underline{J}_i^{TX}$. By (12.1.17), for $1 < p < +\infty$,

$$\left\|\int_0^t \langle b^2 R^{TX}\left(K_i^{TX}, \dot{x}\right)\dot{x}, \delta w\rangle\right\|_p \simeq \left\|\left[\int_0^t \left|b^2 R^{TX}\left(K_i^{TX}, \dot{x}\right)\dot{x}\right|^2 ds\right]^{1/2}\right\|_p. \qquad (12.6.3)$$

As we saw in section 12.5, for $0 < b \leq M, \epsilon \leq t \leq M$, the K_i remain uniformly bounded. Therefore,

$$\int_0^t \left|b^2 R^{TX}\left(K_i^{TX}, \dot{x}\right)\dot{x}\right|^2 ds \leq C_{\epsilon,M} \int_0^t |b\dot{x}|^4 ds. \qquad (12.6.4)$$

By (12.2.1), we get

$$\int_0^t |b\dot{x}|^4 ds = \int_0^t \left|Y^{TX}\right|^4 ds. \qquad (12.6.5)$$

By equation (10.8.27) in Proposition 10.8.2, by (12.6.3)–(12.6.5), we find that for $1 < p < +\infty$, for $0 < b \leq M, \epsilon \leq t \leq M$,

$$\left\|\int_0^t \langle b^2 R^{TX}\left(K_i^{TX}, \dot{x}\right)\dot{x}, \delta w\rangle\right\|_p \leq C_{\epsilon,M,p}\left(1 + |Y^{\mathrm{p}}|^2\right). \qquad (12.6.6)$$

We denote by σ_i either ϖ_i or $\underline{\varpi}_i$. By (12.5.7),

$$\sigma_{i,t} = -\int_0^t R^{TX}\left(K_i^{TX}, \dot{x}\right) ds. \qquad (12.6.7)$$

By (12.2.2), we can rewrite (12.6.7) in the form

$$\sigma_{i,t} = -\int_0^t R^{TX}\left(K_i^{TX}, \dot{w} - b^2\ddot{x}\right) ds. \qquad (12.6.8)$$

Recall that K_i^{TX} is deterministic and smooth, so that

$$\int_0^t R^{TX}\left(K_i^{TX}, \dot{w}\right) ds = \int_0^t R^{TX}\left(K_i^{TX}, \delta w\right). \qquad (12.6.9)$$

Using (12.1.17) and the fact that for $0 < b \leq M, \epsilon \leq t \leq M$, K_i^{TX} is uniformly bounded, the L_p norm of (12.6.9) remains uniformly bounded.

Moreover, if $K_i^{TX} = J_i^{TX}$, using (12.2.2) and the fact that R^{TX} is parallel, we get

$$b^2\int_0^t R^{TX}\left(K_i^{TX}, \ddot{x}\right) ds = bR^{TX}\left(e_i, Y_t^{TX}\right) - \int_0^t R^{TX}\left(b\dot{K}_i^{TX}, Y^{TX}\right) ds. \qquad (12.6.10)$$

If $K_i^{TX} = \underline{J}_i^{TX}$, the first term in the right-hand side of (12.6.10) does not appear.

By equation (10.8.26) in Proposition 10.8.2, under the given conditions on b, t, we get

$$\left\|bR^{TX}\left(e_i, Y_t^{TX}\right)\right\|_p \leq C_{p,M}\left(1 + |Y^{\mathfrak{p}}|\right). \qquad (12.6.11)$$

As we saw after (12.5.5), for $0 < b \leq M, \epsilon \leq t \leq M$, $b\dot{K}_i^{TX}$ remains uniformly bounded on $[0, t]$. Therefore

$$\left|\int_0^t R^{TX}\left(b\dot{K}_i^{TX}, Y^{TX}\right) ds\right| \leq C_{\epsilon,M}\int_0^t \left|Y_s^{TX}\right| ds. \qquad (12.6.12)$$

The same arguments as before show that

$$\left\|\int_0^t R^{TX}\left(b\dot{K}_i^{TX}, Y^{TX}\right) ds\right\|_p \leq C_{\epsilon,M,p}\left(1 + |Y^{\mathfrak{p}}|\right). \qquad (12.6.13)$$

Using (12.6.7)–(12.6.13), for $0 < b \leq M, \epsilon \leq t \leq M$,

$$\|\sigma_{i,t}\|_p \leq C_{\epsilon,M,p}\left(1 + |Y^{\mathfrak{p}}|\right). \qquad (12.6.14)$$

By (12.5.8), (12.6.1), and by (12.6.3)–(12.6.14), we get (12.6.2). The proof of our theorem is completed. $\qquad\square$

Remark 12.6.2. Similar arguments can be used when iterating the integration by parts formula (12.5.8) by the procedure outlined in section 12.5. Bounds similar to the bound in Theorem 12.6.1 can be easily obtained. If k is the order of the differential operator acting on F, in the right-hand side of (12.6.2), $1 + |Y^{\mathfrak{p}}|^2$ is replaced by $1 + |Y^{\mathfrak{p}}|^{2k}$. Since M_t has uniformly controlled L_p norm for $p \in [1, +\infty[$, random variables like M_t can be multiplied and their product still has uniformly controlled L_p norm.

12.7 Uniform rough estimates on $r_{b,t}^X$ for bounded b

Recall that the probability law of Y_t^{TX} is given by $k_{t/b^2}^{TX}(Y, Y') \, dY'$. As we already saw in the proof of Theorem 12.2.1, given $b > 0, t \geq 0$, by (10.7.1), $\exp\left(\left|Y_t^{TX}\right|^2 / 2\right) \in L_2$.

Let $F : \mathcal{X} \to \mathbf{R}$ be a smooth function with compact support.

Theorem 12.7.1. *The following identity holds:*

$$
E\left[\exp\left(\left|Y_t^{TX}\right|^2/2\right)\left\langle dF\left(x_t, Y_t^{TX}\right), \left(c_{x_t}^{TX}, d_{x_t}^{TX}\right)\right\rangle\right]
$$

$$
= \sum_{1 \leq i \leq m} E\left[\exp\left(\left|Y_t^{TX}\right|^2/2\right) F\left(x_t, Y_t^{TX}\right)\left(\left\langle c_{x_t}^{TX}, \tau_t^0 e_i\right\rangle\right.\right.
$$

$$
\int_0^t \left\langle v_i + b^2 R^{TX}\left(J_i^{TX}, \dot{x}\right)\dot{x}, \delta w\right\rangle
$$

$$
\left.\left. + \left\langle d_{x_t}^{TX}, \tau_t^0 e_i\right\rangle \int_0^t \left\langle \underline{v}_i + b^2 R^{TX}\left(\underline{J}_i^{TX}, \dot{x}\right)\dot{x}, \delta w\right\rangle\right)\right]
$$

$$
- \sum_{1 \leq i \leq m} E\left[\exp\left(\left|Y_t^{TX}\right|^2/2\right) F\left(x_t, Y_t^{TX}\right)\left\langle Y_t^{TX}, d_{x_t}^{TX}\right\rangle\right]
$$

$$
- \sum_{1 \leq i \leq m} E\left[\exp\left(\left|Y_t^{TX}\right|^2/2\right) F\left(x_t, Y_t^{TX}\right)\right.
$$

$$
\left.\left(\left\langle c_{x_t}^{TX}, \tau_t^0 \varpi_{i,t} e_i\right\rangle + \left\langle d_{x_t}^{TX}, \tau_t^0 \underline{\varpi}_{i,t} e_i\right\rangle\right)\right]. \quad (12.7.1)
$$

Proof. We replace $F\left(x, Y^{TX}\right)$ by $F\left(x, Y^{TX}\right)\exp\left(\left|Y^{TX}\right|^2/2\right)$ in (12.5.8) and we get (12.7.1). $\square$

Remark 12.7.2. Since $\exp\left(\left|Y_t^{TX}\right|^2/2\right) \in L_2$, in (12.7.1), we may replace this random variable by

$$
\exp\left(\left|Y_t^{TX}\right|^2/2\right)\left(1 + \left|Y_t^{TX}\right|^2\right)^p, p \in \mathbf{N},
$$

and obtain a similar formula. The same is true when replacing $d_{x_t}^{TX}$ by any vertical polynomial vector field like $\nabla_{Y^{TX}}^V$.

By equation (10.8.12) in Proposition 10.8.1, for $0 < b \leq M, \epsilon \leq t \leq M$, since t/b^2 stays away from 0, there exists $q = q_{\epsilon,M}, 1 < q_{\epsilon,M} < 2$ such that for $t \geq \epsilon, 0 < b \leq M, \exp\left(\left|Y_t^{TX}\right|^2/2\right) \in L_q$, and moreover,

$$
\left\|\exp\left(\left|Y_t^{TX}\right|^2/2\right)\right\|_q \leq C_q \exp\left(\left|Y^{\mathfrak{p}}\right|^2/2q\right). \quad (12.7.2)
$$

Let $N_t^{c,d}$ be the expression appearing after $\exp\left(\left|Y_t^{TX}\right|^2/2\right) F\left(x_t, Y_t^{TX}\right)$ in the right-hand side of (12.7.1). Again $N_t^{c,d}$ can be viewed as a linear form

N_t on $T_{x_t} X \times T_{x_t} X$. The same arguments as in the proof of equation (12.6.2) in Theorem 12.6.1 show that for $0 < b \leq M, \epsilon \leq t \leq M$, for $1 < p < +\infty$,

$$\|N_t\|_p \leq C_{\epsilon,M,p} \left(1 + |Y^{\mathfrak{p}}|^2\right). \tag{12.7.3}$$

By (12.7.2), (12.7.3), and by Hölder's inequality, $\exp\left(\left|Y_t^{TX}\right|^2/2\right) N_t \in L_1$, and moreover, there exist $C_{\epsilon,M} > 0, c_{\epsilon,M} \in \,]0,1[$ such that under the given conditions on b, t,

$$\left\|\exp\left(\left|Y_t^{TX}\right|^2/2\right) N_t\right\|_1 \leq C_{\epsilon,M} \exp\left(c_{\epsilon,M} |Y^{\mathfrak{p}}|^2/2\right). \tag{12.7.4}$$

By (12.7.4), we get

$$\exp\left(-|Y^{\mathfrak{p}}|^2/2\right) \left\|\exp\left(\left|Y_t^{TX}\right|^2/2\right) N_t\right\|_1$$
$$\leq C_{\epsilon,M} \exp\left(-(1 - c_{\epsilon,M}) |Y^{\mathfrak{p}}|^2/2\right). \tag{12.7.5}$$

By (11.5.10) and (12.2.10), if $F \in C^{\infty,c}(\mathcal{X}, \mathbf{R})$, we get

$$\int_{\mathcal{X}} r_{b,t}^X \left((x_0, Y^{\mathfrak{p}}), (x', Y^{TX\prime})\right) F\left(x', Y^{TX\prime}\right) dx' dY^{TX\prime}$$
$$= \exp\left(-|Y^{\mathfrak{p}}|^2/2\right) E\left[\exp\left(\left|Y_t^{TX}\right|^2/2\right) F\left(x_t, Y_t^{TX}\right)\right]. \tag{12.7.6}$$

By (12.7.1) and (12.7.6), we obtain

$$\int_{\mathcal{X}} r_{b,t}^X \left((x_0, Y^{\mathfrak{p}}), (x', Y^{TX\prime})\right) \left\langle dF\left(x', Y^{TX\prime}\right), \left(c_{x'}^{TX}, d_{x'}^{TX}\right)\right\rangle dx' dY^{TX\prime}$$
$$= \exp\left(-|Y^{\mathfrak{p}}|^2/2\right) E\left[\exp\left(\left|Y^{TX}\right|^2/2\right) F\left(x_t, Y_t^{TX}\right) N_t^{c,d}\right]. \tag{12.7.7}$$

By proceeding as before, in (12.7.7), $\left\langle dF\left(x', Y^{TX\prime}\right), \left(c_{x'}^{TX}, d_{x'}^{TX}\right)\right\rangle$ can be replaced by the action on F of an arbitrary differential operator in the vector fields c^{TX}, d^{TX} with coefficients that are polynomials in $Y^{TX\prime}$, and we still obtain an analogue of (12.7.7). Moreover, the obvious analogue of (12.7.5) holds.

Now we give an analogue of Definition 9.1.2.

Definition 12.7.3. A function $f : \mathcal{X} \times \mathcal{X} \to \mathbf{R}$ is said to be rapidly decreasing if for any $k \in \mathbf{N}$, $\left(1 + \left|Y^{TX}\right|^k + \left|Y^{TX\prime}\right|^k\right) f\left((x, Y^{TX}), (x', Y^{TX\prime})\right)$ is uniformly bounded. If f also depends on $b > 0, t > 0$, if $D \in \mathbf{R}_+^* \times \mathbf{R}_+^*$, we say that f is uniformly rapidly decreasing for $(b, t) \in D$ if the previous bounds are uniform.

Theorem 12.7.4. For $0 < b \leq M, \epsilon \leq t \leq M$, $r_{b,t}^X \left((x, Y^{TX}), (x', Y^{TX\prime})\right)$ and its covariant derivatives of arbitrary order in $(x', Y^{TX\prime})$ with respect to ∇^{TX} are uniformly rapidly decreasing on $\mathcal{X} \times \mathcal{X}$.

Proof. To establish our theorem, we may as well assume that $(x, Y^{TX}) = (x_0, Y^{\mathfrak{p}})$. Let $u(r) : \mathbf{R}_+ \to [0, 1]$ be a smooth function that is equal to 1 for $r \leq 1$ and to 0 for $r \geq 2$. Take $x' \in X$. We take geodesic coordinates centered at $x' \in X$, and we trivialize TX by parallel transport with respect to ∇^{TX} along such geodesics, so that $\mathcal{X} \sim \mathfrak{p} \times \mathfrak{p}$. Given $\alpha, \beta \in \mathfrak{p}$, set

$$g_{\alpha,\beta}\left(y, Y^{\mathfrak{p}'}\right) = u\left(|y|\right) \exp\left(i \langle \alpha, y \rangle + i \langle \beta, Y^{\mathfrak{p}'} \rangle\right). \tag{12.7.8}$$

Recall that the smooth positive function $\delta : \mathfrak{p} \to \mathbf{R}_+$ was defined in (4.1.11). Set

$$\Phi_{b,t,Y^{\mathfrak{p}},x'}\left(\alpha, \beta\right) = \int_{\mathfrak{p} \times \mathfrak{p}} r_{b,t}^X\left((x_0, Y^{\mathfrak{p}}), (y, Y^{\mathfrak{p}'})\right) g_{\alpha,\beta}\left(y, Y^{\mathfrak{p}'}\right) \delta\left(y\right) dy dY^{\mathfrak{p}'}. \tag{12.7.9}$$

We use equations (12.7.5) and (12.7.7) with F replaced by $g_{\alpha,\beta}$, and their higher order analogues. We find that for any $k \in \mathbf{N}$, there exist $C_{\epsilon,M,k} > 0, c_{\epsilon,M,k} > 0$ such that for $0 < b \leq M, \epsilon \leq t \leq M$,

$$\left(1 + |\alpha|^k + |\beta|^k\right) \left|\Phi_{b,t,Y^{\mathfrak{p}},x'}\left(\alpha, \beta\right)\right| \leq C_{\epsilon,M,k} \exp\left(-c_{\epsilon,M,k} \left|Y^{\mathfrak{p}}\right|^2\right). \tag{12.7.10}$$

Using elementary properties of Fourier transform, we see that for $\epsilon \leq t \leq M, 0 < b \leq M$, $r_{b,t}^X\left((x_0, Y^{\mathfrak{p}}), (x', Y^{TX'})\right)$ and its covariant derivatives of any order in the variable $(x', Y^{TX'}) \in \mathcal{X}$ are not only uniformly bounded, but they also exhibit a Gaussian decay in the variable $Y^{\mathfrak{p}}$. Using (10.8.26) and proceeding as before, the same arguments show that for any $k' \in \mathbf{N}$, these results also apply to $r_{b,t}^X\left((x_0, Y^{\mathfrak{p}}), (x', Y^{TX'})\right) \left(1 + \left|Y^{TX'}\right|^{2k'}\right)$. The proof of our theorem is completed. $\qquad\square$

Remark 12.7.5. In the above argument, we used the fact that X is a symmetric space, because the coordinate system centered at x' can be chosen uniformly on the full X. Also the above proof gives a uniform Gaussian decay of $r_{b,t}^X$ in the variables $Y^{TX}, Y^{TX'}$. We will give a related proof of this fact in Theorem 13.2.4.

Also note that in equation (12.5.2), we chose the control v so as to minimize the norm $|v|_{L_2}$. However, we can instead assume that v vanishes on $[0, t/2]$ and pick the control v such that it minimizes $|v|_{L_2}$ among such controls. This problem is again a problem that was already considered in section 10.3, with the time interval $[t/2, t]$ replacing the time interval $[0, t]$. The considerations of the previous sections still hold in this case. Of course, using the fact that the kernels $s_{b,\cdot}, r_{b,\cdot}$ form semigroups, the interpretation of such manipulations is quite obvious.

12.8 The limit as $b \to 0$

Theorem 12.8.1. *For $t > 0$, as $b \to 0$, we have the pointwise convergence,*

$$r_{b,t}^X\left((x, Y^{TX}), (x', Y^{TX'})\right)$$

$$\to \frac{1}{\pi^{m/2}} p_t\left(x, x'\right) \exp\left(-\frac{1}{2}\left(\left|Y^{TX}\right|^2 + \left|Y^{TX'}\right|^2\right)\right). \tag{12.8.1}$$

Proof. In (12.8.1), we may as well take $x = x_0, Y^{TX} = Y^{\mathfrak{p}} \in \mathfrak{p}$. To establish (12.8.1), we will show that there is a corresponding narrow convergence of measures, which in view of Theorem 12.7.4, immediately leads to (12.8.1). To establish the narrow convergence, we will use a method inspired by Stroock-Varadhan [StV72]. In the whole proof, we will assume that $0 < b \leq 1$.

Let $f : X \to \mathbf{R}$ be a continuous function with derivatives of order ≤ 2 that are uniformly bounded and continuous . Put

$$M_{f,t} = f(x_t) - f(x_0) - b\left(\nabla_{Y_t^{TX}} f(x_t) - \nabla_{Y^{\mathfrak{p}}} f(x_0)\right)$$
$$- \int_0^t \nabla_{Y_s^{TX}}^{TX} \nabla_{Y_s^{TX}} f(x_s) \, ds. \quad (12.8.2)$$

By (12.3.19), $M_{f,t}$ is a martingale given by

$$M_{f,t} = \int_0^t \nabla_{\delta w} f(x_s). \quad (12.8.3)$$

By (12.1.17), (12.8.3), for $0 \leq s \leq t, 1 < p < +\infty$,

$$\|M_{f,t} - M_{f,s}\|_p \leq C_{f,p} |t - s|^{1/2}. \quad (12.8.4)$$

Moreover, if a_s is a real continuous function, for $1 < p < +\infty$,

$$\left|\int_s^t a_u \, du\right|^p \leq \int_s^t |a_u|^p \, du \, (t - s)^{p-1}. \quad (12.8.5)$$

Using (10.8.26) and (12.8.5), we get

$$\left\|\int_s^t \nabla_{Y_u^{TX}}^{TX} \nabla_{Y_u^{TX}} f(x_u) \, du\right\|_p \leq C_{f,p}\left(1 + |Y^{TX}|^2\right)(t - s). \quad (12.8.6)$$

For $s \leq t$, let τ_t^s denote parallel transport from $T_{x_s} X$ into $T_{x_t} X$ along x. with respect to the Levi-Civita connection, and let τ_s^t be its inverse. Then

$$\nabla_{Y_t^{TX}} f(x_t) - \nabla_{Y_s^{TX}} f(x_s)$$
$$= \langle \tau_s^t Y_t^{TX} - Y_s^{TX}, \tau_s^t df(x_t)\rangle + \langle Y_s^{TX}, \tau_s^t df(x_t) - df(x_s)\rangle. \quad (12.8.7)$$

Conditionally on $\mathcal{F}_s$, the probability law of $\tau_s^t Y_t^{TX}$ is a Gaussian with mean $e^{-(t-s)/b^2} Y_s^{TX}$, and variance $\left(1 - e^{-2(t-s)/b^2}\right)/2$, so that using (10.8.26), we get

$$\left\|\tau_s^t Y_t^{TX} - Y_s^{TX}\right\|_p \leq C_p\left(1 - e^{-(t-s)/b^2}\right)(1 + |Y^{TX}|)$$
$$+ C_p\left(1 - e^{-2(t-s)/b^2}\right)^{1/2}. \quad (12.8.8)$$

Moreover,

$$b\left(1 - e^{-(t-s)/b^2}\right) \leq \sqrt{t - s}. \quad (12.8.9)$$

Indeed, (12.8.9) holds for $b \leq \sqrt{t - s}$, and for $b > \sqrt{t - s}$,

$$b\left(1 - e^{-(t-s)/b^2}\right) \leq \frac{t - s}{b} \leq \sqrt{t - s}. \quad (12.8.10)$$

Finally, by the first inequality in (12.8.10), we get

$$b\left(1 - e^{-2(t-s)/b^2}\right)^{1/2} \le \sqrt{2(t-s)}. \qquad (12.8.11)$$

By (12.8.8)–(12.8.11), we obtain

$$b\left\|\tau_s^t Y_t^{TX} - Y_s^{TX}\right\|_p \le C_p\left(1 + \left|Y^{TX}\right|\right)\sqrt{t-s}. \qquad (12.8.12)$$

Also for $s \le t$,

$$\tau_s^t df(x_t) - df(x_s) = \int_s^t \tau_s^u \nabla_{Y_u^{TX}/b}^{TX} df(x_u)\, du. \qquad (12.8.13)$$

By (10.8.26), (12.8.5), and (12.8.13), we obtain

$$\left\|b\left\langle Y_s^{TX}, \tau_s^t df(x_t) - df(x_s)\right\rangle\right\|_p \le C_{f,p}\left(1 + \left|Y^{TX}\right|^2\right)(t-s). \qquad (12.8.14)$$

By (12.8.2), (12.8.4), (12.8.6), (12.8.7), (12.8.12), and (12.8.14), for $1 < p < +\infty, |t - s| \le 1$,

$$\|f(x_t) - f(x_s)\|_p \le C_{f,p}\left(1 + \left|Y^{TX}\right|^2\right)\sqrt{t-s}. \qquad (12.8.15)$$

Let k be a left and right K-invariant real smooth function on G with values in $\mathbf{R}$ that vanishes on K. Then k descends to a smooth function from X into $\mathbf{R}$, which is invariant under the left action of K on X. If $g, g' \in X, pg = x, pg' = x'$, put

$$k(x, x') = k\left(g^{-1}g'\right). \qquad (12.8.16)$$

Then $k(x, x')$ is a smooth function on $X \times X$, which vanishes on the diagonal. We may and we will assume that when $d(x, x') \ge 1$, k is equal to 1.

By using (12.8.15) with $f(y) = k(x_0, y)$, for $0 \le t \le 1$,

$$\|k(x_0, x_t)\|_p \le C_{k,p}\left(1 + \left|Y^{TX}\right|^2\right)\sqrt{t}. \qquad (12.8.17)$$

More generally, using (10.8.26), (12.8.17), and the Markov property of the process $\left(x_., Y^{TX}\right)$, we find that for $0 < s \le t, t - s \le 1$,

$$\|k(x_s, x_t)\|_p \le C_{k,p}\left(1 + \left|Y^{TX}\right|^2\right)\sqrt{t-s}. \qquad (12.8.18)$$

For $t \ge 0$, let $\mathcal{G}_t$ be the σ-algebra on $\mathcal{C}(\mathbf{R}_+, X)$ generated by $x_s, s \le t$, and let $\mathcal{G}_t|_{t \ge 0}$ be the corresponding filtration. Given $T \ge 0$, we use a similar notation over $\mathcal{C}([0, T], X)$. For $T \ge 0$, let $Q_{b,T}$ be the probability law of $x_.$ on $\mathcal{C}([0, T], X)$. Using (12.8.18) and [StV79, page 61, 2.4.2], we deduce from (12.8.18) that for $0 < b \le M$, the $Q_{b,T}$ form a relatively compact set of probability measures for the topology of narrow convergence.

Given $T > 0$, let b_n be a decreasing sequence in $\mathbf{R}_+$ that converges to 0, and is such that $Q_{b_n,T}$ converges to $Q_{0,T}$ for the topology of the narrow convergence. We will show that $Q_{0,T}$ is precisely the probability law of the Brownian motion starting at x_0 on $\mathcal{C}([0, T], X)$.

Let $f : X \to \mathbf{R}$ be a smooth function with compact support. Take s_0, t_0, t such that $0 \le s_0 < t_0 \le t \le T$. Since $M_{f,.}$ is an $\mathcal{F}_.$-martingale, it is also a $\mathcal{G}_{\cdot}$-martingale. Let $H(x_.) : \mathcal{C}([0,T],X) \to \mathbf{R}$ be a bounded continuous function that is $\mathcal{G}_{s_0}$-measurable. Then

$$E^{Q_{b,T}}[H(x_.)M_{f,t}] = E^{Q_{b,T}}[H(x_.)M_{f,t_0}]. \qquad (12.8.19)$$

By (10.8.26), as $n \to +\infty$,

$$E^{Q_{b_n,T}}\left[H(x_.)b_n\left(\nabla_{Y_t^{TX}}f(x_t) - \nabla_{Y_{t_0}^{TX}}f(x_{t_0})\right)\right] \to 0. \qquad (12.8.20)$$

Take $b > 0$ small enough so that $t_0 - b \geq s_0$. To make the notation simpler, we will omit the parallel transport operators along $x_.$. In other words, along $x_.$, TX is trivialized by parallel transport with respect to ∇^{TX}. Then

$$\int_{t_0}^{t} \nabla_{Y_s^{TX}}^{TX} \nabla_{Y_s^{TX}} f(x_s)\,ds = \int_{t_0}^{t} \nabla_{Y_s^{TX}}^{TX} \nabla_{Y_s^{TX}} \left(f(x_s) - f(x_{s-b})\right)\,ds$$

$$+ \int_{t_0-b}^{t-b} \nabla_{Y_{s+b}^{TX}}^{TX} \nabla_{Y_{s+b}^{TX}} f(x_s)\,ds. \qquad (12.8.21)$$

By (10.8.26), (12.8.5), and (12.8.15), for $0 < b \leq 1$, we get

$$\left\| \int_{t_0}^{t} \nabla_{Y_s^{TX}}^{TX} \nabla_{Y_s^{TX}} \left(f(x_s) - f(x_{s-b})\right)\,ds \right\|_p \leq C_{f,p}\left(1 + |Y^{TX}|^4\right)\sqrt{b}.$$

$$(12.8.22)$$

As we saw after (12.8.7), conditionally on $\mathcal{F}_s$, the probability law of Y_{s+b} is a Gaussian centered at $e^{-1/b}Y_s^{TX}$ and with covariance equal to $\left(1 - e^{-2/b}\right)/2$. Since $t_0 - b \geq s_0$, by the above, we get

$$E^{Q_{b,T}}\left[H(x_.)\int_{t_0-b}^{t-b} \nabla_{Y_{s+b}^{TX}}^{TX} \nabla_{Y_{s+b}^{TX}} f(x_s)\,ds\right]$$

$$= e^{-2/b} E\left[H(x_.)\int_{t_0-b}^{t-b} \nabla_{Y_s^{TX}}^{TX} \nabla_{Y_s^{TX}} f(x_s)\,ds\right]$$

$$+ \left(1 - e^{-2/b}\right) E\left[H(x_.)\int_{t_0-b}^{t-b} \frac{1}{2}\Delta^X f(x_s)\,ds\right]. \qquad (12.8.23)$$

We make $b = b_n$ in (12.8.23). By (10.8.26), as $n \to +\infty$, the first term in the right-hand side of (12.8.23) tends to 0. Since $Q_{b_n,T}$ converges to $Q_{0,T}$ in the sense of narrow convergence, as $n \to +\infty$,

$$E^{Q_{b_n,T}}\left[H(x_.)\int_{t_0-b}^{t-b} \frac{1}{2}\Delta^X f(x_s)\,ds\right] \to E^{Q_{0,T}}\left[H(x_.)\int_{t_0}^{t} \frac{1}{2}\Delta^X f(x_s)\,ds\right].$$

$$(12.8.24)$$

By (12.8.21)–(12.8.24), as $n \to +\infty$,

$$E^{Q_{b_n,T}}\left[H(x_.)\int_{t_0}^{t} \nabla_{Y_s^{TX}}^{TX} \nabla_{Y_s^{TX}} f(x_s)\,ds\right]$$

$$\to E^{Q_{0,T}}\left[H(x_.)\int_{t_0}^{t} \frac{1}{2}\Delta^X f(x_s)\,ds\right]. \qquad (12.8.25)$$

Put

$$M_{0,f,t} = f(x_t) - f(x_0) - \int_0^t \frac{1}{2}\Delta^X f(x_s)\, ds. \tag{12.8.26}$$

Using (12.8.2), (12.8.19), (12.8.20), and (12.8.25) we get

$$E^{Q_{0,T}}[H(x.)\, M_{0,t,f}] = E^{Q_{0,T}}[H(x.)\, M_{0,t_0,f}]. \tag{12.8.27}$$

Since (12.8.27) is valid for any bounded continuous function H that is $\mathcal{G}_{s_0}$-measurable with $0 \le s_0 < t_0$, (12.8.27) still holds for $s_0 = t_0$. Therefore $M_{0,f,\cdot}$ is a $\mathcal{G}_\cdot$-martingale. Using a result of Stroock and Varadhan [StV79, chapter 7], $Q_{0,T}$ is just the Brownian measure on $\mathcal{C}([0,T],X)$, with initial value x_0. Since $Q_{0,T}$ is uniquely determined, it follows that as $b \to 0$, $Q_{b,T}$ converges to $Q_{0,T}$.

For $T = +\infty$, we will write Q_b instead of $Q_{b,+\infty}$. It follows from the above that for any $t \ge 0$, as $b \to 0$,

$$E^{Q_b}[f(x_t)] \to E^{Q_0}[f(x_t)]. \tag{12.8.28}$$

Let $F : \mathcal{X} \to \mathbf{R}$ be a smooth function with compact support. Given $t > 0$, and $b > 0$ such that $b \le t$,

$$E\left[F\left(x_t, Y_t^{TX}\right)\right] = E\left[F\left(x_{t-b}, \tau_{t-b}^t Y_t^{TX}\right)\right]$$
$$+ E\left[F\left(x_t, Y_t^{TX}\right) - F\left(x_{t-b}, \tau_{t-b}^t Y_t^{TX}\right)\right]. \tag{12.8.29}$$

Using the above arguments on the probability law of $\tau_{t-b}^t Y_t^{TX}$ conditional on $\mathcal{F}_{t-b}$, we get

$$E\left[F\left(x_{t-b}, \tau_{t-b}^t Y_t^{TX}\right)\right]$$
$$= E\left[\int_{T_{x_{t-b}}X} F\left(x_{t-b}, e^{-1/b} Y_{t-b}^{TX} + \left(1 - e^{-2/b}\right)^{1/2} Z^{TX}\right) \right.$$
$$\left. \exp\left(-|Z^{TX}|^2\right) \frac{dZ^{TX}}{\pi^{m/2}}\right]. \tag{12.8.30}$$

Moreover,

$$\left| F\left(x_{t-b}, e^{-1/b} Y_{t-b}^{TX} + \left(1 - e^{-2/b}\right)^{1/2} Z^{TX}\right) - F\left(x_{t-b}, Z^{TX}\right) \right|$$
$$\le C_F\left(e^{-1/b}|Y_{t-b}^{TX}| + e^{-2/b}|Z^{TX}|\right). \tag{12.8.31}$$

By (10.8.26), (12.8.30), and (12.8.31), for $0 < b \le 1$,

$$E\left[F\left(x_{t-b}, \tau_{t-b}^t Y_t^{TX}\right)\right]$$
$$- E^{Q_b}\left[\int_{T_{x_{t-b}}X} F\left(x_{t-b}, Z^{TX}\right) \exp\left(-|Z^{TX}|^2\right) \frac{dZ^{TX}}{\pi^{m/2}}\right]$$
$$= \mathcal{O}\left(e^{-1/b}(1 + |Y^\mathbf{p}|)\right). \tag{12.8.32}$$

Also since F has compact support, we deduce from (12.8.15) that if in the second term in the left-hand side of (12.8.32), we replace x_{t-b} by x_t, this introduces an error that is $\mathcal{O}\left(\sqrt{b}\left(1+|Y^{\mathfrak{p}}|^2\right)\right)$. Using (12.8.28), (12.8.32), and the above argument, we find that as $b \to 0$,

$$E\left[F\left(x_{t-b}, \tau_{t-b}^t Y_t^{TX}\right)\right]$$
$$\to E^{Q_0}\left[\int_{T_{x_t}X} F\left(x_t, Z^{TX}\right) \exp\left(-\left|Z^{TX}\right|^2\right) \frac{dZ^{TX}}{\pi^{m/2}}\right]. \quad (12.8.33)$$

Now we will estimate the second term in the right-hand side of (12.8.29). We identify F to the smooth function $\widetilde{F} = p^* F$, which maps $G \times \mathfrak{p}$ into $\mathbf{R}$. The function $\widetilde{F}$ is such that if $k \in K$,

$$\widetilde{F}\left(gk, \mathrm{Ad}\left(k^{-1}\right)Y\right) = \widetilde{F}\left(g, Y\right). \quad (12.8.34)$$

Using the notation in (12.2.8), we get

$$F\left(x_t, Y_t^{TX}\right) - F\left(x_{t-b}, \tau_{t-b}^t Y_t^{TX}\right) = \widetilde{F}\left(g_t, Y_t^{\mathfrak{p}}\right) - \widetilde{F}\left(g_{t-b}, Y_t^{\mathfrak{p}}\right). \quad (12.8.35)$$

Recall that θ denotes the Cartan involution. We equip $\mathfrak{g} = \mathfrak{p} \oplus \mathfrak{k}$ with the scalar product $-B\left(\cdot, \theta\cdot\right)$. Also we equip TG with the associated left-invariant metric. We denote by d^G the corresponding distance on G. Set

$$\underline{d}^G = \inf\left(d^G, 1\right). \quad (12.8.36)$$

Since F has compact support, we get

$$\left|\widetilde{F}\left(g_t, Y_t^{\mathfrak{p}}\right) - \widetilde{F}\left(g_{t-b}, Y_t^{\mathfrak{p}}\right)\right| \leq C_F \underline{d}^G\left(g_{t-b}, g_t\right). \quad (12.8.37)$$

Let $u : G \to \mathbf{R}$ be a smooth function that is constant outside of the ball of center 0 and radius 1. Instead of (12.8.2), we now set

$$M_{u,t} = u\left(g_t\right) - u\left(1\right) - b\left(\nabla_{Y_t^{\mathfrak{p}}} u\left(g_t\right) - \nabla_{Y^{\mathfrak{p}}} u\left(1\right)\right) - \int_0^t \nabla_{Y_s^{\mathfrak{p}}} \nabla_{Y_s^{\mathfrak{p}}} u\left(g_s\right) ds. \quad (12.8.38)$$

By (12.3.20),

$$M_{u,t} = \int_0^t \nabla_{\delta w_s} u\left(g_s\right) ds. \quad (12.8.39)$$

By proceeding as in (12.8.4)–(12.8.15), for $0 < b \leq 1, 1 < p < +\infty, s \leq t, |t - s| \leq 1$, we get the analogue of (12.8.15),

$$\left\|u\left(g_t\right) - u\left(g_s\right)\right\|_p \leq C_{u,p}\left(1 + |Y^{\mathfrak{p}}|^2\right)\sqrt{t-s}. \quad (12.8.40)$$

Using (10.8.26) and the Markov property of the process $\left(g_., Y^{\mathfrak{p}}\right)$, under the same conditions as above, we get

$$\left\|u\left(g_s^{-1}g_t\right) - u\left(1\right)\right\|_p \leq C_{u,p}\left(1 + |Y^{\mathfrak{p}}|^2\right)\sqrt{t-s}. \quad (12.8.41)$$

By (12.8.41), we deduce that

$$\left\|\underline{d}_G\left(g_s, g_t\right)\right\|_p \leq C_p\left(1 + |Y^{\mathfrak{p}}|^2\right)\sqrt{t-s}. \quad (12.8.42)$$

By (12.8.35)–(12.8.42), we get

$$\left| E\left[F\left(x_t, Y_t^{TX} \right) - F\left(x_{t-b}, \tau_{t-b}^t Y_t^{TX} \right) \right] \right| \le C_F \sqrt{b}. \tag{12.8.43}$$

By (12.8.29), (12.8.33), (12.8.43), as $b \to 0$,

$$E\left[F\left(x_t, Y_t^{TX} \right) \right] \to E^{Q_0}\left[\int_{T_{x_t} X} F\left(x_t, Z^{TX} \right) \exp\left(-\left| Z^{TX} \right|^2 \right) \frac{dZ^{TX}}{\pi^{m/2}} \right]. \tag{12.8.44}$$

By (12.1.6), (12.2.10), we can rewrite (12.8.44) in the form

$$\int_{\mathcal{X}} F\left(x', Y^{TX\prime} \right) s_{b,t}^X \left(\left(x_0, Y^{\mathfrak{p}} \right), \left(x', Y^{TX\prime} \right) \right) dx' dY^{TX\prime}$$
$$\to \int_{\mathcal{X}} F\left(x', Y^{TX\prime} \right) p_t\left(x_0, x' \right) \exp\left(-\left| Y^{TX\prime} \right|^2 \right) dx' \frac{dY^{TX\prime}}{\pi^{m/2}}. \tag{12.8.45}$$

By (11.5.10) and by Theorem 12.7.4, for $0 < b \le 1$, given $\left(x, Y^{TX} \right)$, as functions of $\left(x', Y^{TX\prime} \right)$, the functions $s_{b,t}^X \left(\left(x_0, Y^{\mathfrak{p}} \right), \left(x', Y^{TX\prime} \right) \right)$ are uniformly bounded on compact subsets of $\mathcal{X}$ as well as their derivatives of arbitrary order in these variables. Using (12.8.45), as $b \to 0$, we get the pointwise convergence

$$s_{b,t}^X \left(\left(x_0, Y^{TX}, \left(x', Y^{TX\prime} \right) \right) \right) \to \pi^{-m/2} p_t\left(x_0, x' \right) \exp\left(-\left| Y^{TX\prime} \right|^2 \right). \tag{12.8.46}$$

By (11.5.10) and (12.8.46), we get (12.8.1). The proof of our theorem is completed. □

Remark 12.8.2. The proof of Theorem 12.8.1 shows that given $\left(x, Y^{TX} \right) \in \mathcal{X}$, the convergence in (12.8.1) is uniform when $\left(x', Y^{TX\prime} \right)$ varies in compact subsets of $\mathcal{X}$, and also that the derivatives of arbitrary order in $\left(x', Y^{TX\prime} \right)$ also converge uniformly on the compact subsets. The proof gives a probabilistic counterpart to arguments of Bismut-Lebeau [BL08, chapter 17].

Establishing the narrow convergence of $Q_{b,T}$ to $Q_{0,T}$ is difficult because, contrary to equations (12.2.1) and (12.2.8), which can be solved pointwise for any $w. \in \mathcal{C}\left(\mathbf{R}_+, \mathfrak{p} \right)$, equations (12.1.3) and (12.1.4) are genuinely stochastic differential equations. Still by proceeding as in [B81b, Théorème 1.2], we can replace the narrow convergence of probability measures by uniform convergence in probability over the probability space $\left(\mathcal{C}\left(\mathbf{R}_+, \mathfrak{p} \right), \mathcal{F}_\infty, P \right)$. This way, one can avoid the apparent duplication of arguments on the martingales $M_{f,.}$ and $M_{u,.}$.

The method used in the proof of Theorem 12.8.1 also shows that as $b \to 0$, the probability law of $g.$ on $\mathcal{C}\left([0, T], G \right)$ converges narrowly to the probability law of the solution of equation (12.1.4). The proof uses the bound in (12.8.42), and the same martingale argument as in the proof of Theorem 12.8.1. Let $P_{\left(g_t, Y_t^{TX} \right)}$ be the probability law of $\left(g_t, Y_t^{TX} \right)$, let $P_{g_t}^0$ be the probability law of g_t in (12.1.4). For $t > 0$, the same argument as in the proof of Theorem 12.8.1 shows that as $b \to 0$, we have the narrow convergence of probability measures on $G \times \mathfrak{p}$,

$$P_{\left(g_t, Y_t^{TX}\right)} \to P_{g_t}^0 \exp\left(-\left|Y^{TX}\right|^2\right) \frac{dY^{TX}}{\pi^{m/2}}. \tag{12.8.47}$$

12.9 The rough estimates as $b \to +\infty$

For $a > 0$, set

$$K_a s\left(x, Y^{TX}\right) = s\left(x, a Y^{TX}\right). \tag{12.9.1}$$

Put

$$\underline{\mathcal{A}}_b^X = K_b \mathcal{A}_b^X K_b^{-1}. \tag{12.9.2}$$

By (11.1.1), we get

$$\underline{\mathcal{A}}_b^X = \frac{1}{2}\left(-\frac{1}{b^4}\Delta^V + \left|Y^{TX}\right|^2 - \frac{m}{b^2}\right) - \nabla_{Y^{TX}}. \tag{12.9.3}$$

For $t > 0$, let $\underline{r}_{b,t}^X\left(\left(x, Y^{TX}\right), \left(x', Y^{TX\prime}\right)\right)$ be the smooth kernel associated with $\exp\left(-t\underline{\mathcal{A}}_b^X\right)$. Then

$$\underline{r}_{b,t}^X\left(\left(x, Y^{TX}\right), \left(x', Y^{TX\prime}\right)\right) = b^m r_{b,t}^X\left(\left(x, b Y^{TX}\right), \left(x', b Y^{TX\prime}\right)\right). \tag{12.9.4}$$

Take $Y^{\mathfrak{p}} \in \mathfrak{p}$. Instead of (12.2.1), we consider the stochastic differential equation,

$$\dot{x} = Y^{TX}, \qquad\qquad \dot{Y}^{TX} = \frac{\dot{w}}{b^2}, \tag{12.9.5}$$

$$x_0 = p1, \qquad\qquad Y_0^{TX} = Y^{\mathfrak{p}}.$$

The first line of (12.9.5) can be written in the form

$$\ddot{x} = \frac{\dot{w}}{b^2}, \qquad\qquad Y^{TX} = \dot{x}. \tag{12.9.6}$$

By (12.9.5), we obtain

$$\tau_0^t Y_t^{TX} = Y^{\mathfrak{p}} + \frac{1}{b^2} w_t. \tag{12.9.7}$$

By (12.9.7), the probability law of $\tau_0^t Y_t^{TX}$ is a Gaussian centered at $Y^{\mathfrak{p}}$, and with variance $\frac{t}{b^4}$. Then we get an analogue of (10.8.10), i.e., for $1 < p < +\infty$, $b > 0, t \geq 0$,

$$\left\|Y_t^{TX}\right\|_p \leq \left|Y^{\mathfrak{p}}\right| + C_p \frac{\sqrt{t}}{b^2}. \tag{12.9.8}$$

Let $F : \mathcal{X} \to \mathbf{R}$ be a smooth function with compact support. By using Proposition 11.5.2 and the Itô calculus as in the proof of Theorem 12.2.1, and the Feynman-Kac formula, for $t > 0$, we get

$$\exp\left(-t\underline{\mathcal{A}}_b^X\right) F\left(x_0, Y^{\mathfrak{p}}\right) = E\left[\exp\left(\frac{mt}{2b^2} - \frac{1}{2}\int_0^t \left|Y^{TX}\right|^2 ds\right) F\left(x_t, Y_t^{TX}\right)\right]. \tag{12.9.9}$$

Observe that when proving (12.9.9), the considerations of integrability in the proof of Theorem 12.2.1 are no longer necessary.

Let k be a multi-index, where the indices run in $1, \ldots, 2m$. Let $\nabla^{TX,k}_{(x',Y^{TX'})}$ be the corresponding covariant derivative of order k with respect to ∇^{TX} in the variables $(x', Y^{\widetilde{TX'}})$.

Theorem 12.9.1. *Given ϵ, M with $0 < \epsilon \le M$, there exists $C_{\epsilon,M} > 0$ such that for $b \ge 1, \epsilon \le t \le M$,*

$$\mathbf{r}^X_{b,t}\left((x, Y^{TX}), (x', Y^{TX'})\right) \le C_{\epsilon,M} b^{4m}. \tag{12.9.10}$$

For $b \ge 1, \epsilon \le t \le M$, for any multi-index k,

$$\left|\nabla^{TX,k}_{(x',Y^{TX'})} \mathbf{r}^X_{b,t}\left((x, Y^{TX}), (x', Y^{TX'})\right)\right| / b^{4m+2|k|}\Big| \tag{12.9.11}$$

is uniformly rapidly decreasing on $\mathcal{X} \times \mathcal{X}$.

Proof. If h, B are taken as in section 12.1, instead of (12.4.3), we consider the system

$$b^2\left(\ddot{J}^{TX} + R^{TX}\left(J^{TX}, \dot{x}\right)\dot{x}\right) = h + \left[B + \varpi, \frac{\delta w}{ds}\right], \tag{12.9.12}$$

$$\dot{\varpi} = \left[J^{TX}, \dot{x}\right], \; J^{TX}_0 = 0, \; \dot{J}^{TX}_0 = 0, \; \varpi_0 = 0.$$

By proceeding as in the proof of Theorem 12.4.1, instead of (12.4.4), we get

$$E\left[\exp\left(-\frac{1}{2}\int_0^t |Y^{TX}|^2 \, ds\right)\left\langle dF\left(x_t, Y^{TX}_t\right), \left(J^{TX}_t, \dot{J}^{TX}_t\right)\right\rangle\right]$$

$$= E\left[\exp\left(-\frac{1}{2}\int_0^t |Y^{TX}|^2 \, ds\right) F\left(x_t, Y^{TX}_t\right)\right.$$

$$\left.\left(\int_0^t \left\langle Y^{TX}, \dot{J}^{TX}\right\rangle ds + \int_0^t \langle h, \delta w\rangle\right)\right]. \tag{12.9.13}$$

Consider the system

$$b^2\left(\ddot{J}^{TX} + R^{TX}\left(J^{TX}, \dot{x}\right)\dot{x}\right) = h, \qquad \dot{\varpi} = \left[J^{TX}, \dot{x}\right], \tag{12.9.14}$$

$$J^{TX}_0 = 0, \dot{J}^{TX}_0 = 0, \qquad\qquad\qquad \varpi_0 = 0.$$

Then (12.9.14) is a special case of (12.9.12), so that (12.9.13) still holds.

As an aside, let us note that the integration by parts formulas in (12.4.4), (12.4.6) and in (12.9.13), (12.9.14) can be shown to be equivalent.

Now we proceed as in [BL08, section 14.3]. Set

$$\psi_s = s^2\left(3 - 2s\right), \qquad \overline{\psi}_s = s^2\left(-1 + s\right). \tag{12.9.15}$$

Then

$$\psi_0 = \psi'_0 = 0, \qquad\qquad \psi_1 = 1, \psi'_1 = 0, \tag{12.9.16}$$

$$\overline{\psi}_0 = \overline{\psi}'_0 = 0, \qquad\qquad \overline{\psi}_1 = 0, \overline{\psi}'_1 = 1.$$

For $e, f \in \mathfrak{p}$, set

$$J_s^{TX} = \psi_{s/t} e + t \overline{\psi}_{s/t} f, \tag{12.9.17}$$

so that

$$J_0^{TX} = 0, \; \dot{J}_0^{TX} = 0, \qquad J_t^{TX} = e, \; \dot{J}_t^{TX} = f. \tag{12.9.18}$$

Incidentally, note that since J_s^{TX} is a polynomial of degree 3 in the variable s, J^{TX} is the solution of a variational problem associated with the functional $H_{\infty,t}(J) = \frac{1}{2} \int_0^t \left| \ddot{J}^{TX} \right|^2 ds$.

By (12.9.13), (12.9.14), and (12.9.18), we get

$$E\left[\exp\left(-\frac{1}{2} \int_0^t \left| Y^{TX} \right|^2 ds \right) \left\langle dF\left(x_t, Y_t^{TX} \right), \tau_t^0 (e, f) \right\rangle \right]$$

$$= E\left[\exp\left(-\frac{1}{2} \int_0^t \left| Y^{TX} \right|^2 ds \right) F\left(x_t, Y_t^{TX} \right)\right.$$

$$\left. \left(\int_0^t \left\langle Y^{TX}, \dot{J}^{TX} \right\rangle ds + b^2 \int_0^t \left\langle \ddot{J}^{TX} + R^{TX} \left(J^{TX}, \dot{x} \right) \dot{x}, \delta w \right\rangle \right) \right]. \tag{12.9.19}$$

Formulas similar to (12.5.8) can also be obtained. In particular, inspection of (12.1.17), (12.9.5), and (12.9.8) shows that for $1 < p < +\infty, b \geq 1, \epsilon \leq t \leq M$, the L_p norm of the relevant term M_t to the right of $F\left(x_t, Y_t^{TX} \right)$ in the right-hand side of (12.9.19) is dominated by $C_{\epsilon,M,p} b^2 \left(1 + |Y^{\mathfrak{p}}|^2 \right)$.

In the integration by parts formula where k derivatives of F appear in the left-hand side, the corresponding bound is given by $C_{\epsilon,M,p,k} b^{2k} \left(1 + |Y^{\mathfrak{p}}|^{2k} \right)$.

Now we use the same notation as in the proof of Theorem 12.7.4. In particular we fix $x' \in X$, and we choose the same coordinate system on $\mathcal{X}$. Also we define $g_{\alpha,\beta}(y, Y^{\mathfrak{p}'})$ as in (12.7.8). Set

$$\underline{\Phi}_{b,t,Y^{\mathfrak{p}},x'}(\alpha, \beta) = \int_{\mathfrak{p} \times \mathfrak{p}} \underline{r}_{b,t}^X \left((x_0, Y^{\mathfrak{p}}), (y, Y^{\mathfrak{p}'}) \right) g_{\alpha,\beta}(y, Y^{\mathfrak{p}'}) \delta(y) \, dy dY^{\mathfrak{p}'}. \tag{12.9.20}$$

By the above, we find that given $k \in \mathbf{N}$, there exists $C_{\epsilon,M} > 0$ such that for $b \geq 1, \epsilon \leq t \leq M$,

$$\left(1 + \left(|\alpha|^k + |\beta|^k \right) / b^{2k} \right) \left| \underline{\Phi}_{b,t,Y^{\mathfrak{p}},x'}(\alpha, \beta) \right| \leq C_{\epsilon,M,k} \left(1 + |Y^{\mathfrak{p}}|^{4k} \right). \tag{12.9.21}$$

Some care has to be given to the derivation of (12.9.21). Indeed, in the coordinates $(y, Y^{\mathfrak{p}'})$, e is a combination with smooth bounded coefficients of $\frac{\partial}{\partial y_i} - \nabla_{\Gamma_y^{TX} \left(\frac{\partial}{\partial y^i} \right) Y^{\mathfrak{p}'}}^V$, where ∇^V denotes differentiation in the variable $Y^{\mathfrak{p}'}$, and Γ^{TX} denotes the connection form for ∇^{TX}. Equivalently, $\frac{\partial}{\partial y^i}$ is a smooth linear combination of (e, f), the coefficients growing at most linearly in $Y^{\mathfrak{p}'}$. This explains the power $4k$ in the right-hand side of (12.9.21) instead of the more natural $2k$. This issue did not appear explicitly in (12.7.10), where polynomials in $Y^{\mathfrak{p}}$ were absorbed by a Gaussian factor.

The critical fact is that the constant $C_{\epsilon,M,k} > 0$ in the right-hand side of (12.7.10) does not depend on $Y^{\mathfrak{p}}, x'$. By taking (12.7.10) with $k = 2m + 1$ and using an inverse Fourier transform, from (12.7.10), we get

$$\underline{r}^X_{b,t}\left((x_0, Y^{\mathfrak{p}}), (x', Y^{TX\prime})\right) \leq C_{\epsilon,M} b^{4m}\left(1 + |Y^{\mathfrak{p}}|^{4(2m+1)}\right). \qquad (12.9.22)$$

Of course (12.9.22) remains valid when replacing $(x_0, Y^{\mathfrak{p}})$ by any $(x, Y^{TX}) \in \mathcal{X}$, so that

$$\underline{r}^X_{b,t}\left((x, Y^{TX}), (x', Y^{TX\prime})\right) \leq C_{\epsilon,M} b^{4m}\left(1 + |Y^{TX}|^{4(2m+1)}\right). \qquad (12.9.23)$$

Using (12.9.8) and proceeding as before, in (12.9.23), for $k' \in \mathbf{N}$, we replace $\underline{r}^X_{b,t}\left((x, Y^{TX}), (x', Y^{TX\prime})\right)$ by $\underline{r}^X_{b,t}\left((x, Y^{TX}), (x', Y^{TX\prime})\right)\left(1 + |Y^{TX\prime}|^{2k'}\right)$, and still obtain an estimate similar to (12.9.22), i.e.,

$$\underline{r}^X_{b,t}\left((x, Y^{TX}), (x', Y^{TX\prime})\right)\left(1 + |Y^{TX\prime}|^{2k'}\right)$$
$$\leq C_{\epsilon,M,k'} b^{4m}\left(1 + |Y^{TX}|^{4(2m+1)+2k'}\right). \qquad (12.9.24)$$

Clearly,

$$\underline{r}^X_{b,t}\left((x, Y^{TX}), (x', Y^{TX\prime})\right) = \int_{\mathcal{X}} \underline{r}^X_{b,t/2}\left((x, Y^{TX}), (z, Z^{TX})\right)$$
$$\underline{r}^X_{b,t/2}\left((z, Z^{TX}), (x', Y^{TX\prime})\right) dz dZ^{TX}. \qquad (12.9.25)$$

By (12.9.22), (12.9.25), we get

$$\underline{r}^X_{b,t}\left((x, Y^{TX}), (x', Y^{TX\prime})\right) \leq C' b^{4m} \int_{\widehat{\mathcal{X}}} \underline{r}^X_{b,t/2}\left((x, Y^{TX}), (z, Z^{TX})\right)$$
$$\left(1 + |Z^{TX}|^{2(2m+1)}\right) dz dZ^{TX}. \qquad (12.9.26)$$

By (10.4.9), (10.4.11), and (12.9.3), we get the crucial identity,

$$\int_{\mathcal{X}} \underline{r}^X_{b,t/2}\left((x_0, Y^{\mathfrak{p}}), (z, Z^{TX})\right)\left(1 + |Z^{TX}|^{2(2m+1)}\right) dz dZ^{TX}$$
$$= \int_{\mathfrak{p}} h^{\mathfrak{p}}_{t/2b^2}\left(bY^{\mathfrak{p}}, Z^{\mathfrak{p}}\right)\left(1 + \left|\frac{Z^{\mathfrak{p}}}{b}\right|^{2(2m+1)}\right) dZ^{\mathfrak{p}}. \qquad (12.9.27)$$

By (10.7.10), for $u < b^2 \coth\left(t/b^2\right)$, we get

$$\int_{\mathfrak{p}} h^{\mathfrak{p}}_{t/b^2}\left(bY^{\mathfrak{p}}, Z^{\mathfrak{p}}\right) \exp\left(u |Z^{\mathfrak{p}}|^2 / 2b^2\right) dZ^{\mathfrak{p}}$$
$$= \left(\frac{e^{t/b^2}}{\cosh\left(t/b^2\right) - u \sinh\left(t/b^2\right)/b^2}\right)^{m/2}$$
$$\exp\left(\frac{1}{2}\frac{u \coth\left(t/b^2\right)/b^2 - 1}{\coth\left(t/b^2\right) - u/b^2} b^2 |Y^{\mathfrak{p}}|^2\right). \qquad (12.9.28)$$

In (12.9.28), we take $u = \frac{1}{2}b^2 \tanh\left(t/b^2\right)$, so that

$$\frac{u \coth\left(t/b^2\right)/b^2 - 1}{\coth\left(t/b^2\right) - u/b^2} b^2 = -\frac{b^2}{2} \frac{\tanh\left(t/b^2\right)}{1 - \frac{1}{2}\tanh^2\left(t/b^2\right)} \leq -b^2 \tanh\left(t/b^2\right),$$

$$(12.9.29)$$

$$\cosh\left(t/b^2\right) - u \sinh\left(t/b^2\right)/b^2 \geq \frac{e^{t/b^2}}{4}.$$

With such a choice of u, we get

$$\int_{\mathfrak{p}} h_{t/b^2}^{\mathfrak{p}}\left(bY^{\mathfrak{p}}, Z^{\mathfrak{p}}\right) \exp\left(u\left|Z^{\mathfrak{p}}\right|^2/2b^2\right) dZ^{\mathfrak{p}} \leq C \exp\left(-u\left|Y^{\mathfrak{p}}\right|^2\right). \quad (12.9.30)$$

For $b \geq 1, \epsilon \leq t \leq M$, u has a positive lower bound. From (12.9.26)–(12.9.28), and (12.9.30), we conclude that there exist $C'_{\epsilon,M} > 0, C''_{\epsilon,M} > 0$ such that in the above range of parameters,

$$\underline{r}_{b,t}^X\left(\left(x_0, Y^{\mathfrak{p}}\right), \left(x', Y^{TX\prime}\right)\right) \leq C'_{\epsilon,M} b^{4m} \exp\left(-C''_{\epsilon,M}\left|Y^{\mathfrak{p}}\right|^2\right). \quad (12.9.31)$$

In (12.9.31), we may replace $\left(x_0, Y^{\mathfrak{p}}\right)$ by any $\left(x, Y^{TX}\right) \in \mathcal{X}$, and so we get (12.9.10).

By (12.9.24), for any $k' \in \mathbf{N}$, the above arguments can be applied to

$$\underline{r}_{b,t}^X\left(\left(x, Y^{TX}\right), \left(x', Y^{TX\prime}\right)\right)\left(1 + \left|Y^{TX\prime}\right|^{2k'}\right).$$

By the analogue of (12.9.31), $\underline{r}_{b,t}^X\left(\left(x, Y^{TX}\right), \left(x', Y^{TX}\right)\right)/b^{4m}$ is uniformly rapidly decreasing on $\mathcal{X} \times \mathcal{X}$, i.e., we obtain the second part of our theorem for $k = 0$.

Let k be a multi-index. By equation (12.9.21) with k replaced by $2m + 1 + |k|$, for $b \geq 1, \epsilon \leq t \leq M$, we get

$$\left|\nabla_{(x', Y^{TX\prime})}^{TX,k} \underline{r}_{b,t}^X\left(\left(x, Y^{TX}\right), \left(x', Y^{TX\prime}\right)\right)\right|$$

$$\leq C_{\epsilon,M,k} b^{4m+2|k|}\left(1 + \left|Y^{TX}\right|^{4(2m+1+|k|)}\right). \quad (12.9.32)$$

By proceeding as in (12.9.25)–(12.9.31), and using (12.9.32), we get an analogue of (12.9.31), in which $\underline{r}_{b,t}^X$ is replaced by $\left|\nabla_{(x', Y^{TX\prime})}^{TX,k} \underline{r}_{b,t}^X\right|$. As before, in the above, $\underline{r}_{b,t}^X$ can be replaced by $\underline{r}_{b,t}^X\left(1 + \left|Y^{TX\prime}\right|^{2k'}\right)$. So we have established the second part of our theorem, which completes the proof. $\square$

Remark 12.9.2. Needless to say, joint derivatives in all variables can be introduced in (12.9.11), by using group invariance, or by exchanging the roles of $\left(x, Y^{TX}\right)$ and $\left(x', Y^{TX\prime}\right)$.

12.10 The heat kernel $\mathfrak{r}_{b,t}^X$ on $\widehat{\mathcal{X}}$

Recall that the scalar operators $\mathfrak{A}_b^X, \mathfrak{B}_b^X$ over $\widehat{\mathcal{X}}$ were defined in (11.6.1)–(11.6.3), and their heat kernels $\mathfrak{r}_{b,t}^X, \mathfrak{s}_{b,t}^X$ in Definition 11.6.1.

In the sequel, $\mathcal{C}\left(\mathbf{R}_+, \mathfrak{p}\right)$ will be replaced by $\mathcal{C}\left(\mathbf{R}_+, \mathfrak{p} \oplus \mathfrak{k}\right)$. However, we will still use the same notation as before for this new space. For example $w. = w_.^{TX} + w_.^N$ is the generic path, and P denotes the Brownian measure on $\left(\mathcal{C}\left(\mathbf{R}_+, \mathfrak{p} \oplus \mathfrak{k}\right), \mathcal{F}_\infty\right)$.

Take $Y \in \mathfrak{g}$. Equation (12.2.1) should be replaced by

$$
\dot{x} = \frac{Y^{TX}}{b}, \qquad\qquad \dot{Y} = -\frac{Y}{b^2} + \frac{\dot{w}}{b}, \qquad\qquad (12.10.1)
$$
$$
x_0 = p1, \qquad\qquad Y_0 = Y.
$$

Let $F : \widehat{\mathcal{X}} \to \mathbf{R}$ be a smooth function with compact support. Using the Itô calculus, instead of (12.2.10) in Theorem 12.2.1, we have the identity

$$
\exp\left(-t\mathfrak{B}_b^X\right) F\left(x_0, Y\right) = E\left[F\left(x_t, Y_t\right)\right]. \qquad (12.10.2)
$$

The considerations of section 12.3 remain valid for the operator $\mathfrak{B}_b^X$.

Let $h. = h^{\mathfrak{p}} + h^{\mathfrak{k}}$ be a bounded process with values in $\mathfrak{g} = \mathfrak{p} \oplus \mathfrak{k}$, which is predictable with respect to the filtration $\mathcal{F}_t|_{t \geq 0}$. We still take $B.$ as in sections 12.1 and 12.4. We define $w_.^\ell$ as in (12.1.8). Of course $w_.^\ell$ takes now its values in $\mathfrak{g}$.

Let $J. = J^{TX} + J^N$ be the solution of the differential equation,

$$
b^2\ddot{J} + \dot{J} + b^2 R^{TX \oplus N}\left(J^{TX}, \dot{x}\right)\left(\dot{x} + \frac{Y^N}{b}\right) = h + \left[B + \varpi, \frac{\delta w}{ds}\right], \quad (12.10.3)
$$
$$
\dot{\varpi} = \left[J^{TX}, \dot{x}\right], \qquad J_0 = 0, \dot{J}_0 = 0, \qquad \varpi_0 = 0.
$$

Then the analogue of (12.4.1), (12.4.2) is just

$$
J_s^{TX} = \frac{\partial}{\partial \ell} x_s^\ell, \qquad\qquad \dot{J}_s = \frac{1}{b}\frac{DY_s}{D\ell}. \qquad (12.10.4)
$$

Also J^{TX}, ϖ still verify equation (12.4.3), with h replaced by $h^{\mathfrak{p}}$, and $w.$ replaced by $w_.^{\mathfrak{p}}$. Note that J^N appears in (12.10.3) only through its first two differentials. If $F : \widehat{\mathcal{X}} \to \mathbf{R}$ is a smooth function with compact support, then the obvious analogue of (12.4.4) holds, i.e.,

$$
E\left[\left\langle dF\left(x_t, Y_t\right), \left(J_t^{TX}, b\dot{J}_t\right)\right\rangle\right] = E\left[F\left(x_t, Y_t\right)\int_0^t \langle h, \delta w\rangle\right]. \qquad (12.10.5)
$$

Instead of (12.4.6), we consider now the differential equation,

$$
b^2\ddot{J} + \dot{J} + b^2 R^{TX \oplus N}\left(J^{TX}, \dot{x}\right)\left(\dot{x} + \frac{Y^N}{b}\right) = h, \quad \dot{\varpi} = \left[J^{TX}, \dot{x}\right], \qquad (12.10.6)
$$
$$
J_0 = 0, \dot{J}_0 = 0, \qquad\qquad\qquad\qquad \varpi_0 = 0.
$$

Then (12.10.6) is a special case of (12.10.3), so that (12.10.5) still holds.

Now L_2 denotes the Hilbert space of square integrable function defined on $[0, t]$ with values in $\mathfrak{g}$. Instead of (12.5.2), for $e \in \mathfrak{p}, f \in \mathfrak{g}$, we consider the differential equation,

$$
b^2\ddot{J} + \dot{J} = v, \qquad\qquad J_0 = 0, \qquad \dot{J}_0 = 0, \qquad (12.10.7)
$$
$$
J_t^{TX} = e, \qquad\qquad\qquad\qquad \dot{J}_t = f/b.
$$

The control problem still consists in finding v with minimal norm $|v|_{L_2}$. Note that in (12.10.7), J_t^N is allowed to vary freely.

This variational problem can be decoupled into a problem on v^{p} associated with (e, f^{p}), and a problem on v^{t} associated with f^{t}. These two problems were considered in section 10.3. By (10.3.11) and (10.3.38), we get

$$\frac{1}{2}|v|_{L_2}^2 = K_{b,t}\left((0,0),(e, f^{\mathrm{p}}/b)\right) + K_{b,t}^*\left(0, f^{\mathrm{t}}/b\right). \tag{12.10.8}$$

Instead of (12.5.6), we get

$$E\left[\langle dF(x_t, Y_t), \tau_t^0(e, f)\rangle\right]$$
$$= E\left[F\left(x_t, Y_t^{TX}\right) \int_0^t \left\langle v + b^2 R^{TX \oplus N}\left(J^{TX}, \dot{x}\right)\left(\dot{x} + \frac{Y^N}{b}\right), \delta w\right\rangle\right]. \tag{12.10.9}$$

We extend Definition 12.7.3 to functions $f : \widehat{\mathcal{X}} \times \widehat{\mathcal{X}} \to \mathbf{R}$. Using the results of chapter 10, and in particular the remarks after (10.3.17) together with the above considerations, it is easy to proceed as in sections 12.5–12.7, and to obtain the following analogue of Theorem 12.7.4.

Theorem 12.10.1. *For $0 < b \le M, \epsilon \le t \le M$, $\mathfrak{r}_{b,t}^X\left((x, Y),(x', Y')\right)$ and its covariant derivatives of arbitrary order in (x', Y') with respect to $\nabla^{TX \oplus N}$ are uniformly rapidly decreasing on $\widehat{\mathcal{X}} \times \widehat{\mathcal{X}}$.*

We now state the analogue of Theorem 12.8.1.

Theorem 12.10.2. *For $t > 0$, as $b \to 0$, we have the pointwise convergence,*

$$\mathfrak{r}_{b,t}^X\left((x, Y),(x', Y')\right) \to \frac{1}{\pi^{(m+n)/2}} p_t(x, x') \exp\left(-\frac{1}{2}\left(|Y|^2 + |Y'|^2\right)\right). \tag{12.10.10}$$

Proof. The proof follows the same lines as the proof of Theorem 12.8.1. Indeed the first part of the proof concerning the probability law of x. is identical. Then we handle the extra component $Y^{\cdot}$ exactly as in that proof. $\qquad \square$

For $a > 0$, recall that K_a was defined in (2.14.2). Put

$$\underline{\mathfrak{A}}_b^X = K_b \mathfrak{A}_b^X K_b^{-1}. \tag{12.10.11}$$

By (11.6.1), (12.10.11), we get

$$\underline{\mathfrak{A}}_b^X = \frac{1}{2}\left(-\frac{1}{b^4}\Delta^{TX \oplus N} + |Y|^2 - \frac{m+n}{b^2}\right) - \nabla_{Y^{TX}}. \tag{12.10.12}$$

For $t > 0$, let $\underline{\mathfrak{r}}_{b,t}^X\left((x, Y),(x', Y')\right)$ be the smooth kernel associated with the operator $\exp\left(-t\underline{\mathfrak{A}}_b^X\right)$. Instead of (12.9.4), we get

$$\underline{\mathfrak{r}}_{b,t}^X\left((x, Y),(x', Y')\right) = b^{m+n}\mathfrak{r}_{b,t}^X\left((x, bY),(x', bY')\right). \tag{12.10.13}$$

Let $w.$ be a Brownian motion valued in $\mathfrak{g} = \mathfrak{p} \oplus \mathfrak{k}$. Instead of equation (12.9.5), we now consider the differential equation

$$\dot{x} = Y^{TX}, \qquad\qquad \dot{Y} = \frac{\dot{w}}{b^2}, \qquad\qquad (12.10.14)$$

$$x_0 = p1, \qquad\qquad Y_0 = Y.$$

Instead of (12.9.9), if $F : \widehat{\mathcal{X}} \to \mathbf{R}$ is a smooth function with compact support, for $t > 0$, we get

$$\exp\left(-t\underline{\mathfrak{A}}_b^X\right) F\left(x_0, Y\right) = E\left[\exp\left(\frac{(m+n)t}{2b^2} - \frac{1}{2}\int_0^t |Y|^2 \, ds\right) F\left(x_t, Y_t\right)\right]. \tag{12.10.15}$$

If k is a multi-index in $1, \ldots, 2m+n$, let $\nabla^{TX\oplus N,k}_{(x',Y')}$ be the covariant derivative of order k with respect to the connection $\nabla^{TX\oplus N}$. We have the following analogue of Theorem 12.9.1.

Theorem 12.10.3. *Given $\epsilon > 0, M > 0$ with $\epsilon \le M$, there exists $C_{\epsilon,M} > 0$ such that for $b \ge 1, \epsilon \le t \le M$,*

$$\mathfrak{r}_{b,t}^X\left((x,Y),(x',Y')\right) \le C_{\epsilon,M} b^{4m+2n}. \tag{12.10.16}$$

For $b \ge 1, \epsilon \le t \le M$, for any multi-index k,

$$\left|\nabla^{TX\oplus N,k}_{(x',Y')}\mathfrak{r}_{b,t}^X\left((x,Y),(x',Y')\right)/b^{4m+2n+2|k|}\right| \tag{12.10.17}$$

is uniformly rapidly decreasing on $\widehat{\mathcal{X}} \times \widehat{\mathcal{X}}$.

Proof. The proof proceeds exactly as the proof of Theorem 12.9.1. $\qquad\qquad\square$

Remark 12.10.4. The considerations of Remark 12.9.2 also apply to the kernel $\mathfrak{r}_{b,t}^X$.

12.11 The heat kernel $\mathsf{r}_{b,t}^X$ on $\widehat{\mathcal{X}}$

Recall that the scalar differential operators $\mathsf{A}_b^X, \mathsf{B}_b^X$ were defined in (11.7.1), (11.7.3), and (11.7.4), and their smooth heat kernels $\mathsf{r}_{b,t}^X, \mathsf{s}_{b,t}^X$ in Definition 11.7.2.

Let $(x., Y.)$ be as in (12.10.1). Let $F : \widehat{\mathcal{X}} \to \mathbf{R}$ be a smooth function with compact support. An application of the Itô calculus and of the Feynman-Kac formula similar to what we did in the proof of equation (12.2.10) in Theorem 12.2.1, in (12.9.9), and in (12.10.15) shows that for $t > 0$,

$$\exp\left(-t\mathsf{B}_b^X\right) F\left(x_0, Y\right) = E\left[\exp\left(-\frac{1}{2}\int_0^t \left|\left[Y^N, Y^{TX}\right]\right|^2 ds\right) F\left(x_t, Y_t\right)\right]. \tag{12.11.1}$$

By comparing (12.10.2) and (12.11.1), we get

$$\mathsf{s}_{b,t}^X \le \mathsf{s}_{b,t}^X. \tag{12.11.2}$$

By (11.6.5), (11.7.29), equation (12.11.2) is equivalent to

$$r_{b,t}^X \leq \mathfrak{r}_{b,t}^X. \tag{12.11.3}$$

The analogue of Theorem 12.10.1 holds. Indeed, as we shall see in the proof of Theorem 12.11.2, the exponential term in (12.11.1) does not create any new difficulty. The analogue of equation (12.10.10) says that as $b \to 0$,

$$r_{b,t}^X \left((x, Y), (x', Y') \right) \to \frac{1}{\pi^{(m+n)/2}} \exp\left(\frac{t}{8} \mathrm{Tr}\left[C^{\mathfrak{e}, \mathfrak{p}} \right] \right)$$
$$p_t(x, x') \exp\left(-\frac{1}{2} \left(|Y|^2 + |Y'|^2 \right) \right). \tag{12.11.4}$$

For a hint on the proof of (12.11.4), we refer to equation (2.16.25), to Proposition 14.10.1, and also to Theorem 14.11.2, where a more difficult result is proved for the operator $\mathcal{L}_{A,b}^X$.

Definition 12.11.1. Set

$$\underline{A}_b^X = K_b A_b^X K_b^{-1}. \tag{12.11.5}$$

By (11.7.2), (12.10.11), and (12.11.5), we get

$$\underline{A}_b^X = \frac{b^4}{2} \left| \left[Y^N, Y^{TX} \right] \right|^2 + \underline{\mathfrak{A}}_b^X. \tag{12.11.6}$$

By (12.10.12), (12.11.6), we obtain

$$\underline{A}_b^X = \frac{b^4}{2} \left| \left[Y^N, Y^{TX} \right] \right|^2 + \frac{1}{2} \left(-\frac{1}{b^4} \Delta^{TX \oplus N} + |Y|^2 - \frac{m+n}{b^2} \right) - \nabla_{Y^{TX}}. \tag{12.11.7}$$

Let $(x_., Y_.)$ be as in (12.10.14). If $F : \widehat{\mathcal{X}} \to \mathbf{R}$ is a smooth function with compact support, instead of (12.10.15), we get

$$\exp\left(-t\underline{A}_b^X \right) F(x_0, Y) = E\left[\exp\left(\frac{(m+n)t}{2b^2} - \frac{b^4}{2} \int_0^t \left| \left[Y^N, Y^{TX} \right] \right|^2 ds \right. \right.$$
$$\left. \left. -\frac{1}{2} \int_0^t |Y|^2 ds \right) F(x_t, Y_t) \right]. \tag{12.11.8}$$

For $t > 0$, let $\underline{r}_{b,t}^X \left((x, Y), (x', Y') \right)$ be the smooth kernel associated with the operator $\exp\left(-t\underline{A}_b^X \right)$.

By (12.10.15), (12.11.8), we get

$$\underline{r}_{b,t}^X \leq \mathfrak{r}_{b,t}^X, \tag{12.11.9}$$

which also follows from (12.11.3). By (12.11.9), we deduce that any upper bound for $\mathfrak{r}_{b,t}^X$ is a fortiori valid for $\underline{r}_{b,t}^X$.

Now we establish an analogue of Theorem 12.10.3 for the kernel $\underline{r}_{b,t}^X$.

Theorem 12.11.2. *Given $\epsilon > 0, M > 0$ with $\epsilon \leq M$, there exists $C_{\epsilon,M} > 0$ such that for $b \geq 1, \epsilon \leq t \leq M$,*

$$\mathfrak{r}_{b,t}^{X} \left((x, Y), (x', Y') \right) \leq C_{\epsilon,M} b^{4m+2n}. \tag{12.11.10}$$

For $b \geq 1, \epsilon \leq t \leq M$, for any multi-index k,

$$\left| \nabla_{(x',Y')}^{TX \oplus N, k} \mathfrak{r}_{b,t}^{X} \left((x, Y), (x', Y') \right) / b^{4m+2n+2|k|} \right| \tag{12.11.11}$$

is uniformly rapidly decreasing on $\widehat{\mathcal{X}} \times \widehat{\mathcal{X}}$.

Proof. By equation (12.10.16) in Theorem 12.10.3 and by (12.11.9), we get (12.11.10). However, this argument cannot be used to get (12.11.11).

To establish (12.11.11), we will use again the Malliavin calculus as in the proof of Theorems 12.9.1 and 12.10.3. We take $(x., Y.)$ as in (12.10.14), and we use (12.11.8). We take $\psi_s, \overline{\psi}_s$ as in (12.9.15). If $e \in \mathfrak{p}, f \in \mathfrak{g} = \mathfrak{p} \oplus \mathfrak{k}$ are unit vectors, as in (12.9.17), put

$$J_s = \psi_{s/t} e + t \overline{\psi}_{s/t} f. \tag{12.11.12}$$

Now we apply to (12.11.8) the method used in the proof of Theorem 12.9.1. We get the following analogue of (12.9.19),

$$E \left[\exp \left(-\frac{b^4}{2} \int_0^t \left| [Y^N, Y^{TX}] \right|^2 ds - \frac{1}{2} \int_0^t |Y|^2 ds \right) \right.$$
$$\left. \left\langle dF(x_t, Y_t), \tau_t^0(e, f) \right\rangle \right]$$
$$= E \left[\exp \left(-\frac{b^4}{2} \int_0^t \left| [Y^N, Y^{TX}] \right|^2 ds - \frac{1}{2} \int_0^t |Y|^2 ds \right) \right.$$
$$\left(b^4 \int_0^t \left\langle [Y^N, Y^{TX}], [\dot{J}^N, Y^{TX}] + [Y^N, \dot{J}^{TX}] \right\rangle ds + \int_0^t \left\langle Y, \dot{J} \right\rangle ds \right.$$
$$\left. + b^2 \int_0^t \left\langle \ddot{J} + R^{TX \oplus N}(J^{TX}, \dot{x})(\dot{x} + Y^N), \delta w \right\rangle \right) F(x_t, Y_t) \right]. \tag{12.11.13}$$

In the right-hand side of (12.11.13), the main new difficulty with respect to the proof of Theorems 12.9.1 and 12.10.3 is the presence of the term starting with b^4 in the right-hand side of (12.11.13). However, note that if K_s is a smooth process with values in $T_x X$,

$$\left| b^2 \int_0^t \left\langle [Y^N, Y^{TX}], K \right\rangle ds \right| \leq \frac{b^4}{2} \int_0^t \left| [Y^N, Y^{TX}] \right|^2 ds + \frac{1}{2} \int_0^t |K|^2 ds. \tag{12.11.14}$$

From (12.11.14), for $b \geq 1$, we get

$$\exp \left(-\frac{b^4}{2} \int_0^t \left| [Y^N, Y^{TX}] \right|^2 ds \right) \left| b^4 \int_0^t \left\langle [Y^N, Y^{TX}], K \right\rangle ds \right|$$
$$\leq b^2 \left(C + \frac{1}{2} \int_0^t |K|^2 ds \right). \tag{12.11.15}$$

Note that the fateful factor b^2 appears in the right-hand side of (12.11.15).

We will use (12.11.15) with

$$K = \left[\dot{j}^N, Y^{TX} \right] + \left[Y^N, \dot{j}^{TX} \right].$$
(12.11.16)

In this case,

$$\int_0^t |K|^2 \, ds \leq C \int_0^t |Y|^2 \, ds.$$
(12.11.17)

Proceeding as in the proof of Theorem 12.9.1 while using (12.11.13)–(12.11.17), the right-hand side of (12.11.13) can be controlled in the same way as the right-hand side of (12.9.19).

We claim that the above argument can be iterated. Let us just explain the case where instead a second order operator acts on F. The term with the factor b^4 is no longer a problem because this factor is acceptable when $|k| = 2$. The only new difficulty comes from the products of terms like the one with b^4, which introduce a factor b^8. By squaring (12.11.14), we get

$$\left| b^2 \int_0^t \left\langle [Y^N, Y^{TX}], K \right\rangle ds \right|^2$$

$$\leq C \left(\left(b^4 \int_0^t [Y^N, Y^{TX}]^2 \, ds \right)^2 + \left(\int_0^t |K|^2 \, ds \right)^2 \right).$$
(12.11.18)

From (12.11.18), for $b \geq 1$, we obtain

$$\exp \left(-\frac{b^4}{2} \int_0^t \left| [Y^N, Y^{TX}] \right|^2 ds \right) \left| b^4 \int_0^t \left\langle [Y^N, Y^{TX}], K \right\rangle ds \right|^2$$

$$\leq b^4 C \left(1 + \left(\int_0^t |K|^2 \, ds \right)^2 \right).$$
(12.11.19)

Again the right weight b^4 appears in the right-hand side of (12.11.19).

The above argument can be easily iterated, and leads to (12.11.11) for arbitrary k. The proof of our theorem is completed. $\square$

Chapter Thirteen

Refined estimates on the scalar heat kernel for bounded b

In this chapter, for bounded $b > 0$, we obtain uniform bounds for the kernels $r_{b,t}^X, \mathfrak{r}_{b,t}^X$, with the proper decay at infinity on $\mathcal{X}$ or $\widehat{\mathcal{X}}$. In chapter 14, these bounds will be used to obtain corresponding bounds for the kernel $q_{b,t}^X$, in order to prove Theorem 4.5.2. The arguments developed in section 12.3, which connect the hypoelliptic heat kernel with the wave equation, play an important role in proving the required estimates.

This chapter is organized as follows. In section 13.1, we establish estimates on the Hessian of the distance function on X.

In section 13.2, we obtain the bounds on the heat kernel $r_{b,t}^X$.

In section 13.3, we establish the bounds on the heat kernel $\mathfrak{r}_{b,t}^X$.

In the whole chapter, we will use the notation and the results of chapter 12.

13.1 The Hessian of the distance function

Clearly, there is $C > 0$ such that if $a \in \mathfrak{g}$,

$$|\mathrm{ad}\,(a)| \le C\,|a|\,. \tag{13.1.1}$$

Recall that the function $d^2\,(x, x')$ is smooth on $X \times X$. We fix $x \in X$. If $x' \in X, x' \ne x$, let $u \in T_x X$ be the unit vector that is the derivative in the time parameter of the geodesic connecting x to x', with speed 1. Then

$$\nabla d\,(x, \cdot) = u. \tag{13.1.2}$$

From (13.1.2), we get

$$\nabla \frac{d^2\,(x, \cdot)}{2} = d\,(x, \cdot)\,u. \tag{13.1.3}$$

Since u is of norm 1, and since u is parallel along geodesics centered at x, from (13.1.2), (13.1.3) we get

$$\nabla^{TX}_{\nabla d^2(x,\cdot)/2} \nabla d^2\,(x, \cdot)\,/2 = \nabla d^2\,(x, \cdot)\,/2. \tag{13.1.4}$$

Also since the function d is convex,

$$\nabla^{TX} \nabla d\,(x, \cdot) \ge 0 \text{ on } X \setminus \{x\}. \tag{13.1.5}$$

Taking the trace in (13.1.5), we get

$$\Delta^X d(x, \cdot) \geq 0 \text{ on } X \setminus \{x\}. \tag{13.1.6}$$

We may and we will assume that $x = p1$. We use the geodesic coordinate system on X centered at x, which is given by $a \in \mathfrak{p} \to x' = e^a x \in X$. We identify TX with $\mathfrak{p}$ by parallel transport along the geodesics $s \in \mathbf{R} \to e^{sa} x \in X$ with respect to the Levi-Civita connection. Note that

$$d(x, x') = |a|. \tag{13.1.7}$$

Proposition 13.1.1. *The following identity holds:*

$$\nabla^{TX} \nabla \frac{d^2(x, \cdot)}{2} = \operatorname{ad}(a) \coth(\operatorname{ad}(a)), \tag{13.1.8}$$

so that

$$1 \leq \nabla^{TX} \nabla \frac{d^2(x, \cdot)}{2} \leq 1 + |\operatorname{ad}(a)|. \tag{13.1.9}$$

Moreover,

$$1 \leq \nabla^{TX} \nabla \frac{d^2(x, \cdot)}{2} \leq 1 + Cd(x, \cdot). \tag{13.1.10}$$

In particular,

$$m \leq \Delta^X \frac{d^2(x, \cdot)}{2} \leq m(1 + Cd(x, \cdot)). \tag{13.1.11}$$

Proof. Let J_s be a Jacobi field along the geodesic $s \in [0,1] \to e^{sa} x \in X$ that vanishes at $s = 0$. By proceeding as in (3.4.19), (3.4.30), we get

$$J_s = \frac{\sinh(sad(a))}{\sinh(\operatorname{ad}(a))} J_1, \tag{13.1.12}$$

so that

$$\dot{J}_1 = \operatorname{ad}(a) \coth(\operatorname{ad}(a)) J_1. \tag{13.1.13}$$

By (13.1.13), we get (13.1.8). Moreover, for $x \in \mathbf{R}$,

$$1 \leq x \coth(x) \leq 1 + |x|. \tag{13.1.14}$$

By (13.1.8), (13.1.14), we get (13.1.9). Also equation (13.1.10) follows from (13.1.1), (13.1.7), and (13.1.9). By taking the trace in (13.1.10), we get (13.1.11). The proof of our proposition is completed. □

Now we will establish corresponding results for the function $d(x, \cdot)$.

Proposition 13.1.2. *The following identity holds on $X \setminus \{x\}$,*

$$\nabla^{TX} \nabla d(x, \cdot) = -\frac{\nabla.d(x, \cdot) \nabla.d(x, \cdot)}{d(x, \cdot)} + \frac{\nabla^{TX} \nabla d^2(x, \cdot)/2}{d(x, \cdot)}. \tag{13.1.15}$$

Moreover,

$$\frac{1 - \nabla.d(x, \cdot) \nabla.d(x, \cdot)}{d(x, \cdot)} \leq \nabla^{TX} \nabla d(x, \cdot)$$

$$\leq \frac{1}{d(x, \cdot)} (1 - \nabla.d(x, \cdot) \nabla.d(x, \cdot) + |\operatorname{ad}(a)|). \tag{13.1.16}$$

There is $C > 0$ such that on $X \setminus \{x\}$,

$$\frac{1 - \nabla.d\,(x, \cdot)\,\nabla.d\,(x, \cdot)}{d\,(x, \cdot)} \leq \nabla^{TX}\nabla d\,(x, \cdot)$$

$$\leq \frac{1}{d\,(x, \cdot)}\left(1 - \nabla.d\,(x, \cdot)\,\nabla.d\,(x, \cdot) + Cd\,(x, \cdot)\right). \quad (13.1.17)$$

In particular, on $X \setminus \{x\}$,

$$\frac{m-1}{d\,(x, \cdot)} \leq \Delta^X d\,(x, \cdot) \leq \frac{m-1}{d\,(x, \cdot)} + Cm. \quad (13.1.18)$$

Finally, on $X \setminus B\,(x, 1)$, the covariant derivatives of order ≥ 1 of $d\,(x, \cdot)$ are uniformly bounded.

Proof. Equation (13.1.15) is trivial. Equations (13.1.16)–(13.1.18) can be derived from (13.1.1), (13.1.2), from Proposition 13.1.1 and from (13.1.15).

By (13.1.2) and (13.1.17), the covariant derivatives of order 1 and 2 of $d\,(x, \cdot)$ are uniformly bounded on $X \setminus B\,(x, 1)$. Let us explain how to handle the higher covariant derivatives. We will first rederive the bound on the second derivatives.

Recall that $x = p1$. In the sequel, the scalar product is taken in $T_x X$. Also we still use the coordinate system $a \in \mathfrak{p} \to e^a x \in X$. By (13.1.7), we get

$$\nabla d\,(x, \cdot) = \frac{a}{|a|}. \quad (13.1.19)$$

We trivialize TX by parallel transport with respect to the connection ∇^{TX} along geodesics centered at x by parallel transport with respect to the connection ∇^{TX}. Let Γ^{TX} denote the associated connection form. By (13.1.19), we obtain

$$\nabla_U^{TX}\nabla d\,(x, \cdot) = \frac{1}{|a|}\left(U - \frac{a}{|a|^2}\langle a, U \rangle\right) + \Gamma_a^{TX}\,(U)\,\frac{a}{|a|}. \quad (13.1.20)$$

By (3.7.5), (13.1.20), we obtain

$$\nabla_U^{TX}\nabla d\,(x, \cdot) = -\langle a, U \rangle\,\frac{a}{|a|^3} + \frac{1}{|a|}\cosh\,(\operatorname{ad}\,(a))\,U. \quad (13.1.21)$$

Using (3.7.7), we can rewrite (13.1.21) in the form

$$\nabla_U^{TX}\nabla d\,(x, \cdot) = -\langle a, V \rangle\,\frac{a}{|a|^3} + \frac{\operatorname{ad}\,(a)}{|a|}\coth\,(\operatorname{ad}\,(a))\,V. \quad (13.1.22)$$

In view of (13.1.8) and (13.1.19), (13.1.22) is just a form of (13.1.15).

By (13.1.1), (13.1.7), (13.1.14), and (13.1.15) or (13.1.22), we recover the fact that the tensor $\nabla^{TX}\nabla d\,(x, \cdot)$ is uniformly bounded outside of $B\,(x, 1)$. Using (3.7.7), (3.7.8), and (13.1.22), the same is true for the higher order covariant derivatives of $d\,(x, \cdot)$. This completes the proof of our proposition. $\square$

13.2 Bounds on the scalar heat kernel on $\mathcal{X}$ for bounded b

Recall that the kernel $r_{b,t}^X \left(\left(x, Y^{TX}\right), \left(x', Y^{TX\prime}\right)\right)$ on $\mathcal{X}$ was defined in Definition 11.5.3.

Theorem 13.2.1. *Given* $\epsilon > 0, M > 0, \epsilon \le M$, *there exist* $C > 0, C' > 0, C'' > 0$ *such that for* $0 < b \le M, \epsilon \le t \le M$, $\left(x, Y^{TX}\right), \left(x', Y^{TX\prime}\right) \in \mathcal{X}$,

$$r_{b,t}^X \left(\left(x, Y^{TX}\right), \left(x', Y^{TX\prime}\right)\right)$$
$$\le C \exp\left(-C'\left(d^2\left(x, x'\right) + \left|Y^{TX}\right|^2 + \left|Y^{TX\prime}\right|^2\right) + C''/b^2\right). \quad (13.2.1)$$

Given $\epsilon > 0, M > 0, \epsilon \le M$, *there exist* $C > 0, C' > 0$ *such that for* $0 < b \le M, \epsilon \le t \le M$, $\left(x, Y^{TX}\right), \left(x', Y^{TX\prime}\right) \in \mathcal{X}$,

$$r_{b,t}^X \left(\left(x, Y^{TX}\right), \left(x', Y^{TX\prime}\right)\right)$$
$$\le C \exp\left(-C'\left(b^2 d^2\left(x, x'\right) + \left|Y^{TX}\right|^2 + \left|Y^{TX\prime}\right|^2\right)\right). \quad (13.2.2)$$

Proof. Clearly,

$$r_{b,t}^X \left(\left(x, Y^{TX}\right), \left(x', Y^{TX\prime}\right)\right) = \int_{\mathcal{X}} r_{b,t/2}^X \left(\left(x, Y^{TX}\right), \left(z, Z^{TX}\right)\right)$$
$$r_{b,t/2}^X \left(\left(z, Z^{TX}\right), \left(x', Y^{TX\prime}\right)\right) dz dZ^{TX}. \quad (13.2.3)$$

By Theorem 12.7.4, under the given conditions on b, t, the kernel $r_{b,t/2}^X$ is uniformly bounded by a constant $C' > 0$. By (13.2.3), we get

$$r_{b,t}^X \left(\left(x, Y^{TX}\right), \left(x', Y^{TX\prime}\right)\right) \le C' \int_{\mathcal{X}} r_{b,t/2}^X \left(\left(x, Y^{TX}\right), \left(z, Z^{TX}\right)\right) dz dZ^{TX}.$$
$$(13.2.4)$$

Recall that $x_0 = p1$. We may and we will assume that $\left(x, Y^{TX}\right) = \left(x_0, Y^{\mathfrak{p}}\right)$, with $Y^{\mathfrak{p}} \in \mathfrak{p}$. Now we proceed as in (12.9.27). By comparing (10.4.1) and (11.1.1), we get

$$\int_{\mathcal{X}} r_{b,t/2}^X \left(\left(x_0, Y^{\mathfrak{p}}\right), \left(z, Z^{TX}\right)\right) dz dZ^{TX} = \int_{\mathfrak{p}} h_{t/2b^2}^{\mathfrak{p}} \left(Y^{\mathfrak{p}}, Z^{\mathfrak{p}}\right) dZ^{\mathfrak{p}}. \quad (13.2.5)$$

By (10.7.12), (13.2.4), and (13.2.5), we obtain

$$r_{b,t}^X \left(\left(x_0, Y^{\mathfrak{p}}\right), \left(x', Y^{TX\prime}\right)\right)$$
$$\le C' \left(\frac{e^{t/2b^2}}{\cosh\left(t/2b^2\right)}\right)^{m/2} \exp\left(-\frac{1}{2}\tanh\left(t/2b^2\right)\left|Y^{\mathfrak{p}}\right|^2\right). \quad (13.2.6)$$

By (13.2.6), there exist $C' > 0, c' > 0$ such that for $0 < b \le M, \epsilon \le t \le M$,

$$r_{b,t}^X \left(\left(x_0, Y^{\mathfrak{p}}\right), \left(x', Y^{TX\prime}\right)\right) \le C' \exp\left(-c'\left|Y^{\mathfrak{p}}\right|^2\right). \quad (13.2.7)$$

More generally, by (13.2.7), we get

$$r_{b,t}^X \left(\left(x, Y^{TX}\right), \left(x', Y^{TX\prime}\right)\right) \le C' \exp\left(-c'\left|Y^{TX}\right|^2\right). \quad (13.2.8)$$

By (11.1.1), the formal adjoint $\mathcal{A}_b^{X*}$ of $\mathcal{A}_b^X$ with respect to the standard L_2 scalar product on $\mathcal{X}$ is obtained from $\mathcal{A}_b^X$ by replacing $-\frac{1}{b}\nabla_{Y^{TX}}$ by $\frac{1}{b}\nabla_{Y^{TX}}$. The operator $\mathcal{A}_b^{X*}$ is of the same type as $\mathcal{A}_b^X$. By proceeding as in the proof of (13.2.8), by exchanging the roles of Y^{TX} and $Y^{TX'}$, we also get

$$r_{b,t}^X\left((x, Y^{TX}), (x', Y^{TX'})\right) \leq C' \exp\left(-c' \left|Y^{TX'}\right|^2\right). \tag{13.2.9}$$

In the integral in (13.2.3), either $d(x, z) \geq d(x, x')/2$ or $d(x', z) \geq d(x, x')/2$. Therefore we can refine (13.2.4) into

$$r_{b,t}^X\left((x, Y^{TX}), (x', Y^{TX'})\right)$$
$$\leq C' \int_{\substack{(z, Z^{TX}) \in \mathcal{X} \\ d(x,z) \geq d(x,x')/2}} r_{b,t/2}^X\left((x, Y^{TX}), (z, Z^{TX})\right) dz dZ^{TX}$$
$$+ C' \int_{\substack{(z, Z^{TX}) \in \mathcal{X} \\ d(x',z) \geq d(x,x')/2}} r_{b,t/2}^X\left((z, Z^{TX}), (x', Y^{TX'})\right) dz dZ^{TX}. \tag{13.2.10}$$

By symmetry, it is enough to estimate the first integral in the right-hand side of (13.2.10). In estimating this integral, we may as well assume that $(x, Y^{TX}) = (x_0, Y^\mathfrak{p})$.

Instead of (12.9.5), we consider the stochastic differential equation on $\mathcal{X}$,

$$\dot{x} = \frac{Y^{TX}}{b}, \qquad\qquad \dot{Y}^{TX} = \frac{\dot{w}}{b}, \tag{13.2.11}$$
$$x_0 = p1, \qquad\qquad Y_0^{TX} = Y^\mathfrak{p},$$

so that the first equation in (12.9.6) still holds.

Let $F : \mathcal{X} \to \mathbf{R}$ be a smooth function with compact support. By proceeding as in (12.9.9), we get

$$\exp\left(-t\mathcal{A}_b^X\right) F(x_0, Y^\mathfrak{p}) = E\left[\exp\left(\frac{mt}{2b^2} - \frac{1}{2b^2}\int_0^t \left|Y^{TX}\right|^2 ds\right) F(x_t, Y_t^{TX})\right]. \tag{13.2.12}$$

By (13.2.12), we obtain

$$\int_{\substack{(z, Z^{TX}) \in \mathcal{X} \\ d(x_0,z) \geq d(x_0,x')/2}} r_{b,t/2}^X\left((x_0, Y^\mathfrak{p}), (z, Z^{TX})\right) dz dZ^{TX}$$
$$= E\left[\exp\left(\frac{mt}{4b^2} - \frac{1}{2b^2}\int_0^{t/2} \left|Y^{TX}\right|^2 ds\right) 1_{d(x_0,x_{t/2}) \geq d(x_0,x')/2}\right]. \tag{13.2.13}$$

Let $x_.$ be a path as in (13.2.11). Clearly

$$\frac{1}{2b^2}\int_0^{t/2} \left|Y^{TX}\right|^2 ds \geq \frac{d^2(x_0, x_{t/2})}{t}. \tag{13.2.14}$$

Take $\alpha \in]0, 1[$. By (13.2.13), (13.2.14), we get

$$\int_{\substack{(z, Z^{TX}) \in \mathcal{X} \\ d(x_0, z) \geq d(x_0, x')/2}} r^X_{b, t/2} \left((x_0, Y^{\mathfrak{p}}), (z, Z^{TX}) \right) dz dZ^{TX}$$

$$\leq \exp \left(- \left(1 - \alpha^2 \right) d^2 \left(x_0, x' \right) / 4t \right)$$

$$E \left[\exp \left(\frac{mt}{4b^2} - \frac{\alpha^2}{2b^2} \int_0^{t/2} |Y^{TX}|^2 ds \right) \right]. \quad (13.2.15)$$

Let $O^{\mathfrak{p}}_\alpha$ be the operator

$$O^{\mathfrak{p}}_\alpha = \frac{1}{2} \left(-\Delta^{\mathfrak{p}} + \alpha^2 |Y^{\mathfrak{p}}|^2 - m \right), \quad (13.2.16)$$

and for $t > 0$, let $h^{\mathfrak{p}}_{\alpha, t} (Y^{\mathfrak{p}}, Z^{\mathfrak{p}})$ be the associated heat kernel. Set

$$Y^{\mathfrak{p}}_t = Y^{\mathfrak{p}} + \frac{1}{b} w_t. \quad (13.2.17)$$

Using again the Feynman-Kac formula, we get

$$\int_{\mathfrak{p}} h^{\mathfrak{p}}_{\alpha, t/2b^2} (Y^{\mathfrak{p}}, Z^{\mathfrak{p}}) dZ^{\mathfrak{p}} = E \left[\exp \left(\frac{mt}{4b^2} - \frac{\alpha^2}{2b^2} \int_0^{t/2} |Y^{\mathfrak{p}}|^2 ds \right) \right]. \quad (13.2.18)$$

Recall that the harmonic oscillator $O^{\mathfrak{p}}$ was defined in (10.4.1). For $a > 0$, we define K_a as in (10.4.7). Clearly,

$$K^{-1}_{\sqrt{\alpha}} O^{\mathfrak{p}}_\alpha K_{\sqrt{\alpha}} = \alpha O^{\mathfrak{p}} - \frac{m (1 - \alpha)}{2}. \quad (13.2.19)$$

By (13.2.19), we get

$$h^{\mathfrak{p}}_{\alpha, t} (Y^{\mathfrak{p}}, Z^{\mathfrak{p}}) = \exp (mt (1 - \alpha) / 2) \alpha^{m/2} h^{\mathfrak{p}}_{\alpha t} \left(\sqrt{\alpha} Y^{\mathfrak{p}}, \sqrt{\alpha} Z^{\mathfrak{p}} \right). \quad (13.2.20)$$

By (10.7.12) and (13.2.20), we obtain

$$\int_{\mathfrak{p}} h^{\mathfrak{p}}_{\alpha, t/2b^2} (Y^{\mathfrak{p}}, Z^{\mathfrak{p}}) dZ^{\mathfrak{p}} = \left(\frac{e^{t/2b^2}}{\cosh (\alpha t / 2b^2)} \right)^{m/2}$$

$$\exp \left(-\frac{1}{2} \alpha \tanh (\alpha t / 2b^2) |Y^{\mathfrak{p}}|^2 / 2 \right). \quad (13.2.21)$$

By (13.2.15), (13.2.18), and (13.2.21), we get

$$\int_{\substack{(z, Z^{TX}) \in \mathcal{X} \\ d(x_0, z) \geq d(x_0, x')/2}} r^X_{b, t/2} \left((x_0, Y^{\mathfrak{p}}), (z, Z^{TX}) \right) dz dZ^{TX}$$

$$\leq \exp \left(- \left(1 - \alpha^2 \right) d^2 \left(x_0, x' \right) / 4t \right) \left(\frac{e^{t/2b^2}}{\cosh (\alpha t / 2b^2)} \right)^{m/2}$$

$$\exp \left(-\frac{1}{2} \alpha \tanh (\alpha t / 2b^2) |Y^{\mathfrak{p}}|^2 / 2 \right). \quad (13.2.22)$$

By taking $\alpha = 1/2$, we deduce from (13.2.22) that there exist $C' > 0, c' > 0$ such that under the given conditions on the parameters b, t,

$$\int_{\substack{(z, Z^{TX}) \in \mathcal{X} \\ d(x_0, z) \geq d(x_0, x')/2}} r^X_{b,t/2} \left((x_0, Y^{\mathfrak{p}}), (z, Z^{TX}) \right) dz dZ^{TX}$$

$$\leq C' \exp\left(-c' d^2(x_0, x') + C''/b^2\right). \quad (13.2.23)$$

For $0 < b \leq 1/2$, we can take $\alpha = (1 - b^2)^{1/2}$ in (13.2.22), and we get the existence of $C' > 0, c' > 0$ such that

$$\int_{\substack{(z, Z^{TX}) \in \mathcal{X} \\ d(x_0, z) \geq d(x_0, x')/2}} r^X_{b,t/2} \left((x_0, Y^{\mathfrak{p}}), (z, Z^{TX}) \right) dz dZ^{TX}$$

$$\leq C' \exp\left(-c' b^2 d^2(x_0, x')\right). \quad (13.2.24)$$

As explained after equation (13.2.10), the bounds in (13.2.23) and (13.2.24) give corresponding bounds for $r^X_{b,t} \left((x, Y^{TX}), (x', Y^{TX'}) \right)$.

By (13.2.8), (13.2.9), (13.2.23) and (13.2.24), we get (13.2.1) and (13.2.2). The proof of our theorem is completed. □

Now we establish a second critical estimate.

Theorem 13.2.2. *Given $\epsilon > 0, M > 0, \epsilon \leq M$, there exist $C > 0, C' > 0$ such that for $0 < b \leq M, \epsilon \leq t \leq M$, $(x, Y^{TX}), (x', Y^{TX'}) \in \mathcal{X}$,*

$$r^X_{b,t} \left((x, Y^{TX}), (x', Y^{TX'}) \right)$$
$$\leq C \left(\exp\left(-C' d^2(x, x')\right) + \exp\left(-C'' d(x, x')/b^2\right) \right). \quad (13.2.25)$$

Proof. We will still use arguments taken from the proof of Theorem 13.2.1, but we will complement them by a new crucial step. By (13.2.2), we may and we will assume that $d(x, x') \geq 4$, and also that $b > 0$ is small.

We still start from equation (13.2.10). As before, we will only estimate the first term in the right-hand side of this equation. Recall that $s^X_{b,t}$ is the smooth kernel associated with $\exp\left(-t\mathcal{B}^X_b\right)$. By (11.5.10), we get

$$\int_{\substack{(z, Z^{TX}) \in \mathcal{X} \\ d(x, z) \geq d(x, x')/2}} r^X_{b,t/2} \left((x, Y^{TX}), (z, Z^{TX}) \right) dz dZ^{TX}$$

$$= \exp\left(-\left|Y^{TX}\right|^2/2\right) \int_{\substack{(z, Z^{TX}) \in \mathcal{X} \\ d(x, z) \geq d(x, x')/2}} s^X_{b,t/2} \left((x, Y^{TX}), (z, Z^{TX}) \right)$$

$$\exp\left(\left|Z^{TX}\right|^2/2\right) dz dZ^{TX}. \quad (13.2.26)$$

Again, we may and we will assume that $(x, Y^{TX}) = (x_0, Y^{\mathfrak{p}})$. Instead of (13.2.11), we consider the stochastic differential equation in (12.2.1). By

(12.2.10) and (13.2.26), we get

$$\int_{\substack{(z,Z^{TX})\in\mathcal{X} \\ d(x_0,z)\geq d(x_0,x')/2}} r^X_{b,t/2}\left((x_0,Y^{\mathfrak{p}}),(z,Z^{TX})\right)dzdZ^{TX}$$

$$= \exp\left(-\left|Y^{\mathfrak{p}}\right|^2/2\right)E\left[1_{d(x_0,x_{t/2})\geq d(x_0,x')/2}\exp\left(\left|Y^{TX}_{t/2}\right|^2/2\right)\right]. \quad (13.2.27)$$

Let $k:\mathbf{R}_+\to\mathbf{R}_+$ be a smooth increasing function such that

$$k(u) = u^2 \text{ for } u\leq 1/2, \quad (13.2.28)$$
$$= u \text{ for } u\geq 1.$$

For $z\in X$, set

$$f(z) = k(d(x_0,z)). \quad (13.2.29)$$

Then f is smooth. Since $d(x_0,x')\geq 4$, if $d(x_0,z)\geq d(x_0,x')/2$, then $f(z) = d(x_0,z)$. Since f and its first derivative vanish at x_0, if $d(x_0,x_{t/2})\geq d(x_0,x')/2$, by (12.3.19), we get

$$d(x_0,x_{t/2}) = -b\nabla_{Y^{TX}_{t/2}}f(x_{t/2}) + \int_0^{t/2}\nabla_{\delta w_s}f(x_s)$$
$$+ \int_0^{t/2}\nabla^{TX}_{Y^{TX}_s}\nabla_{Y^{TX}_s}f(x_s)ds. \quad (13.2.30)$$

By (13.2.30), if $d(x_0,x_{t/2})\geq d(x_0,x')/2$, at least one of the terms in the right-hand side of (13.2.30) is larger than $d(x_0,x')/6$. We will consider the contribution of each term to the expectation in the right-hand side of (13.2.27). This will be in increasing order of difficulty.

By (13.1.2), $\nabla.f$ is uniformly bounded, so that

$$\left|\nabla_{Z^{TX}}f(z)\right|\leq c\left|Z^{TX}\right|. \quad (13.2.31)$$

Therefore if $-b\nabla_{Z^{TX}}f(z)\geq d(x_0,x')/6$, then

$$\left|Z^{TX}\right|\geq \frac{c'}{b}d(x_0,x'). \quad (13.2.32)$$

By an obvious analogue of (13.2.26), (13.2.27), we get

$$\int_{\substack{(z,Z^{TX})\in\mathcal{X} \\ |Z^{TX}|\geq\frac{c'}{b}d(x_0,x')}} r^X_{b,t/2}\left((x_0,Y^{\mathfrak{p}}),(z,Z^{TX})\right)dzdZ^{TX}$$

$$= \exp\left(-\left|Y^{\mathfrak{p}}\right|^2/2\right)E\left[1_{|Y^{TX}_{t/2}|\geq\frac{c'}{b}d(x_0,x')}\exp\left(\left|Y^{TX}_{t/2}\right|^2/2\right)\right]. \quad (13.2.33)$$

By proceeding as in (12.9.27) and in (13.2.5), we obtain

$$\int_{\substack{(z,Z^{TX})\in\mathcal{X} \\ |Z^{TX}|\geq\frac{c'}{b}d(x_0,x')}} r^X_{b,t/2}\left((x_0,Y^{\mathfrak{p}}),(z,Z^{TX})\right)dzdZ^{TX}$$

$$= \int_{\mathfrak{p}} h^{\mathfrak{p}}_{t/2b^2}(Y^{\mathfrak{p}},Z^{\mathfrak{p}})1_{|Z^{\mathfrak{p}}|\geq\frac{c'}{b}d(x_0,x')}dZ^{\mathfrak{p}}. \quad (13.2.34)$$

By Chebyshev's inequality, for $u \geq 0$,

$$\int_{\mathfrak{p}} h_{t/2b^2}^{\mathfrak{p}} (Y^{\mathfrak{p}}, Z^{\mathfrak{p}}) 1_{|Z^{\mathfrak{p}}| \geq \frac{c'}{b} d(x_0, x')} dZ^{\mathfrak{p}} \leq \exp \left(-\frac{u}{2} c'^2 d^2 (x_0, x') / b^2 \right)$$

$$\int_{\mathfrak{p}} h_{t/2b^2}^{\mathfrak{p}} (Y^{\mathfrak{p}}, Z^{\mathfrak{p}}) \exp \left(\frac{u}{2} |Z^{\mathfrak{p}}|^2 \right) dZ^{\mathfrak{p}}. \quad (13.2.35)$$

In (13.2.35), we take

$$u = \frac{1}{2} \tanh \left(\epsilon / 2 M^2 \right). \quad (13.2.36)$$

Note that $u < 1 < \coth \left(t/2b^2 \right)$. Also for $t \geq \epsilon, b \leq M$, $\coth \left(t/2b^2 \right) \leq \coth \left(\epsilon/2M^2 \right)$. Since $0 < u < 1/2$, we get

$$\frac{u \coth \left(t/2b^2 \right) - 1}{\coth \left(t/2b^2 \right) - u} \leq \frac{u \coth \left(\epsilon/2M^2 \right) - 1}{\coth \left(\epsilon/2M^2 \right) - u} = \frac{-u}{1 - 2u^2} \leq -u. \quad (13.2.37)$$

By (10.7.11), (13.2.35)–(13.2.37), we obtain

$$\int_{\mathfrak{p}} h_{t/2b^2}^{\mathfrak{p}} (Y^{\mathfrak{p}}, Z^{\mathfrak{p}}) 1_{|Z^{\mathfrak{p}}| \geq \frac{c'}{b} d(x_0, x')} dZ^{\mathfrak{p}} \leq \left(\frac{2}{1 - \tanh \left(\epsilon/2M^2 \right)/2} \right)^{m/2}$$

$$\exp \left(-\frac{1}{4} \tanh \left(\epsilon/2M^2 \right) \left(|Y^{\mathfrak{p}}|^2 + c'^2 d^2 (x_0, x') / b^2 \right) \right). \quad (13.2.38)$$

By (13.2.31)–(13.2.38), the contribution of the first term in the right-hand side of (13.2.30) to estimating the right-hand side of (13.2.27) is compatible with (13.2.25).

By Hölder's inequality, for $\theta \in]1, +\infty[$,

$$E \left[1_{\int_0^{t/2} \nabla_{\delta w_s} f(x_s) \geq d(x_0, x')/6} \exp \left(\left| Y_{t/2}^{TX} \right|^2 /2 \right) \right]$$

$$\leq \left(E \left[\exp \left(\theta \left| Y_{t/2}^{TX} \right|^2 /2 \right) \right] \right)^{1/\theta}$$

$$P \left[\int_0^{t/2} \nabla_{\delta w_s} f(x_s) \geq d(x_0, x')/6 \right]^{(\theta-1)/\theta}. \quad (13.2.39)$$

Since $\nabla.f$ is uniformly bounded, by a well-known inequality on Itô integrals [StV79, eq. (2.1) in Theorem 2.4.1], there exist $C > 0, C' > 0$ such that for $\epsilon \leq t \leq M$,

$$P \left[\int_0^{t/2} \nabla_{\delta w_s} f(x_s) \geq d(x_0, x')/6 \right] \leq C \exp \left(-C' d^2 (x_0, x') \right). \quad (13.2.40)$$

In the sequel, we take $\theta = 3/2$. By (10.8.12), given $\epsilon > 0$, for $b > 0$ small enough and $t \geq \epsilon$, we get

$$E \left[\exp \left(\theta \left| Y_{t/2}^{TX} \right|^2 /2 \right) \right] \leq C \exp \left(|Y^{\mathfrak{p}}|^2 /2 \right). \quad (13.2.41)$$

By (13.2.39)–(13.2.41), we get

$$\exp\left(-\left|Y^{\mathfrak{p}}\right|^2/2\right) E\left[1_{\int_0^{t/2} \nabla_{\delta w_s} f(x_s) \geq d(x_0, x')/6} \exp\left(\left|Y_{t/2}^{TX}\right|^2/2\right)\right]$$
$$\leq C \exp\left(-C'd^2(x_0, x') - C''\left|Y^{\mathfrak{p}}\right|^2\right). \quad (13.2.42)$$

By (13.2.42), the contribution of the second term in the right-hand side of (13.2.30) to the right-hand side of (13.2.27) is also compatible with (13.2.25).

We now come to the most difficult step in the proof, i.e., to the control of the contribution of the third term in the right-hand side of (13.2.30). We still take $\theta = 3/2$. As in (13.2.39), we get

$$E\left[1_{\int_0^{t/2} \nabla_{Y_s^{TX}}^{TX} \nabla_{Y_s^{TX}} f(x_s) ds \geq d(x_0, x')/6} \exp\left(\left|Y_{t/2}^{TX}\right|^2/2\right)\right]$$
$$\leq \left(E\left[\exp\left(\theta\left|Y_{t/2}^{TX}\right|^2/2\right)\right]\right)^{1/\theta}$$
$$P\left[\int_0^{t/2} \nabla_{Y_s^{TX}}^{TX} \nabla_{Y_s^{TX}} f(x_s) ds \geq d(x_0, x')/6\right]^{(\theta-1)/\theta}. \quad (13.2.43)$$

By (13.1.17), there is $C > 0$ such that

$$\nabla^{TX}_{\cdot}\nabla_{\cdot} f \leq \frac{C}{2}. \quad (13.2.44)$$

From (13.2.44), we get

$$\int_0^{t/2} \nabla_{Y_s^{TX}}^{TX} \nabla_{Y_s^{TX}} f(x_s) ds \leq \frac{C}{2} \int_0^{t/2} \left|Y_s^{TX}\right|^2 ds. \quad (13.2.45)$$

By (13.2.45), using Chebyshev's inequality, we find that for any $\alpha > 0$,

$$P\left[\int_0^{t/2} \nabla_{Y_s^{TX}}^{TX} \nabla_{Y_s^{TX}} f(x_s) ds \geq d(x_0, x')/6\right] \leq \exp\left(-\alpha d(x_0, x')/6b^2\right)$$
$$E\left[\exp\left(\frac{\alpha C}{2b^2} \int_0^{t/2} \left|Y_s^{TX}\right|^2 ds\right)\right]. \quad (13.2.46)$$

In the sequel, we take $\beta, 0 < \beta < 1$ and we choose α given by

$$\alpha = \frac{\beta^2}{C}. \quad (13.2.47)$$

Set

$$P_\beta^{\mathfrak{p}} = \frac{1}{2}\left(-\Delta^{\mathfrak{p}} + 2\nabla_{Y^{\mathfrak{p}}}\right) - \frac{\beta^2}{2}\left|Y^{\mathfrak{p}}\right|^2. \quad (13.2.48)$$

For $t > 0$, let $k_{\beta,t}^{\mathfrak{p}}(Y^{\mathfrak{p}}, Z^{\mathfrak{p}})$ be the smooth kernel associated with $\exp\left(-tP_\beta^{\mathfrak{p}}\right)$. Let $Y_{\cdot}^{\mathfrak{p}}$ be taken as in (12.2.8). By the Feynman-Kac formula, we get

$$E\left[\exp\left(\frac{\beta^2}{2b^2} \int_0^{t/2} \left|Y^{\mathfrak{p}}\right|^2 ds\right)\right] = \int_{\mathfrak{p}} k_{\beta,t/2b^2}^{\mathfrak{p}}(Y^{\mathfrak{p}}, Z^{\mathfrak{p}}) dZ^{\mathfrak{p}}. \quad (13.2.49)$$

Now we use the notation in the proof of Theorem 13.2.1. Let $\rho \in]0,1[$ be given by

$$\rho^2 = 1 - \beta^2. \tag{13.2.50}$$

By (10.4.3), (10.4.4), (13.2.16), and (13.2.48), we get

$$\exp\left(-|Y^{\mathfrak{p}}|^2/2\right) P_{\beta}^{\mathfrak{p}} \exp\left(|Y^{\mathfrak{p}}|^2/2\right) = O_{\rho}^{\mathfrak{p}}. \tag{13.2.51}$$

By (13.2.49) and (13.2.51), we get

$$E\left[\exp\left(\frac{\beta^2}{2b^2}\int_0^{t/2}|Y_s^{\mathfrak{p}}|^2\,ds\right)\right] = \exp\left(|Y^{\mathfrak{p}}|^2/2\right)$$

$$\int_{\mathfrak{p}} h_{\rho,t/2b^2}^{\mathfrak{p}}\left(Y^{\mathfrak{p}}, Z^{\mathfrak{p}}\right) \exp\left(-|Z^{\mathfrak{p}}|^2/2\right) dZ^{\mathfrak{p}}. \tag{13.2.52}$$

By (13.2.20) and (13.2.52), we obtain

$$E\left[\exp\left(\frac{\beta^2}{2b^2}\int_0^{t/2}|Y_s^{\mathfrak{p}}|^2\,ds\right)\right] = \exp\left(mt\left(1-\rho\right)/4b^2\right)\exp\left(|Y^{\mathfrak{p}}|^2/2\right)$$

$$\int_{\mathfrak{p}} h_{\rho t/2b^2}^{\mathfrak{p}}\left(\sqrt{\rho}Y^{\mathfrak{p}}, Z^{\mathfrak{p}}\right) \exp\left(-|Z^{\mathfrak{p}}|^2/2\rho\right) dZ^{\mathfrak{p}}. \tag{13.2.53}$$

Now we use equation (10.7.10), with t replaced by $\rho t/2b^2$, and $u = -1/\rho$, together with (13.2.53), and we get

$$E\left[\exp\left(\frac{\beta^2}{2b^2}\int_0^{t/2}|Y_s^{\mathfrak{p}}|^2\,ds\right)\right] = \exp\left(mt\left(1-\rho\right)/4b^2\right)$$

$$\left[\frac{e^{\rho t/2b^2}}{\cosh\left(\rho t/2b^2\right) + \frac{\sinh(\rho t/2b^2)}{\rho}}\right]^{m/2} \exp\left(\frac{1}{2}\frac{\beta^2 \tanh\left(\rho t/2b^2\right)}{\rho + \tanh\left(\rho t/2b^2\right)}|Y^{\mathfrak{p}}|^2\right). \tag{13.2.54}$$

Incidentally, observe that if we replace β by γb with $\gamma \geq 0$, and if we make $b \to 0$ in (13.2.54), the limit is $\exp\left(mt\gamma^2/8\right)$, which indicates that as $b \to 0$, the probability law of $\int_0^{t/2}|Y_s^{TX}|^2\,ds$ converges to the Dirac mass at $mt/4$. This fact is very important from a probabilistic point of view, and plays a key role in the proof of our main result. We refer to Proposition 14.10.1 for more details on this point.

We now go back to the case where β is fixed with $0 < \beta < 1$. By (13.2.41), (13.2.43), (13.2.46), (13.2.47), (13.2.54), for $b > 0$ small enough and $\epsilon \leq t \leq M$, we obtain

$$\exp\left(-|Y^{\mathfrak{p}}|^2/2\right) E\left[1_{\int_0^{t/2}\nabla_{Y_s^{TX}}^{TX}\nabla_{Y_s^{TX}}f(x_s)ds \geq d(x_0,x')/6}\exp\left(\left|Y_{t/2}^{TX}\right|^2/2\right)\right]$$

$$\leq C'\left[\exp\left(-|Y^{\mathfrak{p}}|^2/2\right)\exp\left(-\frac{\beta^2}{C}d\left(x_0,x'\right)/6b^2\right)\right.$$

$$\exp\left(mt\left(1-\rho\right)/4b^2\right)\exp\left(\frac{1}{2}\frac{\beta^2 \tanh\left(\rho t/2b^2\right)}{\rho + \tanh\left(\rho t/2b^2\right)}|Y^{\mathfrak{p}}|^2\right)\right]^{(\theta-1)/\theta}. \tag{13.2.55}$$

Equivalently,

$$\exp\left(-|Y^{\mathfrak{p}}|^2/2\right) E\left[1_{\int_0^{t/2} \nabla^{TX}_{YTX}\nabla_{YTX}f(x_s)ds\geq d(x_0,x')/6}\exp\left(\left|Y^{TX}_{t/2}\right|^2/2\right)\right]$$

$$\leq C'\left[\exp\left(-\frac{\beta^2}{C}d\left(x_0,x'\right)/6b^2\right)\right.$$

$$\left.\exp\left(mt\left(1-\rho\right)/4b^2\right)\exp\left(-\frac{1}{2}\frac{\rho\left(1+\rho\tanh\left(\rho t/2b^2\right)\right)}{\rho+\tanh\left(\rho t/2b^2\right)}|Y^{\mathfrak{p}}|^2\right)\right]^{(\theta-1)/\theta}.$$

$$(13.2.56)$$

Now observe that since $\rho \leq 1$,

$$\frac{1+\rho\tanh\left(\rho t/2b^2\right)}{\rho+\tanh\left(\rho t/2b^2\right)} \geq 1. \qquad (13.2.57)$$

By (13.2.56), (13.2.57), we get

$$\exp\left(-|Y^{\mathfrak{p}}|^2/2\right) E\left[1_{\int_0^{t/2} \nabla^{TX}_{YTX}\nabla_{YTX}f(x_s)ds\geq d(x_0,x')/6}\exp\left(\left|Y^{TX}_{t/2}\right|^2/2\right)\right]$$

$$\leq C'\left[\exp\left(-\left(\frac{\beta^2}{C}d\left(x_0,x'\right)/6b^2-mt\left(1-\rho\right)/4b^2\right)-\frac{\rho}{2}|Y^{\mathfrak{p}}|^2\right)\right]^{(\theta-1)/\theta}.$$

$$(13.2.58)$$

For $d\left(x_0,x'\right)$ large enough,

$$\frac{\beta^2}{C}d\left(x_0,x'\right)/6-mt\left(1-\rho\right)/4 \geq \frac{\beta^2}{C}d\left(x_0,x'\right)/12. \qquad (13.2.59)$$

By (13.2.2), (13.2.58), and (13.2.59), the contribution of the third term in (13.2.30) to the right-hand side of (13.2.27) is also compatible with the estimate (13.2.25). The proof of our theorem is completed. □

Remark 13.2.3. A complete explanation for the remark that follows equation (13.2.54) will be given in Proposition 14.10.1. Also note that if we only assume $0 < t \leq M, 0 < b \leq M$, using (10.7.11), the bound in (13.2.38) can be replaced by the fact that given $0 < u < 1$,

$$\int_{\mathfrak{p}} h^{\mathfrak{p}}_{t/2b^2}\left(Y^{\mathfrak{p}}, Z^{\mathfrak{p}}\right) 1_{|Z^{\mathfrak{p}}|\geq \frac{c'}{b}d(x_0,x')}dZ^{\mathfrak{p}}$$

$$\leq \left(\frac{2}{1-u}\right)^{m/2}\exp\left(\frac{1}{2}u\left(|Y^{\mathfrak{p}}|^2-\frac{c'^2}{b^2}d^2\left(x_0,x'\right)\right)\right). \qquad (13.2.60)$$

Similarly, when $0 < t \leq M$, there is still a bound like (13.2.42), with a negative constant C'' in the right-hand side.

We will now obtain the estimate we were looking for.

Theorem 13.2.4. *Given* $\epsilon > 0, M > 0, \epsilon \le M$, *there exist* $C > 0, C' > 0$ *such that for* $0 < b \le M, \epsilon \le t \le M$, $\left(x, Y^{TX}\right), \left(x', Y^{TX'}\right) \in \mathcal{X}$,

$$r_{b,t}^{X}\left(\left(x, Y^{TX}\right), \left(x', Y^{TX'}\right)\right)$$
$$\le C \exp\left(-C'\left(d^2(x, x') + \left|Y^{TX}\right|^2 + \left|Y^{TX'}\right|^2\right)\right). \quad (13.2.61)$$

Proof. When $d(x, x')$ remains uniformly bounded, our estimate follows from equation (13.2.2) in Theorem 13.2.1. Also by taking $d(x, x')$ large enough, equation (13.2.61) follows from (13.2.1), (13.2.2), and (13.2.25). $\qquad\square$

13.3 Bounds on the scalar heat kernel on $\widehat{\mathcal{X}}$ for bounded b

Recall that the kernel $\mathfrak{r}_{b,t}^{X}\left((x, Y), (x', Y')\right)$ on $\widehat{\mathcal{X}}$ was defined in Definition 11.6.1. Now we state the obvious extension of Theorem 13.2.4 to the kernel $\mathfrak{r}_{b,t}^{X}$.

Theorem 13.3.1. *Given* $\epsilon > 0, M > 0, \epsilon \le M$, *there exist* $C > 0, C' > 0$ *such that for* $0 < b \le M, \epsilon \le t \le M$, $(x, Y), (x', Y') \in \mathcal{X}$,

$$\mathfrak{r}_{b,t}^{X}\left((x, Y), (x', Y')\right) \le C \exp\left(-C'\left(d^2(x, x') + |Y|^2 + |Y'|^2\right)\right). \quad (13.3.1)$$

Proof. The proof of our theorem will be obtained by following the same steps that led to the proof of Theorem 13.2.4, i.e., by establishing the obvious extensions of Theorems 13.2.1 and 13.2.2. Of course, we will also use the rough estimates for the kernel $\mathfrak{r}_{b,t}^{X}$ that were established in Theorem 12.10.1.

First, we review the proof of Theorem 13.2.1. The obvious analogue of equation (13.2.5) holds, i.e., if $Y \in \mathfrak{g}$,

$$\int_{\widehat{\mathcal{X}}} \mathfrak{r}_{b,t/2}^{X}\left((x_0, Y), (z, Z)\right) dz dZ = \int_{\mathfrak{g}} h_{t/2b^2}^{\mathfrak{p}\oplus\mathfrak{k}}(Y, Z) dZ. \quad (13.3.2)$$

By proceeding as in (13.2.6), we get the obvious analogue of (13.2.8),

$$\mathfrak{r}_{b,t}^{X}\left((x_0, Y), (x', Y')\right) \le C' \exp\left(-c'|Y|^2\right). \quad (13.3.3)$$

The obvious analogue of (13.2.10) holds for $\mathfrak{r}_{b,t}^{X}$. Moreover, if $Y = \left(Y^{\mathfrak{p}}, Y^{\mathfrak{k}}\right)$,

$$\int_{\substack{(z,Z)\in\widehat{\mathcal{X}} \\ d(x_0,z)\ge d(x_0,x')/2}} \mathfrak{r}_{b,t/2}^{X}\left((x_0, Y), (z, Z)\right) dz dZ$$

$$= \int_{\substack{(z,Z^{TX})\in\mathcal{X} \\ d(x_0,z)\ge d(x_0,x')/2}} r_{b,t/2}^{X}\left((x_0, Y^{\mathfrak{p}}), (z, Z^{TX})\right) dz dZ^{TX}$$

$$\int_{\mathfrak{k}} h_{t/2b^2}^{\mathfrak{k}}\left(Y^{\mathfrak{k}}, Z^{\mathfrak{k}}\right) dZ^{\mathfrak{k}}. \quad (13.3.4)$$

Using (10.7.12), (13.2.23), and (13.2.24), we get the analogue of these bounds for the left-hand side of (13.3.4). This completes the proof of the analogue of Theorem 13.2.1.

Now we review the proof of the analogue of Theorem 13.2.2. We combine (10.7.12), (13.2.26) and the subsequent estimates in the proof of Theorem 13.2.2 together with (13.3.4), so that the analogue of Theorem 13.2.2 also holds for $\mathfrak{r}_{b,t}^X$.

Then we can use the arguments of the proof of Theorem 13.2.4 and obtain (13.3.1). The proof of our theorem is completed. $\qquad\square$

Chapter Fourteen

The heat kernel $q_{b,t}^X$ for bounded b

The purpose of this chapter is to establish the estimates of Theorem 4.5.2 for the hypoelliptic heat kernel $q_{b,t}^X\left((x,Y),(x',Y')\right)$ on $\widehat{\mathcal{X}}$. More precisely, we show that for bounded $b > 0$, the heat kernel verifies uniform Gaussian type estimates. Also we study the limit as $b \to 0$ of this heat kernel.

The method consists in using the techniques of chapters 12 and 13 for the scalar heat kernels over $\mathcal{X}$ and $\widehat{\mathcal{X}}$, and to control the kernel $q_{b,t}^X$ by using a Feynman-Kac formula. Still, because the operator $\mathcal{L}_{A,b}^X$ contains matrix terms that themselves diverge as $b \to 0$, we have to be careful. In particular, the contribution of the representation ρ^E will be obtained by using results of chapter 12 applied to the symmetric space attached to the complexification $K_{\mathbf{C}}$ of K.

This chapter is organized as follows. In section 14.1, we give a probabilistic construction of the elliptic heat operators $\exp\left(-t\mathcal{L}_A^X\right)$ on X.

In section 14.2, we extend the wave equation considerations of section 12.3 to the operator $\mathcal{L}_b^X$.

In section 14.3, by a trivial change of variables on $\widehat{\mathcal{X}}$, we obtain a new operator $\mathcal{L}_{A,b}^{X\prime}$.

In section 14.4, we give a probabilistic construction of the heat operators $\exp\left(-t\mathcal{L}_{A,b}^{X\prime}\right)$. A matrix valued process $U.$ acting on $\Lambda^{\cdot}\left(T^*X \oplus N^*\right) \otimes F$ appears, which can be naturally expressed in the form $U. = V. \otimes W..$ The estimation of the heat kernel $q_{b,t}^X$ will be obtained by estimating $|V_t|$ and $|W_t|$.

In section14.5, we estimate $|V_t|$.

In section 14.6, we estimate $|W_t|$ via the introduction of the symmetric space attached to $K_{\mathbf{C}}$.

In section 14.7, when E is the trivial representation, we prove uniform estimates on $q_{b,t}^X$ for bounded $b > 0$.

In section 14.8, we establish such estimates in the general case, i.e., we obtain the first part of Theorem 4.5.2.

In section 14.9, we obtain uniform rough estimates on higher order derivatives of our hypoelliptic heat kernels. Such estimates will be needed in the proof of the existence of their limit as $b \to 0$.

In section 14.10, we study the behavior of the process $V.$ as $b \to 0$. We prove in particular that it converges in probability to an explicit deterministic process.

Finally, in section 14.11, we obtain the limit as $b \to 0$ of the heat kernel

$q_{b,t}^X$, i.e., we establish the second part of Theorem 4.5.2.

14.1 A probabilistic construction of $\exp\left(-t\mathcal{L}_A^X\right)$

We use the same notation as in section 12.1. Let $s \in \mathbf{R}_+ \to w_s \in \mathfrak{p}$ be a Brownian motion with values in $\mathfrak{p}$, with $w_0 = 0$. Let E denote the corresponding expectation operator. Again, we consider equation (12.1.3), i.e.,

$$\dot{x} = \dot{w}, \qquad\qquad x_0 = p1, \qquad\qquad (14.1.1)$$

and also equation (12.1.4),

$$\dot{g} = \dot{w}, \qquad\qquad g_0 = 1, \qquad\qquad (14.1.2)$$

so that as in (12.1.5),

$$x_{.} = pg_{.}. \qquad\qquad (14.1.3)$$

Here, τ_t^0 denotes parallel transport on F with respect to the connection ∇^F from F_{x_0} into F_{x_t}, and τ_0^t is its inverse. Note that since $g_{.}$ is the horizontal lift of $x_{.}$, τ_t^0 is immediately obtained from $g_{.}$.

We will establish the obvious extension of equation (12.1.6).

Proposition 14.1.1. Let $u \in C^{\infty,c}(X, F)$. For any $t \geq 0$,

$$\exp\left(-t\mathcal{L}_A^X\right) u(x_0) = \exp\left(-\frac{t}{8}B^*\left(\kappa^{\mathfrak{g}}, \kappa^{\mathfrak{g}}\right) - \frac{t}{2}C^{\mathfrak{k},F} - tA\right) E\left[\tau_0^t u(x_t)\right].$$

$$(14.1.4)$$

Proof. By (2.13.1)–(2.13.3), and (4.5.1), we get

$$\mathcal{L}_A^X = -\frac{1}{2}\Delta^{H,X} + \frac{1}{8}B^*\left(\kappa^{\mathfrak{g}}, \kappa^{\mathfrak{g}}\right) + \frac{1}{2}C^{\mathfrak{k},F} + A. \qquad (14.1.5)$$

Recall that the last three terms in (14.1.5) are parallel with respect to ∇^F. Equation (14.1.4) now follows from (14.1.5) and from Itô's formula. $\qquad\square$

14.2 The operator $\mathcal{L}_b^X$ and the wave equation

Set

$$\mathcal{M}_{A,b}^X = \exp\left(|Y|^2/2\right)\mathcal{L}_{A,b}^X \exp\left(-|Y|^2/2\right). \qquad (14.2.1)$$

For $A = 0$, we use instead the notation $\mathcal{M}_b^X$.

By (2.13.5), (4.5.1), we get

$$\mathcal{M}_{A,b}^X = \frac{1}{2}\left|\left[Y^N, Y^{TX}\right]\right|^2 + \frac{1}{2b^2}\left(-\Delta^{TX\oplus N} + 2\nabla_Y^V\right) + \frac{N^{\Lambda^{\cdot}(T^*X\oplus N^*)}}{b^2}$$

$$+ \frac{1}{b}\left(\nabla_{Y^{TX}}^{C^\infty(TX\oplus N,\widehat{\pi}^*(\Lambda^{\cdot}(T^*X\oplus N^*)\otimes F))} + \widehat{c}\left(\mathrm{ad}\left(Y^{TX}\right)\right)\right.$$

$$\left. - c\left(\mathrm{ad}\left(Y^{TX}\right) + i\theta\mathrm{ad}\left(Y^N\right)\right) - i\rho^E\left(Y^N\right)\right) + A. \quad (14.2.2)$$

Set

$$\mathcal{R} = \nabla_{YTX}^{C^\infty(TX\oplus N, \hat{\pi}^*(\Lambda^{\cdot}(T^*X\oplus N^*)\otimes F))} + \hat{c}\left(\mathrm{ad}\left(Y^{TX}\right)\right)$$
$$- c\left(\mathrm{ad}\left(Y^{TX}\right) + i\theta\mathrm{ad}\left(Y^N\right)\right) - i\rho^E\left(Y^N\right). \quad (14.2.3)$$

By (14.2.2), (14.2.3), we get

$$\mathcal{M}_b^X = \frac{1}{2}\left|\left[Y^N, Y^{TX}\right]\right|^2 + \frac{1}{2b^2}\left(-\Delta^{TX\oplus N} + 2\nabla_Y^V\right) + \frac{N^{\Lambda^{\cdot}(T^*X\oplus N^*)}}{b^2} + \frac{\mathcal{R}}{b}. \quad (14.2.4)$$

Note that $\mathcal{R}$ maps smooth sections of $\Lambda^{\cdot}(T^*X\oplus N^*)\otimes F$ over X to smooth sections of $\hat{\pi}^*(\Lambda^{\cdot}(T^*X\oplus N^*)\otimes F)$ over $\hat{\mathcal{X}}$.

Now we establish an extension of equation (12.3.4) in Proposition 12.3.1.

Proposition 14.2.1. *If $s \in C^\infty(X, \Lambda^{\cdot}(T^*X\oplus N^*)\otimes F)$, then*

$$\left(\mathcal{M}_b^X - \frac{1}{2}\left|\left[Y^N, Y^{TX}\right]\right|^2\right)\left(1 - b\left(1 + N^{\Lambda^{\cdot}(T^*X\oplus N^*)}\right)^{-1}\mathcal{R}\right)\hat{\pi}^*s$$
$$= \frac{N^{\Lambda^{\cdot}(T^*X\oplus N^*)}}{b^2}\hat{\pi}^*s - \mathcal{R}\left(1 + N^{\Lambda^{\cdot}(T^*X\oplus N^*)}\right)^{-1}\mathcal{R}\hat{\pi}^*s. \quad (14.2.5)$$

Proof. The eigenspace of the fibrewise operator $\frac{1}{2}\left(-\Delta^{TX\oplus N} + 2\nabla_Y^V\right)$ associated with the eigenvalue 0 is generated by the constants, and the linear functions of Y generate the eigenspace associated with the eigenvalue 1. Equation (14.2.5) then follows from (14.2.4) and from the fact that $\mathcal{R}\hat{\pi}^*s$ depends linearly on Y. $\qquad\square$

14.3 Changing Y into $-Y$

Definition 14.3.1. *Let I be the map $s(x, Y) \to s(x, -Y)$. Set*

$$\mathcal{L}_{A,b}^{X\prime} = I\mathcal{L}_{A,b}^X I^{-1}. \quad (14.3.1)$$

By (2.13.5), (4.5.1), and (14.3.1), we get

$$\mathcal{L}_{A,b}^{X\prime} = \frac{1}{2}\left|\left[Y^N, Y^{TX}\right]\right|^2 + \frac{1}{2b^2}\left(-\Delta^{TX\oplus N} + |Y|^2 - m - n\right)$$
$$+ \frac{N^{\Lambda^{\cdot}(T^*X\oplus N^*)}}{b^2} - \frac{1}{b}\left(\nabla_{YTX}^{C^\infty(TX\oplus N, \hat{\pi}^*(\Lambda^{\cdot}(T^*X\oplus N^*)\otimes F))} + \hat{c}\left(\mathrm{ad}\left(Y^{TX}\right)\right)\right.$$
$$\left. - c\left(\mathrm{ad}\left(Y^{TX}\right) + i\theta\mathrm{ad}\left(Y^N\right)\right) - i\rho^E\left(Y^N\right)\right) + A. \quad (14.3.2)$$

For $t > 0$, let $q_{b,t}^{X\prime}((x, Y), (x', Y'))$ be the smooth kernel associated with the operator $\exp\left(-t\mathcal{L}_{A,b}^{X\prime}\right)$. Then

$$q_{b,t}^{X\prime}((x, Y), (x', Y')) = q_{b,t}^X((x, -Y), (x', -Y')). \quad (14.3.3)$$

14.4 A probabilistic construction of $\exp\left(-t\mathcal{L}_{A,b}^{X'}\right)$

Let $s \in \mathbf{R}_+ \to w_s = w_s^{\mathfrak{p}} + w_s^{\mathfrak{k}} \in \mathfrak{g} = \mathfrak{p} \oplus \mathfrak{k}$ be a Brownian motion, with $w_0 = 0$. Instead of (13.2.11), we consider the stochastic differential equation

$$\dot{x} = \frac{Y^{TX}}{b}, \qquad\qquad \dot{Y} = \frac{\dot{w}}{b}, \qquad (14.4.1)$$
$$x_0 = p1, \qquad\qquad Y_0 = Y.$$

Let τ_t^0 denote parallel transport from x_0 to x_t with respect to the connection $\nabla^{\Lambda^{\cdot}(T^*X\oplus N^*)\otimes F}$, and let τ_0^t be its inverse.

Along the path $x.$, we trivialize $\Lambda^{\cdot}(T^*X \oplus N^*), F$ by parallel transport with respect to the connections $\nabla^{\Lambda^{\cdot}(T^*X\oplus N^*)}, \nabla^F$. Consider the differential equation along the path $x.$,

$$\frac{dU}{ds} = U\left[-\frac{N^{\Lambda^{\cdot}(T^*X\oplus N^*)}}{b^2} + \frac{1}{b}\widehat{c}\left(\mathrm{ad}\left(Y^{TX}\right)\right)\right.$$
$$\left. - \frac{1}{b}c\left(\mathrm{ad}\left(Y^{TX}\right) + i\theta\mathrm{ad}\left(Y^N\right)\right) - \frac{i}{b}\rho^E\left(Y^N\right)\right], \qquad (14.4.2)$$
$$U_0 = 1.$$

Theorem 14.4.1. Let $s \in C^{\infty,c}\left(\widehat{\mathcal{X}}, \widehat{\pi}^*\left(\Lambda^{\cdot}\left(T^*X \oplus N^*\right) \otimes F\right)\right)$. For $t > 0$, the following identity holds:

$$\exp\left(-t\mathcal{L}_{A,b}^{X'}\right)s\left(x_0, Y\right) = E\left[\exp\left(\frac{(m+n)t}{2b^2} - \frac{1}{2}\int_0^t\left|[Y^N, Y^{TX}]\right|^2 ds\right.\right.$$
$$\left.\left. - \frac{1}{2b^2}\int_0^t |Y|^2 ds - tA\right)U_t\tau_0^t s\left(x_t, Y_t\right)\right]. \quad (14.4.3)$$

Proof. Using the results of section 11.8, equation (14.4.3) can be proved by the same arguments as equations (12.2.10), (12.9.9), (12.11.1), (12.11.8), and (13.2.12). The only significant difference is that in (14.4.2), (14.4.3), we have used a matrix version of the Feynman-Kac formula. Incidentally, since A is parallel and commutes with $\rho^E\left(Y^N\right)$, its contribution has been factored out in (14.4.3). $\qquad\square$

Definition 14.4.2. Set

$$M_b = -\frac{N^{\Lambda^{\cdot}(T^*X\oplus N^*)}}{b^2} + \frac{1}{b}\widehat{c}\left(\mathrm{ad}\left(Y^{TX}\right)\right) - \frac{1}{b}c\left(\mathrm{ad}\left(Y^{TX}\right) + i\theta\mathrm{ad}\left(Y^N\right)\right). \qquad (14.4.4)$$

As we saw in the proof of Theorem 2.13.2, M_b is self-adjoint.

We use the same trivialization of $\Lambda^{\cdot}(T^*X \oplus N^*), F$ along the path $x.$ as in (14.4.2). Let $V_\cdot, W_\cdot$ be the solutions of the differential equations,

$$\frac{dV}{ds} = V M_b, \qquad\qquad V_0 = 1, \qquad\qquad (14.4.5)$$

$$\frac{dW}{ds} = W \left[-\frac{i}{b} \rho^E \left(Y^N \right) \right], \qquad W_0 = 1.$$

In (14.4.5), $V_\cdot \in \mathrm{Aut}\left(\Lambda^\cdot \left(T^*X \oplus N^* \right)_{x_0} \right)$, and $W_\cdot \in \mathrm{Aut}\left(E_{x_0} \right)$. Then we have the obvious identity

$$U_\cdot = V_\cdot \otimes W_\cdot. \qquad\qquad (14.4.6)$$

14.5 Estimating $V_\cdot$

As before, we identify $\Lambda^\cdot \left(T^*X \oplus N^* \right)$ with $\Lambda^\cdot \left(T^*X \oplus N^* \right) \otimes_{\mathbf{R}} \mathbf{C}$. We still denote by $\langle\ \rangle$ the Hermitian product on $\Lambda^\cdot \left(T^*X \oplus N^* \right)$.

Proposition 14.5.1. *There is $C > 0$ such that if $b > 0, \eta > 0, f \in \Lambda^\cdot \left(T^*X \oplus N^* \right)$,*

$$\langle M_b f, f \rangle \le \frac{1}{2} \left(\frac{C^2}{\eta} + \frac{\eta}{b^2} \left| Y \right|^2 \right) \left| f \right|^2. \qquad (14.5.1)$$

There exists $C' > 0$ such that for $b > 0, f \in \Lambda^\cdot \left(T^*X \oplus N^* \right)$,

$$\langle M_b f, f \rangle \le \frac{C'}{2} \left| Y \right|^2 \left| f \right|^2 - \frac{1}{2b^2} \left\langle N^{\Lambda^\cdot \left(T^*X \oplus N^* \right)} f, f \right\rangle. \qquad (14.5.2)$$

Proof. If $f \in \Lambda^\cdot \left(T^*X \oplus N^* \right)$, then

$$\langle M_b f, f \rangle \le \frac{C}{b} \left| Y \right| \left| f \right|^2. \qquad (14.5.3)$$

By (14.5.3), we get (14.5.1).

Let $e_1, \dots, e_m$ be an orthonormal basis of TX, let $e_{m+1}, \dots, e_{m+n}$ be an orthonormal basis of N. By (2.16.11), we get

$$\left\langle \left(\widehat{c} \left(\mathrm{ad} \left(Y^{TX} \right) \right) - c \left(\mathrm{ad} \left(Y^{TX} \right) \right) \right) f, f \right\rangle$$
$$= -2 \sum_{\substack{1 \le i \le m \\ m+1 \le j \le m+n}} \left\langle \left[Y^{TX}, e_i \right], e_j \right\rangle \mathrm{Re} \left\langle i_{e_i} i_{e_j} f, f \right\rangle. \qquad (14.5.4)$$

Using (14.5.4), we conclude that

$$\left| \left\langle \left(\widehat{c} \left(\mathrm{ad} \left(Y^{TX} \right) \right) - c \left(\mathrm{ad} \left(Y^{TX} \right) \right) \right) f, f \right\rangle \right| \le C \left| Y^{TX} \right| \left| \sqrt{N^{\Lambda^\cdot \left(T^*X \oplus N^* \right)}} f \right| \left| f \right|. \qquad (14.5.5)$$

By (2.16.11), (2.16.12), we get

$$\left| \left\langle c \left(i\theta \mathrm{ad} \left(Y^N \right) \right) f, f \right\rangle \right| \le C \left| Y^N \right| \left| \sqrt{N^{\Lambda^\cdot \left(T^*X \oplus N^* \right)}} f \right| \left| f \right|. \qquad (14.5.6)$$

Also for any $\eta > 0$,

$$\frac{1}{b} \left| Y \right| \left| \sqrt{N^{\Lambda^\cdot \left(T^*X \oplus N^* \right)}} f \right| \left| f \right| \le \frac{1}{2} \left(\frac{\eta}{b^2} \left\langle N^{\Lambda^\cdot \left(T^*X \oplus N^* \right)} f, f \right\rangle + \frac{1}{\eta} \left| Y \right|^2 \left| f \right|^2 \right). \qquad (14.5.7)$$

By (14.4.4), (14.5.5), (14.5.6) and by taking η small enough in (14.5.7), we get (14.5.2). The proof of our proposition is completed. $\qquad\square$

We take $V.$ as in (14.4.5).

Theorem 14.5.2. *There exists $C > 0$ such that for $\eta > 0, b > 0, t > 0$,*

$$|V_t| \leq \exp\left(\frac{1}{2}\frac{C^2}{\eta}t + \frac{\eta}{2b^2}\int_0^t |Y_s|^2\, ds\right). \tag{14.5.8}$$

Moreover, there exists $C' > 0$ such that for any $b > 0, t > 0$,

$$|V_t| \leq \exp\left(\frac{C'}{2}\int_0^t |Y_s|^2\, ds\right). \tag{14.5.9}$$

Proof. Let $V_.^*$ be the adjoint of $V.$. Then $V_.^*$ is the solution of the differential equation

$$\frac{dV^*}{ds} = M_b V^*, \qquad\qquad V_0^* = 1. \tag{14.5.10}$$

If $f \in \Lambda^\cdot (T^*X \oplus N^*)_{x_0}$, by (14.5.10), we get

$$\frac{d}{ds}|V^*f|^2 = 2\langle M_b V^* f, V^* f\rangle. \tag{14.5.11}$$

By (14.5.1), (14.5.2), and (14.5.11), using Gronwall's lemma, we get (14.5.8) and (14.5.9) for $|V_t^*|$, which is equivalent to the corresponding estimates for $|V_t|$. $\qquad\square$

14.6 Estimating $W.$

We take $W.$ as in (14.4.5). A first crude estimate for W_t is the obvious

$$|W_t| \leq \exp\left(\frac{C}{b}\int_0^t |Y^N|\, ds\right). \tag{14.6.1}$$

From (14.6.1), we find that for any $\eta > 0$,

$$|W_t| \leq \exp\left(\frac{1}{2}\frac{C^2}{\eta}t + \frac{\eta}{2b^2}\int_0^t |Y|^2\, ds\right), \tag{14.6.2}$$

which is exactly of the same type as (14.5.8). Still, the estimate (14.6.2) is useless for small $b > 0$.

Let $K_{\mathbf{C}}$ be the complexification of K. Let $\mathfrak{k}_{\mathbf{C}}$ be the Lie algebra of $K_{\mathbf{C}}$. Then $\mathfrak{k}_{\mathbf{C}} = i\mathfrak{k} \oplus \mathfrak{k}$ is the splitting of $\mathfrak{k}_{\mathbf{C}}$ corresponding to the splitting $\mathfrak{g} = \mathfrak{p} \oplus \mathfrak{k}$. The symmetric bilinear form $\mathrm{Re}\, B|_{\mathfrak{k}_{\mathbf{C}}}$ has the same properties as the symmetric bilinear form B on $\mathfrak{g}$.

Let $1_{K_{\mathbf{C}}}$ denote the unit element in $K_{\mathbf{C}}$. Let $X_K = K_{\mathbf{C}}/K$ be the symmetric space associated with the pair $(K_{\mathbf{C}}, K)$. More generally, we will use the same notation for the objects attached to $K_{\mathbf{C}}$ as for the objects attached to G. Temporarily, we denote by 1_G the unit element in G.

Using the unique decomposition of elements in $K_{\mathbf{C}}$ as in (3.2.26), we find that there exists $C > 0$ such that if $h \in K_{\mathbf{C}}$,

$$\left|\rho^E(h)\right| \leq e^{Cd(p1,ph)}, \qquad \left|\rho^E\left(h^{-1}\right)\right| \leq e^{Cd(p1,ph)}. \tag{14.6.3}$$

Consider the differential equation on X_K,

$$\dot{y} = -\frac{iY^N}{b}, \qquad\qquad \dot{Y}^N = \frac{\dot{w}^{\mathfrak{k}}}{b}, \qquad (14.6.4)$$

$$y_0 = p1_{K_{\mathbf{C}}}, \qquad\qquad Y_0^N = Y^N.$$

As in (12.2.8), we may as well consider the differential equation on $h \in K_{\mathbf{C}}$,

$$\dot{h} = \frac{-iY^N}{b}, \qquad\qquad \dot{Y}^N = \frac{\dot{w}^{\mathfrak{k}}}{b}, \qquad (14.6.5)$$

$$h_0 = 1_{K_{\mathbf{C}}}, \qquad\qquad Y_0^N = Y^N,$$

so that in (14.6.4), $y. = ph..$

By (14.4.5) and (14.6.5), we get

$$W_t = \rho^E(h_t). \qquad (14.6.6)$$

By (14.6.3), (14.6.6), we get

$$|W_t| \leq e^{Cd(y_0, y_t)}, \qquad\qquad \left|W_t^{-1}\right| \leq e^{Cd(y_0, y_t)}. \qquad (14.6.7)$$

14.7 A proof of (4.5.3) when E is trivial

Here, we assume ρ^E to be trivial, so that $U. = V..$ Also we make $A = 0$, so that we eliminate the subscript A. By (14.4.3),

$$\left|\exp\left(-t\mathcal{L}_b^{X\prime}\right) s(x_0, Y)\right|$$
$$\leq E\left[\exp\left(\frac{(m+n)t}{2b^2} - \frac{1}{2b^2}\int_0^t |Y|^2\, ds\right) |V_t|\, |s(x_t, Y_t)|\right]. \qquad (14.7.1)$$

By (14.5.9) and (14.7.1), we obtain

$$\left|\exp\left(-t\mathcal{L}_b^{X\prime}\right) s(x_0, Y)\right| \leq E\left[\exp\left(\frac{(m+n)t}{2b^2} - \frac{1}{2b^2}\int_0^t |Y|^2\, ds\right.\right.$$
$$\left.\left. + \frac{C'}{2}\int_0^t |Y|^2\, ds\right) |s(x_t, Y_t)|\right]. \qquad (14.7.2)$$

Recall that the operator scalar $\mathfrak{A}_b^X$ on $\widehat{\mathcal{X}}$ was defined in (11.6.1). Set

$$\mathfrak{A}_b^{X\prime} = \mathfrak{A}_b^X - \frac{C'}{2}|Y|^2. \qquad (14.7.3)$$

As we shall see later, for $b > 0$ small enough and $t > 0$, the heat operator $\exp\left(-t\mathfrak{A}_b^{X\prime}\right)$ and its smooth heat kernel $\mathfrak{r}_{b,t}^{X\prime}((x,Y),(x',Y'))$ are well-defined. By proceeding as in (12.9.9), (13.2.12), and (14.4.3), we find that if $F \in C^{\infty, c}\left(\widehat{\mathcal{X}}, \mathbf{R}\right)$, if $(x., Y.)$ is taken as in (14.4.1), then

$$\exp\left(-t\mathfrak{A}_b^{X\prime}\right) F(x_0, Y) = E\left[\exp\left(\frac{(m+n)t}{2b^2} - \frac{1}{2b^2}\int_0^t |Y|^2\, ds\right.\right.$$
$$\left.\left. + \frac{C'}{2}\int_0^t |Y|^2\, ds\right) F(x_t, Y_t)\right]. \qquad (14.7.4)$$

Then (14.7.2) can be written in the form

$$\left|\exp\left(-t\mathcal{L}_b^{X'}\right)s\left(x_0,Y\right)\right| \le \exp\left(-t\mathfrak{A}_b^{X'}\right)|s|\left(x_0,Y\right). \tag{14.7.5}$$

From (14.7.5), we get

$$\left|q_{b,t}^{X'}\left((x,Y),(x',Y')\right)\right| \le \mathfrak{r}_{b,t}^{X'}\left((x,Y),(x',Y')\right). \tag{14.7.6}$$

For $b > 0$ small enough, let $\alpha_b \in]0,1[$ be such that

$$\alpha_b^2 = 1 - C'b^2. \tag{14.7.7}$$

Then

$$\mathfrak{A}_b^{X'} = \frac{1}{2b^2}\left(-\Delta^{TX\oplus N} + \alpha_b^2|Y|^2 - m - n\right) - \frac{1}{b}\nabla_{Y^{TX}}. \tag{14.7.8}$$

For $a > 0$, we define K_a as in (2.14.2). Then

$$K_{\sqrt{\alpha_b}}^{-1}\mathfrak{A}_b^{X'}K_{\sqrt{\alpha_b}} = \frac{\alpha_b}{2b^2}\left(-\Delta^{TX\oplus N} + |Y|^2 - m - n\right)$$
$$- \frac{1}{b\sqrt{\alpha_b}}\nabla_{Y^{TX}} - \frac{1-\alpha_b}{2b^2}(m+n). \tag{14.7.9}$$

Set

$$\beta_b = \frac{b}{\alpha_b^{3/2}}, \qquad\qquad \tau_b = \frac{1}{\alpha_b^2}. \tag{14.7.10}$$

Then we can rewrite (14.7.9) in the form

$$K_{\sqrt{\alpha_b}}^{-1}\mathfrak{A}_b^{X'}K_{\sqrt{\alpha_b}} = \tau_b\mathfrak{A}_{\beta_b}^X - \frac{1-\alpha_b}{2b^2}(m+n). \tag{14.7.11}$$

By (14.7.11), it is now clear that for $b > 0$ small enough, for $t > 0$, the operator $\exp\left(-t\mathfrak{A}_b^{X'}\right)$ is well-defined, and moreover,

$$\mathfrak{r}_{b,t}^{X'}\left((x,Y),(x',Y')\right) = \exp\left(\frac{1-\alpha_b}{2b^2}(m+n)t\right)\alpha_b^{(m+n)/2}$$
$$\mathfrak{r}_{\beta_b,\tau_b t}^X\left((x,\sqrt{\alpha_b}Y),(x',\sqrt{\alpha_b}Y')\right). \tag{14.7.12}$$

By combining equation (13.3.1) in Theorem 13.3.1 with (14.3.3), (14.7.6), (14.7.7), and (14.7.12), for $b > 0$ small enough, and $\epsilon \le t \le M$, we get (4.5.3).

Take now $\mu > 0, M > 0, \mu \le M$, and assume that $\mu \le b \le M$. Instead of (14.5.9), we will now use (14.5.8), while taking $\eta = \frac{1}{4}$. The same arguments as before still lead to the proof of (4.5.3). The proof of this estimate is then completed.

Remark 14.7.1. Using (14.5.8) with $b > 0$ small would not lead to the proof of (4.5.3).

14.8 A proof of the estimate (4.5.3) in the general case

Note that if μ, M are taken as in section 14.7, if $\mu \leq b \leq M$, using the estimates (14.5.8) and (14.6.2), the proof of equation (4.5.3) in Theorem 4.5.2 proceeds exactly as in section 14.7. Therefore, in the sequel, we may as well assume $b > 0$ to be small.

By (14.4.3), (14.4.6), (14.5.9), and (14.6.7), we obtain

$$
\left| \exp\left(-t\mathcal{L}_{A,b}^{X\prime}\right) s\left(x_0, Y\right) \right| \leq E\left[\exp\left(\frac{(m+n)\,t}{2b^2} - \frac{1}{2b^2} \int_0^t |Y|^2 \, ds \right.\right.
$$
$$
\left.\left. + \frac{C'}{2} \int_0^t |Y|^2 \, ds + Cd\left(y_0, y_t\right) + |A|\,t \right) \left| s\left(x_t, Y_t\right) \right| \right]. \quad (14.8.1)
$$

Theorem 14.8.1. *There exists $b_0 \in]0,1]$ such that given $\epsilon > 0, M > 0, \epsilon \leq M$, there exist $C_{\epsilon,M} > 0, C'_{\epsilon,M} > 0, C''_{\epsilon,M} > 0$ such that for $b \in]0,b_0], t \in [\epsilon, M], p \geq 0$,*

$$
E\left[\exp\left(\frac{(m+n)\,t}{2b^2} - \frac{1}{2b^2} \int_0^t |Y|^2 \, ds \right.\right.
$$
$$
\left.\left. + \frac{C'}{2} \int_0^t |Y|^2 \, ds + pd\left(y_0, y_t\right) \right) \right]
$$
$$
\leq C_{\epsilon,M} \exp\left(-C'_{\epsilon,M} |Y|^2 + C''_{\epsilon,M} p^2 \right). \quad (14.8.2)
$$

Proof. We use the notation of section 14.7. Let $r_{b,t}^{X_K}$ be the analogue of the smooth kernel $r_{b,t}^X$ on the symmetric space X_K. By (14.6.4) and (14.7.12), we get

$$
E\left[\exp\left(\frac{(m+n)\,t}{2b^2} - \frac{1}{2b^2} \int_0^t |Y|^2 \, ds \right.\right.
$$
$$
\left.\left. + \frac{C'}{2} \int_0^t |Y|^2 \, ds + pd\left(y_0, y_t\right) \right) \right]
$$
$$
= \exp\left(\frac{1-\alpha_b}{2b^2}(m+n)\,t \right) \int_{\mathfrak{p}} h_{\tau_b t/\beta_b^2}^{\mathfrak{p}}\left(\sqrt{\alpha_b} Y^{TX}, Y^{TX\prime} \right) dY^{TX\prime}
$$
$$
\int_{X_K} r_{\beta_b,\tau_b t}^{X_K}\left(\left(y_0, -iY^N\right), \left(y, -iY^{N\prime}\right) \right) \exp\left(pd\left(y_0, y\right) \right) dy \, dY^{N\prime}. \quad (14.8.3)
$$

By (10.7.12), we obtain

$$
\int_{\mathfrak{p}} h_{\tau_b t/\beta_b^2}^{\mathfrak{p}}\left(\sqrt{\alpha_b} Y^{TX}, Y^{TX\prime} \right) dY^{TX\prime}
$$
$$
= \left(\frac{e^{\tau_b t/\beta_b^2}}{\cosh\left(\tau_b t/\beta_b^2 \right)} \right)^{m/2} \exp\left(-\frac{\tanh\left(\tau_b t/\beta_b^2 \right)}{2} \alpha_b |Y^{TX}|^2 \right). \quad (14.8.4)
$$

Also, using (4.1.11), (4.1.12), and equation (13.2.61) in Theorem 13.2.4, we find that for $b > 0$ small enough, and $\epsilon \le t \le M$,

$$\int_{\mathcal{X}_K} r_{\beta_b,\tau_b t}^{X_K} \left(\left(y_0, -iY^N \right), \left(y, -iY^{N'} \right) \right) \exp \left(pd \left(y_0, y \right) \right) dy dY^{N'}$$

$$\le C \exp \left(-C' \left| Y^N \right|^2 + C'' p^2 \right). \quad (14.8.5)$$

By (14.7.7) and (14.8.3)–(14.8.5), we get (14.8.2). The proof of our theorem is completed. $\square$

Recall that the operator $\mathfrak{B}_b^X$ on $\widehat{\mathcal{X}}$ was defined in (11.6.2), (11.6.3). Put

$$\mathfrak{B}_b^{X'} = \mathfrak{B}_b^X - \frac{C'}{2} |Y|^2. \quad (14.8.6)$$

By (11.6.2), (14.7.3), and (14.8.6), we get

$$\mathfrak{B}_b^{X'} = \exp \left(|Y|^2 / 2 \right) \mathfrak{A}_b^{X'} \exp \left(- |Y|^2 / 2 \right). \quad (14.8.7)$$

To construct the semigroup $\exp \left(-t \mathfrak{B}_b^{X'} \right)$, we consider again equation (12.10.1) and the suitable modification of (14.6.4), i.e.,

$$\dot{x} = \frac{Y^{TX}}{b}, \qquad \dot{y} = -\frac{iY^N}{b}, \qquad \dot{Y} = -\frac{Y}{b^2} + \frac{\dot{w}}{b}, \quad (14.8.8)$$

$$x_0 = p1_G, \qquad y_0 = p1_{K_C}, \qquad Y_0 = Y.$$

The first equations in (12.2.8) and (14.6.5) still hold, so that

$$\dot{g} = \frac{Y^{\mathfrak{p}}}{b}, \qquad \qquad \dot{h} = -\frac{iY^{\mathfrak{k}}}{b}, \quad (14.8.9)$$

$$g_0 = 1_G, \qquad \qquad h_0 = 1_{K_C}.$$

If $F \in C^{\infty,c} \left(\widehat{\mathcal{X}}, \mathbf{R} \right)$, for $t > 0$,

$$\exp \left(-t \mathfrak{B}_b^{X'} \right) F \left(x_0, Y \right) = E \left[\exp \left(\frac{C'}{2} \int_0^t |Y|^2 \, ds \right) F \left(x_t, Y_t \right) \right]. \quad (14.8.10)$$

Theorem 14.8.2. *There exists $b_0 \in]0,1]$ such that given $\epsilon > 0, M > 0, \epsilon \le M$, there exist $C_{\epsilon,M} > 0, C'_{\epsilon,M} > 0, C''_{\epsilon,M} > 0$ such that for $b \in]0, b_0], t \in [\epsilon, M], p \ge 0$,*

$$\exp \left(- |Y|^2 / 2 \right) E \left[\exp \left(|Y_t|^2 / 2 + \frac{C'}{2} \int_0^t |Y|^2 \, ds + pd \left(y_0, y_t \right) \right) \right]$$

$$\le C_{\epsilon,M} \exp \left(-C'_{\epsilon,M} |Y|^2 + C''_{\epsilon,M} p^2 \right). \quad (14.8.11)$$

There exists $b_0 \in]0,1]$ such that given $\epsilon > 0, M > 0, \epsilon \le M$, there exist $C_{\epsilon,M} > 0, C'_{\epsilon,M} > 0, C''_{\epsilon,M} > 0$ such that for $b \in]0, b_0], t \in [\epsilon, M], s \in [t/2, t], p \ge 0$,

$$\exp \left(- |Y|^2 / 2 \right) E \left[\exp \left(|Y_t|^2 / 2 + pd \left(y_0, y_s \right) \right) \right]$$

$$\le C_{\epsilon,M} \exp \left(-C'_{\epsilon,M} |Y|^2 + C''_{\epsilon,M} p^2 \right). \quad (14.8.12)$$

Proof. We claim that the right-hand side of (14.8.2) is just the left-hand side of (14.8.11). If $p = 0$, this is just a consequence of (14.7.4), (14.8.7), and (14.8.10). When p is nonzero, either one includes the process $y_{\cdot}$ as part of the Markov process $(x_{\cdot}, y_{\cdot}, Y_{\cdot})$ and uses a similar argument. Alternatively, a proper use of the Girsanov transformation leads to the equality we just mentioned. Then (14.8.11) is equivalent to (14.8.2).

Recall that the kernel $k_{\cdot}^{\mathfrak{p} \oplus \mathfrak{k}}$ was defined in section 10.4. For $s \geq 0$, let τ_0^s still denote the parallel transport from $(TX \oplus N)_{x_s}$ into $(TX \oplus N)_{x_0}$ with respect to the connection $\nabla^{TX \oplus N}$. Using the Markov property of the process $(x_{\cdot}, Y_{\cdot})$, for $0 \leq s \leq t$, we get

$$E\left[\exp\left(|Y_t|^2/2 + pd(y_0, y_s)\right)\right]$$
$$= E\left[e^{pd(y_0,y_s)} \int_{\mathfrak{g}} k_{(t-s)/b^2}^{\mathfrak{p}\oplus\mathfrak{k}}(\tau_0^s Y_s, Y) \exp\left(|Y|^2/2\right) dY\right]. \quad (14.8.13)$$

By (10.7.9), (14.8.13), we obtain

$$E\left[\exp\left(|Y_t|^2/2 + pd(y_0, y_s)\right)\right] \leq 2^{(m+n)/2} E\left[\exp\left(|Y_s|^2/2\right) e^{pd(y_0,y_s)}\right].$$
$$(14.8.14)$$

By (14.8.11) with $C' = 0$, and by (14.8.14), we find that for $s \geq t/2$, (14.8.12) holds. The proof of our theorem is completed. $\qquad\square$

Remark 14.8.3. The condition $s \in [t/2, t]$ in (14.8.12) is imposed by the fact that the proof relies on equation (14.8.2) in Theorem 14.8.1. The proof of this equation itself relies on equation (14.8.5), where the pointwise estimates of chapter 13 on the kernel $r_{\beta_b, \tau_b t}^{X_K}$ are used, which require $t > 0$ to stay away from 0. However, the estimate (14.8.5) being in integrated form, it can be obtained by the methods of chapter 13, and more specifically by using equation (13.2.30), without having to use any pointwise estimate on the kernel $r_{\beta_b, \tau_b t}^{X_K}$. Ultimately, in (14.8.12), we can as well take $s \in [0, t]$.

Let $k(u)$ be a smooth function as in (13.2.28). For $y, y' \in X_K$, set

$$f(y, y') = k(d(y, y')). \quad (14.8.15)$$

Clearly,

$$|f(y, y') - d(y, y')| \leq 1. \quad (14.8.16)$$

By (14.8.16), we deduce that if $y, y', y'' \in X_K$, then

$$f(y, y'') \leq f(y, y') + f(y', y'') + 3. \quad (14.8.17)$$

By (14.8.16), in equation (14.8.11), we may as well replace $d(y_0, y)$ by the function $f(y_0, y)$.

We fix $p > 0$. The constants that follow will depend on $p > 0$. The value of p will be determined later.

Take $F \in C^{\infty,c}\left(\widehat{\mathcal{X}}, \mathbf{R}\right)$. We claim that for $b \in]0, b_0], \epsilon \leq t \leq M$, the methods of chapter 12 can be applied to the expression

$$\exp\left(-|Y|^2/2\right) E\left[\exp\left(|Y_t|^2/2 + \frac{C'}{2}\int_0^t |Y|^2\, ds + pf(y_0, y_t)\right) F(x_t, Y_t)\right].$$
(14.8.18)

Indeed, an analogue of equation (12.7.1) can be easily obtained, because, by Proposition 13.1.2, the covariant derivatives of order ≥ 1 in the variable y of $f(y_0, y)$ are uniformly bounded on X_K. The fact that estimates like (14.8.11) hold for any value of $p > 0$ results in that the arguments in the proofs of Theorems 12.7.4 and 12.10.1 can be used without any change. Ultimately, we find that there is a nonnegative smooth kernel $\bar{\mathfrak{r}}_{b,t}^X((x,Y),(x',Y'))$, such that

$$\exp\left(-|Y|^2/2\right) E\left[\exp\left(|Y_t|^2/2 + \frac{C'}{2}\int_0^t |Y|^2\, ds + pf(y_0, y_t)\right) F(x_t, Y_t)\right]$$
$$= \int_{\widehat{\mathcal{X}}} \bar{\mathfrak{r}}_{b,t}^X((x_0, Y),(x',Y'))\, F(x', Y')\, dx' dY'. \quad (14.8.19)$$

Also $\bar{\mathfrak{r}}_{b,t}^X$ and its covariant derivatives are uniformly bounded on $\widehat{\mathcal{X}}$.

Contrary to what the notation suggests, the kernels $\bar{\mathfrak{r}}_{b,t}^X$ do not form a semigroup. Still, using (14.8.17) and the Markov property of the process $(x., y., Y.)$, instead of the equality in (13.2.3), we have the inequality,

$$\bar{\mathfrak{r}}_{b,t}^X((x,Y),(x',Y')) \leq C \int_{\widehat{\mathcal{X}}} \bar{\mathfrak{r}}_{b,t/2}^X((x,Y),(z,Z))$$
$$\bar{\mathfrak{r}}_{b,t/2}^X((z,Z),(x',Y'))\, dz dZ. \quad (14.8.20)$$

Using (14.8.20) and proceeding as in the proofs of Theorems 13.2.1, 13.2.2, 13.2.4, and 13.3.1, the analogue of (13.3.1) holds, i.e., there exist $C > 0, C' > 0$ such that for $0 < b \leq b_0, \epsilon \leq t \leq M$,

$$\bar{\mathfrak{r}}_{b,t}^X((x,Y),(x',Y')) \leq C \exp\left(-C'\left(d^2(x,x') + |Y|^2 + |Y'|^2\right)\right). \quad (14.8.21)$$

If $C > 0$ is the constant that appears in (14.8.1), we take $p = C$. In the right-hand side of (14.8.1), by (14.8.16), we can as well replace $d(y_0, y_t)$ by $f(y_0, y_t)$ and still get an inequality with an extra constant in the right-hand side. Using (14.8.1), by proceeding as in the beginning of the proof of Theorem 14.8.2, and using (14.8.19), instead of (14.7.5), we get

$$\left|\exp\left(-t\mathcal{L}_{A,b}^{X'}\right) s(x_0, Y)\right| \leq C \int_{\widehat{\mathcal{X}}} \bar{\mathfrak{r}}_{b,t}^X((x_0, Y),(x',Y'))\, |s(x', Y')|\, dx' dY'.$$
(14.8.22)

By (14.8.22), instead of (14.7.6), we obtain

$$\left|q_{b,t}^{X'}((x,Y),(x',Y'))\right| \leq \bar{\mathfrak{r}}_{b,t}^X((x,Y),(x',Y')). \quad (14.8.23)$$

By (14.3.3), (14.8.21), and (14.8.23), we get (4.5.3). This completes the proof of equation (4.5.3) in the general case.

14.9 Rough estimates on the derivatives of $q_{b,t}^{X'}$ for bounded b

Equation (4.5.3) gives a uniform estimate on $q_{b,t}^{X'}\left((x,Y),(x',Y')\right)$. Now we briefly explain how to obtain uniform bounds on the covariant derivatives of arbitrary order.

To make the argument simpler, we will always assume that $b_0 \in]0,1]$ is taken as in Theorem 14.8.2, and we will assume that $0 < b \le b_0$. The case of b bounded and bounded away from 0 can be dealt with by the same methods. Set

$$\mathcal{M}_{A,b}^{X'} = \exp\left(|Y|^2/2\right) \mathcal{L}_{A,b}^{X'} \exp\left(-|Y|^2/2\right). \tag{14.9.1}$$

Note that $\mathcal{M}_{A,b}^{X'}$ can be obtained from $\mathcal{M}_{A,b}^{X}$ in (14.2.2) by conjugation by I.

We consider again the stochastic differential equation (12.10.1). We define U as in (14.4.2). If $s \in C^{\infty,c}\left(\widehat{\mathcal{X}}, \widehat{\pi}^*\left(\Lambda^{\cdot}\left(T^*X \oplus N^*\right) \otimes F\right)\right)$, instead of (14.4.3), we get

$$\exp\left(-t\mathcal{M}_{A,b}^{X'}\right) s\left(x_0,Y\right)$$
$$= E\left[\exp\left(-\frac{1}{2}\int_0^t \left|[Y^N, Y^{TX}]\right|^2 ds - tA\right) U_t \tau_0^t s\left(x_t, Y_t\right)\right]. \tag{14.9.2}$$

The idea is now to apply the techniques of the Malliavin calculus to (14.9.2). We will proceed as in chapter 12, and more precisely in sections 12.4–12.7 and in section 12.10.

Also, instead of obtaining bounds of the type (12.7.3), we will show directly an inequality that is slightly stronger than (12.7.5). Namely, we will show that for $\epsilon > 0, M > 0, \epsilon \le M, p \ge 1$, if $b \in]0,b_0], t \in [\epsilon,M]$, if N_t is one of the random variables produced by integration by parts, then

$$\exp\left(-|Y|^2/2\right) \left\|\exp\left(|Y_t|^2/2\right)|N_t|^p\right\|_1 \le C_{\epsilon,M,p} \exp\left(-C'_{\epsilon,M,p}|Y|^2\right). \tag{14.9.3}$$

Needless to say, as shown in (12.7.3)–(12.7.5), such bounds are verified by the random variables that appear in chapter 12. The need for these stronger estimates appears because by (14.6.7), $|W_t|$ is big compared to the terms that appeared in chapter 12.

In relation with the above, we will systematically use controls v that vanish identically on the interval $[0,t/2]$, while still picking the optimal control on the interval $[t/2,t]$. This possibility was already explored in Remarks 12.7.5 and 14.8.3.

As explained in the proof of Theorem 12.11.2, the term

$$\exp\left(-\frac{1}{2}\int_0^t \left|[Y^N, Y^{TX}]\right|^2 ds\right)$$

does not create any difficulty with respect to what was done in sections 12.4–12.7, no more than τ_0^t. Here we will concentrate on the term U_t. Of course we still write U_t as in (14.4.6).

First, we explain how to obtain the derivative of $V.\tau_0^1$ with respect to the parameter ℓ at $\ell = 0$, when replacing $w.$ by $w_.^\ell$ as in (12.1.8). Here, as in section 12.10, $w_.^\ell$ takes its values in $\mathfrak{g} = \mathfrak{p} \oplus \mathfrak{k}$.

Set

$$\mathbf{V}. = \frac{\partial}{\partial \ell} V.\dot{\tau_0}|_{\ell=0}. \tag{14.9.4}$$

We use the trivialization of $\Lambda^\cdot (T^*X \oplus N^*)$ by parallel transport with respect to the connection $\nabla^{TX \oplus N}$. Also we use the notation in (12.10.3). By (14.4.5), we get

$$\frac{d}{ds}\mathbf{V} = \mathbf{V}M_b + V\left(\frac{D}{D\ell}M_b + R^{TX \oplus N}\left(J^{TX}, \dot{x}\right)\right), \qquad \mathbf{V}_0 = 0. \tag{14.9.5}$$

From (14.9.5), we deduce that

$$\mathbf{V}_t = \int_0^t V_s\left(\frac{D}{D\ell}M_b + R^{TX \oplus N}\left(J^{TX}, \dot{x}\right)\right)V_s^{-1}dsV_t. \tag{14.9.6}$$

By (14.4.4), we get

$$\frac{D}{D\ell}M_b = \widehat{c}\left(\mathrm{ad}\left(\dot{j}^{TX}\right)\right) - c\left(\mathrm{ad}\left(\dot{j}^{TX} + i\theta\mathrm{ad}\left(\dot{j}^N\right)\right)\right). \tag{14.9.7}$$

By (14.9.7), we get

$$\left|\frac{D}{D\ell}M_b\right| \leq C\left|\dot{j}\right|. \tag{14.9.8}$$

For $s \leq t$, by (14.5.9) applied to the interval $[s,t]$, we get

$$\left|V_s^{-1}V_t\right| \leq \exp\left(\frac{C'}{2}\int_s^t |Y_u|^2\,du\right). \tag{14.9.9}$$

We take unit vectors $e \in \mathfrak{p}, f \in \mathfrak{k}$, and we take $J.$ as in (12.10.7), so that $J_0 = 0, J_t^{TX} = e, \dot{J}_t = f/b$. As explained before, we choose the control v that minimizes $|v|_{L_2}$ among the controls that vanish on $[0, t/2]$. In particular $J.$ also vanishes on $[0, t/2]$.

As we saw in (10.3.17), (10.3.48) and in the considerations that follow these two equations, when $b \to 0, \epsilon \leq t \leq M, \dot{J}$ remains uniformly bounded in L_2. By (14.9.8), (14.9.9), we conclude that for $0 < b \leq 1$,

$$\left|\int_0^t V_s\frac{D}{D\ell}M_bV_s^{-1}dsV_t\right| \leq C\exp\left(\frac{C'}{2}\int_0^t |Y_s|^2\,ds\right). \tag{14.9.10}$$

We will now estimate the contribution of the second term in (14.9.6) by the same technique as in (14.9.7)–(14.9.10). Let $\mathbf{v}.$ be the solution of the differential equation

$$\frac{d}{ds}\mathbf{v} = \mathbf{v}M_b + VR^{TX \oplus N}\left(J^{TX}, \dot{x}\right), \qquad \mathbf{v}_0 = 0. \tag{14.9.11}$$

By (2.1.10) and (12.10.1), we get

$$R^{TX \oplus N}\left(J^{TX}, \dot{x}\right) = -\mathrm{ad}\left(\left[J^{TX}, \frac{Y^{TX}}{b}\right]\right). \tag{14.9.12}$$

Let $\mathcal{M}_b \in \mathrm{End}\,(\Lambda^{\cdot}\,(T^*X \oplus N^*) \oplus \Lambda^{\cdot}\,(T^*X \oplus N^*))$ be given by

$$\mathcal{M}_b = \begin{bmatrix} M_b & -\mathrm{ad}\,\left(\left[J^{TX}, \frac{Y^{TX}}{b}\right]\right) \\ 0 & M_b \end{bmatrix}. \tag{14.9.13}$$

Consider the differential equation

$$\frac{d}{ds}\mathcal{V} = \mathcal{V}\mathcal{M}_b, \qquad\qquad \mathcal{V}_0 = 1. \tag{14.9.14}$$

Then $\mathcal{V}$ is given by

$$\mathcal{V} = \begin{bmatrix} V & \mathbf{v} \\ 0 & V \end{bmatrix}. \tag{14.9.15}$$

To estimate $\mathbf{v}, \mathbf{V}$, we will instead estimate $\mathcal{V}$.

By the considerations that follow (10.3.47), for $0 < b \leq 1$, J^{TX} remains pointwise uniformly bounded. This is also a consequence of the fact that $\dot{J}^{TX}$ is uniformly bounded in L_2. Also when acting on $\Lambda^{\cdot}\,(T^*X \oplus N^*)$, $\mathrm{ad}\left[J^{TX}, \frac{Y^{TX}}{b}\right]$ preserves the total degree and vanishes on $\Lambda^0\,(T^*X \oplus N^*)$. It follows that if $f, f' \in \Lambda^{\cdot}\,(T^*X \oplus N^*)$,

$$\left|\left\langle \mathrm{ad}\,\left(\left[J^{TX}, \frac{Y^{TX}}{b}\right]\right) f, f' \right\rangle\right|$$
$$\leq C\left|\frac{Y^{TX}}{b}\right|\left|\left|\sqrt{N^{\Lambda^{\cdot}\,(T^*X \oplus N^*)}}f\right|\right|\left|\left|\sqrt{N^{\Lambda^{\cdot}\,(T^*X \oplus N^*)}}f'\right|\right|. \tag{14.9.16}$$

By (14.5.2), (14.9.13), and (14.9.16), we conclude that

$$\mathrm{Re}\,\langle \mathcal{M}_b\,(f, f'), (f, f')\rangle \leq \frac{C'}{2}\,|Y|^2\,|(f, f')|^2. \tag{14.9.17}$$

By proceeding as in the proof of (14.5.9) in Theorem 14.5.2, we deduce from (14.9.17) that for $0 < b \leq 1, \epsilon \leq t \leq M$,

$$|\mathcal{V}_t| \leq \exp\left(\frac{C'}{2}\int_0^t |Y|^2\,ds\right). \tag{14.9.18}$$

Therefore we have a corresponding estimate for $\mathbf{v}_t$. By combining (14.9.10) with this estimate, we get a corresponding estimate for $\mathbf{V}_t$.

Now we will apply the same ideas to $W_{\cdot}$. Set

$$\mathbf{W}_{\cdot} = \frac{\partial}{\partial \ell}W_{\cdot}\tau_0'|_{\ell=0}. \tag{14.9.19}$$

Using the second equation in (14.4.5) and proceeding as in (14.9.5), we get

$$\frac{d}{ds}\mathbf{W} = \mathbf{W}\rho^E\left(-iY^N/b\right) + W\rho^E\left(-i\dot{J}^N + R^N\left(J^{TX}, \dot{x}\right)\right), \qquad \mathbf{W}_0 = 0. \tag{14.9.20}$$

Using (2.1.10) and (14.9.20), we get

$$\mathbf{W}_t = \int_0^t W_s\rho^E\left(-i\dot{J}^N - \left[J^{TX}, \dot{x}\right]\right)W_s^{-1}ds\,W_t. \tag{14.9.21}$$

Recall that $J.$ vanishes on $[0, t/2]$. Using (14.6.7) and the fact that $\dot{J}$ remains uniformly bounded in L_2, we get

$$\left| \int_0^t W_s \rho^E \left(-i \dot{J}^N \right) W_s^{-1} ds W_t \right|$$

$$\leq C' \exp \left(C d \left(y_0, y_t \right) \right) \left(1 + \int_{t/2}^t \exp \left(4 C d \left(y_0, y_s \right) \right) ds \right). \quad (14.9.22)$$

By (14.8.12), (14.9.22), and using Hölder's inequality, we find that for $b \in]0, b_0], \epsilon \leq t \leq M, p > 1,$

$$\exp \left(- |Y|^2 / 2 \right) E \left[\exp \left(|Y_t|^2 / 2 \right) \left| \int_0^t W_s \rho^E \left(-i \dot{J}^N \right) W_s^{-1} ds W_t \right|^p \right]$$

$$\leq C_{\epsilon, M, p} \exp \left(- C'_{\epsilon, M} |Y|^2 \right). \quad (14.9.23)$$

Recall that TX has been trivialized by parallel transport with respect to the connection ∇^{TX} along $x.$. By (14.8.8),

$$\int_0^t W_s \rho^E \left([J^{TX}, \dot{x}] \right) W_s^{-1} ds W_t = \int_0^t W_s \rho^E \left([J^{TX}, -b^2 \ddot{x} + \dot{w}^p] \right) W_s^{-1} ds W_t. \quad (14.9.24)$$

Moreover,

$$\int_0^t W_s \rho^E \left([J^{TX}, \dot{w}^p] \right) W_s^{-1} ds = \int_0^t W_s \rho^E \left([J^{TX}, \delta w^p] \right) W_s^{-1}, \quad (14.9.25)$$

i.e., (14.9.25) is an Itô integral. By (14.6.7),

$$\exp \left(- |Y|^2 / 2 \right) E \left[\exp \left(|Y_t|^2 / 2 \right) \left| \int_0^t W_s \rho^E \left([J^{TX}, \delta w^p] \right) W_s^{-1} W_t \right|^p \right]$$

$$\leq \exp \left(- |Y|^2 / 2 \right) \left(E \left[\exp \left(|Y_t|^2 / 2 \right) e^{2 C p d(y_0, y_t)} \right] \right.$$

$$\left. + E \left[\exp \left(|Y_t|^2 / 2 \right) \left| \int_0^t W_s \rho^E \left([J^{TX}, \delta w^p] \right) W_s^{-1} \right|^{2p} \right] \right). \quad (14.9.26)$$

By (14.8.11), we can dominate the first term in the right-hand side of (14.9.26) by an expression similar to the right-hand side of (14.9.23). Take $\theta = 3/2$. By Hölder's inequality, we get

$$\exp \left(- |Y|^2 / 2 \right) E \left[\exp \left(|Y_t^2 / 2| \right) \left| \int_0^t W_s \rho^E \left([J^{TX}, \delta w^p] \right) W_s^{-1} \right|^{2p} \right]$$

$$\leq \left[\exp \left(- |Y|^2 / 2 \right) E \left[\exp \left(\theta |Y_t|^2 / 2 \right) \right] \right]^{1/\theta}$$

$$\left[\exp \left(- |Y|^2 / 2 \right) E \left[\left| \int_0^t W_s \rho^E \left([J^{TX}, \delta w^p] \right) W_s^{-1} \right|^{2p\theta/(\theta-1)} \right] \right]^{(\theta-1)/\theta}.$$

$$(14.9.27)$$

By (10.8.12), given $\epsilon > 0$, for $b > 0$ small enough, and for $t \geq \epsilon$, the first term in the right-hand side of (14.9.27) is uniformly bounded. Moreover, using (12.1.17), (14.6.7), and the fact that J_s^{TX} vanishes for $s \leq t/2$, for $p > 1$, we obtain

$$E\left[\left|\int_0^t W_s \rho^E\left([J^{TX}, \delta w^{TX}]\right) W_s^{-1}\right|^{2p\theta/(\theta-1)}\right]$$
$$\leq C_p E\left[\left[\int_{t/2}^t e^{4Cd(y_0, y_s)} ds\right]^{p\theta/(\theta-1)}\right]. \quad (14.9.28)$$

Moreover, for $\epsilon \leq t \leq M$,

$$\left[\int_{t/2}^t e^{4Cd(y_0, y_s)} ds\right]^{p\theta/(\theta-1)} \leq C_{M,p} \int_{t/2}^t e^{4Cpd(y_0, y_s)\theta/(\theta-1)} ds. \quad (14.9.29)$$

By (14.8.11), (14.9.28), and (14.9.29), we obtain

$$\exp\left(-|Y|^2/2\right) E\left[\left|\int_0^t W_s \rho^E\left([J^{TX}, \delta w^{TX}]\right) W_s^{-1}\right|^{2p\theta/(\theta-1)}\right]$$
$$\leq C_{M,p} \exp\left(-|Y|^2/2\right) E\left[\int_{t/2}^t e^{4Cpd(y_0, y_s)\theta/(\theta-1)} ds\right]$$
$$\leq C_{\epsilon,M,p} \exp\left(-C'_{\epsilon,M,p}|Y|^2\right). \quad (14.9.30)$$

By (14.9.26)–(14.9.30), the left-hand side of (14.9.26) can be also dominated by an expression similar to the right-hand side of (14.9.23).

Moreover, since $J_t^{TX} = e$, using (14.4.5) and (14.8.8), we get

$$\int_0^t W_s \rho^E\left([J^{TX}, -b^2\ddot{x}]\right) W_s^{-1} ds W_t = b W_t \rho^E\left([e, -Y_t^{TX}]\right)$$
$$+ \int_{t/2}^t W_s \rho^E\left([-iY^N, [J^{TX}, Y^{TX}]] + [b\dot{j}^{TX}, Y^{TX}]\right) W_s^{-1} ds W_t. \quad (14.9.31)$$

By (14.6.7), for $0 \leq b \leq 1$,
$$|b W_t \rho^E\left([e, -Y_t^{TX}]\right)| \leq c\left(e^{2Cd(y_0, y_t)} + |Y_t|^2\right). \quad (14.9.32)$$

Take $\theta = 3/2$. For $p \geq 1$, using Hölder's inequality, we obtain

$$E\left[\exp\left(|Y_t|^2/2\right)|Y_t|^{2p}\right]$$
$$\leq \left[E\left[\exp\left(\theta|Y_t|^2/2\right)\right]\right]^{1/\theta} E\left[|Y_t|^{2p\theta/(\theta-1)}\right]^{(\theta-1)/\theta}. \quad (14.9.33)$$

By (10.8.12), (10.8.26), and (14.9.33), given $\epsilon > 0$, for $b > 0$ small enough and $t \geq \epsilon$,
$$\exp\left(-|Y|^2/2\right) E\left[\exp\left(|Y_t|^2/2\right)|Y_t|^{2p}\right] \leq C_{\epsilon,p} \exp\left(-C'_{\epsilon,p}|Y|^2\right). \quad (14.9.34)$$

By (14.6.7), (14.8.12), (14.9.32), and (14.9.34), for $b > 0$ small enough, $t \in [\epsilon, M]$, we get

$$\exp\left(-|Y|^2/2\right) E\left[\exp\left(|Y_t|^2/2\right) |bW_t \rho^E\left([e, -Y_t^{TX}]\right)|^p\right]$$
$$\leq C_{\epsilon,M,p} \exp\left(-C'_{\epsilon,M,p} |Y|^2\right). \quad (14.9.35)$$

Using (14.6.7) and the fact that J^{TX} remains uniformly bounded, we get

$$\left|\int_{t/2}^t W_s \rho^E\left([-iY^N, [J^{TX}, Y^{TX}]]\right) W_s^{-1} ds W_t\right| \leq ce^{2Cd(y_0, y_t)}$$
$$+ c\int_{t/2}^t e^{8Cd(y_0, y_s)} ds + c\int_{t/2}^t |Y_s|^8 ds. \quad (14.9.36)$$

The first two terms in the right-hand side of (14.9.36) can be dealt with as (14.9.22), (14.9.23). We will now explain how to deal with the last term in the right-hand side of (14.9.36). We take again $\theta = 3/2$. By Hölder's inequality, for $\epsilon \leq t \leq M$,

$$E\left[\exp\left(|Y_t|^2/2\right) \int_{t/2}^t |Y_s|^{8p} ds\right] \leq C_M \left[E\left[\exp\left(\theta |Y_t|^2/2\right)\right]\right]^{1/\theta}$$
$$\left[E\left[\int_{t/2}^t |Y_s|^{8p\theta/(\theta-1)} ds\right]\right]^{(\theta-1)/\theta}. \quad (14.9.37)$$

By (10.8.12), (10.8.26), and (14.9.37), given $\epsilon > 0$, for $b > 0$ small enough and $\epsilon \leq t \leq M, p > 1$, we obtain

$$\exp\left(-|Y|^2/2\right) E\left[\exp\left(|Y_t|^2/2\right) \int_{t/2}^t |Y_s|^{8p} ds\right]$$
$$\leq C_{\epsilon,M,p} \exp\left(-C'_{\epsilon,M,p} |Y|^2\right). \quad (14.9.38)$$

By (14.9.36) and the considerations which follow this equation, and by (14.9.38), we get

$$\exp\left(-|Y|^2/2\right) E\left[\exp\left(|Y_t|^2/2\right)\right.$$
$$\left.\left|\int_{t/2}^t W_s \rho^E\left([-iY^N, [J^{TX}, Y^{TX}]]\right) W_s^{-1} ds W_t\right|^p\right]$$
$$\leq C_{\epsilon,M,p} \exp\left(-C'_{\epsilon,M,p} |Y|^2\right). \quad (14.9.39)$$

Moreover, since $\dot{J}^{TX}$ remains uniformly bounded in L_2, the last term in the right-hand side of (14.9.31) can be treated by the same methods as in (14.9.22), (14.9.36).

By (14.9.24)–(14.9.39), given $\epsilon > 0$, for $b > 0$ small enough, $\epsilon \le t \le M$, and $p > 1$,

$$\exp\left(-|Y|^2/2\right) E\left[\exp\left(|Y_t|^2/2\right)\left|\int_0^t W_s \rho^E\left([J^{TX}, \dot{x}]\right) W_s^{-1} ds W_t\right|^p\right]$$

$$\le C_{\epsilon,M,p} \exp\left(-C'_{\epsilon,M}|Y|^2\right). \quad (14.9.40)$$

Ultimately, we have shown that the methods of chapter 12 can be applied to the semigroup $\exp\left(-t\mathcal{L}_{A,b}^{X'}\right)$. We obtain this way a uniform control on the first covariant derivatives of $q_{b,t}^{X'}\left((x, Y), (x', Y')\right)$ in the variables (x', Y').

The principle of iteration of the above to higher derivatives in the variables (x', Y') should be clear. We leave the details to the reader.

By exchanging the roles of (x, Y) and (x', Y'), similar results can be proved also for the covariant derivatives in the variables (x, Y). They will not be needed here.

Remark 14.9.1. We assumed the control v to vanish on $[0, t/2]$ in order to use the estimate (14.8.12) in Theorem 14.8.2, which requires $s \in [t/2, t]$. However, as explained in Remark 14.8.3, this estimate remains valid also for $s \in [0, t]$. Ultimately, this restriction on v is unnecessary.

14.10 The behavior of V. as $b \to 0$

Let $\mathbf{1} \in \mathrm{End}(\mathfrak{g})$ be the identity. We identify $\mathbf{1}$ to the scalar product on $\mathfrak{g} = \mathfrak{p} \oplus \mathfrak{k}$. Also, if $Y \in \mathfrak{g}$, we denote by $Y \otimes Y$ the quadratic form $u \in \mathfrak{g} \to \langle u, Y \rangle^2$.

Let $(V, \|\,\|)$ be a Banach space. For $b > 0$, let U_b be a random variable with values in V. Recall that U_b is said to converge to 0 in probability as $b \to 0$ if for any $\epsilon > 0$,

$$P\left[\|U_b\| \ge \epsilon\right] \to 0. \quad (14.10.1)$$

We still assume equations (14.8.8) and (14.8.9) to be verified.

Proposition 14.10.1. *Given $\alpha > 1/2, M > 0$, as $b \to 0$, the process $|\log(b)|^{-\alpha} Y$. converges uniformly on $[0, M]$ to 0 in probability.*

For any $M > 0$, as $b \to 0$, the process $\int_0^t Y_s \otimes Y_s ds$ converges uniformly on $[0, M]$ to $\frac{t}{2}\mathbf{1}$ in probability, and the process $\int_0^t \left|[Y^{\mathfrak{p}}, Y^{\mathfrak{k}}]\right|^2 ds$ converges uniformly on $[0, M]$ to $-\frac{1}{4}\mathrm{Tr}\left[C^{\mathfrak{k},\mathfrak{p}}\right] t$ in probability.

Proof. As in (10.8.3), by (14.8.8), we get

$$Y_t = e^{-t/b^2} Y + \frac{1}{b} e^{-t/b^2} \int_0^t e^{s/b^2} \delta w_s. \quad (14.10.2)$$

By (14.10.2), to establish the first part of our proposition, we may as well assume that $Y_0 = 0$. By a time change argument, there is a Brownian motion B. with values in $\mathfrak{g} = \mathfrak{p} \oplus \mathfrak{k}$ such that $B_0 = 0$, and moreover,

$$\int_0^t e^{s/b^2} \delta w_s = \frac{b}{\sqrt{2}} B_{e^{2t/b^2} - 1}. \quad (14.10.3)$$

By (14.10.2), (14.10.3), we get

$$Y_t = \frac{e^{-t/b^2}}{\sqrt{2}} B_{e^{2t/b^2}-1}.$$

(14.10.4)

By (14.10.4), we conclude that the uniform convergence result takes place as long as we assume that $0 \le t \le b^2$.

Now we assume that $t \ge b^2$. It is well-known that $B'_s = sB_{1/s}$ is still a Brownian motion starting at 0. We can write (14.10.4) in the form

$$Y_t = \sqrt{2} \sinh\left(t/b^2\right) B'_{\left(e^{2t/b^2}-1\right)^{-1}}.$$

(14.10.5)

Since $t \ge b^2$, $\left(e^{2t/b^2}-1\right)^{-1}$ remains uniformly bounded.

Moreover, by the law of iterated logarithm [ReY99, Theorem II (1.9)], we get

$$\limsup_{t \to 0} \frac{B'_t}{\sqrt{2t \log|\log t|}} = 1.$$

(14.10.6)

Also observe that

$$\log\left(e^{2t/b^2}-1\right) \le 2t/b^2,$$

(14.10.7)

so that for $b > 0$ small enough and $b^2 \le t \le M$,

$$\left|\log\left(e^{2t/b^2}-1\right)\right| \le 2M/b^2.$$

(14.10.8)

By (14.10.8), for $b > 0$ small enough and $b^2 \le t \le M$,

$$\log\left|\log\left(e^{2t/b^2}-1\right)\right| \le \log\left(2M/b^2\right).$$

(14.10.9)

By (14.10.5)–(14.10.9), we obtain the first part of our proposition.

For the solution of the third equation in (14.8.8) with $b = 1$, we use the notation $Z. = Y.$. Set

$$Y_{b,t} = Z_{t/b^2}.$$

(14.10.10)

As we saw after (10.8.25), for $b > 0$, the probability law of the process $Y_{b,.}$ coincides with the probability law of $Y.$ for the given b. Then

$$\int_0^t Y_{b,s} \otimes Y_{b,s} ds = b^2 \int_0^{t/b^2} Z_s \otimes Z_s ds.$$

(14.10.11)

The ergodic theorem can be applied to the Markov process $Z.$, whose invariant measure is given by $\exp\left(-|Y|^2\right) dY/\pi^{(m+n)/2}$. As $t \to +\infty$, we get the almost sure convergence,

$$\frac{1}{t} \int_0^t Z_s \otimes Z_s ds \to \frac{1}{2}.$$

(14.10.12)

In the sequel, we assume that (14.10.12) is verified. Given $\epsilon > 0$, there exists $A_\epsilon > 0$ such that for $t \ge A_\epsilon$,

$$\left|\frac{1}{t} \int_0^t Z_s \otimes Z_s ds - \frac{1}{2}\right| \le \frac{\epsilon}{M}.$$

(14.10.13)

By (14.10.11), (14.10.13), we deduce that if $b^2 A_\epsilon \le t \le M$,

$$\left| \int_0^t Y_{b,s} \otimes Y_{b,s} ds - t\frac{1}{2} \right| \le \epsilon. \tag{14.10.14}$$

Moreover, by (14.10.11), there exists $b_\epsilon > 0$ such that for $0 < b \le b_\epsilon$, and $t \le b^2 A_\epsilon$, (14.10.14) also holds. It follows that given $\epsilon > 0$, there exists $b_\epsilon > 0$ such that for $0 < b \le b_\epsilon$, (14.10.14) holds for any $t \in [0, M]$. Therefore, almost surely, as $b \to 0$, $\int_0^t Y_{b,s} \otimes Y_{b,s} ds$ converges uniformly on $[0, M]$ to $t\frac{1}{2}$. Since $Y.$ and $Y_{b,\cdot}$ have the same probability law, this implies the convergence in probability that is mentioned in our proposition.

By (14.10.10), we get

$$\int_0^t \left| \left[Y_{b,s}^{\mathrm{p}}, Y_{b,s}^{\mathrm{f}} \right] \right|^2 ds = b^2 \int_0^{t/b^2} \left| \left[Z_s^{\mathrm{p}}, Z_s^{\mathrm{f}} \right] \right|^2 ds. \tag{14.10.15}$$

By the ergodic theorem, we know that as $t \to +\infty$, we have the almost sure convergence,

$$\frac{1}{t} \int_0^t \left| \left[Z_s^{\mathrm{p}}, Z_s^{\mathrm{f}} \right] \right|^2 ds \to -\frac{1}{4} \mathrm{Tr} \left[C^{\mathrm{f}, \mathrm{p}} \right]. \tag{14.10.16}$$

Using (14.10.16) and proceeding as before, we get the last part of our proposition. $\qquad\square$

Recall that $R(Y)$ was defined in (2.16.10). By (14.4.4), we get

$$M_b = -\frac{N^{\Lambda^\cdot (T^* X \oplus N^*)}}{b^2} + \frac{R(Y)}{b}. \tag{14.10.17}$$

Using (14.4.5), we obtain

$$V_t = \exp\left(-t N^{\Lambda^\cdot (T^* X \oplus N^*)}/b^2 \right)$$
$$+ \int_0^t V_s \frac{R(Y_s)}{b} \exp\left(-(t-s) N^{\Lambda^\cdot (T^* X \oplus N^*)}/b^2 \right) ds. \tag{14.10.18}$$

As in section 2.16, $\mathbf{P}$ denotes the projection operator from $\Lambda^\cdot (T^* X \oplus N^*)$ on $\Lambda^0 (T^* X \oplus N^*) = \mathbf{R}$, and $\mathbf{P}^\perp = 1 - \mathbf{P}$.

Proposition 14.10.2. *For any $p > 2, M > 0$, there exist $b_p \in]0, 1], C_{p,M} > 0, C' > 0$ such that for $b \in]0, b_p]$,*

$$\left\| \sup_{0 \le t \le M} \left(V_t - \exp\left(-t N^{\Lambda^\cdot (T^* X \oplus N^*)}/b^2 \right) \right) \mathbf{P}^\perp \right\|_p$$
$$\le C_{p,M} b^{(p-2)/p} (1 + |Y|) \exp\left(\frac{C' b^2}{2} |Y|^2 \right). \tag{14.10.19}$$

For any $\epsilon > 0, M > 0, \epsilon \le M$, $V.\mathbf{P}^\perp$ converges uniformly to 0 on $[\epsilon, M]$ in probability.

Proof. Clearly,

$$\left| \int_0^t V_s \frac{R(Y_s)}{b} \exp\left(-(t-s) N^{\Lambda^{\cdot}(T^*X \oplus N^*)}/b^2 \right) ds \mathbf{P}^\perp \right|$$

$$\leq \int_0^t \left| V_s \frac{R(Y_s)}{b} \exp\left(-(t-s) N^{\Lambda^{\cdot}(T^*X \oplus N^*)}/b^2 \right) \mathbf{P}^\perp \right| ds. \quad (14.10.20)$$

Let $p > 2$ and let q be such that $\frac{1}{p} + \frac{1}{q} = 1$. By Hölder's inequality, for $0 \leq t \leq M$, we get

$$\int_0^t \left| V_s \frac{R(Y_s)}{b} \exp\left(-(t-s) N^{\Lambda^{\cdot}(T^*X \oplus N^*)}/b^2 \right) \mathbf{P}^\perp \right| ds$$

$$\leq b^{(p-2)/p} \left[\int_0^M |V_s R(Y_s)|^p \, ds \right]^{1/p}$$

$$\left\{ \frac{1 - \exp\left(-qMN^{\Lambda^{\cdot}(T^*X \oplus N^*)}/b^2 \right)}{qN^{\Lambda^{\cdot}(T^*X \oplus N^*)}} \mathbf{P}^\perp \right\}^{1/q}. \quad (14.10.21)$$

Moreover,

$$\left\{ E\left[\int_0^M |V_s R(Y_s)|^p \, ds \right] \right\}^{1/p} \leq \left\{ E\left[\int_0^M |V_s|^{2p} \, ds \right] \right\}^{1/2p}$$

$$\left\{ E\left[\int_0^M |R(Y_s)|^{2p} \, ds \right] \right\}^{1/2p}. \quad (14.10.22)$$

By equation (10.8.26) in Proposition 10.8.2, we get

$$\left\{ E\left[\int_0^M |R(Y_s)|^{2p} \, ds \right] \right\}^{1/2p} \leq C_{p,M} (1 + |Y|). \quad (14.10.23)$$

Let C' be the positive constant that appears in (14.5.9). Then

$$\left\{ E\left[\int_0^M |V_s|^{2p} \, ds \right] \right\}^{1/2p} \leq C_{p,M} \left\| \exp\left(\frac{C'}{2} \int_0^M |Y|^2 \, ds \right) \right\|_{2p}. \quad (14.10.24)$$

By proceeding as in (14.7.7)–(14.7.12) and using the notation in these equations, and by (14.8.7), for $b^2 < 1/C'$, we get

$$E\left[\exp\left(\frac{C'}{2} \int_0^t |Y|^2 \, ds \right) \right] = \exp\left(\frac{1 - \alpha_b}{2b^2} (m+n) t \right)$$

$$\exp\left(\frac{|Y|^2}{2} \right) \int_{\mathfrak{g}} h_{\alpha_b t/b^2}^{\mathfrak{p} \oplus \mathfrak{k}} (\sqrt{\alpha_b} Y, Y') \exp\left(-\frac{|Y'|^2}{2\alpha_b} \right) dY'. \quad (14.10.25)$$

By (10.7.10), we obtain

$$
\int_{\mathfrak{g}} h^{\mathfrak{p}\oplus\mathfrak{k}}_{\alpha_b t/b^2} \left(\sqrt{\alpha_b}Y, Y'\right) \exp\left(-\frac{|Y'|^2}{2\alpha_b}\right) dY'
$$

$$
= \left(\frac{e^{\alpha_b t/b^2}}{\cosh\left(\alpha_b t/b^2\right) + \sinh\left(\alpha_b t/b^2\right)/\alpha_b}\right)^{(m+n)/2}
$$

$$
\exp\left(-\frac{\alpha_b}{2}\frac{\coth\left(\alpha_b t/b^2\right) + \alpha_b}{\coth\left(\alpha_b t/b^2\right)\alpha_b + 1}|Y|^2\right). \quad (14.10.26)
$$

By (14.10.25), (14.10.26), we get

$$
E\left[\exp\left(\frac{C'}{2}\int_0^t |Y|^2\, ds\right)\right] = \left(\frac{e^{t/b^2}}{\cosh\left(\alpha_b t/b^2\right) + \sinh\left(\alpha_b t/b^2\right)/\alpha_b}\right)^{(m+n)/2}
$$

$$
\exp\left(\frac{1}{2}\frac{C'b^2}{\coth\left(\alpha_b t/b^2\right)\alpha_b + 1}|Y|^2\right)
$$

$$
\leq \exp\left((1-\alpha_b)\, t\, (m+n)\, /2b^2\right) \exp\left(\frac{C'b^2}{2}|Y|^2\right). \quad (14.10.27)
$$

By (14.10.24) and (14.10.27), for $b \in]0,1]$ small enough,

$$
\left\{E\left[\int_0^M |V_s|^{2p}\, ds\right]\right\}^{1/2p} \leq C_{p,M} \exp\left(\frac{C'b^2}{2}|Y|^2\right). \quad (14.10.28)
$$

By (14.10.18), (14.10.20)–(14.10.23), and (14.10.28), we get (14.10.19).
Clearly,

$$
\left|\exp\left(-tN^{\Lambda^{\cdot}(T^*X\oplus N^*)}/b^2\right)\mathbf{P}^{\perp}\right| \leq \exp\left(-t/b^2\right). \quad (14.10.29)
$$

By (14.10.19) and (14.10.29), we get the second part of our proposition. $\quad\square$

Now we establish a path integral version of equation (14.2.5) in Proposition 14.2.1.

Proposition 14.10.3. *The following identity holds:*

$$
V_t\left(1 + b\left(1 + N^{\Lambda^{\cdot}(T^*X\oplus N^*)}\right)^{-1}R\left(Y_t\right)\right)
$$

$$
= 1 + b\left(1 + N^{\Lambda^{\cdot}(T^*X\oplus N^*)}\right)^{-1}R\left(Y_0\right)
$$

$$
+ \int_0^t V_s\left(-\frac{N^{\Lambda^{\cdot}(T^*X\oplus N^*)}}{b^2} + R\left(Y_s\right)\left(1 + N^{\Lambda^{\cdot}(T^*X\oplus N^*)}\right)^{-1}R\left(Y_s\right)\right) ds
$$

$$
+ \int_0^t V_s\left(1 + N^{\Lambda^{\cdot}(T^*X\oplus N^*)}\right)^{-1}R\left(\delta w\right). \quad (14.10.30)
$$

Proof. By (14.10.17), the first equation in (14.4.5) can be written in the form

$$\frac{dV}{ds} = V\left(-\frac{N^{\Lambda^{\cdot}(T^*X\oplus N^*)}}{b^2} + \frac{R(Y)}{b}\right).\tag{14.10.31}$$

Using (14.8.8) and (14.10.31), we get (14.10.30). $\qquad\square$

By (14.10.30), if $f \in \Lambda^0(T^*X \oplus N^*) = \mathbf{R}$, we get

$$V_t\left(1 + b\left(1 + N^{\Lambda^{\cdot}(T^*X\oplus N^*)}\right)^{-1}R(Y_t)\right)f$$

$$= \left(1 + b\left(1 + N^{\Lambda^{\cdot}(T^*X\oplus N^*)}\right)^{-1}R(Y_0)\right)f$$

$$+ \int_0^t V_s R(Y_s)\left(1 + N^{\Lambda^{\cdot}(T^*X\oplus N^*)}\right)^{-1}R(Y_s)f\,ds$$

$$+ \int_0^t V_s\left(1 + N^{\Lambda^{\cdot}(T^*X\oplus N^*)}\right)^{-1}R(\delta w)f.\tag{14.10.32}$$

In the sequel, we simply take $f = 1$.

Proposition 14.10.4. *Given $M > 0$, as $b \to 0$, the following processes converge uniformly over $[0, M]$ in probability,*

$$\int_0^t V_s\mathbf{P}^\perp R(Y_s)\left(1 + N^{\Lambda^{\cdot}(T^*X\oplus N^*)}\right)^{-1}R(Y_s)f\,ds \to 0,\tag{14.10.33}$$

$$\int_0^t V_s\left(1 + N^{\Lambda^{\cdot}(T^*X\oplus N^*)}\right)^{-1}R(\delta w)f \to 0.$$

Proof. Clearly, for $0 \le t \le M$,

$$\left|\int_0^t V_s\mathbf{P}^\perp R(Y_s)\left(1 + N^{\Lambda^{\cdot}(T^*X\oplus N^*)}\right)^{-1}R(Y_s)f\,ds\right|$$

$$\le C\int_0^M \left|V_s\mathbf{P}^\perp\right||Y_s|^2\,ds.\tag{14.10.34}$$

Take $p > 2$, and let q be such that $\frac{1}{p} + \frac{1}{q} = 1$. By Hölder's inequality, we get

$$E\left[\int_0^M \left|V_s\mathbf{P}^\perp\right||Y_s|^2\,ds\right]$$

$$\le \left\{E\left[\int_0^M \left|V_s\mathbf{P}^\perp\right|^p\,ds\right]\right\}^{1/p}\left\{E\left[\int_0^M |Y_s|^{2q}\,ds\right]\right\}^{1/q}.\tag{14.10.35}$$

By equation (10.8.26) in Proposition 10.8.2, we get

$$\left\{E\left[\int_0^M |Y_s|^{2q}\,ds\right]\right\}^{1/q} \le C_{M,q}\left(1 + |Y|^2\right).\tag{14.10.36}$$

By equation (14.10.19) in Proposition 14.10.2, and by (14.10.34)–(14.10.36), to establish the first part of our theorem, we may as well replace $V_s \mathbf{P}^\perp$ in the right-hand side of (14.10.35) by $\exp\left(-s N^{\Lambda^{\cdot}(T^*X \oplus N^*)}/b^2\right) \mathbf{P}^\perp$, which concludes the proof of the first convergence in (14.10.33).

By (2.16.14), $\mathbf{P} R(Y) \mathbf{P} = 0$, and so

$$\int_0^t V_s \left(1 + N^{\Lambda^{\cdot}(T^*X \oplus N^*)}\right)^{-1} R(\delta w) f$$

$$= \int_0^t V_s \mathbf{P}^\perp \left(1 + N^{\Lambda^{\cdot}(T^*X \oplus N^*)}\right)^{-1} R(\delta w) f. \quad (14.10.37)$$

Take $p > 1$. By (12.1.17),

$$\left\| \sup_{0 \le t \le M} \left| \int_0^t V_s \mathbf{P}^\perp \left(1 + N^{\Lambda^{\cdot}(T^*X \oplus N^*)}\right)^{-1} R(\delta w) f \right| \right\|_p$$

$$\le C_p \left\| \left[\int_0^M \left| V_s \mathbf{P}^\perp \right|^2 ds \right]^{1/2} \right\|_p. \quad (14.10.38)$$

Moreover, for $p > 2$,

$$\left\| \left[\int_0^M \left| V \mathbf{P}^\perp \right|^2 ds \right]^{1/2} \right\|_p \le C_{p,M} \left\{ E \left[\int_0^M \left| V \mathbf{P}^\perp \right|^p ds \right] \right\}^{1/p}. \quad (14.10.39)$$

By using again equation (14.10.19) in Proposition 14.10.2, we get the second part of our proposition. $\qquad \square$

Recall that for $Y \in TX \oplus N$, $S(Y)$ was defined in Definition 2.16.2, and was evaluated in Proposition 2.16.3. Note $S(Y)$ is a nonnegative scalar multiple of $\mathbf{P}$, and that this scalar depends quadratically on Y.

Definition 14.10.5. Put

$$\underline{V}_t = \exp \left(\int_0^t S(Y_s) \, ds \right). \quad (14.10.40)$$

Theorem 14.10.6. *For $\epsilon > 0, M > 0, \epsilon \le M$, as $b \to 0$, $V. - \underline{V}.\mathbf{P}$ converges uniformly to 0 on $[\epsilon, M]$ in probability.*

Proof. By Proposition 14.10.2, to establish our theorem, we may as well replace $V.$ by $V.\mathbf{P}$. Set

$$v_t = V_t \mathbf{P} - \mathbf{P} - \int_0^t V_s \mathbf{P} S(Y_s) \, ds. \quad (14.10.41)$$

By (14.10.41), we get

$$V_t \mathbf{P} = \underline{V}_t \mathbf{P} + v_t + \int_0^t v_s S(Y_s) \underline{V}_s^{-1} ds \underline{V}_t. \quad (14.10.42)$$

By (14.10.42), since $S(Y_s)$ is nonnegative, we get

$$|V_t \mathbf{P} - \underline{V}_t \mathbf{P}| \le |v_t| + |\underline{V}_t - 1| \sup_{0 \le s \le M} |v_s|. \tag{14.10.43}$$

By (14.5.9) and (14.10.27), given $p > 1$, for $b > 0$ small enough,

$$\left\| \sup_{0 \le t \le M} |V_t| \right\|_p \le C_{M,p} \exp\left(\frac{C'b^2}{2} |Y|^2 \right). \tag{14.10.44}$$

By Proposition 14.10.1 and by (14.10.44), as $b \to 0$,

$$bV_t \left(1 + N^{\Lambda^\cdot (T^*X \oplus N^*)} \right)^{-1} R(Y_t)$$

converges to 0 uniformly on $[0, M]$ in probability. Using (2.16.15), (14.10.32), and Proposition 14.10.4, as $b \to 0$, v. converges uniformly to 0 on $[0, M]$ in probability.

Moreover, it is obvious that $\underline{V}$. also satisfies (14.10.44). By (14.10.43), we deduce that as $b \to 0$, $V.\mathbf{P} - \underline{V}.\mathbf{P}$ converges to 0 uniformly on $[0, M]$ in probability. This completes the proof of our theorem. □

Recall that $\delta \in \mathbf{R}$ was defined in Definition 2.16.4.

Theorem 14.10.7. *For $\epsilon > 0, M > 0, \epsilon \le M$, as $b \to 0$, V_t converges uniformly to $\exp(\delta t) \mathbf{P}$ on $[\epsilon, M]$ in probability.*

Proof. By Theorem 14.10.6, to establish our theorem, it is enough to prove that as $b \to 0$, $\underline{V}_t$ converges to $\exp(\delta t)$ uniformly on $[0, M]$ in probability. By (14.10.40), we have to show that $\int_0^t S(Y_s)\, ds$ converges to δt uniformly on $[0, M]$ in probability. This is a consequence of Proposition 2.16.5, of (2.16.23), and of Proposition 14.10.1. □

14.11 The limit of $q_{b,t}^{X'}$ as $b \to 0$

Recall that given $b > 0$, W. is the solution of the second equation in (14.4.5). As before if $f \in \mathfrak{k}^\mathbf{C}$, we identify f with the corresponding left-invariant vector field on $K_\mathbf{C}$.

Note that h. in (14.6.5) is the obvious analogue for $K_\mathbf{C}$ of g. for G in (12.2.8).

Definition 14.11.1. Let $\left(w = (w^\mathfrak{p}, w^\mathfrak{k})\right)$ be a Brownian motion with values in $\mathfrak{g} = \mathfrak{p} \oplus \mathfrak{k}$. Let $g_{0,\cdot}, h_{0,\cdot} \in K_\mathbf{C}$ be the solutions of the stochastic differential equations,

$$dg_{0,\cdot} = dw^\mathfrak{p}, \qquad \dot{h}_{0,\cdot} = -idw^\mathfrak{k}, \tag{14.11.1}$$

$$g_{0,0} = 1_G, \qquad h_{0,0} = 1_{K_\mathbf{C}}.$$

Put

$$x_{0,\cdot} = pg_{0,\cdot}, \qquad y_{0,\cdot} = ph_{0,\cdot}, \qquad W_{0,\cdot} = \rho^E(h_{0,\cdot}). \tag{14.11.2}$$

Then $W_{0,\cdot}$ is the solution of the stochastic differential equation

$$dW_{0,\cdot} = W_{0,\cdot} \rho^E\left(-idw^\mathfrak{k} \right), \qquad W_{0,0} = 1. \tag{14.11.3}$$

Recall that the kernel $q_{0,t}^X \left((x, Y), (x', Y') \right)$ was defined in (4.5.2).

Theorem 14.11.2. *Let* $s \in C^{\infty,c}\left(\widehat{\mathcal{X}}, \widehat{\pi}^*\left(\Lambda^{\cdot}\left(T^*X \oplus N^*\right) \otimes F\right)\right)$. *Then for* $t > 0, (x, Y) \in \widehat{\mathcal{X}}$, *as* $b \to 0$,

$$\int_{\widehat{\mathcal{X}}} q_{b,t}^X\left((x, Y), (x', Y')\right) s\left(x', Y'\right) dx' dY'$$

$$\to \int_{\widehat{\mathcal{X}}} q_{0,t}^X\left((x, Y), (x', Y')\right) s\left(x', Y'\right) dx' dY'. \quad (14.11.4)$$

Proof. To establish our theorem, we may and we will assume that $x = x_0$. Take $(x., Y.)$ as in (12.10.1). By (14.1.4), (14.3.1), (14.4.6), (14.9.1), and (14.9.2), we need to show that for $t > 0$, as $b \to 0$,

$$E\left[\exp\left(-\frac{1}{2}\int_0^t \left|\left[Y^N, Y^{TX}\right]\right|^2 ds - tA\right) V_t \otimes W_t \tau_0^t s\left(x_t, Y_t\right)\right]$$

$$\to \pi^{-(m+n)/2} \exp\left(-\frac{t}{8}B^*\left(\kappa^{\mathfrak{g}}, \kappa^{\mathfrak{g}}\right) - \frac{t}{2}C^{\mathfrak{k},F} - tA\right)$$

$$E\left[\tau_0^t \int_{(TX \oplus N)_{x_t}} \mathbf{P}s\left(x_{0,t}, Y\right) \exp\left(-|Y|^2\right) dY\right]. \quad (14.11.5)$$

1. The case where ρ^E is trivial.

   First, we will assume that ρ^E is trivial, so that we can disregard W_t and $C^{\mathfrak{k},F}$ in (14.11.5).

   By Theorem 14.10.7, we know that for $t > 0$, as $b \to 0$, V_t converges in probability to $\exp(\delta t)\mathbf{P}$. By (14.10.44), to establish (14.11.5), we may as well replace V_t by $\exp(\delta t)\mathbf{P}$. Similarly, by the last part of Proposition 14.10.1, we may as well replace $\frac{1}{2}\int_0^t \left|\left[Y^N, Y^{TX}\right]\right|^2 ds$ by $-\frac{1}{8}\mathrm{Tr}\left[C^{\mathfrak{k},\mathfrak{p}}\right] t$. Using (2.16.26), to prove (14.11.5), we only need to show that if $F \in C^{\infty,c}\left(\widehat{\mathcal{X}}, \mathbf{R}\right)$, as $b \to 0$,

   $$E\left[F\left(x_t, Y_t\right)\right]$$

   $$\to \pi^{-(m+n)/2} E\left[\int_{(TX \oplus N)_{x_{0,t}}} F\left(x_{0,t}, Y\right) \exp\left(-|Y|^2\right) dY\right],$$

   $$(14.11.6)$$

   which is a consequence of (11.6.5), (12.1.6), (12.10.2), and of Theorems 12.10.1 and 12.10.2. Needless to say, we need much less than the convergence of kernels in Theorem 12.10.2, but just the intermediate narrow convergence result given in the proof of Theorem 12.8.1, and its analogue over $\widehat{\mathcal{X}}$.

2. The case of a general ρ^E

   By (14.8.11), given $\epsilon > 0, M > 0$, for $t \in [\epsilon, M], p > 0$,

   $$E\left[\exp\left(pd\left(y_0, y_t\right)\right)\right] \le C_{\epsilon,M} \exp\left(-C'_{\epsilon,M}|Y|^2 + C''_{\epsilon,M}p^2\right). \quad (14.11.7)$$

Using (14.6.7), (14.10.44), the considerations in the first part of the proof, and (14.11.7), to establish (14.11.5), we only need to show that if $s \in C^{\infty,c}\left(\widehat{\mathcal{X}}, \widehat{\pi}^* F\right)$, as $b \to 0$,

$$E\left[W_t \tau_0^t s\left(x_t, Y_t\right)\right] \to \pi^{-(m+n)/2} \exp\left(-\frac{t}{2}C^{\mathfrak{k},F}\right)$$

$$E\left[\tau_0^t \int_{(TX \oplus N)_{x_{0,t}}} s\left(x_{0,t}, Y\right) \exp\left(-|Y|^2\right) dY\right]. \quad (14.11.8)$$

Let $H : K_{\mathbf{C}} \to \operatorname{End}\left[E_{x_0}\right]$ be a smooth bounded function. By equation (12.8.47) in Remark 12.8.2 applied to G and $K_{\mathbf{C}}$, for $t > 0$, as $b \to 0$, we get

$$E\left[H\left(h_t\right) \tau_0^t s\left(x_t, Y_t\right)\right] \to E\left[H\left(h_{0,t}\right)\right]$$

$$\pi^{-(m+n)/2} E\left[\tau_0^t \int_{(TX \oplus N)_{x_{0,t}}} s\left(x_{0,t}, Y\right) \exp\left(-|Y|^2\right) dY\right]. \quad (14.11.9)$$

By (14.6.6), (14.6.7), and (14.11.7), equation (14.11.9) can be used with the unbounded $H = \rho^E$, so that $H\left(h_t\right) = W_t$. When $H = \rho^E$, by (14.11.2),

$$H\left(h_{0,t}\right) = W_{0,t}. \quad (14.11.10)$$

By transforming equation (14.11.3) for $W_{0,\cdot}$ into an Itô stochastic differential equation, we get

$$dW_{0,\cdot} = -\frac{1}{2}W_{0,\cdot}C^{\mathfrak{k},E}dt + W_{0,\cdot}\rho^E\left(-i\delta w^{\mathfrak{k}}\right), \quad W_{0,0} = 1, \quad (14.11.11)$$

so that

$$E\left[W_{0,t}\right] = \exp\left(-\frac{t}{2}C^{\mathfrak{k},E}\right). \quad (14.11.12)$$

From (14.11.9)–(14.11.12), we get (14.11.8). The proof of our theorem is completed.

□

Given $(x, Y) \in \widehat{\mathcal{X}}$, equation (4.5.4) in Theorem 4.5.2 is an obvious consequence of the uniform estimates on $q_{b,t}^{X\prime}\left((x, Y), (x', Y')\right)$ and its covariant derivatives of arbitrary order in the variable $(x', Y') \in \widehat{\mathcal{X}}$ that were established in section 14.9, and of Theorem 14.11.2.

Chapter Fifteen

The heat kernel $q_{b,t}^X$ for b large

The purpose of this chapter is to establish the estimates of Theorem 9.1.1 on the hypoelliptic heat kernel $\underline{q}_{b,t}^X$. More precisely, we show that as $b \to +\infty$, $\underline{q}_{b,t}^X$ exhibits the proper decay away from $\widehat{\mathcal{F}}_\gamma = \widehat{i}_a \mathcal{N}\left(k^{-1}\right) \subset \widehat{\mathcal{X}}$.

To avoid being overburdened with technicalities at the very beginning, first, we prove similar estimates on scalar hypoelliptic heat kernels over $\mathcal{X}$. Such estimates are then extended to scalar hypoelliptic heat kernels over $\widehat{\mathcal{X}}$. Ultimately, we extend the estimates to the kernel $\underline{q}_{b,t}^X$ using the Feynman-Kac formula. The term $\frac{1}{2}\left|\left[Y^N, Y^{TX}\right]\right|^2$ in the right-hand side of equation (2.13.5) for $\mathcal{L}_b^X$ plays a crucial role in proving the estimates.

This chapter is organized as follows. In section 15.1, we establish uniform Gaussian estimates on the scalar heat kernel $\underline{r}_{b,t}^X$ over $\mathcal{X}$.

In section 15.2, we establish important estimates on the deviation of the solution of the differential equation (12.9.5) from the geodesic flow as $b \to +\infty$. Such estimates are easy. Unavoidably the hyperbolic nature of the geodesic flow appears in the estimates. The main purpose of the chapter is to convert such pointwise estimates on the trajectories to estimates on the hypoelliptic heat kernels.

In section 15.3, we show that as $b \to \infty$, $\underline{r}_{b,1}^X\left(\left(x, Y^{TX}\right), \gamma\left(x, Y^{TX}\right)\right)$ decays in the proper way away from $i_a X\left(\gamma\right)$. The estimates are based on the properties of the pseudodistance on $\mathcal{X}$ established in section 3.9, and also on the results of section 15.2. The argument goes roughly as follows. Away from $\mathcal{F}_\gamma = i_a X\left(\gamma\right)$, $\varphi_{1/2}\left(x, Y^{TX}\right)$ and $\varphi_{-1/2}\gamma\left(x, Y^{TX}\right)$ are far away. On the other hand, by the results of section 15.2, the hypoelliptic heat kernel propagates more and more along the geodesic flow. The point is to ultimately deduce from these two facts the proper decay of the hypoelliptic heat kernel.

In section 15.4, we obtain uniform Gaussian estimates for $\underline{r}_{b,1}^X$ near $i_a X\left(\gamma\right)$.

In section 15.5, we obtain corresponding results for the scalar heat kernel $\underline{r}_{b,1}^X$ on $\widehat{\mathcal{X}}$ away from $\widehat{i}_a \mathcal{N}\left(k^{-1}\right)$. Because $\underline{r}_{b,1}^X$ is dominated by $\underline{r}_{b,1}^X$, some of these estimates are proved instead for $\underline{r}_{b,1}^X$.

In section 15.6, we obtain uniform estimates for $\underline{r}_{b,1}^X$ near $\widehat{i}_a \mathcal{N}\left(k^{-1}\right)$. These estimates are non-Gaussian.

In section 15.7, we prove Theorem 9.1.1 by using the estimates on the scalar heat kernel $\underline{r}_{b,t}^X$ to obtain corresponding estimates for $\underline{q}_{b,t}^X$.

In section 15.8, we establish Theorem 9.1.3, i.e., we obtain uniform estimates for the covariant derivatives of $\underline{q}_{b,t}^X$.

In section 15.9, we prove Theorem 9.5.6.

Finally, in section 15.10, we establish Theorem 9.11.1.

In the present chapter, we use the notation of the previous chapters of the book. In particular the semisimple element $\gamma \in G$ is taken as in (3.3.2).

15.1 Uniform estimates on the kernel $r^X_{b,t}$ over $\mathcal{X}$

We use the notation of section 12.9. For $t > 0$, $r^X_{b,t}\left(\left(x, Y^{TX}\right), \left(x', Y^{TX\prime}\right)\right)$ is the smooth kernel associated with $\exp\left(-t\mathcal{A}^X_b\right)$. Set

$$r^X_b = r^X_{b,1}. \tag{15.1.1}$$

Theorem 15.1.1. *Given $\epsilon > 0, M > 0, \epsilon \le M$, there exist $C_{\epsilon,M} > 0, C'_{\epsilon,M} > 0$ such that for $b \ge 1, \epsilon \le t \le M, \left(x, Y^{TX}\right), \left(x', Y^{TX\prime}\right) \in \mathcal{X}$,*

$$r^X_{b,t}\left(\left(x, Y^{TX}\right), \left(x', Y^{TX\prime}\right)\right)$$
$$\le C_{\epsilon,M} b^{4m} \exp\left(-C'_{\epsilon,M}\left(d^2\left(x, x'\right) + \left|Y^{TX}\right|^2 + \left|Y^{TX\prime}\right|^2\right)\right). \tag{15.1.2}$$

Proof. We follow the proof of Theorem 13.2.1. Using the uniform bounds on the kernel $r^X_{b,t}$ that were given in equation (12.9.10) in Theorem 12.9.1, and using also the analogue of (13.2.3), we get the analogue of (13.2.4),

$$r^X_{b,t}\left(\left(x, Y^{TX}\right), \left(x', Y^{TX\prime}\right)\right)$$
$$\le C_{\epsilon,M} b^{4m} \int_{\mathcal{X}} r^X_{b,t/2}\left(\left(x, Y^{TX}\right), \left(z, Z^{TX}\right)\right) dz dZ^{TX}. \tag{15.1.3}$$

Recall that $x_0 = p1$. We may and we will assume that $\left(x, Y^{TX}\right) = \left(x_0, Y^{\mathfrak{p}}\right)$, with $Y^{\mathfrak{p}} \in \mathfrak{p}$. By proceeding as in (13.2.5), we get

$$\int_{\mathcal{X}} r^X_{b,t/2}\left(\left(x_0, Y^{\mathfrak{p}}\right), \left(z, Z^{TX}\right)\right) dz dZ^{TX} = \int_{\mathfrak{p}} h^{\mathfrak{p}}_{t/2b^2}\left(bY^{\mathfrak{p}}, Z^{\mathfrak{p}}\right) dZ^{\mathfrak{p}}. \tag{15.1.4}$$

By (10.7.12), we obtain

$$\int_{\mathfrak{p}} h^{\mathfrak{p}}_{t/2b^2}\left(bY^{\mathfrak{p}}, Z^{\mathfrak{p}}\right) dZ^{\mathfrak{p}} = \left(\frac{e^{t/2b^2}}{\cosh\left(t/2b^2\right)}\right)^{m/2}$$
$$\exp\left(-\frac{1}{2}b^2 \tanh\left(t/2b^2\right)\left|Y^{\mathfrak{p}}\right|^2\right). \tag{15.1.5}$$

By (15.1.3)–(15.1.5), we get

$$r^X_{b,t}\left(\left(x, Y^{TX}\right), \left(x', Y^{TX\prime}\right)\right) \le C_{\epsilon,M} b^{4m} \exp\left(-C'_{\epsilon,M}\left|Y^{TX}\right|^2\right). \tag{15.1.6}$$

Also we get a corresponding estimate by exchanging the roles of $\left(x, Y^{TX}\right)$ and $\left(x', Y^{TX\prime}\right)$.

By proceeding as in (13.2.10) and using again the uniform bounds on $\underline{r}^X_{b,t}$ of Theorem 12.9.1, we get

$$\underline{r}^X_{b,t}\left(\left(x,Y^{TX}\right),\left(x',Y^{TX\prime}\right)\right)$$

$$\leq C_{\epsilon,M}b^{4m}\int_{\substack{(z,Z^{TX})\in\mathcal{X}\\d(x,z)\geq d(x,x')/2}}\underline{r}^X_{b,t/2}\left(\left(x,Y^{TX}\right),\left(z,Z^{TX}\right)\right)dzdZ^{TX}$$

$$+\,C_{\epsilon,M}b^{4m}\int_{\substack{(z,Z^{TX})\in\mathcal{X}\\d(x',z)\geq d(x,x')/2}}\underline{r}^X_{b,t/2}\left(\left(z,Z^{TX}\right),\left(x',Y^{TX\prime}\right)\right)dzdZ^{TX}.$$

$$(15.1.7)$$

We only need to estimate the first term in the right-hand side of (15.1.7).

By (13.2.22), we get

$$\int_{\substack{(z,Z^{TX})\in\mathcal{X}\\d(x_0,z)\geq d(x_0,x')/2}}\underline{r}^X_{b,t/2}\left(\left(x_0,Y^{\mathfrak{p}}\right),\left(z,Z^{TX}\right)\right)dzdZ^{TX}$$

$$\leq \exp\left(-\left(1-\alpha^2\right)d^2\left(x_0,x'\right)/4t\right)$$

$$\left(\frac{e^{t/2b^2}}{\cosh\left(\alpha t/2b^2\right)}\right)^{m/2}\exp\left(-\frac{1}{2}\alpha b^2\tanh\left(\alpha t/2b^2\right)|Y^{\mathfrak{p}}|^2/2\right).\quad(15.1.8)$$

By (15.1.7), and by taking $\alpha=\frac{1}{\sqrt{2}}$ in (15.1.8), we find that the estimate we obtain is compatible with (15.1.2). By (15.1.6) and (15.1.8), we get (15.1.2). The proof of our theorem is completed. □

Remark 15.1.2. The estimate (15.1.2) is not so useful, because it diverges as $b\to+\infty$. However, given the structure of the operator $\underline{\mathcal{A}}^X_b$, this divergence is unavoidable. In the sequel, we will study the kernel $\underline{r}^X_{b,t}$ in more detail.

15.2 The deviation from the geodesic flow for large b

We introduce the notation

$$h=\frac{1}{b^2}.\quad(15.2.1)$$

By (12.9.3), the operator $\underline{\mathcal{A}}^X_b$ can be written in the form

$$\underline{\mathcal{A}}^X_b=\frac{1}{2}\left(-h^2\Delta^V+\left|Y^{TX}\right|^2-hm\right)-\nabla_{Y^{TX}}.\quad(15.2.2)$$

The operator $\underline{\mathcal{A}}^X_\infty$ is well-defined and given by

$$\underline{\mathcal{A}}^X_\infty=\frac{1}{2}\left|Y^{TX}\right|^2-\nabla_{Y^{TX}}.\quad(15.2.3)$$

We consider equation (12.9.5), in which the starting point $\left(x,Y^{TX}\right)\in\mathcal{X}$ is now arbitrary. This equation takes the form

$$\dot{x}=Y^{TX},\qquad\qquad\dot{Y}^{TX}=h\dot{w},\qquad(15.2.4)$$

$$x_0=x,\qquad\qquad Y^{TX}_0=Y^{TX}.$$

When $h = 0$, we will write $(\underline{x}_., \underline{Y}_.^{TX})$ instead of $(x., Y^{TX})$, so that

$$\dot{\underline{x}} = \underline{Y}^{TX}, \hspace{4cm} \underline{\dot{Y}}^{TX} = 0, \hspace{1.5cm} (15.2.5)$$
$$\underline{x}_0 = x, \hspace{4cm} \underline{Y}_0^{TX} = Y^{TX}.$$

Recall that $\varphi_t|_{t \in \mathbf{R}}$ denotes the geodesic flow on $\mathcal{X}$. Then in (15.2.5),

$$\left(\underline{x}_t, \underline{Y}_t^{TX}\right) = \varphi_t\left(x, Y^{TX}\right). \hspace{2cm} (15.2.6)$$

As we saw in section 12.2 after (12.2.7), equation (15.2.4) can be solved for any continuous path $w.$, and the solution depends continuously on the path $w..$

Recall that the pseudodistance δ on $\mathcal{X}$ was defined in (3.8.1).

Theorem 15.2.1. *Given $M > 0$, there exists $\eta_M > 0$ such that for $h \in [0, 1], t \in [0, M], (x, Y^{TX}) \in \mathcal{X}$,*

$$\delta\left(\left(\underline{x}_t, \underline{Y}_t^{TX}\right), \left(x_t, Y_t^{TX}\right)\right) \leq \exp\left(\eta_M\left(\left|Y^{TX}\right| + h\int_0^t |w_s|\, ds + h|w_t|\right)\right)$$
$$h\left(\int_0^t |w_s|\, ds + |w_t|\right). \hspace{1cm} (15.2.7)$$

Proof. We may and we will assume that $(x, Y^{TX}) = (p1, Y^{\mathfrak{p}})$. We will proceed as in (12.2.8), (12.2.9). Let $w.$ be a Brownian motion with values in $\mathfrak{p}$. Instead of equation (12.2.8), and in view of (15.2.4), we consider the differential equation on $G \times \mathfrak{p}$,

$$\dot{g} = Y^{\mathfrak{p}}, \hspace{4cm} \dot{Y}^{\mathfrak{p}} = h\dot{w}, \hspace{1.5cm} (15.2.8)$$
$$g_0 = 1, \hspace{4cm} Y_0^{\mathfrak{p}} = Y^{\mathfrak{p}}.$$

We can rewrite the first equation in (15.2.8) in the form

$$\dot{g}_t = Y^{\mathfrak{p}} + hw_t, \hspace{3cm} g_0 = 1. \hspace{1.5cm} (15.2.9)$$

As in (12.2.9), we get

$$x_t = pg_t, \hspace{2cm} \left(x_t, Y_t^{TX}\right) = \left(g_t, Y_t^{\mathfrak{p}}\right). \hspace{1cm} (15.2.10)$$

Also the solution of (15.2.8) depends smoothly on h.

Put

$$\underline{\omega}_s^{\mathfrak{g}} = \omega^{\mathfrak{g}}\left(\frac{\partial}{\partial h} g_s\right). \hspace{2cm} (15.2.11)$$

By (2.1.8), we get

$$\underline{\dot{\omega}}^{\mathfrak{g}} = -[Y^{\mathfrak{p}}, \underline{\omega}^{\mathfrak{g}}] + w, \hspace{2cm} \underline{\omega}_0^{\mathfrak{g}} = 0. \hspace{1cm} (15.2.12)$$

Of course, we can split $\underline{\omega}^{\mathfrak{g}}$ into its $\mathfrak{p}$ and $\mathfrak{k}$ components $\underline{\omega}^{\mathfrak{p}}, \underline{\omega}^{\mathfrak{k}}$, so that

$$\underline{\omega}^{\mathfrak{g}} = \underline{\omega}^{\mathfrak{p}} + \underline{\omega}^{\mathfrak{k}}. \hspace{2cm} (15.2.13)$$

Using (15.2.12) and Gronwall's lemma, there exists $\eta > 0$ such that

$$\left|\underline{\omega}_t^{\mathfrak{g}}\right| \leq \exp\left(\eta\int_0^t |Y_s^{\mathfrak{p}}|\, ds\right)\int_0^t |w_s|\, ds. \hspace{1.5cm} (15.2.14)$$

By (15.2.14), for $0 \le h \le 1$, we get

$$\left|\omega_t^{\mathfrak{g}}\right| \le \exp\left(\eta\left(t\,|Y^{\mathfrak{p}}| + h\int_0^t |w_s|\,ds\right)\right)\int_0^t |w_s|\,ds. \qquad (15.2.15)$$

By (15.2.15), for $0 \le h \le 1$,

$$d\left(\underline{x}_t, x_t\right) \le \exp\left(\eta\left(t\,|Y^{\mathfrak{p}}| + h\int_0^t |w_s|\,ds\right)\right) h\int_0^t |w_s|\,ds. \qquad (15.2.16)$$

Let $\frac{D}{Dh}$ denote the covariant derivative with respect to the Levi-Civita connection. Then

$$\frac{D}{Dh}Y_t^{\mathfrak{p}} = \left[\omega_t^{\mathfrak{k}}, Y_t^{\mathfrak{p}}\right] + w_t. \qquad (15.2.17)$$

Let $\underline{\nabla}_{\underline{x}_t}^{TX}$ be the flat connection on TX that is associated with the trivialization of TX along radial geodesics centered at $\underline{x}_t$ by parallel transport with respect to ∇^{TX}. Set

$$\Gamma^{TX} = \nabla^{TX} - \underline{\nabla}_{\underline{x}_t}^{TX}. \qquad (15.2.18)$$

Using (3.7.8), we get

$$\left|\Gamma^{TX}\right| \le C. \qquad (15.2.19)$$

In (15.2.19), the norm of $\Gamma_x^{TX} \in T^*X \otimes \mathrm{End}\,(TX)$ is calculated with respect to the natural metric of this vector bundle.

Let $\frac{D'}{Dh}$ denote covariant differentiation with respect to $\underline{\nabla}_{\underline{x}_t}^{TX}$. Clearly,

$$\frac{D'}{Dh}Y_t^{\mathfrak{p}} = \frac{D}{Dh}Y_t^{\mathfrak{p}} - \Gamma_{\underline{x}_t}^{TX}\left(\frac{\partial x_t}{\partial h}\right)Y_t^{\mathfrak{p}}. \qquad (15.2.20)$$

By (15.2.14)–(15.2.20), we get

$$\left|\frac{D'}{Dh}Y_t^{\mathfrak{p}}\right| \le \exp\left(\eta\left(t\,|Y^{\mathfrak{p}}| + h\int_0^t |w_s|\,ds\right)\right)\int_0^t |w_s|\,ds$$
$$(|Y^{\mathfrak{p}}| + h\,|w_t|) + |w_t|. \qquad (15.2.21)$$

By (15.2.16), (15.2.21), we get (15.2.7). The proof of our theorem is completed. $\qquad\square$

15.3 The scalar heat kernel on $\mathcal{X}$ away from $\mathcal{F}_\gamma = i_a X\,(\gamma)$

Recall that the displacement function d_γ was defined in (3.1.3).

Theorem 15.3.1. *Given* $\beta > 0, \epsilon > 0, M > 0, \epsilon \le M$, *there exist* $\eta_M > 0, C_{\epsilon,M} > 0, C'_{\epsilon,M} > 0, C''_{\gamma,\beta,M} > 0$ *such that for* $b \ge 1, \epsilon \le t \le M$, $\left(x, Y^{TX}\right) \in \mathcal{X}$, *if* $d\left(x, X\,(\gamma)\right) \ge \beta$,

$$\underline{r}_{b,t}^X\left(\left(x, Y^{TX}\right), \gamma\left(x, Y^{TX}\right)\right)$$

$$\le C_{\epsilon,M}b^{4m}\exp\left(-C'_{\epsilon,M}\left(d_\gamma^2\,(x) + \left|Y^{TX}\right|^2\right)\right.$$

$$\left. - C''_{\gamma,\beta,M}\exp\left(-2\eta_M\left|Y^{TX}\right|\right)b^4\right). \qquad (15.3.1)$$

There exists $\eta > 0$ such that given $\beta > 0, \mu > 0$, there exist $C > 0, C' > 0, C''_{\gamma,\beta,\mu} > 0$ such that for $b \geq 1, (x, Y^{TX}) \in \mathcal{X}$, if $d(x, X(\gamma)) \leq \beta, |Y^{TX} - a^{TX}| \geq \mu$,

$$\underline{r}_b^X \left((x, Y^{TX}), \gamma(x, Y^{TX}) \right)$$
$$\leq C b^{4m} \exp\left(-C' |Y^{TX}|^2 - C''_{\gamma,\beta,\mu} \exp\left(-2\eta |Y^{TX}| \right) b^4 \right). \quad (15.3.2)$$

Given $\beta > 0, \epsilon > 0, M > 0, \epsilon \leq M, \nu > 0$, there exist $C_{\epsilon,M} > 0, C'_{\epsilon,M} > 0, C''_{\beta,M,\nu} > 0$ such that for $b \geq 1, \epsilon \leq t \leq M$, if $(x, Y^{TX}), (x', Y^{TX\prime}) \in \mathcal{X}$, and $|Y^{TX}| \leq \nu, |Y^{TX\prime}| \leq \nu$, if $\delta\left((\underline{x}_t, \underline{Y}_t^{TX}), (x', Y^{TX\prime}) \right) \geq \beta$, then

$$\underline{r}_{b,t}^X \left((x, Y^{TX}), (x', Y^{TX\prime}) \right) \leq C_{\epsilon,M} b^{4m}$$
$$\exp\left(-C'_{\epsilon,M} \left(d^2(x, x') + |Y^{TX}|^2 + |Y^{TX\prime}|^2 \right) - C''_{\beta,M,\nu} b^4 \right). \quad (15.3.3)$$

Proof. As in (3.9.5), set

$$\left(x', Y^{TX\prime} \right) = \gamma\left(x, Y^{TX} \right). \quad (15.3.4)$$

Then

$$\left| Y^{TX\prime} \right| = \left| Y^{TX} \right|. \quad (15.3.5)$$

Under the assumptions of the first part of our theorem, by equation (3.9.54) in Theorem 3.9.2, for $0 \leq t \leq M$,

$$\delta\left(\varphi_{t/2}(x, Y^{TX}), \varphi_{-t/2}(x', Y^{TX\prime}) \right) \geq C_{\gamma,\beta,M}. \quad (15.3.6)$$

Set

$$\varphi_t\left(x, Y^{TX} \right) = \left(\underline{x}_t, \underline{Y}_t^{TX} \right), \qquad \varphi_{-t}\left(x', Y^{TX\prime} \right) = \left(\underline{x}'_t, \underline{Y}_t^{TX\prime} \right). \quad (15.3.7)$$

Then

$$\left| \underline{Y}_t^{TX} \right| = \left| \underline{Y}_t^{TX\prime} \right| = \left| Y^{TX} \right| = \left| Y^{TX\prime} \right|. \quad (15.3.8)$$

Equation (15.3.6) can be rewritten in the form

$$\delta\left(\left(\underline{x}_{t/2}, \underline{Y}_{t/2}^{TX} \right), \left(\underline{x}'_{t/2}, \underline{Y}_{t/2}^{TX\prime} \right) \right) \geq C_{\gamma,\beta,M}. \quad (15.3.9)$$

Recall that if $x_0 \in X$, the distance $d_{x_0}\left((x, f), (x', f') \right)$ on $\mathcal{X}$ was defined in (3.8.3). As in (3.9.33) let $\underline{z}_t$ denote the middle point on the geodesic segment connecting $\underline{x}_t$ and $\underline{x}'_t$. Then (15.3.9) can be written in the form

$$d_{\underline{z}_{t/2}}\left(\left(\underline{x}_{t/2}, \underline{Y}_{t/2}^{TX} \right), \left(\underline{x}'_{t/2}, \underline{Y}_{t/2}^{TX\prime} \right) \right) \geq C_{\gamma,\beta,M}. \quad (15.3.10)$$

By (15.3.10), if $(z, Z^{TX}) \in \mathcal{X}$, either $d_{\underline{z}_{t/2}}\left(\left(\underline{x}_{t/2}, \underline{Y}_{t/2}^{TX} \right), (z, Z^{TX}) \right) \geq C_{\gamma,\beta,M}/2$ or $d_{\underline{z}_{t/2}}\left(\left(\underline{x}'_{t/2}, \underline{Y}_{t/2}^{TX\prime} \right), (z, Z^{TX}) \right) \geq C_{\gamma,\beta,M}/2$. Also by (3.8.9),

$$d_{\underline{z}_{t/2}}\left(\left(\underline{x}_{t/2}, \underline{Y}_{t/2}^{TX} \right), (z, Z^{TX}) \right)$$
$$\leq \delta\left(\left(\underline{x}_{t/2}, \underline{Y}_{t/2}^{TX} \right), (z, Z^{TX}) \right) + C d\left(\underline{x}_{t/2}, z \right) \left| Y^{TX} \right|. \quad (15.3.11)$$

By (12.9.25), we get

$$\underline{r}^X_{b,t}\left(\left(x,Y^{TX}\right),\left(x',Y^{TX\prime}\right)\right)$$

$$\leq \int_{\substack{(z,Z^{TX})\in\mathcal{X} \\ d_{\underline{z}_{t/2}}((\underline{x}_{t/2},\underline{Y}^{TX}_{t/2}),(z,Z^{TX}))\geq C_{\gamma,\beta,M}/2}} \underline{r}^X_{b,t/2}\left(\left(x,Y^{TX}\right),\left(z,Z^{TX}\right)\right)$$
$$\underline{r}^X_{b,t/2}\left(\left(z,Z^{TX}\right),\left(x',Y^{TX\prime}\right)\right)dzdZ^{TX}$$

$$+ \int_{\substack{(z,Z^{TX})\in\mathcal{X} \\ d_{\underline{z}_{t/2}}((\underline{x}'_{t/2},\underline{Y}^{TX\prime}_{t/2}),(z,Z^{TX}))\geq C_{\gamma,\beta,M}/2}} \underline{r}^X_{b,t/2}\left(\left(x,Y^{TX}\right),\left(z,Z^{TX}\right)\right)$$
$$\underline{r}^X_{b,t/2}\left(\left(z,Z^{TX}\right),\left(x',Y^{TX\prime}\right)\right)dzdZ^{TX}. \quad (15.3.12)$$

By equation (15.1.2) in Theorem 15.1.1, for $\epsilon \leq t \leq M, b \geq 1$,

$$\int_{\substack{(z,Z^{TX})\in\mathcal{X} \\ d_{\underline{z}_{t/2}}((\underline{x}_{t/2},\underline{Y}^{TX}_{t/2}),(z,Z^{TX}))\geq C_{\gamma,\beta,M}/2}} \underline{r}^X_{b,t/2}\left(\left(x,Y^{TX}\right),\left(z,Z^{TX}\right)\right)$$
$$\underline{r}^X_{b,t/2}\left(\left(z,Z^{TX}\right),\left(x',Y^{TX\prime}\right)\right)dzdZ^{TX}$$

$$\leq C_{\epsilon,M}b^{4m}\int_{\substack{(z,Z^{TX})\in\mathcal{X} \\ d_{\underline{z}_{t/2}}((\underline{x}_{t/2},\underline{Y}^{TX}_{t/2}),(z,Z^{TX}))\geq C_{\gamma,\beta,M}/2}} \underline{r}^X_{b,t/2}\left(\left(x,Y^{TX}\right),\left(z,Z^{TX}\right)\right)$$
$$dzdZ^{TX}. \quad (15.3.13)$$

We may and we will assume that $x = x_0$, with $x_0 = p1$. By (12.9.9),

$$\int_{\substack{(z,Z^{TX})\in\mathcal{X} \\ d_{\underline{z}_{t/2}}((\underline{x}_{t/2},\underline{Y}^{TX}_{t/2}),(z,Z^{TX}))\geq C_{\gamma,\beta,M}/2}}$$
$$\underline{r}^X_{b,t/2}\left(\left(x,Y^{TX}\right),\left(z,Z^{TX}\right)\right)dzdZ^{TX}$$

$$= E\left[\exp\left(\frac{mt}{4b^2} - \frac{1}{2}\int_0^{t/2}\left|Y^{TX}\right|^2 ds\right)\right.$$
$$\left. 1_{d_{\underline{z}_{t/2}}\left(\left(\underline{x}_{t/2},\underline{Y}^{TX}_{t/2}\right),\left(x_{t/2},Y^{TX}_{t/2}\right)\right)\geq C_{\gamma,\beta,M}/2}\right]. \quad (15.3.14)$$

By (15.3.11), if $d_{\underline{z}_{t/2}}\left(\left(\underline{x}_{t/2},\underline{Y}^{TX}_{t/2}\right),\left(x_{t/2},Y^{TX}_{t/2}\right)\right) \geq C_{\gamma,\beta,M}/2$, then

$$C_{\gamma,\beta,M}/2 \leq \delta\left(\left(\underline{x}_{t/2},\underline{Y}^{TX}_{t/2}\right),\left(x_{t/2},Y^{TX}_{t/2}\right)\right) + Cd\left(\underline{x}_{t/2},x_{t/2}\right)\left|Y^{TX}\right|. \quad (15.3.15)$$

By (15.2.7), (15.3.15), for $b \geq 1, 0 \leq t \leq M$, there exist $\eta_M > 0, D_{\gamma,\beta,M} > 0$ such that for $b \geq 1, t \in [0,M]$, under the above conditions,

$$D_{\gamma,\beta,M} \leq \exp\left(\eta_M\left(\left|Y^{TX}\right| + \frac{1}{b^2}\left(\int_0^{t/2}\left|w_s\right|ds + \left|w_{t/2}\right|\right)\right)\right)$$
$$\frac{1}{b^2}\left(\int_0^{t/2}\left|w_s\right|ds + \left|w_{t/2}\right|\right). \quad (15.3.16)$$

By (15.3.14), (15.3.16), we get

$$\int_{\substack{(z,Z^{TX})\in\mathcal{X} \\ d_{\underline{z}_{t/2}}((\underline{x}_{t/2},\underline{Y}_{t/2}^{TX}),(z,Z^{TX}))\geq C_{\gamma,\beta,M}/2}} \underline{r}_{b,t/2}^X\left((x,Y^{TX}),(z,Z^{TX})\right)dzdZ^{TX}$$

$$\leq C_M P\left[\exp\left(\frac{\eta_M}{b^2}\left(\int_0^{t/2}|w_s|\,ds+|w_{t/2}|\right)\right)\right.$$

$$\left.\frac{1}{b^2}\left(\int_0^{t/2}|w_s|\,ds+|w_{t/2}|\right)\geq D_{\gamma,\beta,M}\exp\left(-\eta_M\left|Y^{TX}\right|\right)\right]. \quad (15.3.17)$$

For $0\leq t\leq M$,

$$\int_0^{t/2}|w_s|\,ds+|w_{t/2}|\leq C_M\sup_{0\leq s\leq t/2}|w_s|. \quad (15.3.18)$$

Note that given $d>0$, there exists $c_{d,M}>0$ such that if $x\geq 0, 0\leq y\leq d$ are such that

$$\exp\left(\eta_M x\right)x>y, \quad (15.3.19)$$

then

$$x>c_{d,M}y. \quad (15.3.20)$$

By (15.3.18)–(15.3.20), for $0\leq t\leq M$, we obtain

$$P\left[\exp\left(\frac{\eta_M}{b^2}\left(\int_0^{t/2}|w_s|\,ds+|w_{t/2}|\right)\right)\frac{1}{b^2}\left(\int_0^{t/2}|w_s|\,ds+|w_{t/2}|\right)\right.$$

$$\left.\geq D_{\gamma,\beta,M}\exp\left(-\eta_M\left|Y^{TX}\right|\right)\right]$$

$$\leq P\left[\sup_{0\leq s\leq M/2}|w_s|\geq D'_{\gamma,\beta,M}\exp\left(-\eta_M\left|Y^{TX}\right|\right)b^2\right]. \quad (15.3.21)$$

By [ReY99, Proposition II (1.8)], there exist $c>0, c'>0$ such that for $t\geq 0, \ell\geq 0$,

$$P\left[\sup_{0\leq s\leq t}|w_s|\geq \ell\right]\leq c\exp\left(-c'\ell^2/2t\right). \quad (15.3.22)$$

By (15.3.22), there exists $D''_{\gamma,\beta,M}>0$ such that

$$P\left[\sup_{0\leq s\leq M/2}|w_s|\geq D'_{\gamma,\beta,M}\exp\left(-\eta_M\left|Y^{TX}\right|\right)b^2\right]$$

$$\leq c\exp\left(-C''_{\gamma,\beta,M}\exp\left(-2\eta_M\left|Y^{TX}\right|\right)b^4\right). \quad (15.3.23)$$

By (15.3.13), (15.3.17), (15.3.21), and (15.3.23), we get

$$
\int_{\substack{(z,Z^{TX})\in\mathcal{X} \\ d_{\underline{z}_{t/2}}((\underline{x}_{t/2},\underline{Y}_{t/2}^{TX}),(z,Z^{TX}))\geq C_{\gamma,\beta,M}/2}} r_{b,t/2}^{X}\left((x,Y^{TX}),(z,Z^{TX})\right)
$$
$$
r_{b,t/2}^{X}\left((z,Z^{TX}),(x',Y^{TX\prime})\right)dzdZ^{TX}
$$
$$
\leq C_{\epsilon,M}b^{4m}\exp\left(-C_{\gamma,\beta,M}''\exp\left(-2\eta_M\left|Y^{TX}\right|\right)b^4\right). \quad (15.3.24)
$$

Using (15.3.5) and interchanging the roles of (x, Y^{TX}) and $(x', Y^{TX\prime})$, we get a similar estimate for the second term in the right-hand side of (15.3.12). By combining equation (15.1.2) in Theorem 15.1.1 with the above estimates, we get (15.3.1).

Now we establish (15.3.2), which is an estimate for $t = 1$. Instead of equation (3.9.54) in Theorem 3.9.2, we use equation (3.9.58) in Theorem 3.9.3, and we proceed exactly as before.

Finally, we will establish (15.3.3). We no longer assume that $(x', Y^{TX\prime}) = \gamma\left(x, Y^{TX}\right)$, but otherwise the previous notation will still be in force. For $\rho > 0$, set

$$
\mathcal{A}_\rho = \Bigg\{ (z, Z^{TX}) \in \mathcal{X}, \delta\left(\left(\underline{x}_{t/2},\underline{Y}_{t/2}^{TX}\right),(z, Z^{TX})\right) < \rho,
$$
$$
\delta\left(\left(\underline{x}_{t/2}',\underline{Y}_{t/2}^{TX\prime}\right),(z, Z^{TX})\right) < \rho\Bigg\}. \quad (15.3.25)
$$

We will show that under the assumptions of the third part of our theorem, for $\rho > 0$ small enough, $\mathcal{A}_\rho$ is empty. Take $(z, Z^{TX}) \in \mathcal{A}_\rho$. By (3.8.4), (15.3.25), we get

$$
d_z\left(\left(\underline{x}_{t/2},\underline{Y}_{t/2}^{TX}\right),\left(\underline{x}_{t/2}',\underline{Y}_{t/2}^{TX\prime}\right)\right) \leq 2\rho. \quad (15.3.26)
$$

Moreover,

$$
\varphi_{t/2}\left(\underline{x}_{t/2},\underline{Y}_{t/2}^{TX}\right) = \left(\underline{x}_t,\underline{Y}_t^{TX}\right), \qquad \varphi_{t/2}\left(\underline{x}_{t/2}',\underline{Y}_{t/2}^{TX\prime}\right) = \left(x',Y^{TX\prime}\right).
$$
$$(15.3.27)$$

Also by (15.3.27), we get

$$
\left|\underline{Y}_{t/2}^{TX}\right| = \left|Y^{TX}\right|, \qquad \left|\underline{Y}_{t/2}^{TX\prime}\right| = \left|Y^{TX\prime}\right|. \quad (15.3.28)
$$

Under the assumptions of the third part of our theorem, we know that $\left|Y^{TX}\right| \leq \nu, \left|Y^{TX\prime}\right| \leq \nu$. By (3.8.10), (15.3.27), and (15.3.28), there exists $\rho_{\beta,M,\nu} > 0$ such that if $\rho = \rho_{\beta,M,\nu}$, if for a given $t \in [0, M]$, (15.3.26) holds, then

$$
\delta\left((\underline{x}_t,\underline{Y}_t^{TX}),(x',Y^{TX\prime})\right) \leq \frac{\beta}{2}. \quad (15.3.29)
$$

By (15.3.29), under the assumptions of the third part of our theorem, $\mathcal{A}_{\rho_{\beta,M,\nu}}$ is empty.

By (12.9.25), as in (15.3.12), we get

$$\underline{r}_{b,t}^X \left(\left(x, Y^{TX} \right), \left(x', Y^{TX\prime} \right) \right)$$

$$\leq \int_{\substack{(z,Z^{TX}) \in \mathcal{X} \\ \delta((\underline{x}_{t/2}, \underline{Y}_{t/2}^{TX}), (z,Z^{TX})) \geq \rho_{\beta, M, \nu}}} \underline{r}_{b,t/2}^X \left(\left(x, Y^{TX} \right), \left(z, Z^{TX} \right) \right)$$

$$\underline{r}_{b,t/2}^X \left(\left(z, Z^{TX} \right), \left(x', Y^{TX\prime} \right) \right) dz dZ^{TX}$$

$$+ \int_{\substack{(z,Z^{TX}) \in \mathcal{X} \\ \delta((\underline{x}_{t/2}', \underline{Y}_{t/2}^{TX\prime}), (z,Z^{TX})) \geq \rho_{\beta, M, \nu}}} \underline{r}_{b,t/2}^X \left(\left(x, Y^{TX} \right), \left(z, Z^{TX} \right) \right)$$

$$\underline{r}_{b,t/2}^X \left(\left(z, Z^{TX} \right), \left(x', Y^{TX\prime} \right) \right) dz dZ^{TX}. \quad (15.3.30)$$

Using (15.3.30) and proceeding as in (15.3.12)–(15.3.24), we get (15.3.3). The proof of our theorem is completed. $\qquad\square$

Remark 15.3.2. A consequence of equation (15.3.3) in Theorem 15.3.1 and of its proof is that as should be the case, as $b \to +\infty$, the heat flow for $\underline{A}_b^X$ propagates more and more along the geodesic flow.

15.4 Gaussian estimates for $\underline{r}_b^X$ near $i_a X(\gamma)$

In the sequel, for $x \in X$, we will write a^{TX} instead of a_x^{TX}.

Theorem 15.4.1. *Given $\nu > 0$, there exist $C > 0, C' > 0, C_\nu' > 0$ such that for $b \geq 1, f \in \mathfrak{p}^\perp(\gamma), |f| \leq 1, x = \rho_\gamma(1, f), Y^{TX} \in T_x X, \left| Y^{TX} \right| \leq \nu$,*

$$\underline{r}_b^X \left(\left(x, Y^{TX} \right), \gamma \left(x, Y^{TX} \right) \right) \leq C b^{4m}$$

$$\exp \left(-C' \left| Y^{TX} \right|^2 - C_\nu' \left(|f|^2 + \left| Y^{TX} - a^{TX} \right|^2 \right) b^4 \right). \quad (15.4.1)$$

Proof. Let C_ν be the positive constant appearing in Theorem 3.9.4. By equations (15.3.1) and (15.3.2) in Theorem 15.3.1, as long as $|f|$ or $\left| Y^{TX} - a^{TX} \right|$ stay away from 0, (15.4.1) holds. So we may as well assume that $|f| + \left| Y^{TX} - a^{TX} \right| \leq c$, with c such that $0 < c C_\nu < 1$.

We use the notation in the first part of the proof of Theorem 15.3.1. By equation (3.9.60) in Theorem 3.9.4, instead of (15.3.6), we have

$$\delta \left(\varphi_{1/2} \left(x, Y^{TX} \right), \varphi_{-1/2} \left(x', Y^{TX\prime} \right) \right) \geq C_\nu \left(|f| + \left| Y^{TX} - a^{TX} \right| \right). \quad (15.4.2)$$

The proof of our theorem continues exactly as the proof of the first part of Theorem 15.3.1. $\qquad\square$

15.5 The scalar heat kernel on $\widehat{\mathcal{X}}$ away from $\widehat{\mathcal{F}}_\gamma = \widehat{i}_a \mathcal{N}\left(k^{-1} \right)$

We use the notation of sections 12.10 and 12.11. In particular, for $t > 0$, $\underline{\mathfrak{r}}_{b,t}^X \left((x, Y), (x', Y') \right)$ is the smooth kernel on $\widehat{\mathcal{X}}$ associated with $\exp\left(-t \underline{\mathfrak{A}}_b^X \right)$, and $\underline{r}_{b,t}^X \left((x, Y), (x', Y') \right)$ is the smooth kernel associated with $\exp\left(-t \underline{A}_b^X \right)$.

Set

$$\mathfrak{r}_b^X = \mathfrak{r}_{b,1}^X, \qquad\qquad \mathfrak{r}_b^X = \mathfrak{r}_{b,1}^X. \qquad (15.5.1)$$

By (12.11.9), (15.1.1), and (15.5.1), we get

$$\mathfrak{r}_b^X \le \mathfrak{r}_b^X. \qquad (15.5.2)$$

Theorem 15.5.1. *Given* $\epsilon > 0, M > 0, \epsilon \le M$, *there exist* $C_{\epsilon,M} > 0, C'_{\epsilon,M} > 0$ *such that for* $b \ge 1, \epsilon \le t \le M, (x,Y), (x',Y') \in \widehat{\mathcal{X}}$,

$$\mathfrak{r}_{b,t}^X((x,Y),(x',Y'))$$

$$\le C_{\epsilon,M} b^{4m+2n} \exp\left(-C'_{\epsilon,M}\left(d^2(x,x') + |Y|^2 + |Y'|^2\right)\right). \qquad (15.5.3)$$

Proof. We use the uniform bounds on the kernel $\mathfrak{r}_{b,t}^X$ that were established in Theorem 12.10.3, and we proceed as in the proof of Theorem 15.1.1.   □

As in (15.2.1), we still take $h = 1/b^2$. We consider equation (12.10.14), in which the starting point $(x,Y) \in \widehat{\mathcal{X}}$ is arbitrary. Let $w. = w.^{TX} \oplus w.^N$ be a Brownian motion valued in $T_x X \oplus N_x$. Equation (12.10.14) takes the form

$$\dot{x} = Y^{TX}, \qquad\qquad \dot{Y} = h\dot{w}, \qquad (15.5.4)$$

$$x_0 = x, \qquad\qquad Y_0 = Y.$$

We denote by $(\underline{x}., \underline{Y}.)$ the solution of (15.5.4) for $h = 0$.

Recall that the pseudodistance δ on $\widehat{\mathcal{X}}$ was defined in section 3.8. As usual, we write $Y = (Y^{TX}, Y^N)$. We have the following extension of Theorem 15.2.1.

Theorem 15.5.2. *Given* $M > 0$, *there exists* $\eta_M > 0$ *such that for* $h \in [0,1], t \in [0,M], (x,Y) \in \widehat{\mathcal{X}}$,

$$\delta((\underline{x}_t, \underline{Y}_t),(x_t,Y_t)) \le \exp\left(\eta_M\left(|Y^{TX}| + h\int_0^t |w_s|\,ds + h|w_t|\right)\right)$$

$$h\left(\int_0^t |w_s|\,ds + |w_t|\right)\left(1 + |Y^N|\right). \qquad (15.5.5)$$

Proof. With respect to Theorem 15.2.1, we only have to deal with the extra component Y^N of Y. We use the notation on the proof of Theorem 15.2.1. We may and we will assume that $x_0 = p1$. Instead of (15.2.8), we now have

$$\dot{g} = Y^{\mathfrak{p}}, \qquad\qquad \dot{Y} = h\dot{w}, \qquad (15.5.6)$$

$$g_0 = 1, \qquad\qquad Y_0 = Y.$$

In (15.5.6), we split $Y.$ in the form

$$Y. = Y.^{\mathfrak{p}} + Y.^{\mathfrak{k}}. \qquad (15.5.7)$$

Let $\frac{D}{dh}$ also denote covariant differentiation with respect to the connection ∇^N. As in (15.2.17), we get

$$\frac{D}{Dh}Y_t^{\mathfrak{k}} = [\omega_t^{\mathfrak{k}}, Y_t^{\mathfrak{k}}] + w_t^{\mathfrak{k}}. \qquad (15.5.8)$$

Let $\underline{\nabla}_{\underline{x}_t}^N$ be the flat connection on N that is associated with the trivialization of N along radial geodesics centered at $\underline{x}_t$ by parallel transport with respect to the connection ∇^N. Set

$$\Gamma^N = \nabla^N - \underline{\nabla}_{\underline{x}_t}^N. \tag{15.5.9}$$

A formula for Γ^N is given in (3.7.9). As in (15.2.19), by (3.7.8), we obtain

$$\left| \Gamma^N \right| \le C. \tag{15.5.10}$$

As in (15.2.20), we get

$$\frac{D'}{Dh} Y^{\mathfrak{e}} = \frac{D}{Dh} Y^{\mathfrak{e}} - \Gamma_{x_t}^N \left(\frac{\partial x_t}{\partial h} \right) Y_t^{\mathfrak{e}}. \tag{15.5.11}$$

By equation (15.2.7) in Theorem 15.2.1, by (15.2.15), and (15.5.8)–(15.5.11), we get (15.5.5). The proof of our theorem is completed. $\qquad\square$

We use the same conventions as before Theorems 3.9.5 and 9.1.1. In particular, $\mathrm{Ad}\left(k^{-1}\right)$ acts on the fibres on N. Also we still write a^{TX} instead of a_x^{TX}.

Theorem 15.5.3. *Given $\beta > 0, \epsilon > 0, M > 0, \epsilon \le M$, there exist $\eta_M > 0, C_{\epsilon,M} > 0, C'_{\epsilon,M} > 0, C''_{\gamma,\beta,M} > 0$ such that for $b \ge 1, \epsilon \le t \le M, (x,Y) \in \widehat{\mathcal{X}}$, if $d\left(x, X\left(\gamma\right)\right) \ge \beta$,*

$$\mathfrak{r}_{b,t}^X \left((x,Y), \gamma\left(x,Y\right) \right) \le C_{\epsilon,M} b^{4m+2n} \exp\left(-C'_{\epsilon,M} \left(d_\gamma^2\left(x\right) + |Y|^2 \right) \right.$$

$$\left. - C''_{\gamma,\beta,M} \exp\left(-2\eta_M \left| Y^{TX} \right| \right) b^4 \right). \tag{15.5.12}$$

There exists $\eta > 0$ such that given $\beta > 0, \mu > 0$, there exist $C > 0, C' > 0, C''_{\gamma,\beta,\mu} > 0$ such that for $b \ge 1, (x,Y) \in \widehat{\mathcal{X}}$, if $d\left(x, X\left(\gamma\right)\right) \le \beta, \left| Y^{TX} - a^{TX} \right| \ge \mu$,

$$\mathfrak{r}_b^X \left((x,Y), \gamma\left(x,Y\right) \right)$$
$$\le C b^{4m+2n} \exp\left(-C' |Y|^2 - C''_{\gamma,\beta,\mu} \exp\left(-2\eta \left| Y^{TX} \right| \right) b^4 \right). \tag{15.5.13}$$

Given $\nu > 0, \rho > 0$, there exist $C > 0, C' > 0, C''_{\gamma,\nu,\rho} > 0$ such that for $\beta \in]0,1]$ small enough, $b \ge 1, f \in \mathfrak{p}^\perp\left(\gamma\right), |f| \le \beta, x = \rho_\gamma\left(1, f\right), \left| Y^{TX} - a^{TX} \right| \le \beta, \left| Y^N \right| \ge \rho$, if $\left| \left(\mathrm{Ad}\left(k^{-1}\right) - 1 \right) Y^N \right| \ge \nu \left| Y^N \right|$, or if $\left| \left[a^{TX}, Y^N \right] \right| \ge \nu \left| Y^N \right|$, then

$$\mathfrak{r}_b^X \left((x,Y), \gamma\left(x,Y\right) \right) \le C b^{4m+2n}$$
$$\exp\left(-C' \left| Y^N \right|^2 - C''_{\gamma,\nu,\rho} \left(1 + \left| Y^N \right|^2 \right)^{-1} b^4 \right). \tag{15.5.14}$$

Given $\beta > 0, \epsilon > 0, M > 0, \epsilon \leq M, \nu > 0$, there exist $C_{\epsilon,M} > 0, C'_{\epsilon,M} > 0, C''_{\beta,M,\nu} > 0$ such that for $b \geq 1, \epsilon \leq t \leq M$, if $(x,Y),(x',Y') \in \widehat{\mathcal{X}}$, and $|Y| \leq \nu, |Y'| \leq \nu$, if $\delta\left((\underline{x}_t,\underline{Y}_t),(x',Y')\right) \geq \beta$, then

$$\mathfrak{r}^X_{b,t}\left((x,Y),(x',Y')\right)$$
$$\leq C_{\epsilon,M} b^{4m+2n} \exp\left(-C'_{\epsilon,M}\left(d^2\left(x,x'\right)+|Y|^2+|Y'|^2\right)-C''_{\beta,M,\nu}b^4\right). \tag{15.5.15}$$

Proof. As in (3.9.61), set

$$(x',Y') = \gamma\left(x,Y\right). \tag{15.5.16}$$

To obtain (15.5.12), we proceed exactly as in the proof of Theorem 15.3.1. Instead of (15.3.12), we get

$$\mathfrak{r}^X_{b,t}\left((x,Y),(x',Y')\right)$$
$$\leq \int_{\substack{(z,Z)\in\widehat{\mathcal{X}} \\ d_{\underline{z}_{t/2}}((\underline{x}_{t/2},\underline{Y}^{TX}_{t/2}),(z,Z^{TX}))\geq C_{\gamma,\beta,M}/2}} \mathfrak{r}^X_{b,t/2}\left((x,Y),(z,Z)\right)$$
$$\mathfrak{r}^X_{b,t/2}\left((z,Z),(x',Y')\right)dzdZ$$
$$+ \int_{\substack{(z,Z^{TX})\in\widehat{\mathcal{X}} \\ d_{\underline{z}_{t/2}}((\underline{x}'_{t/2},\underline{Y}^{TX'}_{t/2}),(z,Z^{TX}))\geq C_{\gamma,\beta,M}/2}} \mathfrak{r}^X_{b,t/2}\left((x,Y),(z,Z)\right)$$
$$\mathfrak{r}^X_{b,t/2}\left((z,Z),(x',Y')\right)dzdZ. \tag{15.5.17}$$

By equation (15.5.3) in Theorem 15.5.1, instead of (15.3.13), we get

$$\int_{\substack{(z,Z)\in\widehat{\mathcal{X}} \\ d_{\underline{z}_{t/2}}((\underline{x}_{t/2},\underline{Y}^{TX}_{t/2}),(z,Z^{TX}))\geq C_{\gamma,\beta,M}/2}} \mathfrak{r}^X_{b,t/2}\left((x,Y),(z,Z)\right)$$
$$\mathfrak{r}^X_{b,t/2}\left((z,Z),(x',Y')\right)dzdZ$$
$$\leq C_{\epsilon,M} b^{4m+2n} \int_{\substack{(z,Z)\in\widehat{\mathcal{X}} \\ d_{\underline{z}_{t/2}}((\underline{x}_{t/2},\underline{Y}^{TX}_{t/2}),(z,Z))\geq C_{\gamma,\beta,M}/2}}$$
$$\mathfrak{r}^X_{b,t/2}\left((x,Y),(z,Z)\right)dzdZ^{TX}. \tag{15.5.18}$$

Now we use equation (12.10.15) instead of equation (12.9.9), and by proceeding as in the proof of Theorem 15.3.1, we get an estimate for (15.5.18) similar to (15.3.24), with b^{4m} replaced by b^{4m+2n}. The second term in the right-hand side of (15.5.18) can also be estimated by the same method. Combining equation (15.5.3) with these estimates, we get (15.5.12).

By proceeding as in the proof of Theorem 15.3.1, we also obtain (15.5.13). Note that in the proof of (15.5.12) and (15.5.13), only Theorem 15.2.1 has been used, and not the more general Theorem 15.5.2.

Now we concentrate on the proof of (15.5.14). Under the first set of assumptions, we will prove the estimate (15.5.14) in which $\mathfrak{r}^X_b$ is replaced by $\mathfrak{r}^X_b$.

Using (15.5.2), this will also give (15.5.14) for $\underline{\mathfrak{r}}_b^X$. We still use the notation in (15.5.16). Set

$$\varphi_t(x, Y) = (\underline{x}_t, \underline{Y}_t), \qquad \varphi_t(x', Y') = (\underline{x}'_t, \underline{Y}'_t). \qquad (15.5.19)$$

In particular $\underline{Y}^N_\cdot, \underline{Y}^{N'}_\cdot$ denote the components in N of $\underline{Y}_\cdot, \underline{Y}'_\cdot$.

Take $\nu > 0, \rho > 0$, and assume that $\left|Y^N\right| \geq \rho, \left|\left(\mathrm{Ad}\left(k^{-1}\right) - 1\right) Y^N\right| \geq \nu \left|Y^N\right|$. By equation (3.9.63) in Theorem 3.9.5, there exists $\beta_{\gamma,\nu} > 0$ such that for $|f| \leq \beta_{\gamma,\nu}, \left|Y^{TX} - a^{TX}\right| \leq \beta_{\gamma,\nu}$, we get

$$\left|\tau^{\underline{x}'_{1/2}}_{\underline{x}_{1/2}} \underline{Y}^{N'}_{1/2} - \underline{Y}^N_{1/2}\right| \geq \frac{\nu}{2} \left|Y^N\right| \geq \frac{\nu}{2}\rho. \qquad (15.5.20)$$

Let $\underline{z}_{1/2}$ denote the middle point on the geodesic segment connecting $\underline{x}_{1/2}$ and $\underline{x}'_{1/2}$. We can rewrite (15.5.20) in the form

$$\left|\tau^{\underline{x}'_{1/2}}_{\underline{z}_{1/2}} \underline{Y}^{N'}_{1/2} - \tau^{\underline{x}_{1/2}}_{\underline{z}_{1/2}} \underline{Y}^N_{1/2}\right| \geq \frac{\nu}{2} \left|Y^N\right| \geq \frac{\nu}{2}\rho. \qquad (15.5.21)$$

By (15.5.21), we get

$$\int_{(z,Z)\in\widehat{\mathcal{X}}} \mathfrak{r}^X_{b,1/2}\left((x,Y),(z,Z)\right) \mathfrak{r}^X_{b,1/2}\left((z,Z),(x',Y')\right) dzdZ$$

$$\leq \int_{\substack{(z,Z)\in\widehat{\mathcal{X}} \\ \left|\tau^z_{\underline{z}_{1/2}} Z^N - \tau^{\underline{x}_{1/2}}_{\underline{z}_{1/2}} \underline{Y}^N_{1/2}\right| \geq \nu\rho/4}} \mathfrak{r}^X_{b,1/2}\left((x,Y),(z,Z)\right)$$

$$\mathfrak{r}^X_{b,1/2}\left((z,Z),(x',Y')\right) dzdZ$$

$$+ \int_{\substack{(z,Z)\in\widehat{\mathcal{X}} \\ \left|\tau^z_{\underline{z}_{1/2}} Z^N - \tau^{\underline{x}'_{1/2}}_{\underline{z}_{1/2}} \underline{Y}^{N'}_{1/2}\right| \geq \nu\rho/4}} \mathfrak{r}^X_{b,1/2}\left((x,Y),(z,Z)\right)$$

$$\mathfrak{r}^X_{b,1/2}\left((z,Z),(x',Y')\right) dzdZ. \qquad (15.5.22)$$

As explained in section 3.8, equations (3.8.5)–(3.8.10) are also valid when $f, f' \in N$, so that

$$\left|\tau^z_{\underline{z}_{1/2}} Z^N - \tau^{\underline{x}_{1/2}}_{\underline{z}_{1/2}} \underline{Y}^N_{1/2}\right| \leq \left|\tau^z_{\underline{x}_{1/2}} Z^N - \underline{Y}^N_{1/2}\right| + Cd\left(\underline{x}_{1/2}, z\right) \left|Y^N\right|. \qquad (15.5.23)$$

By (15.5.5), (15.5.23), and keeping in mind that $\left|Y^{TX}\right|$ remains uniformly bounded, we obtain

$$\left|\tau^{\underline{x}_{1/2}}_{\underline{z}_{1/2}} Y^N_{1/2} - \tau^{\underline{x}_{1/2}}_{\underline{z}_{1/2}} \underline{Y}^N_{1/2}\right| \leq C \exp\left(\frac{\eta}{b^2}\left(\int_0^{1/2} |w_s|\, ds + |w_{1/2}|\right)\right)$$

$$\frac{1}{b^2}\left(\int_0^{1/2} |w_s|\, ds + |w_{1/2}|\right)\left(1 + \left|Y^N\right|\right). \qquad (15.5.24)$$

By (12.10.15), (15.5.24), and proceeding as in (15.3.17)–(15.3.23), we get

$$
\int_{\substack{(z,Z)\in\widehat{\mathcal{X}} \\ \left|\tau^z_{\pm1/2}Z^N - \tau^{\pm1/2}_{\pm1/2}\underline{Y}^N_{1/2}\right|\geq\nu\rho/4}} \mathfrak{r}^X_{b,1/2}\left((x,Y),(z,Z)\right)
$$

$$
\mathfrak{r}^X_{b,1/2}\left((z,Z),(x',Y')\right)dzdZ
$$

$$
\leq Cb^{4m+2n}\exp\left(-C''_{\gamma,\nu,\rho}\left(1+\left|Y^N\right|^2\right)^{-1}b^4\right). \quad (15.5.25)
$$

The second integral in the right-hand side of (15.5.22) can be dominated in the same way. Combining these bounds with Theorem 15.5.1, we obtain equation (15.5.14) for $\mathfrak{r}^X_b$, from which, as explained before, (15.5.14) follows for $\underline{\mathfrak{r}}^X_b$.

Now we will establish (15.5.14) under our second set of assumptions. We still use the notation in (15.5.16). By the analogue of equation (12.9.25) for the kernel $\mathfrak{r}^X_{b,t}$ and by equation (12.11.10) in Theorem 12.11.2, we get

$$
\mathfrak{r}^X_b\left((x,Y),(x',Y')\right)\leq Cb^{4m+2n}\int_{\widehat{\mathcal{X}}}\mathfrak{r}^X_{b,1/2}\left((x,Y),(z,Z)\right)dzdZ. \quad (15.5.26)
$$

Given $\nu>0$, for $\beta>0$ small enough, if $\left|\left[a^{TX},Y^N\right]\right|\geq\nu\left|Y^N\right|$ and $\left|Y^{TX}-a^{TX}\right|\leq\beta$, then

$$
\left|\left[Y^N,Y^{TX}\right]\right|\geq\frac{\nu}{2}\left|Y^N\right|. \quad (15.5.27)
$$

In the sequel, we may assume that (15.5.27) holds, and we will estimate the right-hand side of (15.5.26).

We may and we will assume that $x=x_0$. By (12.11.8), we obtain

$$
\int_{\widehat{\mathcal{X}}}\mathfrak{r}^X_{b,1/2}\left((x,Y),(z,Z)\right)dzdZ
$$

$$
=E\left[\exp\left(\frac{(m+n)}{4b^2}-\frac{1}{2}\int_0^{1/2}|Y|^2\,ds-\frac{b^4}{2}\int_0^{1/2}\left|\left[Y^N,Y^{TX}\right]\right|^2ds\right)\right].
$$

$$
(15.5.28)
$$

Since $x=x_0$, we can take $Y^{\mathfrak{g}}\in\mathfrak{g}$ with $Y^{\mathfrak{g}}=Y^{\mathfrak{p}}+Y^{\mathfrak{k}},Y^{\mathfrak{p}}\in\mathfrak{p},Y^{\mathfrak{k}}\in\mathfrak{k}$, so that $Y^{\mathfrak{g}},Y^{\mathfrak{p}},Y^{\mathfrak{k}}$ represent Y,Y^{TX},Y^N. By (15.5.27), we get

$$
\left|\left[Y^{\mathfrak{k}},Y^{\mathfrak{p}}\right]\right|\geq\frac{\nu}{2}\left|Y^{\mathfrak{k}}\right|. \quad (15.5.29)
$$

Let $w^{\mathfrak{g}}=w^{\mathfrak{p}}+w^{\mathfrak{k}}$ be a Brownian motion in $\mathfrak{g}=\mathfrak{p}\oplus\mathfrak{k}$ with $w_0=0$. Let E still denote the corresponding expectation. Put

$$
Y^{\mathfrak{g}}_s=Y^{\mathfrak{g}}+\frac{w^{\mathfrak{g}}_s}{b^2}. \quad (15.5.30)
$$

The process $Y^{\mathfrak{g}}$ splits as

$$
Y^{\mathfrak{g}}=Y^{\mathfrak{p}}+Y^{\mathfrak{k}}. \quad (15.5.31)
$$

Also equation (15.5.28) can be written in the form

$$\int_{\mathcal{X}} \mathfrak{r}_{b,1/2}^X \left((x,Y), (z,Z) \right) dz dZ$$

$$= E \left[\exp \left(\frac{(m+n)}{4b^2} - \frac{1}{2} \int_0^{1/2} |Y_s^{\mathfrak{g}}|^2 \, ds - \frac{b^4}{2} \int_0^{1/2} \left| [Y_s^{\mathfrak{k}}, Y_s^{\mathfrak{p}}] \right|^2 ds \right) \right].$$

$$(15.5.32)$$

By (15.5.30), since $|Y^{\mathfrak{p}}|$ remains uniformly bounded, we get

$$\left| [Y_s^{\mathfrak{k}}, Y_s^{\mathfrak{p}}] \right| \geq \left| [Y^{\mathfrak{k}}, Y^{\mathfrak{p}}] \right| - \frac{C'}{b^2} |Y^{\mathfrak{k}}| |w_s^{\mathfrak{p}}| - \frac{C'}{b^2} |w_s^{\mathfrak{k}}| - \frac{C'}{b^4} |w_s^{\mathfrak{p}}| |w_s^{\mathfrak{k}}|. \quad (15.5.33)$$

By (15.5.29), (15.5.33), we obtain

$$\left| [Y_s^{\mathfrak{k}}, Y_s^{\mathfrak{p}}] \right| \geq \left(\frac{\nu}{2} - \frac{C'}{b^2} |w_s^{\mathfrak{p}}| \right) |Y^{\mathfrak{k}}| - \frac{C'}{b^2} |w_s^{\mathfrak{k}}| - \frac{C'}{b^4} |w_s^{\mathfrak{p}}| |w_s^{\mathfrak{k}}|. \quad (15.5.34)$$

By (15.3.22), there exists $C'' > 0$ such that

$$P \left[\sup_{0 \leq s \leq 1/2} |w_s^{\mathfrak{p}}| \geq \frac{b^2 \nu}{4C'} \right] \leq c \exp \left(-C'' b^4 \right). \quad (15.5.35)$$

Moreover, if $\sup\limits_{0 \leq s \leq 1/2} |w_s^{\mathfrak{p}}| < \frac{b^2 \nu}{4C'}$, if $|Y^{\mathfrak{k}}| \geq \rho$, by (15.5.34), for $0 \leq s \leq 1/2$, we get

$$\left| [Y_s^{\mathfrak{k}}, Y_s^{\mathfrak{p}}] \right| \geq \frac{\nu}{4} \rho - \frac{C'}{b^2} |w_s^{\mathfrak{k}}| - \frac{C'}{b^4} |w_s^{\mathfrak{p}}| |w_s^{\mathfrak{k}}|. \quad (15.5.36)$$

By still using (15.3.22), (15.5.35), and (15.5.36), there exist $c > 0, C_{\nu,\rho} > 0$ such that

$$P \left[\inf_{0 \leq s \leq 1/2} \left| [Y_s^{\mathfrak{k}}, Y_s^{\mathfrak{p}}] \right| \leq \frac{\nu \rho}{8} \right] \leq c \exp \left(-C_{\nu,\rho} b^4 \right). \quad (15.5.37)$$

By (15.5.28) and (15.5.37), we conclude that given $\nu > 0, \rho > 0$, there exist $c > 0, C_{\nu,\rho} > 0$ such that for $\beta \in]0,1]$ small enough, under the conditions of the second part of the statement leading to (15.5.14),

$$\int_{\mathcal{X}} \mathfrak{r}_{b,1/2}^X \left((x,Y), (z,Z) \right) dz dZ \leq c \exp \left(-C_{\nu,\rho} b^4 \right). \quad (15.5.38)$$

By equation (15.5.3) in Theorem 15.5.1, by (15.5.26), and by (15.5.38), we get a stronger statement than (15.5.14) under the second set of assumptions, since in this case, the factor $\left(1 + |Y^N|^2 \right)^{-1}$ is unnecessary.

As to the proof of equation (15.5.15), using Theorem 15.5.2 instead of Theorem 15.2.1, it is the same as the proof of equation (15.3.3) in Theorem 15.3.1. The proof of our theorem is completed. □

Remark 15.5.4. The conclusions of Remark 15.3.2 remain valid for the operator $\mathfrak{A}_b^X$, if we replace the geodesic flow in $\mathcal{X}$ by the geodesic flow in $\widehat{\mathcal{X}}$. By (12.11.9), this is still true for the operator $\underline{A}_b^X$.

Note that for $b \geq 1$,

$$C' \left|Y^N\right|^2 + C''_{\gamma,\nu,\rho} \left(1 + \left|Y^N\right|^2\right)^{-1} b^4 \geq c_{\gamma,\nu,\rho} \left(\left|Y^N\right|^2 + b^2\right). \quad (15.5.39)$$

Indeed, (15.5.39) is trivially true for $\left|Y^N\right|^2 \geq b^2$ and for $\left|Y^N\right|^2 < b^2$. By (15.5.39), we can then give a weaker version of (15.5.14). The estimates (15.5.14) and (15.5.15) will not be used in the sequel.

15.6 Estimates on the scalar heat kernel on $\widehat{\mathcal{X}}$ near $\widehat{i}_a \mathcal{N}\left(k^{-1}\right)$

Now, we will give an upper bound for $\mathrm{r}_b^X\left(\left(x, Y\right), \gamma\left(x, Y\right)\right)$ near $\widehat{i}_a \mathcal{N}\left(k^{-1}\right)$. Recall that we write a^{TX} instead of a_x^{TX}.

Theorem 15.6.1. *There exist $c > 0, C > 0, C'_\gamma > 0$ such that for $b \geq 1, f \in \mathfrak{p}^\perp\left(\gamma\right), |f| \leq 1, x = \rho_\gamma\left(1, f\right), Y \in \left(TX \oplus N\right)_x, \left|Y^{TX} - a^{TX}\right| \leq 1,$*

$$\mathrm{r}_b^X\left(\left(x, Y\right), \gamma\left(x, Y\right)\right) \leq cb^{4m+2n} \exp\Bigg(-C\left|Y^N\right|^2$$
$$- C'_\gamma \left(|f|^2 + \left|Y^{TX} - a^{TX}\right|^2\right) b^4 - C'_\gamma \left|\left(\mathrm{Ad}\left(k^{-1}\right) - 1\right) Y^N\right| b^2$$
$$- C'_\gamma \left|\left[a^{TX}, Y^N\right]\right| b^2\Bigg). \quad (15.6.1)$$

Proof. We can handle f and Y^{TX} by the same methods as in the proof of Theorem 15.4.1. So we concentrate on how to deal with Y^N. We will use arguments that were already used in the proof of Theorem 15.5.3.

Let $C > 0$ be a fixed constant. We claim that if

$$\left|\left(\mathrm{Ad}\left(k^{-1}\right) - 1\right) Y^N\right| \leq C \left(|f| + \left|Y^{TX} - a^{TX}\right|\right) \left|Y^N\right|, \quad (15.6.2)$$

then the third term in the right-hand side of (15.6.1) can be easily obtained. Indeed, by (15.6.2), we get

$$b^2 \left|\left(\mathrm{Ad}\left(k^{-1}\right) - 1\right) Y^N\right| \leq C' \left(b^4 \left(|f|^2 + \left|Y^{TX} - a^{TX}\right|^2\right) + \left|Y^N\right|^2\right), \quad (15.6.3)$$

so that the first two exponential terms in the right-hand side of (15.6.1) already incorporate the third term. Therefore in the sequel, we may as well assume that for a given $C > 0$,

$$\left|\left(\mathrm{Ad}\left(k^{-1}\right) - 1\right) Y^N\right| > C \left(|f| + \left|Y^{TX} - a^{TX}\right|\right) \left|Y^N\right|. \quad (15.6.4)$$

Let $c_\gamma > 0$ be the constant that appears in Theorem 3.9.5. By taking $C = 2c_\gamma$ in (15.6.4), by equation (3.9.63) in Theorem 3.9.5, there exists $C' > 0$ such that for $|f| \leq 1, \left|Y^{TX} - a^{TX}\right| \leq 1$, we get

$$\left|\tau_{\pm 1/2}^{x'_{1/2}} \underline{Y}_{1/2}^{N'} - \underline{Y}_{1/2}^N\right| \geq C' \left|\left(\mathrm{Ad}\left(k^{-1}\right) - 1\right) Y^N\right|. \quad (15.6.5)$$

Using (15.6.5), we can proceed as in the proofs of Theorems 15.3.1 and 15.5.3, and we get the estimate,

$$\underline{r}_b^X \left((x,Y), (x',Y') \right) \le C b^{4m+2n}$$
$$\exp \left(-C' \left| \left(\mathrm{Ad} \left(k^{-1} \right) - 1 \right) Y^N \right|^2 \left(1 + \left| Y^N \right|^2 \right)^{-1} b^4 \right). \qquad (15.6.6)$$

Note that as in (15.5.25), the term $\left(1 + \left| Y^N \right|^2 \right)^{-1}$ is forced upon us by equation (15.5.5).

If

$$1 + \left| Y^N \right|^2 > \left| \left(\mathrm{Ad} \left(k^{-1} \right) - 1 \right) Y^N \right| b^2, \qquad (15.6.7)$$

then the presence of the term $\left| Y^N \right|^2$ in the right-hand side of (15.6.1) can be made to incorporate the term $\left| \left(\mathrm{Ad} \left(k^{-1} \right) - 1 \right) Y^N \right| b^2$. If

$$1 + \left| Y^N \right|^2 \le \left| \left(\mathrm{Ad} \left(k^{-1} \right) - 1 \right) Y^N \right| b^2, \qquad (15.6.8)$$

the term (15.6.6) is enough to produce the required exponential in the right-hand side of of (15.6.1).

We will now obtain the remaining term in (15.6.1), which will be more difficult. Take $C > 0$. We claim that when

$$\left| \left[a^{TX}, Y^N \right] \right| \le C \left| \left[Y^{TX} - a^{TX}, Y^N \right] \right|, \qquad (15.6.9)$$

then the remaining estimate in (15.6.1) follows easily. Indeed from (15.6.9), we get

$$b^2 \left| \left[a^{TX}, Y^N \right] \right| \le C' \left(b^4 \left| Y^{TX} - a^{TX} \right|^2 + \left| Y^N \right|^2 \right). \qquad (15.6.10)$$

Therefore, the first two terms in the right-hand side of (15.6.1) can be made to absorb the last term.

In the sequel, we may as well assume that $C > 0$ is a fixed constant and that

$$\left| \left[a^{TX}, Y^N \right] \right| > C \left| \left[Y^{TX} - a^{TX}, Y^N \right] \right|. \qquad (15.6.11)$$

Recall that $x = \rho_\gamma(1, f)$. We use the notation of equation (15.5.4). Then

$$Y_s^{TX} = Y^{TX} + \frac{w_s^{TX}}{b^2}, \qquad Y_s^N = Y^N + \frac{w_s^N}{b^2}, \qquad (15.6.12)$$
$$Y_s = Y_s^{TX} + Y_s^N.$$

Equation (15.5.28) still holds.

Observe that for $u, v \in \mathbf{R}^n, \alpha \ge 1$,

$$|u + v|^2 \ge (1 - 1/\alpha) |u|^2 - (\alpha - 1) |v|^2. \qquad (15.6.13)$$

In particular, for $1 \le \alpha \le 2$,

$$|u + v|^2 \ge (\alpha - 1) \left(\frac{1}{2} |u|^2 - |v|^2 \right). \qquad (15.6.14)$$

Set

$$\|w_\cdot^{TX}\| = \sup_{0 \le s \le 1/2} |w_s^{TX}|, \qquad \|w_\cdot^N\| = \sup_{0 \le s \le 1/2} |w_s^N|. \qquad (15.6.15)$$

Let $\eta \in]0, 1]$ be a constant whose precise value will be chosen later. Set

$$\alpha = 1 + \frac{\eta}{1 + \|w_\cdot^{TX}\|^2 + \|w_\cdot^N\|^2}. \qquad (15.6.16)$$

By (15.6.12), (15.6.14), we get

$$|[Y_s^{TX}, Y_s^N]|^2 \ge (\alpha - 1) \left(\frac{1}{2} |[a^{TX}, Y^N]|^2 \right.$$

$$\left. - \left| [Y^{TX} - a^{TX}, Y^N] + \left[Y^{TX}, \frac{w_s^N}{b^2} \right] + \left[\frac{w_s^{TX}}{b^2}, Y^N \right] + \left[\frac{w_s^{TX}}{b^2}, \frac{w_s^N}{b^2} \right] \right|^2 \right).$$

$$(15.6.17)$$

By (15.6.17), we obtain

$$\int_0^{1/2} |[Y_s^{TX}, Y_s^N]|^2 \, ds$$

$$\ge \frac{\alpha - 1}{2} \left(\frac{1}{2} |[a^{TX}, Y^N]|^2 - C' |[Y^{TX} - a^{TX}, Y^N]|^2 \right)$$

$$- \frac{C'}{2} (\alpha - 1) \left(\frac{1}{b^4} |Y^N|^2 \|w_\cdot^{TX}\|^2 + \frac{1}{b^4} |Y^{TX}|^2 \|w_\cdot^N\|^2 \right.$$

$$\left. + \frac{1}{b^8} \|w_\cdot^{TX}\|^2 \|w_\cdot^N\|^2 \right). \qquad (15.6.18)$$

By choosing the constant $C > 0$ adequately in (15.6.11), we deduce from (15.6.19) that we may as well assume that

$$b^4 \int_0^{1/2} |[Y_s^{TX}, Y_s^N]|^2 \, ds \ge \frac{1}{8} (\alpha - 1) |[a^{TX}, Y^N]|^2 b^4$$

$$- \frac{C'}{2} (\alpha - 1) \left(|Y^N|^2 \|w_\cdot^{TX}\|^2 + |Y^{TX}|^2 \|w_\cdot^N\|^2 \right.$$

$$\left. + \frac{1}{b^4} \|w_\cdot^{TX}\|^2 \|w_\cdot^N\|^2 \right). \qquad (15.6.19)$$

We will use the estimate (15.6.19) to obtain an upper bound for the right-hand side of (15.5.28). We will control the four terms in the right-hand side of (15.6.19).

The first term in the second line of (15.6.19) is the most annoying, and this is where the right choice of α in (15.6.16) is crucial. Indeed, let $C > 0$ be the positive constant that appears in (15.6.1). Then

$$C |Y^N|^2 - \frac{C'}{2} (\alpha - 1) |Y^N|^2 \|w_\cdot^{TX}\|^2 \ge \left(C - \frac{C'}{2} \eta \right) |Y^N|^2. \qquad (15.6.20)$$

By taking $\eta \in]0,1]$ small enough, we deduce from (15.6.20) that

$$C \left| Y^N \right|^2 - \frac{C'}{2} (\alpha - 1) \left| Y^N \right|^2 \left\| w_\cdot^{TX} \right\|^2 \geq \frac{C}{2} \left| Y^N \right|^2 . \qquad (15.6.21)$$

Combining equation (15.5.28) with (15.6.21) indicates that the first term in the second line of (15.6.19) can be absorbed by the quadratic term $\left| Y^N \right|^2$ already appearing in the right-hand side of (15.6.1). Of course, this fixes the constant η once and for all.

Recall that $\left| Y^{TX} \right|$ remains uniformly bounded, so that

$$(\alpha - 1) \left| Y^{TX} \right|^2 \left\| w_\cdot^N \right\|^2 \leq C''. \qquad (15.6.22)$$

Therefore the second term in the second line of (15.6.19) is irrelevant in our estimate.

Ultimately, in the right-hand side of (15.6.19), what remains to control is

$$\frac{1}{2} (\alpha - 1) \left(\frac{1}{4} \left| [a^{TX}, Y^N] \right|^2 b^4 - \frac{C'}{b^4} \left\| w_\cdot^{TX} \right\|^2 \left\| w_\cdot^N \right\|^2 \right). \qquad (15.6.23)$$

Using [ReY99, Proposition II (1.8)] as in (15.3.22), for $y \geq 0$, we get

$$P \left[\left\| w_\cdot^{TX} \right\| \left\| w_\cdot^N \right\| \geq y \right] \leq P \left[\left\| w_\cdot^{TX} \right\| \geq \sqrt{y} \right]$$
$$+ P \left[\left\| w_\cdot^N \right\| \geq \sqrt{y} \right] \leq c \exp \left(-c' y \right). \qquad (15.6.24)$$

By (15.6.24), given $d > 0$, we get

$$P \left[\left\| w_\cdot^{TX} \right\| \left\| w_\cdot^N \right\| \geq d \left| [a^{TX}, Y^N] \right| b^4 \right] \leq c \exp \left(-c'd \left| [a^{TX}, Y^N] \right| b^4 \right). \qquad (15.6.25)$$

By (15.6.25), we find that to find an upper bound for (15.5.28) compatible with (15.6.1), we may as well replace the expression in (15.6.23) by

$$\frac{1}{16} (\alpha - 1) \left| [a^{TX}, Y^N] \right|^2 b^4. \qquad (15.6.26)$$

By still using (15.3.22), we obtain

$$P \left[\left(\left\| w_\cdot^{TX} \right\| + \left\| w_\cdot^N \right\| \right) \geq \left| [a^{TX}, Y^N] \right|^{1/2} b \right] \leq c \exp \left(-c' \left| [a^{TX}, Y^N] \right| b^2 \right). \qquad (15.6.27)$$

By (15.6.16), (15.6.26), we have to find a proper lower bound for (15.6.26) on the set $\left(\left\| w_\cdot^{TX} \right\| + \left\| w_\cdot^N \right\| \right) < \left| [a^{TX}, Y^N] \right|^{1/2} b$. Using (15.6.16), on this set,

$$(\alpha - 1) \left| [a^{TX}, Y^N] \right|^2 b^4 \geq \eta \left(\left| [a^{TX}, Y^N] \right| b^2 - 1 \right). \qquad (15.6.28)$$

Equation (15.6.28) still produces the last exponential term in the right-hand side of (15.6.1). The proof of our theorem is completed. $\qquad\qquad\qquad$ □

Remark 15.6.2. It is remarkable that in the uniform estimate (15.6.1), two important terms are non-Gaussian.

15.7 A proof of Theorem 9.1.1

Theorem 9.1.1 is just the analogue of Theorems 15.5.1, 15.5.3, and 15.6.1 for the heat kernel $q_{b,t}^X$. To prove this theorem, we will combine the methods of section 14.4 with the above results.

Recall that the operator $\mathcal{L}_{A,b}^X$ on $\widehat{\mathcal{X}}$ is given by (9.1.3), and that for $t > 0$, $q_{b,t}^X((x, Y), (x', Y'))$ is the smooth kernel associated with $\exp\left(-t\mathcal{L}_{A,b}^X\right)$. Also when $t = 1$, we write q_b^X instead of $q_{b,1}^X$.

A first step in the proof of Theorem 9.1.1 is to obtain an analogue of equation (14.4.3) for $\exp\left(-t\mathcal{L}_{A,b}^X\right)$.

Take $(x., Y.)$ as in (12.10.14). Instead of equation (14.4.2), we consider the differential equation

$$\frac{dU}{ds} = U\left[-\frac{N^{\Lambda^{\cdot}(T^*X\oplus N^*)}}{b^2} + \widehat{c}\left(\text{ad}\left(Y^{TX}\right)\right)\right.$$

$$\left. - c\left(\text{ad}\left(Y^{TX}\right) + i\theta\text{ad}\left(Y^N\right)\right) - i\rho^E\left(Y^N\right)\right], \tag{15.7.1}$$

$$U_0 = 1.$$

Now we give an analogue of Theorem 14.4.1.

Theorem 15.7.1. Let $s \in C^{\infty,c}\left(\widehat{\mathcal{X}}, \widehat{\pi}^*\left(\Lambda^{\cdot}\left(T^*X \oplus N^*\right) \otimes F\right)\right)$. For $t > 0$, the following identity holds:

$$\exp\left(-t\mathcal{L}_{A,b}^X\right)s\left(x_0, Y\right) = E\left[\exp\left(\frac{(m+n)t}{2b^2} - \frac{b^4}{2}\int_0^t |[Y^N, Y^{TX}]|^2\,ds\right.\right.$$

$$\left.\left. - \frac{1}{2}\int_0^t |Y|^2\,ds - tA\right)U_t\tau_0^t s\left(x_t, Y_t\right)\right]. \tag{15.7.2}$$

Proof. The proof of our theorem is the same as the proof of Theorem 14.4.1. □

By proceeding as in the proof of Proposition 14.5.1 and of Theorem 14.5.2, we find that for $b \geq 1$,

$$|U_t| \leq \exp\left(ct + c'\int_0^t |Y|\,ds\right). \tag{15.7.3}$$

By (15.7.3), for $0 \leq t \leq M$,

$$|U_t| \leq C_M \exp\left(\frac{1}{4}\int_0^t |Y|^2\,ds\right). \tag{15.7.4}$$

By (15.7.2), (15.7.4), we obtain

$$\left| \exp\left(-t\underline{\mathcal{L}}_{A,b}^X\right) s\left(x_0, Y\right) \right|$$

$$\leq C_M E\left[\exp\left(\frac{(m+n)t}{2b^2} - \frac{b^4}{2}\int_0^t \left| [Y^N, Y^{TX}] \right|^2 ds \right.\right.$$

$$\left.\left. - \frac{1}{4}\int_0^t |Y|^2\, ds \right) \left| s\left(x_t, Y_t\right) \right| \right]. \qquad (15.7.5)$$

Let $\underline{\mathsf{A}}_b^{X\prime}$ be the scalar operator on $\widehat{\mathcal{X}}$,

$$\underline{\mathsf{A}}_b^{X\prime} = \frac{b^4}{2}\left| [Y^N, Y^{TX}] \right|^2 + \frac{1}{2}\left(-\frac{1}{b^4}\Delta^{TX\oplus N} + \frac{1}{2}|Y|^2 - \frac{m+n}{b^2} \right) - \nabla_{Y^{TX}}.$$
$$(15.7.6)$$

For $t > 0$, let $\underline{\mathsf{r}}_{b,t}^{X\prime}\left((x,Y),(x',Y')\right)$ be the smooth kernel associated with the operator $\exp\left(-t\underline{\mathsf{A}}_b^{X\prime}\right)$. Using the Itô calculus as in (12.11.8), if $F \in C^{\infty,c}\left(\widehat{\mathcal{X}}, \mathbf{R}\right)$, we get

$$\exp\left(-t\underline{\mathsf{A}}_b^{X\prime}\right) F\left(x_0, Y\right) = E\left[\exp\left(\frac{(m+n)t}{2b^2} - \frac{b^4}{2}\int_0^t \left| [Y^N, Y^{TX}] \right|^2 ds \right.\right.$$

$$\left.\left. - \frac{1}{4}\int_0^t |Y|^2\, ds \right) F\left(x_t, Y_t\right) \right]. \qquad (15.7.7)$$

By (15.7.5), (15.7.7), we obtain

$$\left| \exp\left(-t\underline{\mathcal{L}}_{A,b}^X\right) s\left(x_0, Y\right) \right| \leq C_M \exp\left(-t\underline{\mathsf{A}}_b^{X\prime}\right) |s|\left(x_0, Y\right). \qquad (15.7.8)$$

By (15.7.8), we find that

$$\left| q_{b,t}^X\left((x_0,Y),(x',Y')\right) \right| \leq C_M \underline{\mathsf{r}}_{b,t}^{X\prime}\left((x_0,Y),(x',Y')\right). \qquad (15.7.9)$$

Of course, in (15.7.9), (x_0, Y) can be replaced by an arbitrary $(x, Y) \in \widehat{\mathcal{X}}$.

Comparison of equations (12.11.7) and (15.7.6) for $\underline{\mathsf{A}}_b^X$ and $\underline{\mathsf{A}}_b^{X\prime}$ shows that only the coefficient of $|Y|^2$ differ, this coefficient being $\frac{1}{2}$ for $\underline{\mathsf{A}}_b^X$ and $\frac{1}{4}$ for $\underline{\mathsf{A}}_b^{X\prime}$. However, a quick inspection of the arguments in the present chapter shows that the arguments we gave to estimate $\underline{\mathsf{r}}_{b,t}^X$ remain valid for $\underline{\mathsf{r}}_{b,t}^{X\prime}$. By Theorems 15.5.1, 15.5.3, and 15.6.1 applied to $\underline{\mathsf{r}}_{b,t}^{X\prime}$, and by (15.7.9), we get Theorem 9.1.1.

15.8 A proof of Theorem 9.1.3

Now we establish Theorem 9.1.3. This result is just the analogue of Theorems 12.9.1 and 12.10.3 for the kernel $\underline{q}_{b,t}^X$. The proof consists of using the same methods in the proofs of these results, combined with the methods of the

Malliavin calculus as in section 14.9, while still using the estimates (15.7.3) and (15.7.4). Because here $b \geq 1$, no sophisticated arguments like the ones used in section 14.9 are needed, the problem in that section being that as $b \to 0$, some of the matrix terms became singular.

15.9 A proof of Theorem 9.5.6

The proof of Theorem 9.5.6 is essentially the same as the proof of the corresponding inequalities (9.1.10) and (9.2.12) for q_b^X, which were established in section 15.7. Let us now give more details.

We use the notation of chapter 9. First, note that by (9.5.3), with respect to $\mathcal{L}_{A,b}^X$, $\mathfrak{L}_{A,b}^X$ contains the extra term α. By equation (9.4.1) in Proposition 9.4.1, the norm of this term is uniformly bounded, so that it can easily be handled by the Feynman-Kac arguments of sections 14.4 and 15.7.

Recall that the flat connection $\nabla^{\Lambda^{\cdot}(T^*X \oplus N^*), f*, \widehat{f}}$ on $\Lambda^{\cdot}(T^*X \oplus N^*)$ was defined in Definition 2.4.1. By Theorem 2.13.2, equation (2.4.5) gives the splitting of the operator $\nabla_{YTX}^{C^\infty\left(TX \oplus N, \widehat{\pi}^*\left(\Lambda^{\cdot}(T^*X \oplus N^*)\widehat{\otimes}F\right)\right), f*, \widehat{f}}$ into its skew-adjoint and self-adjoint components with respect to the Hermitian product $\langle \rangle$ in (2.13.6). However, the restriction of $\nabla^{\Lambda^{\cdot}(T^*X \oplus N^*), f, \widehat{f}}$ to the extra $\Lambda^{\cdot}\left((\underline{T}X \oplus \underline{N})(\gamma)^*\right)$ does not preserve its metric.

Let $\nabla^{(TX \oplus N)(\gamma), f}$ be the obvious flat connection on $(TX \oplus N)(\gamma)$ that one obtains via the isomorphism $(TX \oplus N)(\gamma) = \mathfrak{z}(\gamma)$, and let $\nabla^{\Lambda^{\cdot}(\underline{T}X \oplus \underline{N})(\gamma)^*, f}$ be the corresponding flat connection on $\Lambda^{\cdot}\left((\underline{T}X \oplus \underline{N})(\gamma)^*\right)$.

Let Q be the orthogonal projection from $TX \oplus N$ on $(TX \oplus N)(\gamma)$ with respect to the scalar product of $TX \oplus N$. Let $\nabla^{(TX \oplus N)(\gamma)}$ be the projection of $\nabla^{TX \oplus N}$ on $(TX \oplus N)(\gamma)$ with respect to this scalar product, and let $\nabla^{\Lambda^{\cdot}((\underline{T}X \oplus \underline{N})(\gamma)^*)}$ be the induced connection on $\Lambda^{\cdot}\left((\underline{T}X \oplus \underline{N})(\gamma)^*\right)$. If $e \in TX$, set

$$\mathrm{ad}_Q(e) = Q\mathrm{ad}(e)Q. \qquad (15.9.1)$$

Then $\mathrm{ad}_Q(e) \in \mathrm{End}\left((TX \oplus N)(\gamma)\right)$ is self-adjoint.

By projecting the two sides of (2.2.2) on $(TX \oplus N)(\gamma)$, we get

$$\nabla^{(TX \oplus N)(\gamma), f} = \nabla^{(TX \oplus N)(\gamma)} + \mathrm{ad}_Q(\cdot). \qquad (15.9.2)$$

Let $\underline{\mathrm{ad}}_Q(e)$ be the obvious action of $-\widetilde{\mathrm{ad}}_Q(e)$ on $\Lambda^{\cdot}\left((\underline{T}X \oplus \underline{N})(\gamma)^*\right)$. Then $\underline{\mathrm{ad}}_Q(e)$ is self-adjoint. By (15.9.2), we get

$$\nabla^{\Lambda^{\cdot}(\underline{T}X \oplus \underline{N})(\gamma)^*, f} = \nabla^{\Lambda^{\cdot}(\underline{T}X \oplus \underline{N})(\gamma)^*} + \underline{\mathrm{ad}}_Q(\cdot). \qquad (15.9.3)$$

Let $\nabla^{\Lambda^{\cdot}(T^*X \oplus N^*)\widehat{\otimes}\Lambda^{\cdot}((\underline{T}X \oplus \underline{N})(\gamma)^*), f*, \widehat{f}}$ be the connection on

$$\Lambda^{\cdot}(T^*X \oplus N^*)\widehat{\otimes}\Lambda^{\cdot}\left((\underline{T}X \oplus \underline{N})(\gamma)^*\right)$$

that is induced by $\nabla^{\Lambda^{\cdot}(T^X \oplus N^*), f*, \widehat{f}}$ and $\nabla^{\Lambda^{\cdot}((\underline{T}X \oplus \underline{N})(\gamma)^*), f}$. By equations (2.4.5), (15.9.3), we get

$$\nabla^{\Lambda^{\cdot}(T^*X \oplus N^*)\widehat{\otimes}\Lambda^{\cdot}((\underline{T}X \oplus N)(\gamma)^*), f*, \widehat{f}} = \nabla^{\Lambda^{\cdot}(T^*X \oplus N^*)\widehat{\otimes}\Lambda^{\cdot}((\underline{T}X \oplus N)(\gamma)^*)}$$

$$- c(\mathrm{ad}(\cdot)) + \widehat{c}(\mathrm{ad}(\cdot)) + \underline{\mathrm{ad}}_Q(\cdot). \qquad (15.9.4)$$

Using (9.5.3) and (15.9.4), we find that $\mathfrak{L}_{A,b}^X$ is of exactly the same form as $\underline{\mathcal{L}}_{A,b}^X$. The same arguments as in the proof of (9.2.12) lead us to (9.5.7). The proof of Theorem 9.5.6 is completed.

15.10 A proof of Theorem 9.11.1

We will only briefly sketch the proof of Theorem 9.11.1, which can be split into two main steps:

1. If $Y_0^{\mathfrak{k}} \in \mathfrak{k}(\gamma)$, $(y, Y) \in \mathfrak{p} \oplus \mathfrak{g}$, for $b \geq 1$, $\mathfrak{p}_{a,b,Y_0^{\mathfrak{k}}}^X ((y, Y), (y', Y'))$ and its derivatives of arbitrary order in $(y', Y') \in \mathfrak{p} \times \mathfrak{g}$ are uniformly bounded on compact subsets of $\mathfrak{p} \times \mathfrak{g}$. Here the bounds may well depend on $Y_0^{\mathfrak{k}}$.

2. If $s(y', Y') \in C^{\infty,c}\left(\mathfrak{p} \times \mathfrak{g}, \Lambda^{\cdot}(\mathfrak{g}^*) \widehat{\otimes} \Lambda^{\cdot}(\mathfrak{z}(\gamma)^*) \otimes E\right)$, then

$$\exp\left(-\mathcal{P}_{a,A,b,Y_0^{\mathfrak{k}}}^X\right) s(y, Y) \to \exp\left(-\mathcal{P}_{a,A,\infty,Y_0^{\mathfrak{k}}}^X\right) s(y, Y). \quad (15.10.1)$$

From (1) and (2), we get (9.11.3), i.e., we have established Theorem 9.11.1.

Let us explain how to get step (1). Take $\beta > 0$. By Theorem 9.1.3, we know that for $b \geq 1, \epsilon \leq t \leq M$, if $|y| \leq \beta b^2, |y'| \leq \beta b^2$,

$$b^{-4m-2n} \underline{q}_{b,t}^X \left(e^{-a}\left(e^{y/b^2}, Y/b^2\right), \left(e^{y'/b^2}, Y'/b^2\right)\right)$$

together with its derivatives of any order in the variables (y', Y') are uniformly bounded.

We fix $Y_0^{\mathfrak{k}} \in \mathfrak{k}(\gamma), N > 0$. We claim that for $b \geq 1, \epsilon \leq t \leq M, |y| \leq N, |y'| \leq N$,

$$b^{-4m-2n} \underline{q}_{b,t}^X \left(e^{-a}\left(e^{y/b^2}, a^{TX} + Y_0^{\mathfrak{k}} + Y/b^2\right), \left(e^{y'/b^2}, a^{TX} + Y_0^{\mathfrak{k}} + Y'/b^2\right)\right)$$
$$(15.10.2)$$

and its derivatives of any order in the variables (y', Y') remain uniformly bounded by constants depending on $Y_0^{\mathfrak{k}}$. Indeed $a^{TX} + Y_0^{\mathfrak{k}}$ is now viewed as a section of $TX \oplus N$, which is uniformly bounded together with its covariant derivatives of arbitrary order in the region $|y'| \leq \beta$, the bounds still depending on the choice of $Y_0^{\mathfrak{k}}$. Since in the second term in (15.10.2), this section is evaluated at y'/b^2, covariant derivatives introduce the right power of b^{-2}.

Recall that the kernel $\mathfrak{q}_{b,t}^X$ was defined in Definition 9.5.3, and that $\mathfrak{q}_b^X = \mathfrak{q}_{b,1}^X$. We claim that the considerations we made before also apply to the kernel $\mathfrak{q}_{b,t}^X$. The argument is exactly the same as in the proof of Theorem 9.5.6 that was given in section 15.9. It uses the estimate (9.4.1) in Proposition 9.4.1, and also equation (9.5.3) for $\mathfrak{L}_{A,b}^X$.

By combining (9.9.7) and the above arguments, step (1) is completed.

We move to step (2). We start from equation (9.8.5) for $\mathcal{O}_{a,A,b}^X$, from which we get a corresponding equation for $\mathcal{P}_{a,A,b,Y_0^{\mathfrak{k}}}^X$ in (9.9.5).

As indicated in section 9.9, the vector bundles TX, N have been trivialized by an Euclidean trivialization, and identified with $\mathfrak{p}, \mathfrak{k}$. On the other hand, the coordinate system $y \in \mathfrak{p} \to e^y p1 \in X$ provides another trivialization of TX, which does not preserve the metric. Using the coordinate system y, the canonical identification of the fibre TX with the tangent bundle to X is induced by an isomorphism $r_y : \mathfrak{p} \to \mathfrak{p}$, which coincides with the identity at $y = 0$.

Let $\Gamma^{TX \oplus N}$ be the connection form for $\nabla^{TX \oplus N}$ in the Euclidean trivialization of $TX \oplus N$. Then $\Gamma^{TX \oplus N}$ is a 1-form with values in antisymmetric sections of $\mathrm{End}\,(\mathfrak{p} \oplus \mathfrak{k})$ that preserve $\mathfrak{p}, \mathfrak{k}$. In the coordinates (y, Y), the scalar operator $\nabla_{Y^{TX}}$ can be written in the form

$$\nabla_{Y^{TX}} = \nabla^H_{rY^{\mathfrak{p}}} - \nabla^V_{\Gamma^{TX \oplus N}(rY^{\mathfrak{p}})Y}. \tag{15.10.3}$$

When making the change of variables indicated in (9.9.4), (9.9.5), the operator in (15.10.3) is changed into the operator

$$\nabla^H_{r_{y/b^2}Y^{\mathfrak{p}}} - \nabla^V_{\Gamma^{TX \oplus N}_{y/b^2}(r_{y/b^2}Y^{\mathfrak{p}})(Y^{\mathfrak{k}}_0 + Y/b^2)}. \tag{15.10.4}$$

Let $\mathcal{P}^{X,s}_{a,A,b,Y^{\mathfrak{k}}_0}$ be the scalar part of $\mathcal{P}^X_{a,A,b,Y^{\mathfrak{k}}_0}$. By (9.8.5), (9.9.4), (9.9.5), and (9.10.8), we get

$$\mathcal{P}^{X,s}_{a,A,b,Y^{\mathfrak{k}}_0} = \frac{1}{2} \left| \left[Y^{\mathfrak{k}}_0 + Y^{\mathfrak{k}}/b^2, b^2 a^{TX}_{y/b^2} + Y^{\mathfrak{p}} \right] \right|^2$$

$$+ \frac{1}{2} \left(-\Delta^{\mathfrak{p} \oplus \mathfrak{k}} + \left| a^{TX}_{y/b^2} + Y^{\mathfrak{k}}_0 + Y/b^2 \right|^2 - \frac{1}{b^2}(m+n) \right) - \nabla^H_{r_{y/b^2}Y^{\mathfrak{p}}}$$

$$+ \nabla^V_{\Gamma^{TX \oplus N}_{y/b^2}(r_{y/b^2}Y^{\mathfrak{p}})(Y^{\mathfrak{k}}_0 + Y/b^2)} + \nabla^V_{\left[a^N_{y/b^2}, 2Y^{\mathfrak{p}} + Y^{\mathfrak{k}} + b^2 Y^{\mathfrak{k}}_0 + b^2 a^{TX}_{y/b^2} \right]}. \tag{15.10.5}$$

To the operator $\mathcal{P}^{X,s}_{a,A,b,Y^{\mathfrak{k}}_0}$, we associate the stochastic differential equation in the variables (y, Y),

$$\dot{y}_b = r_{y_b/b^2}Y^{\mathfrak{p}}_b,$$

$$\dot{Y}_b = -\Gamma^{TX \oplus N}_{y/b^2}\left(r_{y_b/b^2}Y^{\mathfrak{p}}_b \right)\left(Y^{\mathfrak{k}}_0 + Y_b/b^2 \right) \tag{15.10.6}$$

$$- \left[a^N_{y_b/b^2}, 2Y^{\mathfrak{p}}_b + Y^{\mathfrak{k}}_b + b^2 Y^{\mathfrak{k}}_0 + b^2 a^{TX}_{y_b/b^2} \right] + \dot{w},$$

$$y_{b,0} = y, \qquad Y_{b,0} = Y.$$

Equation (15.10.6) is just the coordinate version of a stochastic differential equation over $\widehat{\mathcal{X}}$, for which existence is already granted by the results of section 12.2.

An application of the Feynman-Kac formula shows that if $F : \mathfrak{p} \times \mathfrak{g} \to \mathbf{R}$ is smooth with compact support,

$$\exp\left(-t\mathcal{P}^{X,s}_{a,A,b,Y^{\mathfrak{k}}_0} \right) F(y, Y) = \exp\left(\frac{(m+n)t}{2b^2} \right)$$

$$E\left[\exp\left(-\frac{1}{2}\int_0^t \left| \left[Y^{\mathfrak{k}}_0 + Y^{\mathfrak{k}}_b/b^2, b^2 a^{TX}_{y_b/b^2} + Y^{\mathfrak{p}}_b \right] \right|^2 ds \right. \right.$$

$$\left. \left. - \frac{1}{2}\int_0^t \left| a^{TX}_{y_b/b^2} + Y^{\mathfrak{k}}_0 + Y_b/b^2 \right|^2 ds \right) F(y_{b,t}, Y_{b,t}) \right]. \tag{15.10.7}$$

By proceeding as in (12.2.7), we rewrite (15.10.6) in the form

$$y_{b,t} = y + \int_0^t r_{y_b/b^2} Y_b^{\mathfrak{p}} ds,$$

$$Y_{b,t} = Y - \int_0^t \Gamma_{y_b/b^2}^{TX \oplus N} \left(r_{y_b/b^2} Y_b^{\mathfrak{p}} \right) \left(Y_0^{\mathfrak{k}} + Y_b/b^2 \right) ds \qquad (15.10.8)$$

$$- \int_0^t \left[a_{y_b/b^2}^N, 2Y_b^{\mathfrak{p}} + Y_b^{\mathfrak{k}} \right] ds - \int_0^t \left[a_{y_b/b^2}^N, b^2 Y_0^{\mathfrak{k}} + b^2 a_{y_b/b^2}^{TX} \right] ds + w_t.$$

By (9.10.4), (9.10.9), we find easily that as $b \to +\infty$, $(y_{b,\cdot}, Y_{b,\cdot})$ converge uniformly over compact sets to $(y_{\infty,\cdot}, Y_{\infty,\cdot})$, which is a solution of

$$y_{\infty,t} = y + \int_0^t Y_\infty ds, \quad Y_{\infty,t} = Y + \int_0^t \left[a + Y_0^{\mathfrak{k}}, [a, y_\infty] \right] ds + w_t. \quad (15.10.9)$$

By (9.10.5), (9.10.6), and by the above convergence result, as $b \to \infty$,

$$E\left[\exp\left(-\frac{1}{2} \int_0^t \left| \left[Y_0^{\mathfrak{k}} + Y_b^{\mathfrak{k}}/b^2, b^2 a_{y/b^2}^{TX} + Y_b^{\mathfrak{p}} \right] \right|^2 ds \right. \right.$$

$$\left. - \frac{1}{2} \int_0^t \left| a_{y/b^2}^{TX} + Y_0^{\mathfrak{k}} + Y_b/b^2 \right|^2 ds \right) F(y_t, Y_t) \right] \to \exp\left(-\frac{t}{2} \left(|a|^2 + |Y_0^{\mathfrak{k}}|^2 \right) \right)$$

$$E\left[\exp\left(-\frac{t}{2} \int_0^t \left| [Y_\infty^{\mathfrak{k}}, a] + [Y_0^{\mathfrak{k}}, Y_\infty^{\mathfrak{p}}] \right|^2 ds \right) F(y_{\infty,t}, Y_{\infty,t}) \right]. \quad (15.10.10)$$

Let $\mathcal{P}_{a,A,\infty,Y_0^{\mathfrak{k}}}^{X,s}$ be the scalar part of the operator $\mathcal{P}_{a,A,\infty,Y_0^{\mathfrak{k}}}^{X}$ in (9.10.1). By (15.10.7) (15.10.10), and using again the Feynman-Kac formula, we can rewrite (15.10.10) in the form

$$\exp\left(-t\mathcal{P}_{a,A,b,Y_0^{\mathfrak{k}}}^{X,s} \right) F(y, Y) \to \exp\left(-t\mathcal{P}_{a,A,\infty,Y_0^{\mathfrak{k}}}^{X,s} \right) F(y, Y). \quad (15.10.11)$$

For $t = 1$, (15.10.11) is just the scalar version of (15.10.1).

Let us now establish equation (15.10.1) itself. Let $U_{a,b,\cdot}$ be the solution of the differential equation

$$\frac{dU_{a,b}}{ds} = U_{a,b} \left[-\frac{N^{\Lambda^\cdot(\mathfrak{p}^* \oplus \mathfrak{k}^*)}}{b^2} - \alpha - c\left(\mathrm{ad}\left(Y_b^{\mathfrak{p}}/b^2 \right) \right) + \widehat{c}\left(\mathrm{ad}\left(Y_b^{\mathfrak{p}}/b^2 \right) \right) \right.$$

$$- c\left(\mathrm{ad}\left(a_{y_b/b^2}^{TX} - a_{y_b/b^2}^N \right) \right) + \widehat{c}(\mathrm{ad}(a)) - c\left(i\theta \mathrm{ad}\left(Y_0^{\mathfrak{k}} + Y^{\mathfrak{k}}/b^2 \right) \right)$$

$$\left. - \rho^E \left(iY_0^{\mathfrak{k}} + iY_b^{\mathfrak{k}}/b^2 - a_{y_b/b^2}^N \right) \right], \qquad (15.10.12)$$

$$U_{a,b,0} = 1.$$

By (2.4.5), (9.8.5), and (9.9.5), and proceeding as in the proofs of Theorems 14.4.1 and 15.7.1, if $s(x, Y)$ is taken as in (15.10.1), instead of (15.10.7), we get

$$\exp\left(-t\mathcal{P}^X_{a,A,b,Y_0^\mathfrak{t}}\right) s\left(y,Y\right) = \exp\left(\frac{(m+n)\,t}{2b^2}\right)$$

$$E\left[\exp\left(-\frac{1}{2}\int_0^t \left|\left[Y_0^\mathfrak{t}+Y_b^\mathfrak{t}/b^2, b^2 a^{TX}_{y_b/b^2}+Y_b^\mathfrak{p}\right]\right|^2\right.\right.$$

$$\left.\left.-\frac{1}{2}\int_0^t \left|a^{TX}_{y_b/b^2}+Y_0^\mathfrak{t}+Y_b/b^2\right|^2 - tA\right) U_{a,b,t}\tau_0^t s\left(y_{b,t},Y_{b,t}\right)\right]. \quad (15.10.13)$$

By (15.10.12), as $b \to +\infty$, $U_{a,b,\cdot}$ converges uniformly over compact sets to $U_{a,\infty,\cdot}$ which is given by

$$\frac{dU_{a,\infty}}{ds} = U_{a,\infty}\left[-\alpha + \widehat{c}\left(\operatorname{ad}\left(a\right)\right) - c\left(\operatorname{ad}\left(a\right) + i\theta\operatorname{ad}\left(Y_0^\mathfrak{t}\right)\right) - i\rho^E\left(Y_0^\mathfrak{t}\right)\right],$$

$$(15.10.14)$$

$$U_{a,\infty,0} = 1.$$

By (9.10.1), (15.10.12)–(15.10.14), and proceeding as in (15.10.11), we find that as $b \to +\infty$,

$$\exp\left(-t\mathcal{P}^X_{a,A,b,Y_0^\mathfrak{t}}\right) s\left(y,Y\right) \to \exp\left(-t\mathcal{P}^X_{a,A,\infty,Y_0^\mathfrak{t}}\right) s\left(y,Y\right). \quad (15.10.15)$$

For $t = 1$, we get (15.10.1).

The proof of Theorem 9.11.1 is completed.

Bibliography

[ABo67]   M. F. Atiyah and R. Bott. A Lefschetz fixed point formula for elliptic complexes. I. *Ann. of Math. (2)*, 86:374–407, 1967.

[ABo68]   M. F. Atiyah and R. Bott. A Lefschetz fixed point formula for elliptic complexes. II. Applications. *Ann. of Math. (2)*, 88:451–491, 1968.

[ABoP73]   M. Atiyah, R. Bott, and V. K. Patodi. On the heat equation and the index theorem. *Invent. Math.*, 19:279–330, 1973.

[AS68a]   M. F. Atiyah and I. M. Singer. The index of elliptic operators. I. *Ann. of Math. (2)*, 87:484–530, 1968.

[AS68b]   M. F. Atiyah and I. M. Singer. The index of elliptic operators. III. *Ann. of Math. (2)*, 87:546–604, 1968.

[BaGSc85] W. Ballmann, M. Gromov, and V. Schroeder. *Manifolds of nonpositive curvature*, volume 61 of *Progress in Mathematics*. Birkhäuser Boston Inc., Boston, MA, 1985.

[BeVe83]   N. Berline and M. Vergne. Zéros d'un champ de vecteurs et classes caractéristiques équivariantes. *Duke Math. J.*, 50(2):539–549, 1983.

[B81a]   J.-M. Bismut. Martingales, the Malliavin calculus and hypoellipticity under general Hörmander's conditions. *Z. Wahrsch. Verw. Gebiete*, 56(4):469–505, 1981.

[B81b]   J.-M. Bismut. *Mécanique aléatoire*, volume 866 of *Lecture Notes in Mathematics*. Springer-Verlag, Berlin, 1981. With an English summary.

[B84]   J.-M. Bismut. *Large deviations and the Malliavin calculus*, volume 45 of *Progress in Mathematics*. Birkhäuser Boston Inc., Boston, MA, 1984.

[B05]   J.-M. Bismut. The hypoelliptic Laplacian on the cotangent bundle. *J. Amer. Math. Soc.*, 18(2):379–476 (electronic), 2005.

[B08a]   J.-M. Bismut. The hypoelliptic Dirac operator. In *Geometry and dynamics of groups and spaces*, volume 265 of *Progr. Math.*, pages 113–246. Birkhäuser, Basel, 2008.

[B08b]   J.-M. Bismut. The hypoelliptic Laplacian on a compact Lie group. *J. Funct. Anal.*, 255(9):2190–2232, 2008.

[B08c]   J.-M. Bismut. Loop spaces and the hypoelliptic Laplacian. *Comm. Pure Appl. Math.*, 61(4):559–593, 2008.

[B09a]   J.-M. Bismut. A survey of the hypoelliptic Laplacian. *Astérisque*, (322):39–69, 2008. Géométrie différentielle, physique mathématique, mathématiques et société. II.

[B09b]   J.-M. Bismut. Laplacien hypoelliptique et intégrales orbitales. *C. R. Acad. Sci. Paris Sér. I Math.*, 347:1189–1195, 2009.

[BL91]   J.-M. Bismut and G. Lebeau. Complex immersions and Quillen metrics. *Inst. Hautes Études Sci. Publ. Math.*, (74):ii+298 pp. (1992), 1991.

[BL08]   J.-M. Bismut and G. Lebeau. *The hypoelliptic Laplacian and Ray-Singer metrics*, volume 167 of *Annals of Mathematics Studies*. Princeton University Press, Princeton, NJ, 2008.

[BLo95]   J.-M. Bismut and J. Lott. Flat vector bundles, direct images and higher real analytic torsion. *J. Amer. Math. Soc.*, 8(2):291–363, 1995.

[BZ92]   J.-M. Bismut and W. Zhang. An extension of a theorem by Cheeger and Müller. *Astérisque*, (205):235, 1992. With an appendix by François Laudenbach.

[BuGu70]   D. L. Burkholder and R. F. Gundy. Extrapolation and interpolation of quasi-linear operators on martingales. *Acta Math.*, 124:249–304, 1970.

[CP81]   J. Chazarain and A. Piriou. *Introduction à la théorie des équations aux dérivées partielles linéaires.* Gauthier-Villars, Paris, 1981.

[DH82]   J. J. Duistermaat and G. J. Heckman. On the variation in the cohomology of the symplectic form of the reduced phase space. *Invent. Math.*, 69(2):259–268, 1982.

[DH83]   J. J. Duistermaat and G. J. Heckman. Addendum to: "On the variation in the cohomology of the symplectic form of the reduced phase space." *Invent. Math.*, 72(1):153–158, 1983.

[Do53]   J. L. Doob. *Stochastic processes.* John Wiley & Sons Inc., New York, 1953.

[E96]   P. B. Eberlein. *Geometry of nonpositively curved manifolds.* Chicago Lectures in Mathematics. University of Chicago Press, Chicago, IL, 1996.

[Fr84]   I. B. Frenkel. Orbital theory for affine Lie algebras. *Invent. Math.*, 77(2):301–352, 1984.

[Fri86]   D. Fried. The zeta functions of Ruelle and Selberg. I. *Ann. Sci. École Norm. Sup. (4)*, 19(4):491–517, 1986.

[Fri88]   D. Fried. Torsion and closed geodesics on complex hyperbolic manifolds. *Invent. Math.*, 91(1):31–51, 1988.

[Ge86]   E. Getzler. A short proof of the local Atiyah-Singer index theorem. *Topology*, 25(1):111–117, 1986.

[Gi73]   P. B. Gilkey. Curvature and the eigenvalues of the Laplacian for elliptic complexes. *Advances in Math.*, 10:344–382, 1973.

[Gi84]   P. B. Gilkey. *Invariance theory, the heat equation, and the Atiyah-Singer index theorem*, volume 11 of *Mathematics Lecture Series*. Publish or Perish Inc., Wilmington, DE, 1984.

[GlJ87]   J. Glimm and A. Jaffe. *Quantum physics*. Springer-Verlag, New York, second edition, 1987. A functional integral point of view.

[He78]   S. Helgason. *Differential geometry, Lie groups, and symmetric spaces*, volume 80 of *Pure and Applied Mathematics*. Academic Press Inc. [Harcourt Brace Jovanovich Publishers], New York, 1978.

[Hi74]   N. Hitchin. Harmonic spinors. *Advances in Math.*, 14:1–55, 1974.

[Hö67]   L. Hörmander. Hypoelliptic second order differential equations. *Acta Math.*, 119:147–171, 1967.

[Hö83]   L. Hörmander. *The analysis of linear partial differential operators. I*, volume 256 of *Grundlehren der Mathematischen Wissenschaften*. Springer-Verlag, Berlin, 1983. Distribution theory and Fourier analysis.

[Hör85a]   L. Hörmander. *The analysis of linear partial differential operators. III*. Grundl. Math. Wiss. Band 274. Springer-Verlag, Berlin, 1985. Pseudodifferential operators.

[Hör85b]   L. Hörmander. *The analysis of linear partial differential operators. IV*, volume 275 of *Grundlehren der Mathematischen Wissenschaften*. Springer-Verlag, Berlin, 1985. Fourier integral operators.

[KSh91]   I. Karatzas and S. E. Shreve. *Brownian motion and stochastic calculus*, volume 113 of *Graduate Texts in Mathematics*. Springer-Verlag, New York, second edition, 1991.

[Ka79]   T. Kawasaki. The Riemann-Roch theorem for complex V-manifolds. *Osaka J. Math.*, 16(1):151–159, 1979.

[Kn86]   A. W. Knapp. *Representation theory of semisimple groups*, volume 36 of *Princeton Mathematical Series*. Princeton University Press, Princeton, NJ, 1986. An overview based on examples.

[Koh73]   J. J. Kohn. Pseudo-differential operators and hypoellipticity. In *Partial differential equations (Proc. Sympos. Pure Math., Vol. XXIII, Univ. California, Berkeley, Calif., 1971)*, pages 61–69. Amer. Math. Soc., Providence, RI, 1973.

[Ko76]   B. Kostant. On Macdonald's η-function formula, the Laplacian and generalized exponents. *Advances in Math.*, 20(2):179–212, 1976.

[Ko97]   B. Kostant. Clifford algebra analogue of the Hopf-Koszul-Samelson theorem, the ρ-decomposition $C(\mathfrak{g}) = \mathrm{End}\, V_\rho \otimes C(P)$, and the $\mathfrak{g}$-module structure of $\bigwedge \mathfrak{g}$. *Adv. Math.*, 125(2):275–350, 1997.

[L05]   G. Lebeau. Geometric Fokker-Planck equations. *Port. Math. (N.S.)*, 62(4):469–530, 2005.

[M78]   P. Malliavin. Stochastic calculus of variation and hypoelliptic operators. In *Proceedings of the International Symposium on Stochastic Differential Equations (Res. Inst. Math. Sci., Kyoto Univ., Kyoto, 1976)*, pages 195–263. Wiley, New York, 1978.

[M97]   P. Malliavin. *Stochastic analysis*, volume 313 of *Grundlehren der Mathematischen Wissenschaften*. Springer-Verlag, Berlin, 1997.

[MaQ86]   V. Mathai and D. Quillen. Superconnections, Thom classes, and equivariant differential forms. *Topology*, 25(1):85–110, 1986.

[McK72]   H. P. McKean. Selberg's trace formula as applied to a compact Riemann surface. *Comm. Pure Appl. Math.*, 25:225–246, 1972.

[McKS67]   H. P. McKean, Jr. and I. M. Singer. Curvature and the eigenvalues of the Laplacian. *J. Differential Geometry*, 1(1):43–69, 1967.

[MoSt91]   H. Moscovici and R. J. Stanton. R-torsion and zeta functions for locally symmetric manifolds. *Invent. Math.*, 105(1):185–216, 1991.

[RS71]   D. B. Ray and I. M. Singer. R-torsion and the Laplacian on Riemannian manifolds. *Advances in Math.*, 7:145–210, 1971.

[ReY99]   D. Revuz and M. Yor. *Continuous martingales and Brownian motion*, volume 293 of *Grundlehren der Mathematischen Wissenschaften*. Springer-Verlag, Berlin, third edition, 1999.

[St81a]   D. W. Stroock. The Malliavin calculus and its applications. In *Stochastic integrals (Proc. Sympos., Univ. Durham, Durham, 1980)*, volume 851 of *Lecture Notes in Math.*, pages 394–432. Springer-Verlag, Berlin, 1981.

[St81b]   D. W. Stroock. The Malliavin calculus, a functional analytic approach. *J. Funct. Anal.*, 44(2):212–257, 1981.

[StV72]   D. W. Stroock and S. R. S. Varadhan. On the support of diffusion processes with applications to the strong maximum principle. In *Proceedings of the Sixth Berkeley Symposium on Mathematical Statistics and Probability (Univ. California, Berkeley, Calif., 1970/1971), Vol. III: Probability theory*, pages 333–359. Univ. California Press, Berkeley, CA, 1972.

[StV79]   D. W. Stroock and S. R. S. Varadhan. *Multidimensional diffusion processes*, volume 233 of *Grundlehren der Mathematischen Wissenschaften*. Springer-Verlag, Berlin, 1979.

[T81]   M. E. Taylor. *Pseudodifferential operators*, volume 34 of *Princeton Mathematical Series*. Princeton University Press, Princeton, NJ, 1981.

[W88]   N. R. Wallach. *Real reductive groups. I*, volume 132 of *Pure and Applied Mathematics*. Academic Press Inc., Boston, MA, 1988.

[Wi82]   E. Witten. Supersymmetry and Morse theory. *J. Differential Geom.*, 17(4):661–692 (1983), 1982.

[Y68]   K. Yosida. *Functional analysis*. Second edition, volume 123 of *Grundlehren der mathematischen Wissenschaften*. Springer-Verlag New York, New York, 1968.

Subject Index

Index of Notation